Hidden Treasures

This new and exciting observing guide spotlights an original selection of 109 deep-sky objects that will appeal to sky-watchers worldwide. Stephen James O'Meara's 'hidden treasures' include a wonderful assortment of galaxies, open clusters, planetary nebula and more, all of which have been carefully chosen based on their popularity and ease of observing. None of these objects are included in either the Messier or the Caldwell catalogs, and all are visible in a 4-inch telescope under dark skies, Stunning photographs and beautiful drawings accompany detailed visual descriptions of the objects, which include their rich histories and astrophysical significance. The author's original finder charts are designed to help observers get to their targets fast and efficiently.

STEPHEN JAMES O'MEARA has spent much of his career on the editorial staff of Sky & Telescope magazine. His many astronomical achievements include being the first person to sight Halley's Comet on its 1985 return, he noticing the dark "spokes" in Saturn's B ring before the Voyager 1 spacecraft imaged them, and determining the rotation period of the distant planet Uranus. He received The Texas Star Party Omega Centauri Award for "advancing astronomy through observation, writing, and promotion, and for his love of the sky," and the International Astronomical Union named asteroid 3637 O'Meara in his honor.

DEEP-SKY COMPANIONS

Hidden Treasures

STEPHEN JAMES O'MEARA

SKY & TELESCOPE

CAMBRIDGE UNIVERSITY PRESS

PUBLISHED BY CAMBRIDGE UNIVERSITY PRESS
The Edinburgh Building, Cambridge CB2 2RU, UK www.cambridge.org

AND BY SKY PUBLISHING
49 Bay State Road, Cambridge, Massachusetts 02138-1200, USA SkyandTelescope.com

CAMBRIDGE UNIVERSITY PRESS
Cambridge, New York, Madrid, Cape Town, Singapore, São Paulo

Published in the United States of America by Cambridge University Press, New York

Information on this title: www.cambridge.org/9780521837040

First published 2007

Printed and Bound in Malaysia by Imago

A catalog record for this publication is available from the British Library

ISBN-13 978-0-521-83704-0 hardback (Cambridge University Press)
ISBN-10 0-521-83704-9 hardback (Cambridge University Press)
ISBN-13 978-1-931559-45-4 (Sky Publishing)
ISBN-10 1-931559-45-7 (Sky Publishing)

To Donna,
My jewel of the night. Your beauty is incomparable.

To Daisy Duke such a joy,
My little pot of gold at the end of the rainbow.

And in memory of Milky Way, Miranda-Pyewacket, and Pele.
If time is a treasure, we have been robbed. Our love was priceless.

Who wouldst not leave him in his wandering
To seek for treasure in the jewelled skies . . . ?

Edgar Allan Poe
Sonnet – To Science
(Written in youth)

Contents

Preface

WE ARE ALL TREASURE hunters – storybook pirates searching for riches in the endless sea above. Taking the helms of our telescopes, we lay a course among the stars with the sails of our imaginations open. And what wonders await us as we make our way through the charted territories of the Milky Way: rich, open clusters of hot, young stars, some still swaddled in their nascent nebulosities; ancient globular clusters, the senior citizens of our galaxy, whose teeming suns are packed together like gold doubloons in a sea chest; there are galaxies too numerous to mention lurking beyond our forest of stars, living out their lives in various stages of evolution; and then there are the ghosts – the smoky shells of dying stars, whose very nature reminds us of the ultimate fate of our life-giving Sun.

These celestial treasures cannot be plundered. They can only make us feel, as Joseph Conrad writes in his 1902 adventure novel, *Heart of Darkness*, "meditative, and fit for nothing but placid staring." And there are jewels in the night almost too numerous to mention, some of which rarely get viewed by amateur astronomers. That is why novel lists of neglected deep-sky objects are becoming increasingly popular among observers. And that is why I created this book.

Hidden Treasures is the third title in my *Deep-Sky Companions* series – the other two books are *Deep-Sky Companions: The Messier Objects* and *Deep-Sky Companions: The Caldwell Objects*. This latest work fills an important void. It introduces skywatchers to a new list of 109 deep-sky objects (all visible in small telescopes) not included in the Messier or Caldwell catalogs; 20 additional objects are also listed in Appendix C.

Some objects in the *Hidden Treasures* list are surprisingly bright – bright enough to make one ponder why they have been overlooked. For instance, neither the Messier nor the Caldwell lists include NGC 2451, a 3rd-magnitude "open star cluster" in Puppis known since the seventeenth century; I place the words "open star cluster" in quotes because recent research has shown that this classic object is not a star cluster but an asterism, a chance grouping of stars (see Hidden Treasure 42). Nor do the Messier and Caldwell lists include the 2nd-magnitude Coma Berenices Cluster, which Ptolemy cataloged as a "nebula" in the sixteenth century. And as early as AD 964, Al Sufi described what we now call Brocchi's "Cluster," the famous Coathanger in Sagitta (yet another "cluster" turned asterism). To understand why these objects were not considered, we need to understand the purpose of the Messier and Caldwell catalogs.

Despite popular belief, the Messier catalog is not a list of the "brightest and best" deep-sky objects for small telescopes. It is a list of objects that Messier believed could be confused with a comet that was "just beginning to shine." Its purpose, then, was to help the astronomers of his day, not the celestial tourists of today. Similarly, despite popular belief, the Caldwell catalog does not "pick up" where the Messier catalog left off. In fact, the intent of the Caldwell catalog – to challenge the celestial tourists of today – is just the opposite of the Messier catalog. What's more, Caldwell list's creator, Sir Patrick Caldwell-Moore, selected the objects not because they were the "brightest" and "best" non-Messier objects in the heavens, but because he thought they were either aesthetically pleasing, visually challenging, or astrophysically interesting. And while the Caldwell catalog does include some of the most celebrated non-Messier objects in the night sky – the Double Cluster in Perseus, the Eta Carinae Nebula, the great globular clusters Omega Centauri and 47 Tucanae, and more – it also includes some very dim ones: IC 1613 (an irregular barred dwarf galaxy), NGC 7635 (the Bubble Nebula), and Sharpless 2–155 (the Cave Nebula); all targets more suited for astrophotographers or visual observers with large-aperture telescopes.

That the Messier and Caldwell catalogs are different is good. Both challenge modern observers in their own ways. But, as I have said, there are more celestial treasures hidden among the stars, within the grasp of small-aperture telescopes. Clearly then, there is room for another list of 109 objects that will inspire us to get out and observe, to move beyond the Messier and Caldwell catalogs, and to enrich our observing experience.

WHAT IS "BEST?"

Some experienced deep-sky observers and several astronomical societies have created lists of what they believe are the "finest" or "best" deep-sky objects in the night sky. Many of these lists include objects in the Messier and Caldwell catalogs, as well as some other very stunning and overwhelmingly popular objects not on these lists. But once you remove these objects, few of these "best-of" lists agree on what is "best" to observe. The fact is if everyone agreed on which objects are the "finest," we wouldn't need more than one list. The choices would be clear. But they are not.

When you look up at the sky, what is your favorite star? Which is the "best" constellation? Will others agree? I have many trees and flowers in my yard. Some are my favorites. My wife, Donna, has her favorites too. But do we agree? What about a professional baseball team? If we exclude all the big-name sluggers and pitchers, who's the "best" of the lot? What's your criteria for choosing? Does everyone use the same criteria? Who's your best friend? How do you make such a choice? Will you ever change your mind? Why?

So you see the problem. Saying that someone or something is the "best" is highly subjective, if not extremely personal. The reason the Messier catalog has been so popular with observers worldwide for centuries is because all of the objects are visible in small telescopes. That's not the case with many of the current "best of" or "finest" lists available today. The irony of the Messier catalog, and we come full circle with this point, is that, despite its wide popular appeal, it is *not* a list of the "brightest" and "best" deep-sky objects.

Hidden Treasures is *not* a list of the "brightest," the "finest," or the "best" deep-sky objects. *Hidden Treasures* is a list of 109 deep-sky objects not included in the Messier or Caldwell catalogs. All are visible in a 4-inch telescope. Many are bright. And many have gotten two "thumbs up" from observers around the world. Indeed, many of the objects are not *my* favorites. I selected them because they appear on several "best of" lists, and, when I examined them through my 4-inch telescope, were deemed worthy of inclusion. I went through great strides to make certain that the selection of objects in *Hidden Treasures* consists, in large, of objects that do not require Herculean efforts to be seen in small telescopes.

HOW THE 109 HIDDEN TREASURES WERE SELECTED

Hidden Treasures is original in its creation and scope. The project's roots extend back to 1994, when I began making observations for *Deep-Sky Companions: The Caldwell Objects*; that's also when I began making systematic sweeps of the night sky for comets with my 4-inch f/5 Tele Vue refractor at 23×. As all comet hunters know, the night sky is filled with masqueraders. Whenever I encountered an object not in the Messier or Caldwell lists, I recorded it in a special log set aside for "Additional Objects" – ones that would form the foundation of the 20 additional objects to be spotlighted in *Deep-Sky Companions: The Caldwell Objects*.

The name "hidden treasures" evolved over time. Its origins date to April 30, 1997, when I swept up globular cluster NGC 5286 in Centaurus, and noted that it was "hidden" in the glare of M Centauri. Two months later, I encountered another globular cluster, NGC 6441 in Scorpius, "hidden" in the glare of G Scorpii. Then, on the evening of December 27, 1997, I found the galaxy NGC 404 "hidden" in the glare of Beta Andromedae. Suddenly the idea for a new catalog of objects emerged; that evening I sat down and began drafting a short list of potential "hidden treasures"; what started as an investigation in thought turned into a 10-year treasure hunt in the sky.

I decided to call them hidden treasures for four primary reasons. (1) Some of the objects, like NGC 404 (Hidden Treasure 5) or NGC 2024 (Hidden Treasure 34) are bright enough to be seen in binoculars but can be easily overlooked because they lie so close to a bright star. (2) Other objects are overshadowed by a more popular (Messier or Caldwell) object nearby; the naked-eye open cluster NGC 1647 (Hidden Treasure 27) lies between the Hyades (Caldwell 41) and Messier 1 (the Crab Nebula). (3) Some bright nebulae are small, and they appear on star charts as tiny green boxes (sometimes unnumbered), so they are easily overlooked; NGC 1333 (Hidden Treasure 15) in Perseus is a beautiful reflection nebula that appears on many popular star charts as a box, no bigger than the symbol for a 4th-magnitude star, and NGC 2163 (Hidden Treasure 35), aside from being totally obscure, is a stunning, bipolar planetary nebula in Orion visible in a 2-inch refractor. And (4) some objects – such as the "peculiar planetary" NGC 1360 in Fornax – just seem to lie in regions of sky that are, for one reason or another, neglected by observers due to the lack of bright stars nearby. *Hidden Treasures*, then, will open up a whole new window on the universe for small-telescope users who want to move beyond the Messier and Caldwell catalogs.

When I began my treasure hunt in December 1997, I recorded only objects that were easy to see (meaning visible at a glance), or ones that had immediate "*wow* appeal." The list grew from about two dozen objects in 1998 to about 125 objects by the end of 2001. That's when I became curious as to how popular these objects I was finding were to other observers. I began researching the World Wide Web and other sources for popular deep-sky-observing lists from around the world. The ones I selected for comparison are listed below in alphabetical order:

- "Best Objects in the New General Catalog," A. J. Crayon and Steve Coe, Saguaro Astronomy Club (Phoenix, Arizona).
- "Best Sky Objects: from South African Astronomical Observatory Latitude," John Caldwell, SAAO (South Africa).
- *Celestial Harvest: 300-Plus Showpieces of the Heavens for Telescope Viewing and Contemplation*, Mullaney, James: Mineola, NY: Dover Publications, 2002; formerly *The Finest Deep-Sky Objects*, Mullaney, James and Wallace McCall, Cambridge: Sky Publishing, 1978.
- "Finest N.G.C. Objects," Alan Dyer (Royal Astronomical Society of Canada).
- "Herschel 400," Ancient City Astronomy Club (St. Augustine, Fl).
- "Index Caldwell Catalog," Ian Cooper (Palmerston North, New Zealand).
- "Jack Bennett Catalogue," Jack Bennett (South Africa).
- "Supplementary Catalogue of Bright Deep-Sky Objects," Richard J. Smith (Plymouth).
- TAAS 200, The Albuquerque Astronomical Society (New Mexico).

I suspected that once all the Messier and Caldwell objects were removed from these lists, the overlap would not be overwhelming. I was right. The reasons are many. Some of the lists are for Southern Hemisphere observers, others for Northern Hemisphere observers. The lists also vary depending on the observer's latitude: the list created by the Albuquerque Astronomical Society, for instance, includes more southern objects than does that for the Royal Astronomical Society of Canada. Some of the lists include double and multiple stars, others do not. It also became clear that few of these lists were created solely for the benefit of the small-telescope user.

The challenge I faced was to try and make a compelling list of new deep-sky objects that do not appear in the Messier and Caldwell catalogs, are about 10° above the horizon from latitude 40° north, and can be seen in a 4-inch telescope under a dark sky. To achieve that goal, I went through each list and removed all the Messier and Caldwell objects, followed by the 75 objects I had in my personal hidden treasures list. Next, I removed any objects that could not be seen from my latitude criteria. That left me with 75 objects from my original hidden treasures list and some 80 additional objects from the other lists.

I needed only 129 objects (the 109 hidden treasures plus 20 additional objects), so I further refined the list by including only those objects (with a few exceptions) that appeared in two or more of the selected lists. Since I did not know what their objects looked like through a small telescope (many of the lists used in this comparison were created for observers using larger scopes), I went out and observed all of them with my 4-inch telescope over the course of a year, and removed from the list any objects I could not see, or found very difficult to see.

In time the concept of the Hidden Treasures list evolved further. After sifting through the Messier and Caldwell lists, I realized that they contain a wealth of deep-sky objects discovered prior to September 1782, when William Herschel began his great deep-sky survey. I wondered how difficult it would be to complete that historical list, to include in *Deep-Sky Companions: Hidden Treasures* all the other deep-sky objects that did not make it into those lists. It turned out that the task could be accomplished with minimal effort, as long as I discounted all double or multiple stars mistaken for nebulae, any questionable or missing objects, and the Large and Small Magellanic Clouds. Therefore, anyone who buys the first three volumes of *Deep-Sky Companions* will have at their fingertips the history and astrophysics of every deep-sky object discovered prior to 1782 – up to and including the discoveries made by Caroline Herschel. Six of these objects, all discovered by Abbé Nicolas Louis de Lacaille, cannot be seen from latitude 40° north, but they are bright southern-sky objects. Therefore, only about 103 objects will be 10° or higher in the sky as seen from a latitude of 40° north; I say, "about," because one object, NGC 1291, a fantastic galaxy in Eridanus, is only 1° south of that cut-off point.

THE FINAL LIST

The final Hidden Treasures list did not end with the addition of the historical objects. In fact, the list changed more than two dozen times in the course of 10 years. Sometimes I would either come across a new object in a comet sweep, other times I would learn of a neglected object in a literature sweep. It was incredibly difficult at times to decide which objects should be included in the main list. For instance, shortly before I finalized my list, I decided to pull two of my favorite galaxies – NGC 134, in Sculptor, and NGC 5907, in Draco – because I thought they might be a tad too faint for some small-telescope users. I replaced them with two unanimously "better" and brighter objects. Actually, creating this book was a great learning experience for me, because, thanks to the "best-of" lists mentioned above, I got to see for the first time so many new and wondrous objects. So the Hidden Treasures list is really, in large part, one created by the collective "you." To illustrate this point, I have, in Appendix D, included a deep-sky list comparison table, so you can see at first hand which individuals, or astronomical societies, thought any particular object was worthy of your time.

The final 109 hidden treasures comprise 38 open star clusters, 35 galaxies, 14 planetary nebulae, eight globular star clusters, eight bright nebulae, one dark nebula, one star of high-proper-motion, and four officially recognized asterisms (two of which were formerly known as clusters and one of which I had independently discovered). In the Messier catalog, M7 is the most southerly object with a declination of 35°. Again, all but six of the entries culminate 10° above the southern horizon from a latitude 40° north. All are visible in a 4-inch telescope under a clear, dark sky. In fact, the vast majority of them can be seen (some with effort) in a nineteenth-century, brass spyglass or 7 × 50 binoculars under the same dark-sky conditions.

The following table compares the number and types of deep-sky objects covered in all three catalogs.

Object comparison table

Object type	Messier catalog	Caldwell catalog	Hidden Treasures catalog
Open star clusters	27	28	38
Galaxies	39	35	35
Globular star clusters	28	18	8
Bright nebulae	7	12	8
Planetary nebulae	4	13	14
Dark nebulae	0	1	1
Supernova remnants	1	2 segments of 1	0
Star clouds	1	0	0
High-proper-motion stars	0	0	1
Double stars	1	0	0
Asterisms	1	0	4

Owners of all three titles in the *Deep-Sky Companions* series will have the most up-to-date astrophysical and visual information on nearly 330 deep-sky objects, with ancillary data on many more. And since the astrophysical, visual, and tabular data in *Hidden Treasures* have been gleaned from many of the same sources in *Deep-Sky Companions: The Messier Objects* and *Deep-Sky Companions: The Caldwell Objects*, and since all the objects have been viewed with the same 4-inch telescope under the same sky conditions, observers can compare the physical properties and visual descriptions of any two objects in the three titles with complete confidence. No other series of books to my knowledge offers observers such consistent data.

The Hidden Treasures list includes the brightest, smallest, most unusual, and, arguably, *the* most fascinating planetary nebulae in the night sky. It also includes a double cluster of stars in Perseus, i.e. not *the* Double Cluster, bubbling cauldrons of vapor that mark the sites of intense star formation, starburst galaxies that can manufacture suns at the phenomenal rate of hundreds of millions per year, open clusters so young that the earliest ancestors of humans could have seen them just beginning to shine, and globular clusters so old that they just may be as old as the universe itself. There's a pair of interacting galaxies that are pivotal to Halton Arp's redshift argument (which challenges the almost universally held belief that the large redshifts of quasars and other active galaxies are due to cosmic expansion), and a galaxy whose core may harbor a 10-million-solar-mass black hole. Pink planetaries, yellow globulars, blue clusters, black clouds, they're all here.

As with *Deep-Sky Companions: The Messier Objects* and *Deep-Sky Companions: The Caldwell Objects*, the purpose of this book is to offer new and fresh perspectives on the history and visual appearance of each object; to help you find each object; and to summarize the latest research findings on each. In Chapter 1, "About this book," I discuss the telescopes I used to observe the hidden treasures, my observing sites and methods, helpful observing hints, and more. Since the history, astrophysics, and visual descriptions of many of these objects have never been described at length in any other popular work, this chapter also explains my approach to presenting the information. I detail the 109 hidden treasures in Chapter 2. In many cases the essays describe recent observations from the Hubble Space Telescope, the world's largest ground-based telescopes, and a fleet of spacecraft that now peer (or have peered) into the universe with X-ray and infrared-sensitive "eyes." The essays are also flush with historical anecdotes, including solutions to some outstanding mysteries (including the debate over whether M102 is NGC 5866); observational challenges (naked eye and telescopic); and descriptions of other interesting objects.

Several appendices complete the work. Appendix A is an informative biography of Caroline Herschel written by the esteemed deep-sky observer Barbara Wilson of Houston, Texas. Its intent is not only to complement Larry Mitchell's biography of William Herschel that appears in Appendix C of *Deep-Sky Companions: The Caldwell Objects*, but also to demonstrate how important Caroline's discoveries were to astronomy. The biography is fresh, written from a woman's perspective, and includes several surprises – namely it removes from Caroline Herschel's list of discoveries some objects we have long admired and replaces them with objects that, until recently, no one knew she had discovered. So it is an exciting read. Appendix B tabulates each hidden treasure's position, type, angular size, and apparent magnitude. Appendix C does the same for the 20 additional hidden treasures. Appendix D is a table that lists each hidden treasure and shows which individuals or astronomical societies considered it to be one of the finest in the night sky, or whether it was included for purely historical reasons. At the end of the book, "The treasure chest," is a place for you to "store" each hidden treasure you find. It is, in essence, a checklist. It includes spaces for you to write down important information, such as the date observed, your location, the telescope and magnification used, atmospheric seeing and transparency, and any other special notes you want to record. It is a personal log that you can return to weeks, months, or years later to see how you are progressing as an observer.

Deep-Sky Companions: Hidden Treasures is more, much more, than a valuable resource, or a companion to *Deep-Sky Companions: The Messier Objects* and *Deep-Sky Companions: The Caldwell Objects*; it is *your* companion under the stars. Its aim is to encourage you to observe, to help you find a particular object, and to entertain you, especially on cloudy nights, with the fascinating histories, mysteries, and discoveries that led to our understanding of these objects – our objects. The beauty of the hidden treasures list is that it includes a little something of us all. If there is one thing we amateurs know how to do it is to share – our thoughts, our experiences, and, most of all, our passion for the stars. So go forth into the night and look in wonder at these unsung heroes; they deserve your applause.

Acknowledgments

JUST AS NO PIRATE SAILS ALONE, NO author works in complete isolation. I, in fact, had a band of loyal compatriots assisting me in this project. First I'd like to fire a salute to all the observers and amateur astronomy organizations who promote deep-sky lists that go beyond the Messier objects. These lists helped me to steer my telescope toward objects of popular desire. And thank you Al Nagler of Tele Vue Optics for creating such a wonderful "star ship," which has taken me on countless adventures. Special thanks to Ian Cooper, New Zealand's premier deep-sky observer, for educating me on some of the more popular deep-sky objects visible from the southern skies, and James Mullaney for keeping me abreast of the finest northern deep-sky wonders.

Brent Archinal has been my first mate on every voyage I've embarked on with the *Deep-Sky Companions* series. His insight into the history of astronomy and the discovery of deep-sky objects (especially of star clusters) has proven invaluable. Archinal also reviewed the science in my open cluster essays. Likewise, very special thanks go out to Sun Kwok for reviewing the science in my planetary nebula essays and to William Harris for his reviewing of my globular cluster essays. Charles Lada scanned my coverage of the diffuse nebulae and Larry Mitchell reviewed, and contributed to, my essays on galaxies. Thank you, all of you, for your valuable time and patience. The book would not be the same without your critiques.

A 20-cannon salute goes out to my long-distance friend and fellow sky pirate Barbara Wilson, whose fresh and well-researched history of Caroline Herschel gave me a renewed respect for this great woman. Wilson's history of Caroline had the added benefit of being reviewed by astronomical historian Michael Hoskin, who had recently solved many of the mysteries surrounding objects discovered by, or supposedly discovered by, Caroline Herschel. Hoskin also supplied Barbara with Caroline's original observing records, which led to further surprises about her observations and some new insights into her discoveries.

A special thanks goes out to my longtime friend Daniel W. E. Green who helped me to track down some obscure volumes of astronomical writings. Dan also reviewed Wilson's summary of Caroline Herschel's comet discoveries. *Sky & Telescope* senior editor Dennis di Cicco helped me in my research on the Northern Coalsack; he and fellow senior editor Roger W. Sinnott also provided me with the discovery dates and descriptions of several deep-sky objects found by John Herschel.

Sky & Telescope's rights and permissions manager Sally MacGillivray helped me to get in touch with Sue Tritton at the Royal Observatory in Edinburgh, who, in turn, granted me permission to use the wonderful images of southern deep-sky objects taken with the UK Schmidt Telescope. And a special thank you goes to *Sky & Telescope*'s president/publisher Susan B. Lit for granting me permission to use material from two of my "Eye on the Sky" columns in this book. Much of my essay on Hidden Treasure 31 (the Lost Jewel of Orion) first appeared in the January 2005 issue of *Sky & Telescope* (p. 93); and much of my essay on Hidden Treasure 75 (NGC 5866) first appeared in the March 2005 issue of *Sky & Telescope* (p. 78) under the title "M102: Mystery Solved."

A deep bow goes out to my friend and fellow photographer Charlene Meyers, who took the lovely photograph of me and my telescope on the slopes of Mauna Kea. I also thank my friend and astrophotographer Adriana Sherman of Guatemala, who donated her fine images of NGC 2903 in Leo and the field around NGC 2024 in Orion.

No thank you is long enough, or great enough, to acknowledge my loving wife, Donna, and Daisy Duke, our adorable papillon, for tolerating this rogue sailor who would suddenly cast off on many a clear night and not return 'til morning. They understood the cause was noble. I have been a celestial privateer for Simon Mitton at Cambridge University Press, amassing treasures for a book that promised to enrich the lives of observers around the globe. The journey would not have been possible were it not for his generous support. Thank you Simon for believing in the project and publishing the work. And thank you Jacqueline Garget, my editor at CUP, for taking the project through to completion. Of course, I have had the final say in all matters regarding the accuracy of the text. If you uncover some hideous editorial or scientific crime and demand an execution, I am the pirate you have to capture and hang.

CHAPTER 1
About this book

Seeing is in some respects an art that must be learnt. William Herschel

EVERY MOMENT SPENT under the stars is a treasure hunt – a visual journey that leads us to endless riches in the heavens above. And I've loved each adventure from the beginning. When I was young, perhaps age eight, I set out on one of my first deep-sky adventures – to hunt down the great globular cluster M13 in Hercules. I had seen a full-page photo of it in *Planets, Stars, and Space* (first published in 1957 by Creative Educational Society in cooperation with the American Museum of Natural History, New York), which my father kept on the lower shelf of a bookcase set up in the living room. The book's authors, Joseph Miles Chamberlain and Thomas D. Nicholson, described the cluster as a "huge ball of stars . . . so numerous that the center . . . resembled a brilliant mass of light." My Golden Guide, *The Sky Observer's Guide*, went one step further, saying that this "globular may have 100,000 [stars]." It also said it may be seen with the naked eye.

It seemed incredible to me at the time that if I could first find the Keystone of Hercules among the multitude of stars overhead, I could then search for a citadel of 100,000 suns – one so distant and so tightly packed together (yet so bright) that I could see it with the naked eye as a hazy star.

I was still learning the constellations at the time, so I had to be certain that I had the right keystone. The search was glamorous in itself, because it made me feel the thrill of voyage. I was a flesh-and-blood Jim Hawkins (the protagonist in Robert Louis Stevenson's *Treasure Island*) in search of "buried" cache.

When I finally convinced myself that I had the right keystone, I could scarce believe that the dim smudge I was seeing was the great cluster. So I dashed inside the house, grabbed my father's binoculars, and returned to the chill of the night. After a few minutes of letting my eyes readjust to the darkness, I set out again to search for the keystone. This time, when I spied it, and the little smudge within, I raised the binoculars to the spot and began "bobbing" for the cluster – I was still learning how to use binoculars, so they were quite unsteady. It took a while, but in time I saw a globe of white light flash through the field of view. I narrowed my bobs until I held the object steady. The cluster was small, misty, and round. I saw no teeming masses of suns, no glints of scintillating starlight, just a tiny cloud of light. But I was not disappointed. To a kid just beginning to fathom the universe, the mist of M13 was one of the most majestic sights

The circular depression near the middle of this frame is the summit crater of Kilauea, called Halemaumau, which spans 1000 feet (305 meters).

imaginable. The wonder, it turned out, was not so much finding it, but beholding it, imagining it in ways my eyes could not see.

I've never lost that childhood wonder. In fact, that same fascination guided me in the making of this book.

THE TELESCOPE AND SITE

All the observations for this book were made from the Big Island of Hawaii, where I live. My observing site is near the 4,200-foot-high summit of Kilauea, a gently sloping shield volcano that has been in near-continuous eruption since January 1983. This is the same general area (as the Hawaiian hoary bat flies) where I made the observations for *Deep-Sky Companions: The Messier Objects* and *Deep-Sky Companions: The Caldwell Objects.* But after the latter title was published, I decided to change my observing site because tourist traffic was becoming a nuisance. My new site is more secluded, so I observe fewer car headlights through my old Tele Vue 4-inch f/5 Genesis refractor. Since clouds could blow in unexpectedly at any time, I needed to be migratory; on occasion I had to drive a few thousand feet to get above the clouds to observe a hidden treasure. The photograph above is an aerial shot showing the summit area of Kilauea volcano where the observations were made.

After I completed the observations for *Deep-Sky Companions: The Caldwell Objects*, I called Tele Vue founder, Al Nagler, and asked if his company could, once again, clean up an old friend. (If you've read *Deep-Sky Companions: The Caldwell Objects*, you'll know that after four years of exposure to the acidic gases escaping from Kilauea volcano, I had to have the telescope's optics cleaned, which, when Nagler saw them, almost caused him to faint.) After a long moment of silence – the kind one expects when asked to pray for the loss of a loved one – Al conceded to operate. I got my Genesis back, this time with a

card from the optician who wondered if there was any way, in the future, I could not place the telescope in the direct path of any sulfuric acid clouds. I promised to try.

For the *Hidden Treasures* project, I used three eyepieces (also made by Tele Vue): a 22-mm Panoptic, a 7-mm Nagler, and a 4.8-mm Nagler. On the Genesis these eyepieces provide magnifications of 23×, 72×, and 105×, respectively. A 1.8× Barlow lens gave me additional magnifications of 41×, 130×, and 189×, while a 3× Barlow gave me high powers of 216× and 315×. In addition, I now own a Nagler 3–6-mm zoom eyepiece, which when used together with my Barlow lenses offers me a wide variety of magnifications ranging from 84× to 504×. As a finder I use a Tele Vue Qwik Point (it's like a laser pointer). The Genesis offers me a field of view near 3° wide when I use the 22-mm Panoptic. The telescope sits in the cradle of a sturdy Gibraltar mount, and the entire set up can be broken down in two minutes in case I need to be mobile.

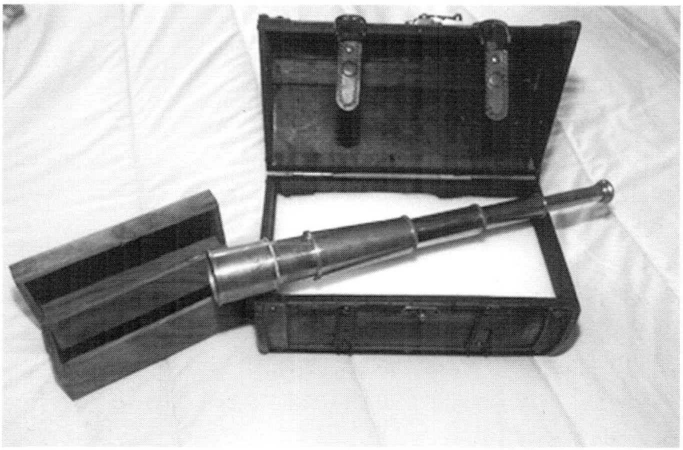

In addition to the Tele Vue 4-inch, I also used a beautiful nineteenth-century brass telescope made by Ross of London. It was a Christmas present from my wife, Donna, who bought it while we were in South Africa during a solar-eclipse cruise in December 2002. I was, at the time, visiting the site where John Herschel had made his great survey of the southern sky, and Donna was off exploring the Capetown area. That's when she found an antique shop and this brass telescope, which I simply refer to as "the antique telescope" in Chapter 2. The tube measures $17\frac{1}{8}$ inches when open, and $7\frac{1}{4}$ inches when closed. The $1\frac{3}{4}$-inch objective is in excellent condition. I found the views through it enlightening. I observed many of the hidden treasures with this telescope. Doing so made me appreciate the views of our astronomical ancestors, especially Caroline Herschel who made her first sweeps of the skies in 1783 with a telescope very similar to this one.

HOW TO USE THIS BOOK

To find a hidden treasure, first locate the object's position on the wide-field finder chart that opens each entry. In keeping with the book's theme of hidden treasure, "X" marks the spot; actually "HT X" marks the spot. (Note that the "X" will most likely not be centered on the map, so you have to hunt a bit to locate it among the stars.) Take note of the brightest stars (those with Greek letters or Flamsteed numbers) near your hidden treasure. Next, locate those stars on the detailed finder chart that accompanies the object's photograph and text (like the wide-field finder chart, the detailed finder charts are oriented with north up and west to the right). Finally, find the part of the text that describes how to locate the object and simply follow the directions.

After the full-page wide-field finder chart, each object's entry in Chapter 2 opens with a photograph of the object (oriented with north up and west to the right, unless otherwise noted) and a list of essential data: hidden treasure number; common name(s), if any; object type; constellation; equinox 2000.0 coordinates; apparent magnitude; angular size or dimensions; surface brightness in magnitudes per square arcminute (for galaxies); distance; and the object's discoverer and discovery date. The text includes a history of the object's discovery; recent research findings; naked-eye or binocular impressions; the object's appearance through the 4-inch Genesis refractor at various magnifications; descriptions by other observers using larger instruments; a visual challenge or two; and brief notes on any interesting objects in the same region of sky.

A "drawing" also accompanies the text, so you can compare your view of any hidden treasure with my own. The views may be very dissimilar, but that's okay; we all see things differently.

You'll find William Herschel's original published description of the object or, if William did not observe the object, his son John's. If neither observer discovered or observed the object, that section is blank. Larry Mitchell, a member of the Houston Astronomical Society, supplied me with William's original notes, which he drew from his original catalogs as they appeared in the *Philosophical Transactions* of the Royal Society of England.

John Herschel's quotations have been gleaned from those given by the "Deepsky Observer's Companion" (www.fortunecity.com/roswell/borley/49/), which was created by the Astronomical Society of South Africa to promote its Deepsky Observing section. The quotes are from John Herschel's original observations, published in 1847 as "results of Astronomical Observations made during the years 1834, 5, 6, 7 [and] 8, at the Cape of Good Hope; Being the completion of a telescopic survey of the whole surface of the visible heavens, commenced in 1825." During his stay in South Africa, John Herschel often made several observations of each object. The quotes used in this book's tables, however, refer only to his first observation; a date is given only if the junior Herschel discovered the object. At the end of each Herschel description is a code contained in parentheses ("H I-156," for instance, or "h 242"). This code dates to a classification system created and used by the Herschels. "H" stands for the elder Herschel and "h" for his son.

The Roman numeral in William Herschel's system identifies the class into which he placed each object:

I. Bright nebulae
II. Faint nebulae
III. Very faint nebulae
IV. Planetary nebulae: stars with burs, with milky chevelure, with short rays, remarkable shapes, etc.
V. Very large nebulae
VI. Very compressed and rich clusters of stars
VII. Pretty much compressed clusters of large or small stars
VIII. Coarsely scattered clusters of stars

The Arabic numeral that follows is simply the order in which that object appears in that class. So H I-156 is the 156th object in Herschel Class I (bright nebulae).

The original 1888 *New General Catalogue* (*NGC*) description, or a description from the supplemental *Index Catalogues*, follows each Herschel description.

SOURCES OF DATA AND INFORMATION

The data and information in this book were drawn from a variety of modern sources. Many of these sources were used in *Deep-Sky Companions: The Messier Objects* and *Deep-Sky Companions: The Caldwell Objects*, so you can compare the properties of the respective catalogs' objects with confidence. Generally speaking, recent research findings on the physical nature of these objects were gleaned from the *Astronomical Journal* or the *Astrophysical Journal*; citations are frequently given. From each object's apparent diameter and distance, I calculated its physical dimensions using the formulas that appear on p. 35 of *Deep-Sky Companions: The Messier Objects*. Other information, such as constellation lore; properties of stars; and object's positions, apparent magnitudes, angular sizes, and surface brightnesses, come from the following excellent sources (primary sources are listed first).

Star names, constellations, and mythology

Allen, Richard Hinckly. *Star Names: Their Lore and Meaning*. New York: Dover Publications, 1963.

Staal, Julius D. W. *The New Patterns in the Sky: Myths and Legends of the Stars*. Blacksburg, VA: McDonald and Woodward, 1988.

Motz, Lloyd and Carol Nathanson. *The Constellations: An Enthusiast's Guide to the Night Sky*. New York: Doubleday, 1988.

Ridpath, Ian. *Star Tales*. New York: Universe Books, 1988.

Stellar magnitudes and spectra

Hirshfeld, Alan, Roger W. Sinnott, and Francois Ochsenbein, eds., *Sky Catalogue 2000.0*, Vol. 1, 2nd edn. Cambridge: Cambridge University Press/Cambridge, MA: Sky Publishing Corp., 1991.

CDS. *SIMBAD* (*Set of Identifications, Measurements, and Bibliography for Astronomical Data*). Strasbourg, France: Centre de Donnees Astronomiques de Strasbourg, http://simbad.u-strasbg.fr/ and http://simbad. harvard.edu/.

Stellar distances

ESA. *The Hipparcos and Tycho Catalogues*. Noordwijk, The Netherlands: European Space Agency, 1997.

Double stars

USNO. *The Washington Double Star Catalog*. Washington, DC: Astrometry

Department, US Naval Observatory, http://ad.usno.navy.mil/ad/wds/wds.html.
Couteau, Paul. *Observing Visual Double Stars*. Cambridge, MA: MIT Press, 1982.
Hirshfeld, Alan and Roger W. Sinnott, eds., *Sky Catalogue 2000.0*, Vol. 2. Cambridge, MA: Sky Publishing Corp./Cambridge: Cambridge University Press, 1985.
Luginbuhl, Christian B. and Brian A. Skiff. *Observing Handbook and Catalogue of Deep-Sky Objects*. Cambridge: Cambridge University Press, 1989.

Variable stars

Hirshfeld, Alan and Roger W. Sinnott, eds., *Sky Catalogue 2000.0*, Vol. 2. Cambridge, MA: Sky Publishing Corp./Cambridge: Cambridge University Press, 1985.
Luginbuhl, Christian B. and Brian A. Skiff. *Observing Handbook and Catalogue of Deep-Sky Objects*. Cambridge: Cambridge University Press, 1989.
American Association of Variable Star Observers, http://www.aavso.org/.
Levy, David. *Observing Variable Stars: A Guide for the Beginner.* Cambridge: Cambridge University Press, 1998.

Open star clusters

Archinal, Brent A. and Steven J. Hynes. *Star Clusters.* Richmond, VA: Willmann-Bell, Inc., 2000.

Globular star clusters

Harris, William E. *Catalog of Parameters for Milky Way Globular Clusters.* Hamilton, ON: McMaster University, http://physwww.physics.mcmaster.ca/%7Eharris/mwgc.dat *Star Clusters.*
Skiff, Brian A. Observational data for galactic globular clusters. *Webb Society Quarterly Journal* **99**: 7 (1995) updated May 2, 1999.

Planetary nebulae

Skiff, Brian A. Precise positions for the NGC/IC Planetary Nebulae. *Webb Society Quarterly Journal* **105**: 15 (1996). (See positions.)
Luginbuhl, Christian B. and Brian A. Skiff. *Observing Handbook and Catalogue of Deep-Sky Objects*. Cambridge: Cambridge University Press, 1989. (See dimensions and central star magnitudes.)
Cragin, Murray, James Lucyk, and Barry Rappaport. *The Deep-Sky Field Guide to Uranometria 2000.0*, 1st edn. Richmond, VA: Willmann-Bell, Inc., 1993.
Acker, Agnes and Francis Gleizes. *Catalogue of the Central Stars of True and Possible Planetary Nebulae.* Strasbourg: Observatory of Strasbourg, France, 1982.
Hynes, Steven J. *Planetary Nebulae: A Practical Guide and Handbook for Amateur Astronomers.* Richmond, VA: Willmann-Bell, Inc., 1991.

Planetary-nebula distances generally were gleaned from the professional literature or from the World Wide Web page of the Space Telescope Science Institute (www.stsci.edu).

Diffuse nebulae

Cragin, Murray, James Lucyk, and Barry Rappaport. *The Deep-Sky Field Guide to Uranometria 2000.0*, 1st edn. Richmond, VA: Willmann-Bell, Inc., 1993.
Hirshfeld, Alan and Roger W. Sinnott, eds., *Sky Catalogue 2000.0*, Vol. 2. Cambridge, MA: Sky Publishing Corp./Cambridge: Cambridge University Press, 1985.

Diffuse-nebula distances were gleaned from the professional literature.

Galaxies

Tully, R. Brent. *Nearby Galaxies Catalog.* Cambridge: Cambridge University Press, 1988. (See types, angular sizes, distances, masses, and luminosities.)

Cragin, Murray, James Lucyk, and Barry Rappaport. *The Deep-Sky Field Guide to Uranometria 2000.0,* 1st edn. Richmond, VA: Willmann-Bell, Inc., 1993. (See apparent magnitudes and surface brightness.)

NASA. *The Extragalactic Database.* Pasadena, CA: Infrared Processing and Analysis Center, http://nedwww.ipac.caltech.edu/. (See positions and, when not available from the above references, other parameters.)

Extragalactic supernovae

IAU. *List of Supernovae.* Cambridge, MA: Central Bureau for Astronomical Telegrams, http://cfa-www.harvard.edu/iau/lists/Supernovae.html.

Historical objects

Glyn Jones, Kenneth. *The Search for the Nebulae.* Bucks: Alpha Academic, 1975.

Information on pirates

Cordingly, David. *Under the Black Flag: The Romance and the Reality of Life Among the Pirates.* Orlando, FL: Harcourt Brace & Company, 1995.

Marine Research Society. *The Pirates' Own Book: Authentic Narratives of the Most Celebrated Sea Robbers.* New York: Dover Publications, 1993.

Note that the World Wide Web Uniform Reference Locators, or URLs, are subject to change. The dimensions, magnitudes, and positions of all other additional deep-sky objects in this book were taken from *The Deep-Sky Field Guide to Uranometria 2000.0.*

As in *Deep-Sky Companions: The Messier Objects,* the data in this book differ from those appearing in older but popular references. This book contains the most up-to-date astronomical data and accurate historical and observational information about each object in the Caldwell catalog than you'll find in any other book in the popular literature.

THE FINDER CHARTS

The wide-field and detailed star charts in Chapter 2 are unlike any you've seen. They are of my own design. I created them to look, somewhat, like a treasure map. The hidden treasure number and proper name of each object appears at the top of the chart. An astronomical compass rose can be found at the bottom (north is always up and west to the right). A magnitude scale is at left, and a handy reference to the corresponding star chart(s) in Wil Tirion's *Sky Atlas 2000.0* and the *Uranometria 2000.0* – two of the most popular star atlases used by observers today – are at lower left.

Of course, since this is a "treasure" map, "X" (actually "HT X") marks the location of the object you want to find. Note that often the "X" will not be centered on the map, so you may have to hunt a bit to locate it; it may even be close to the margins of the map. If two hidden treasures are discussed together in the same essay, both objects will

Hidden Treasures 1 & 2

NGC 189 & 225

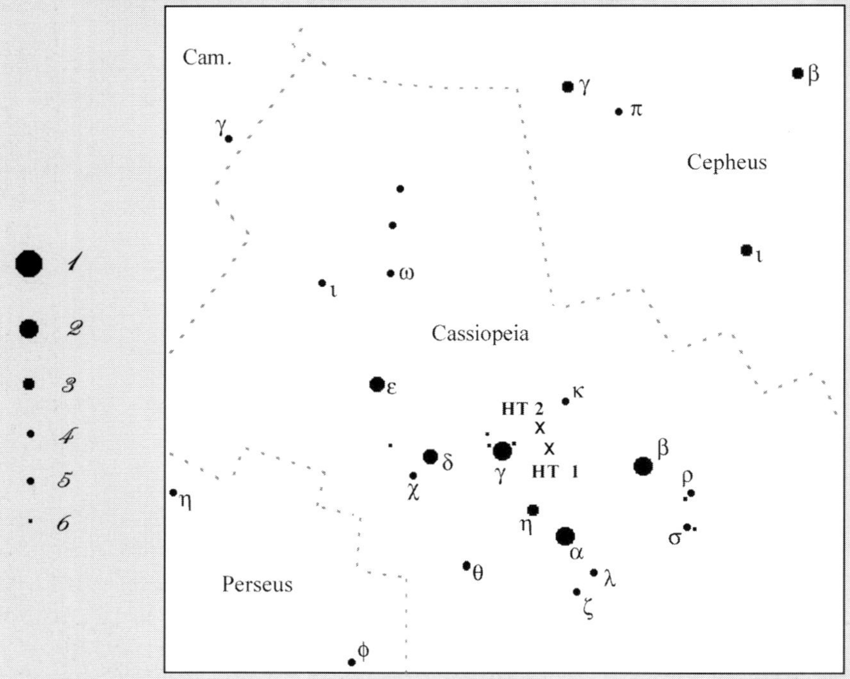

© Stephen James O'Meara

Tirion: Chart 1

Uranometria: Charts 16 & 36

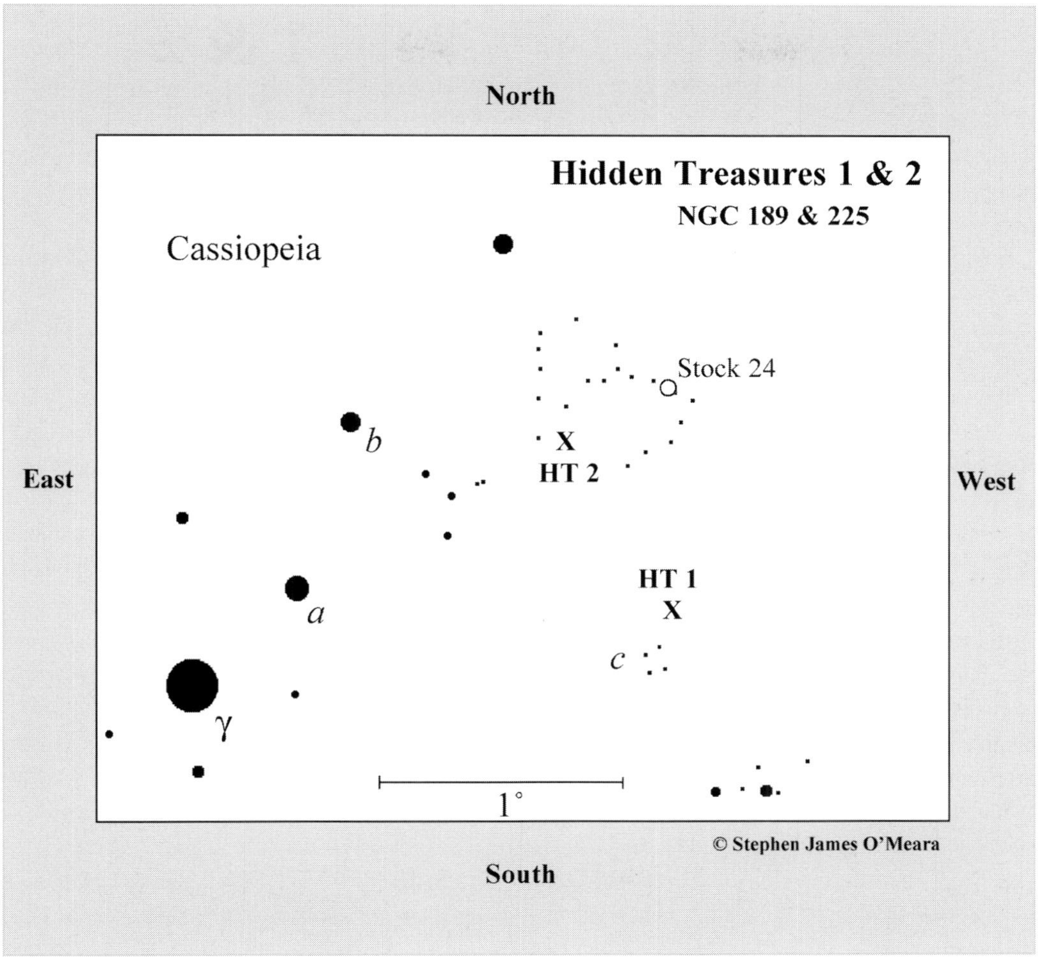

be marked on the same wide-field finder chart with an "X," but they will be differentiated by their hidden treasure number, say, "HT 1" and "HT 2."

The purpose of the wide-field finder charts is simply to show you the brightest constellations or starfields around the hidden treasure. Each show stars roughly to magnitude 6, but generally only in the region around the hidden treasure. Faint stars are more prevalent if the constellation is dim, like Fornax. In creating these charts my philosophy was to get rid of the peripheral noise. Why clutter the view with lots of dim stars and other objects when all you want to do is see where in the sky the hidden treasure lies and which bright stars are nearest to the object?

The brightest stars in each constellation shown have been labeled with either a Bayer (Greek) letter or a Flamsteed number. Some stars may have an italicized lower-case letter, like *a* or *b*; these are additional guide stars, which you'll find in the text as Star *a* or Star *b*, etc. The detailed finder charts have the same orientations and work on the same principles, only on a smaller scale; a scale bar appears at the bottom of each chart. One symbol, a circle, is used to mark the location of other interesting deep-sky objects nearby, a label with the object's proper name accompanies each circle. Note, however, that in the case of *NGC* objects, only the numbers are shown, not the "NGC" prefix. Note too that in the detailed finder charts the italicized letters may also refer to asterisms described in the text.

To find a hidden treasure, then, all you need to do is first locate the object's position on the wide-field finder chart. Next, take note of the brightest stars nearest to the object and locate them on the detailed finder chart. Finally, find the part of the text that describes how to locate the object and simply follow the directions. Nearly all the deep-sky objects mentioned in the text are plotted on the charts, as are named or numbered stars. Sometimes, though, such companion objects or stars being noted are labeled on the photographs rather than on the star charts, especially when the area of interest is small and crowded, as is the case with NGC 281 (Hidden Treasure 3).

THE PHOTOGRAPHS

As in my *Deep-Sky Companions: The Messier Objects* and *Deep-Sky Companions: The Caldwell Objects* all the photographs in this book are reproduced in black and white, with north up and west to the right. The vast majority of deep-sky images in this book are reproduced digitized photographs taken by enormous Schmidt telescopes in both hemispheres. These photos have been made available to astronomers and scientists worldwide by the visionary architects of the Digitized Sky Survey (DSS), which can be perused on the World Wide Web at http://archive.stsci.edu/dss/. (The copyright for the DSS photos of objects used in this book rests with the Anglo-Australian Observatory Board, the United Kingdom Particle Physics and Astronomy Research Council, the California Institute of Technology, and the Associated Universities for Research in Astronomy; they are used here with permission.) Detailed credits appear in Appendix E.

In several cases we have also reproduced images from the Hubble Space Telescope and large telescopes like the Canada–France–Hawaii Telescope atop nearly 14,000-foot-high Mauna Kea volcano on the Big Island of Hawaii. These images were used for the sole purpose of inspiring you to use your imagination. You certainly will not see anything like these images when you look through your telescope, but how else can you fully appreciate what it is you are seeing? So do not be discouraged, be enlightened.

THE DRAWINGS

All the sketches in Chapter 2 are composites of field drawings I made at various magnifications. They are shown approximately with north up and west to the right, unless otherwise noted. Whenever possible, the orientation matches that of the corresponding photograph. Scale bars are included to help you size up each object in your own telescope.

The composites show details visible at low, medium, and high power. For instance, I might have seen a galaxy's faint halo readily at low power but not at high power. On the other hand, a small knot in that galaxy's arm might have been obvious at high power but inconspicuous at low power. Both the halo and the knot will appear in my sketch.

The drawings also show details seen in fits and starts over several hours observing from a dark-sky site. Unless otherwise noted, the drawings all represent views obtained from the area around the 4,200-foot-high summit of Kilauea volcano, an extremely dark location. Under these conditions my 4-inch Genesis will perform as well as an 8- to 10-inch Schmidt-Cassegrain from a suburban site. This means that the details I portray in my drawings should be within the visual grasp of everyone using the large reflectors and catadioptric telescopes that are so popular today.

It's important for you to know that I have not always lived under dark skies. In fact, I spent most of my life – between the ages of 6 and 32 – observing under city lights. I was born in Cambridge, Massachusetts, and lived and observed in that city and its surroundings until I moved to Hawaii in December 1994. I know what it is like not to be able to see deep-sky objects even with large telescopes in city environments. Yet I was never discouraged. I never gave in to city lights. I always tried to make the best of every moment I had with my telescope. If I could only see the bright nucleus of a galaxy and not its arms, I sat behind whatever telescope I had and studied that nucleus until my eyes dripped with fatigue.

At no time did I ever feel inferior to someone seeing spectacular details in a deep-sky object that I could barely see from the city. No, those observing reports inspired me, they strengthened my resolve because I knew that some day I would have an opportunity to see what these great visual observers were detecting from their dark skies. In *Deep-Sky Companions: The Messier Objects*, I mention one of the most eye-opening moments of my life. After 30 years of never having seen the galaxy M74, many of them spent looking for it with the 9-inch f/12 Clark refractor at Harvard College Observatory, I got my first look at it shortly after I moved to Hawaii – in 7 × 35 binoculars. I can hardly begin to tell you how low my jaw dropped that evening. For the first time in my life I could now understand these words by Charles Messier: "This nebula . . . is quite broad, very dim, and extremely difficult to observe."

I did not need any prodding to get out under the stars. I had so much to see. A new window on the universe had opened before my eyes, and a new chapter in my life began. I sit behind the telescope and squeeze every photon from every object I observe because like a man who's crossed the desert without water, I am thirsty . . . only my thirst is for starlight. What's interesting is that I went from using a 9-inch refractor on a regular basis under city lights to using a 4-inch refractor under darks, and I'm seeing more in the deep-sky than ever before.

My *Deep-Sky Companions* series is not so much a set of books as it is a set of personal diaries, one that I have opened up to you. I want you to know what the possibilities are for you to see under a dark night sky. The irony of it all is that when I was younger, it seemed as if the vast majority of amateurs were living under dark skies except for me. Now, light pollution has become so rampant that it seems as if the vast majority of amateurs are observing under city skies, except for me. It's been like taking a rocket ship back in time.

As for now, I am living under natural skies. I feel like an explorer who has been given an opportunity to study the last great tract of rainforest. I spend my time documenting everything I see, because I know that there will come a day when everything I see will no longer be regarded as reality but as a myth. I hope that day never comes, but until it does, I'll be out there seeing what I can see.

Although what I see in my 4-inch is what should be visible to anyone under similar natural dark-sky conditions, I do spend a lot of time behind the eyepiece observing each object, which, I suppose to some, might seem unnatural. But it's not, it's a choice. If you want to learn how to make the most out of every observation, I suggest you read on.

THE PIRATES OF POWER

When we embark on a search for an object – especially one we've never seen before – it's generally best to use the lowest magnification available and a wide field of view. I say generally because the eye's ability to detect an object depends not only on where you place it on your eye's retina but also on the difference in contrast between the object's brightness and the brightness of the background sky; angular size also plays an important part. (A rule of thumb: low power for large objects; high power for small objects.)

Consider how difficult it is to detect Venus with the unaided eye in broad daylight, as opposed to how easily it is seen on the same night. In both cases, the planet's brightness is the same. The reason we see Venus more easily at night is because the background illumination is dimmer; an object appears brighter if its surroundings are dark. The point is, if you live in or near light-polluted skies, using too low a power will hinder your ability to detect faint stars or low-surface brightness objects, if the contrast difference between the object and sky background is too low.

The best way to find the most efficient low powers for you is to experiment with different eyepieces under different sky conditions. Do not be fooled by the seeming exactitude of some equations dealing with object visibility; these equations are general guides at best and do not reflect the full range of possibilities. The only way to determine what you can see is to venture forth and explore. There is no substitute for personal experience, which changes with sky conditions and training. The more you observe, the more adept you will become at seeing finer and finer details; remember, though, that your ability as an observer depends not so much on what you observe, or the number of objects you observe, but how you observe. I'll explain.

Let's imagine we've just embarked on a treasure hunt with our telescopes. Sweeping some 40′ northwest of Beta (β) Canes Venaticorum with low power, we spy the ghostly form of NGC 4490 – the Cocoon Galaxy (Hidden Treasure 63). What should we do next?

That depends on what kind of observer you are, or want to be. Since this book deals with "hidden treasures," I thought you'd enjoy the following analogy between the observing habits of two distinct groups of amateur astronomers and the hunting habits of two distinct groups of real-life pirates.

Some observers pursue their targets like the Barbary pirates of old, whose essence of attack was to "hit and run." In *Under the Black Flag: The Romance and the Reality of Life Among the Pirates*, David Cordingly notes, "it is [also] a characteristic of most of today's pirate attacks in Indonesian waters, when pirates may be on board the victim's ship for no more than nine or ten minutes."

Astronomical observers of the Barbary type are primarily interested in finding and accumulating as many deep-sky objects as possible. They enjoy the challenge of the hunt and are satisfied when their plan of attack succeeds. They are not interested in spending time "on board" each target, examining it carefully and diligently. After the capture, they figuratively toss the treasure on deck, kick it aside, and sail right on to the next.

In the case of NGC 4490, Barbary-type observers would revel in their find by holding it in their gaze for a few minutes, then log it as found, before they set sail for more booty. The more treasures they collect, the bigger their bounty, and the happier they become. Astronomy clubs and organizations further entice this type of activity by promising rewards for their hauls – with pins, certificates, and medals. And that's perfectly fine.

On the other side of the doubloon, however, is the romantic, of which I am one. For us, the hunt is part of a larger adventure. When we "capture" a deep-sky object, we do spend time "on board." We feel the need to sift deeper, to plumb the depths of each "hold," knowing that if we do, if we remain patient, we will be rewarded by the sight of even more riches.

We are like the pirates who sailed the Caribbean and North American waters during the golden age of piracy, whose essence of attack was a mix of lust and leisure. Consider the experience of Captain Samuel Carey, who was commander of the merchant ship *Samuel* on July 13, 1720, when it was attacked by pirates 40 miles east of the banks of Newfoundland. The pirates, Cordingly explains, "swarmed on board the *Samuel* and began taking the ship apart. They tore open the hatches and attacked the cargo like madmen . . . When they finished looting the *Samuel*, the pirates turned their attention to the crew." The attack lasted 48 hours, and it was typical of dozens like it that occurred in the Caribbean and North American waters in the early eighteenth century. Observers who are like the pirates of the Caribbean do not, metaphorically speaking, seek treasures but treasure chests. The real excitement for us is to "break the lock and open the lid," to go beyond what's superficially visible, and to rifle through what's inside. Here's an example of how I visually "tear apart" my targets.

When I capture a new object, I spend a few minutes making an initial inspection of it at low power, then take a break. I am in no hurry to plunder. I stand up, stretch my arms over my head, take a deep breath, and let myself feel the satisfaction of success. Once my body is relaxed, I reposition myself behind the eyepiece (low power) and relax my gaze. Then, while inhaling and exhaling slowly through my nose, I start my first critical examination.

I begin by placing the object in my eye's favored hot spot. Doing so allows me to gain an appreciation of the object's brightness, overall shape, and, if it is elongated, orientation. I assess how difficult the object is to see. I look for interesting objects nearby: for instance, in the case of NGC 4490, a tiny 12th-magnitude galaxy (NGC 4485) lies just north of NGC 4490's northwest flank; is it visible at low power? Everything I see, or don't see, with my peripheral vision, I record.

Some observers would move on to higher powers at this point, but I don't. While I have gained an excellent impression of what the object looks like in my eye's periphery, I now want to know how much detail is visible in

NGC 4490 at low power. To do that I must move the object between the foveal and peripheral regions of the retina. Throughout this process I ask myself questions: Does the galaxy have a bright nucleus or just a core of brightness? Is the halo elongated? Can I see any spiral structure or H-II regions? Are stars superimposed on the galaxy? Is one of them a supernova? And so on.

Again, I impose no time constraints on myself. I will observe an object at low power until I feel I have exhausted my vision's potential. Of course, examining something like NGC 4490, a 10th-magnitude barred spiral galaxy some 6′ in diameter, does not require as much time to observe as something like Messier 31 (M31), a 3rd-magnitude spiral galaxy spanning 3° of sky. But that's about all one can say about time. Once I have completed my inspection at low power, I repeat the process again at medium power (asking myself the same questions), then again at high magnification.

With every increase in power, the peripheral view begins to look more and more like a foveal view at low power, meaning that only the brightest regions of the object are obvious. It will also appear that the contrast between the sky background and the object increases, but it does not; the contrast remains the same; magnification is not selective. The reason has to do, once again, with the physiology of the eye and how it perceives brightness. Brightness is not a physical value but a sensation. "This sensation is roughly, and only roughly," as Cambridge University psychologist Richard L. Gregory explains, "related to the intensity of the light entering the eye."

A point source of a certain brightness appears more intense to the eye than a diffuse source of the same brightness. That is why a 10th-magnitude star is easier to see at a given power than, say, a 10th-magnitude galaxy several arcminutes in diameter. Higher powers exacerbate this effect. When a diffuse object is magnified, its surface brightness dims because the light is spread over a larger area.

In the broadest sense, surface brightness is an object's magnitude divided by its area. The typical surface brightness of a galaxy is 13.5, meaning that each square arcminute of its area is as bright as a magnitude 13.5 star – or about as bright as Pluto at opposition. But Pluto is a point source in small telescopes; so imagine defocusing Pluto until its light covers an area 1′ in diameter. Now imagine magnifying that diffused light from 20× to 200×. That feeble 1′ disk of light is now spread over an area 10 times larger. The eye's ability to detect that light diminishes in proportion. Now take our example galaxy, NGC 4490, which has a surface brightness of 13.0 (a little brighter than normal). The galaxy is also noted for its bright nucleus, jutting bars, and prominent H-II regions. What happens to the view as we increase magnification? Increasing magnification, of course, helps us to separate individual features crowded in the core. At low power, light from the nucleus, the bars, and the H-II regions can blend to form a single intense core of light. Magnification will help to separate the individual features from the nucleus, better to reveal the nature of the galaxy. Of course, the most feeble portions of its disk will fade away with the highest of powers, leaving only the most compact of details to stand out prominently. (That is also why it is important to incorporate views at all powers in your drawings.) The most skilled observers will also know how to scrutinize these remaining filaments of light further by moving them across the critical zone of the retina between the periphery and the fovea.

As you can see, to be an observer of pirate-of-the-Caribbean persuasion requires much time and dedication. So it is understandable that some observers believe observing in this way is too much work. They much prefer to hunt and find, not sit and analyze. Again, that's okay. Regardless of what type of observer you are, the beauty of astronomy is that it can satisfy the appetite of all manner of celestial pirates. Just remember, though, if you consider yourself a Barbary pirate, do not expect to see the subtle details often described by the pirates of the Caribbean. And if you're a pirate of the Caribbean, do not expect to compete in a race against time with a Barbary pirate . . . and win – they'll "not be beat by man nor devil." Of course, there are always exceptions to the rule.

Hidden Treasures 1 & 2

NGC 189 & 225

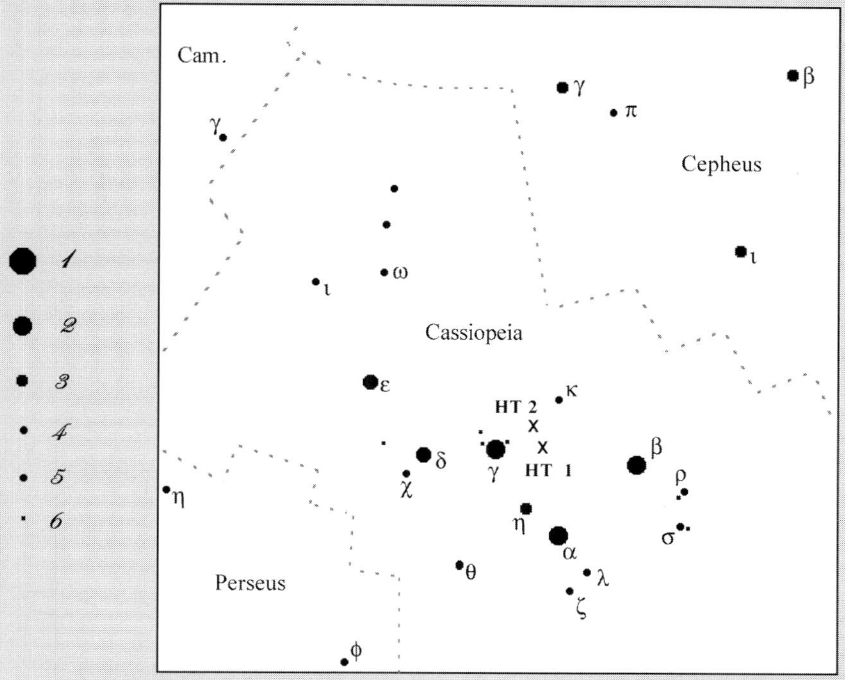

© Stephen James O'Meara

Tirion: Chart 1

Uranometria: Charts 16 & 36

1

NGC 189
Type: Open Cluster
Con: Cassiopeia

RA: 00^h 39.6^m
Dec: +61° 06′
Mag: 8.8
Diam: 5.0′
Dist: 2,400 light-years
Disc: Caroline Herschel, 1783

J. HERSCHEL: Cluster, pretty large, round, stars of 11th to 15th magnitude. (h 36)

NGC: Cluster, pretty large, round, stars of 11th to 15th magnitude.

2

Sailboat Cluster, Broken Heart Cluster
NGC 225
Type: Open Cluster
Con: Cassiopeia

RA: 00^h 43.6^m
Dec: +61° 46′
Mag: 7.0
Diam: 15.0′
Dist: 2,000 light-years
Disc: Caroline Herschel, 1783

W. HERSCHEL: [Observed November 26, 1788] A cluster of very coarsely scattered [bright] stars taking up 15′ or 20′, C. H. discovered 178[3]. (H VIII-78)

NGC: Cluster, large, little compressed, stars from magnitude 9 to 10.

IF THE MESSIER CATALOG OPENS with a bang (M1, the Crab supernova remnant in Taurus) and the Caldwell catalog opens with a whisper (NGC 188, an old and dim open cluster in Cepheus), *Hidden Treasures* opens with a surprise: NGC 189, an 8th-magnitude open cluster in Cassiopeia. The surprise is that indefatigable Caroline Herschel discovered this dim collection of suns in 1783 – a fact that had gone unrecognized for more than two centuries until British astronomical historian Michael Hoskin introduced the world to this fact in the November 2005 *Journal for the History of Astronomy.*

While reviewing Caroline Herschel's original observing notes, Hoskin scrutinized the following description of an object Caroline discovered on September 27, 1783, shortly after she had discovered the open cluster NGC 225 (Hidden Treasure 2) on the same evening:

> About 1 south of the above cluster [NGC 225] a faint nebula surrounded with a great number of both large and small stars. There are more large stars in the field than are marked here [in a diagram] but I took particular notice of the two between which the nebula is situated . . . Mess[ier] has them not.

Hoskin realized that since NGC 225 precedes Gamma (γ) Cassiopeia, this new object must precede it also. The only deep-sky object in the vicinity of Caroline's description is the 8th-magnitude open cluster NGC 189, which lies nearly 1° southwest of NGC 225. "But it is William's VIII.64 (NGC 381) that is credited to Caroline," Hoskin says, "even though this cluster follows Gamma Cas."

Hoskin's find put my mind at ease. It has long been my contention that NGC 381 is too faint for Caroline to have noticed in a sweep, even with her new 4.2-inch telescope, which had a speculum-metal mirror, at 22×. NGC 381 is small (7′), dim (magnitude 9.3), and has a low surface brightness (13.5). I failed to pick it up in a sweep with my 4-inch at 23×. The cluster lies in an extremely rich Milky Way field full of "fuzzy knots." I could see NGC 381 as a dull enhancement in the Milky Way only when I knew *exactly* where to look. Caroline did not have such an advantage.

NGC 189 is not only [0.5]-magnitude brighter than NGC 381, but more condensed (5′); it also has a greater surface brightness (12.3). And while NGC 189 is near Dolidze 12 (a slight concentration of Milky Way), Caroline's cluster is clearly isolated from it. The region is also veiled by obscuring dust, making NGC 225 and NGC 189 stand out more prominently from the otherwise rich background. NGC 189 was, in fact, quite easy to pick up in a comet sweep with my 4-inch at 23× in astronomical twilight.

Hoskin's research clearly dismisses NGC 381 as being one of Caroline's discoveries and makes NGC 189 the outstanding solution. "It would seem therefore," Hoskin says, "that this attribution [of NGC 381] to Caroline is a mistake (perhaps because [it] is positioned between two 6th-magnitude stars 1° apart), and that in fact she should be credited with NGC 189, an object that William never saw." Following Hoskin's recommendation, I removed NGC 381 from the *Hidden Treasures* list and replaced it with NGC 189. I have placed both NGC 189 and NGC 225 in the same write-up, not only because Caroline Herschel discovered them both on the same night, but, just as Caroline's description of NGC 189 suggests, the best way to find NGC 189 is to first locate brighter and larger NGC 225.

You'll find NGC 225 just 2° northwest of Gamma Cas, the central gem of the Celestial W. It is an elegant but neglected open star cluster. Upon her discovery of it, Caroline Herschel wrote,

> About 2 degrees from gamma Cassiopeae making an Isosceles triangle with gamma & kappa, a small cluster of stars, seeming to be intermixed with nebulosity . . . Messier has [it] not.

Caroline swept up the cluster four more times in the next five months. Her brother, William, classified it as a Class VIII object (coarsely scattered clusters of stars) and included it in his catalog of nebulae and clusters.

Under a dark sky, NGC 225 lies in a kidney-bean-shaped bay in the river Milky Way, exactly halfway between Gamma Cas and 4th-magnitude Kappa (κ) Cas. It is also very close to the galactic equator, the dense mid plane of our galaxy, where intervening dust dims the cluster's apparent brightness by about 0.2 magnitude. No matter, the 7th-magnitude glow is easily spied in 7 × 50 binoculars if you know exactly where to look. It is a fine sight in my antique telescope. Look for a bright chevron of stars with a wavy line of a half-dozen or so stars, oriented northwest–southeast, to the west. The cluster is a "milkweed pod" of hazy suns just east of that wavy line. With imagination it looks as if the pod has just burst open, releasing a flurry of seeds with their silken parachutes.

To star hop to NGC 225, start at Gamma Cas and move your scope about 35′ northwest to a golden 5th-magnitude sun (*a*). Next, look for a solitary 6th-magnitude star (*b*) about 40′ to the north-northwest of Star *a*. NGC 225 is almost exactly 1° due west of Star *b*. At 23× in the 4-inch telescope, the field is extremely rich. When I sweep my eye across the nearly 3° field of view, I see NGC 225 marking the southeastern tip of a 40′-wide asterism of 9th- and 10th-magnitude stars that forms a Valentine Heart – two ellipses of starlight, slightly angled with respect to one another that come to a point at NGC 225.

The dim 9th-magnitude open cluster Stock 24 can be seen about 30′ northwest of NGC 225. It marks the northwest edge of the Valentine Heart. Stock 24 is a puny cluster, measuring 5′ across, and though it is 180 stars rich, its brightest member shines at a diminutive 13th magnitude. If you can fit the two clusters in the same field of view, take a moment to ponder their true physical natures. NGC 225 and Stock 24 each span about 9 light-years of space. Stock 24 looks smaller and fainter than NGC 225 because it is nearly three times farther away, which is a shame, because Stock 24 contains more than twice as many stars as does NGC 225. Although the stars of NGC 225 are spread out, making it appear old – as if its members are losing their gravitational grip on one another – it is, in fact, a rather young cluster with an age of about 120 million years, making it almost twice as old as the Pleiades.

NGC 225 is easily resolved in a 4-inch telescope. The brightest member shines at magnitude 9.3. The cluster's faintest suns dip sharply into the rich Milky Way background. At 23× the cluster's shape mimics that of the larger Valentine Heart of stars surrounding it, though NGC 225 looks more like a broken valentine – a heart-shaped cluster with one lobe filled with stars of near uniform brightness (the "milkweed pod") and the other lobe being little more than an empty shell.

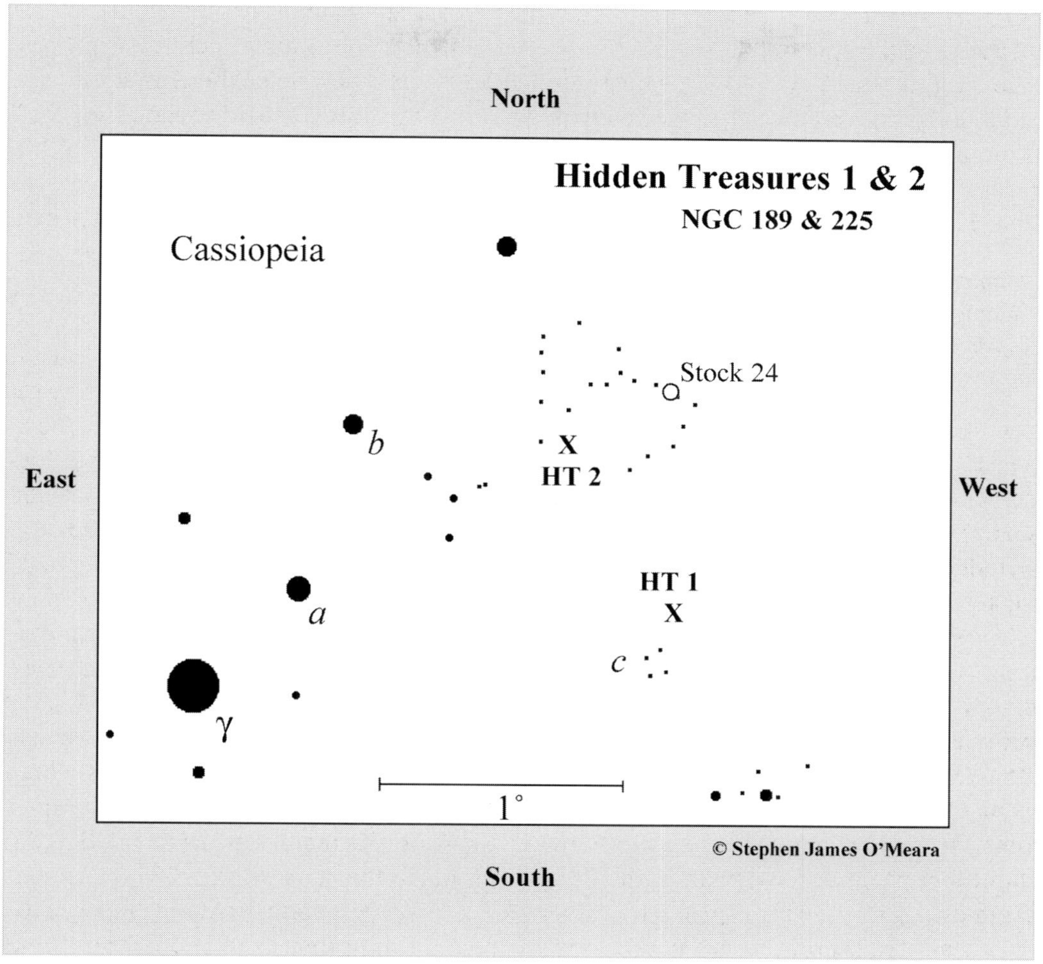
North
Hidden Treasures 1 & 2
NGC 189 & 225
Cassiopeia
Stock 24
b
X
HT 2
East
West
a
HT 1
X
c
γ
1°
© Stephen James O'Meara
South

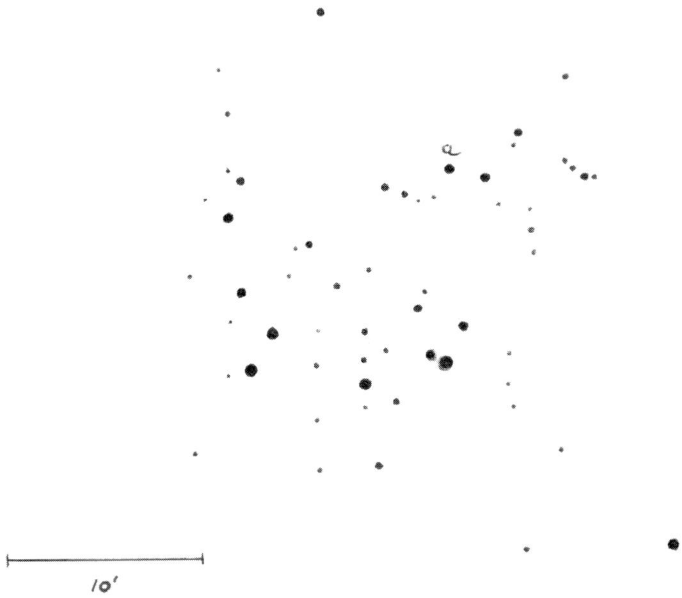

Interestingly, while Smyth notes that no nebulosity exists here, Caroline Herschel, in her discovery description, suspected it did. And it does. The brightest section surrounds the 11th-magnitude star BD +61 154 at the northern fringe of the cluster and is clearly visible in the photograph above. Smyth probably did not see it because he was using a long-focus refractor. Seeing the dim nebula may require a rich-field telescope, like the one that Caroline used, under dark skies – one that will condense the faint

At least that's the way it appears in small apertures.

At 72×, the cluster's stars spread out and lose their appeal. But before you return to low power, take a moment to relax your gaze and scrutinize the scene with a fresh eye. Do you see how that, though the cluster is loose and scattered, its stars are arranged in two distinct regions – a rich section to the southeast (like a lemon slice) and a smaller and sparser congregation to the northwest? If you defocus the view ever so slightly and tap the tube, you should see that these two regions are separated by a meandering lane of darkness. The nineteenth-century British observer, Adm. William Henry Smyth, noted this effect as well, writing that the object is a "small double star, in a loose cluster of about thirty of the 9th and 10th magnitudes, occupying all the field; but there being no stardust, or nebulosity intermixed, the firmament appears unusually dark between them."

light. What is the smallest telescope that will show it? Chuck Layton of Tacoma, Washington, writes that he could not see it in an 8-inch f/6 Newtonian from suburban skies; and William L. Schart of Killeen, Texas, also failed with an 8-inch Schmidt-Cassegrain telescope at 77×. At times I suspected it in the 4-inch at 23×, and at other times not. Perhaps the use of a nebula filter is the solution.

Regardless of the nebula's visibility, NGC 225 is one cluster that plays with the imagination. With averted vision in the antique telescope, the cluster resembles the Greek letter omega (ω), or, better yet, it mimics the larger Celestial W, in which it lies – an aspect first noted by the Rev. Thomas W. Webb in the nineteenth century. In more modern terms, Rod Pommier nicknamed it the Sailboat Cluster, and Jim Anderson of Phelps Lake, North Carolina, notes that when seen simply inverted, the "grouping forms a figure that resembles a stooping falcon with wings arched upward and the pinions extending downward, the head and legs extended to strike its prey . . . Let your imagination run amuck." If you want that "visual punch," keep the stars, which cover an area the size of a quarter Moon, tightly packed at low power. Increased magnification will only darken, or diminish, the view . . . and your imagination.

By the way, Smyth does astutely note that the most conspicuous object within the cluster is a fine double star, whose components shine at magnitudes 8.5 and 11 and are separated by 12″. This pairing can be found just southwest of what appears to be the cluster's visual center. But other dimmer pairs are scattered about.

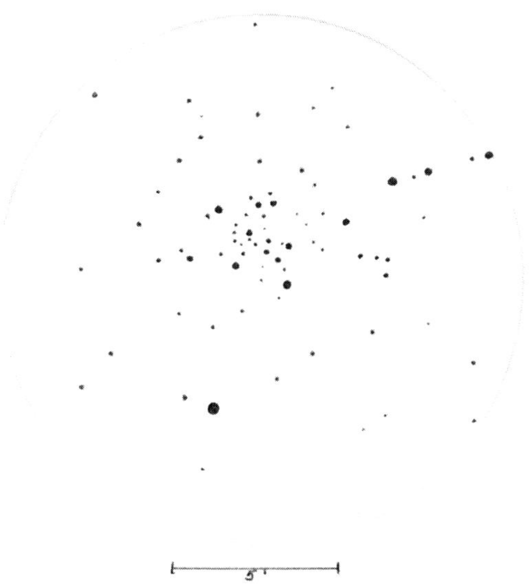

To find NGC 189 look only 50′ to the southwest; it will fit in the same low-power field of view as NGC 225 in most telescopes. Compared to large and scattered NGC 225, NGC 189 is small and round. It also has a beautiful bluish cometlike sheen that stands out distinctly from the background, especially when seen together with NGC 225. NGC 189 lies immediately north of a 7′-wide trapezoid of four 9th- and 10th-magnitude suns. In photographs, the cluster resides within a comparatively rich region bordering the heavily obscured sack of dark nebulosity within which NGC 225 is located. A study of the region by George Alter reveals that the background density is, however, about half that of the other clusters in the neighborhood so that some absorption might justifiably be assumed. NGC 189 is only 4,000 light-years more distant than NGC 225, but it is six times younger. When NGC 189 was formed 20 million years ago, fantastic mountain ranges – the Cordilleras, the Andes, and the Himalayas – were forming as pressure mounted along the margins of colliding crustal plates on Earth, forcing inland seas to shrink and the climate to waver through extremes of hot and cold.

The cluster can be viewed in 7 × 50 binoculars with attention. With 15 × 80 binoculars, British observer D. Branchett calls it a "bright, large group increasing in size with averted vision; scattered stars." NGC 225 is easy to see in my antique telescope. The cluster definitely swells with averted vision in the 4-inch at 23×, when the cluster partially resolves into a sphere of light that shimmers like fireflies in a glass globe. At 72×, the cluster resolves into two distinct classes of stars, namely a splash of about a half-dozen 10th- and 11th-magnitude stars set against a backdrop of some 20 or so dimmer suns with a central concentration that gradually unwinds into a diffuse halo of starlight. Several long and sinuous arms (which may or may not be related to the cluster) extend from this tight core, giving the outer regions a loose spiral structure. With averted vision the round central core of dim suns sits inside a little scoop of bright stars with a handle that extends to the northwest. With imagination, it looks as if some demigod has dipped a ladle into the Milky Way and scooped up some celestial broth.

The view in larger scopes is almost the same. Steve Coe of the Saguaro Astronomy Club, Arizona, says that through his 13.1-inch f/5.6 reflector and 100× the cluster is "gradually much brighter [in the] middle, somewhat mottled[;] averted vision makes it grow." In their book *Star Clusters*, Brent A. Archinal and Steven J. Hynes list 90 members as being associated with the cluster, though it's hard to judge which are.

Before leaving the area, return to Gamma Cas, a famous irregular variable star. Prior to 1910, Gamma shined at magnitude 2.2. By 1937 it had brightened to magnitude 1.6. The star then dipped to a low of 3rd magnitude in 1940, before it slowly brightened again. Today it once again hovers around magnitude 2.2, varying erratically by 0.6 magnitude. Gamma is also a close visual and spectroscopic binary, as well as an X-ray source, which suggests that Gamma could be accreting matter onto a compact companion, such as a neutron star or degenerate dwarf. Keep an eye on Gamma Cas, because, who knows, one day the star could flare up again and alter the appearance of one of the skies most familiar asterisms.

Hidden Treasure 3

NGC 281

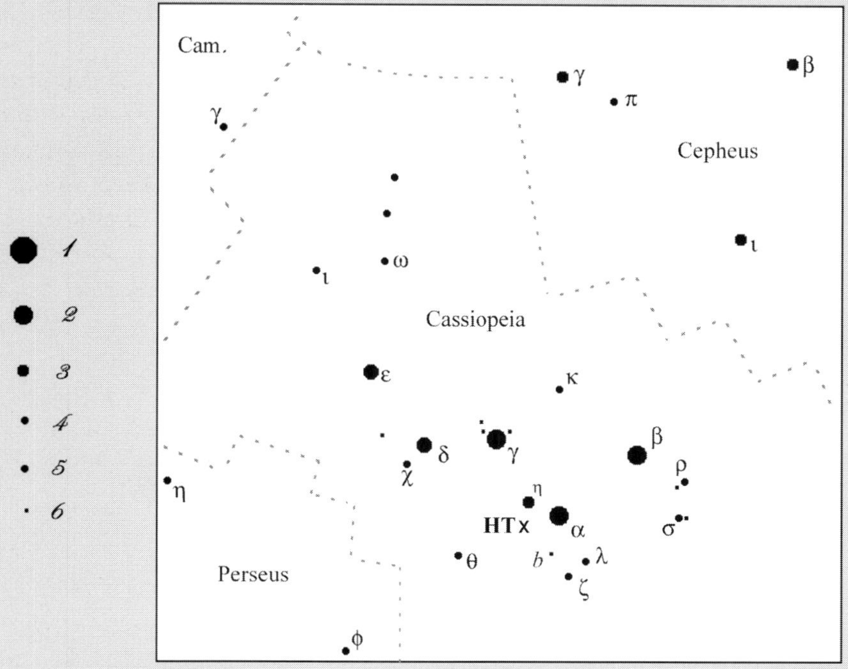

© Stephen James O'Meara

Tirion: Chart 1

Uranometria: Chart 36

3

Pacman Nebula, Lafitte's Grand Isle
NGC 281
Type: Emission Nebula
Con: Cassiopeia

RA: 00^h 52.8^m
Dec: $+56°$ $37'$
Mag: 7.8 (nebula)
Dim: $35' \times 30'$ (nebula)
Dist: ~9,400 light-years
Disc: Edward Emerson Barnard discovered the nebula in 1881; Guillaume Bigourdan discovered the cluster before 1891

HERSCHEL: None.

NGC (NEBULA): Faint, very large, diffused, [faint] triple star on north preceding edge.

NGC 281 IN CASSIOPEIA IS A BIT OF an enigma: *Sky Atlas 2000.0* (second edn), *Uranometria 2000.0*, and the *Millennium Star Atlas* all depict it as an open cluster and a nebula, and it is listed as such in *NGC 2000.0*, the *Observing Handbook and Catalogue of Deep-Sky Objects*, the *Deep-Sky Field Guide*, and *Sky Catalogue 2000.0* (Vol. 2). But, as Brent A. Archinal and Steven J. Hynes note in their book *Star Clusters*, NGC 281 is not a star cluster, but rather the nebulosity involved with the star cluster IC 1590. The hidden treasure you seek, then, is NGC 281 (the nebula). IC 1590 (the cluster) is a visual bonus – like the inlaid gems and precious stones that adorn the Taj Mahal's majestic white marble edifice.

Edward Emerson Barnard (1857–1923) discovered NGC 281 visually on November 26, 1881. Recently married and caring for his invalid mother, Barnard was, at the time, trying to strengthen his financial situation. During the day he worked at Poole's Photograph Gallery in Nashville, Tennessee. At night, he used a 5-inch Byrne refractor to scan the skies for comets. He was also trying his hand at raising chickens. The discovery of NGC 281 came while he was feverishly in quest of the US $200 Warner prize; in 1881 H. H. Warner (a wealthy patron of astronomy from Rochester, New York) announced that he would award a gold medal and US $200 in cash for the first American discoverer of each new comet. Barnard quickly sniffed out a road to success. On the evening of September 17, 1881, he discovered an 8th-magnitude comet low in the western sky. Although the comet did not become a visual spectacle, it did bring him his first Warner prize – part of which he used to purchase a tiny plot of land in Nashville, on which he then financed a home. And it was from this new home, "on a beautiful rising ground" with a clear horizon, that he discovered the new nebula.[1]

The discovery of NGC 281, of course, was a financial disappointment; new comets, not new nebulae, paid. No matter, Barnard went on to discover 16 comets, five of which earned him Warner prizes – enough to pay off the mortgage of his new home, which he named "Comet House." Although new nebulae did not bring any financial rewards, they still intrigued him, especially since no one could yet explain their nature.

Despite the great sky surveys conducted by the Herschels and other celestial treasure hunters over the last century and more, nebulae visible in small telescopes were still being discovered. "While some nebulae could easily be recognized by their fantastic forms," William Sheehan writes in *The Immortal Fire Within: The Life and Work of Edward Emerson Barnard,* "the majority were, at least in the 5-inch telescope Barnard used, 'roundish patches of foggy matter, extremely like comets in appearance'." NGC 281 is no exception; with its curious parabolic shape, the 7th-magnitude glow looks very much like the head of a new comet just beginning to grow a tail.

The question is how could such an object, one that is visible in binoculars today, have gone unnoticed for so long? The answer is twofold. First, the nebula is large ($35' \times 30'$), larger than the full Moon, and its surface brightness is low (23.6), so the great William Herschel hadn't much of a chance to spy it in his narrow 15′ field of view (which is what he used during his sweeps). Second, it's possible that comet hunters prior to Barnard did not encounter NGC 281 because it lies outside the comet "haystack" – that region of sky (within 60° of the Sun) where comets are most likely to be discovered by visual observers using small telescopes. But Barnard did not follow the adopted routine of his predecessors; he believed the entire sky was the comet seeker's domain. The comet seeker, he said, "must examine every portion of [the sky] time and again. . . Everything is interesting and numberless objects are beautiful in the extreme. There is nothing commonplace in the sky."

Seeing the ghostly form of NGC 281 among the rich stellar folds of the Cassiopeia Milky Way must have bordered on the spiritual for Barnard, a man of strong religious conviction. Of all things, Barnard was especially fond of the Milky Way, the "jewel house of the Maker." When we sweep there, he said, "our soul mounts up, up to that wonderful Creator, and we adore the hand that scattered the jewels of heaven so lavishly in this one vast region. No pen, can describe the wonderful scene that the swinging tube reveals as it sweeps among that vast starry array of suns." Yes, NGC 281 was a financial disappointment, but how hard is it to imagine that the discovery of this object in this man's most revered corridor of the heavens would be anything but spiritually uplifting.

Who would blame Barnard for having such feelings? Just look at the stunning photograph of NGC 281 taken with the Canda–France–Hawaii Telescope (CFHT)

[1] NGC 281 was Barnard's second discovery of a deep-sky object; he found his first in July 1881: the magnitude 11.4 galaxy NGC 5584 in Virgo. By 1886, he had discovered 21 nebulae.

atop Mauna Kea in Hawaii (shown here). Here is a complex tapestry of glowing gas forming the backdrop to a tattered curtain of obscuring dust. The entire eastern front of the nebula appears in turmoil. Bulbous masses of dark and light seemingly battle for prominence. It is as if all manner of supernatural shapes and forms are trying to materialize from the bubbling surface of a witch's brew. Among the tumult, needle-thin protuberances and dense nebulous spires jut from the cloud with three-dimensional clarity; some sport at their tips tiny black dots – the mystifying Bok Globules. These dense knots of dust and gas are the black pools into which we peer helplessly, to dream our dreams of creation, for these are the secret birth sites of new stars. How Barnard's heart would have fluttered had he lived to see such an image. It would have been like looking into the mind of his Creator.

Today, we know NGC 281 is an emission nebula – a vast cloud of ionized hydrogen gas in the Perseus arm of the Milky Way. It is excited by a compact trapezium-like cluster of *OB* stars (IC 1590), which, some recent studies show, formed only about 3.5 million years ago. If this is true, its stars are a bit older than those in the Orion Nebula (1 million years) and much younger than the age of our Sun (4.5 billion years). The cluster's brightest member – the *O*5 star HD 5005 (a triple star that shines at magnitude 9.0) – contributes most of the light to the nebula. The glowing gas and dark obscuring clouds are actually part of a larger complex of atomic and molecular clouds forming an 880-light-year-wide ring around NGC 281, which is expanding away from the nebula at 22 kilometers per second. The entire complex was probably first formed by supernovae explosions, which triggered subsequent episodes of star formation, including those ongoing today.

While we will never see the drama captured in the CFHT photograph, we can at least imagine it. To the eye, NGC 281 is an irregular patch of glowing gas a mere $1\frac{3}{4}^\circ$ east of Alpha (α) Cassiopeia (Shedir); its Moon-size glow forms the southeastern apex of a triangle with Alpha and Eta (η) Cassiopeia. I have photographed the nebula numerous times, but never intentionally. It simply appears in the constellation photos I've taken of Cassiopeia with a 35-mm camera and a 50-mm lens. I even knew of its photographic appearance when I was a youngster living in Cambridge, Massachusetts. I probably would have looked for

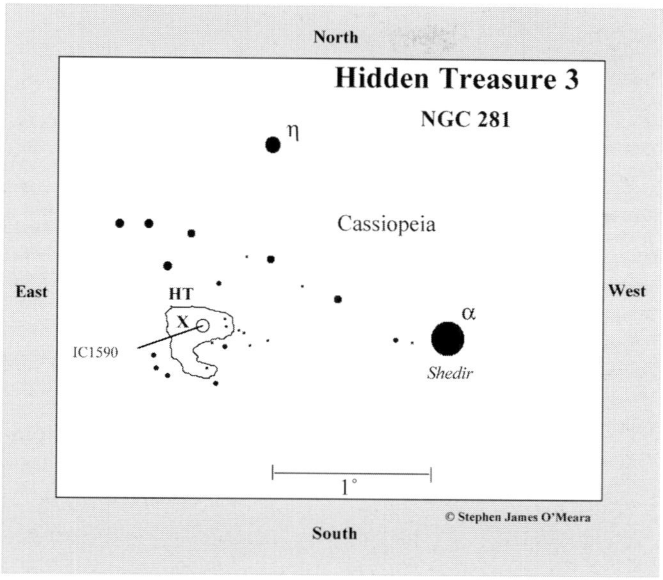

lying back in my car seat, with my head resting comfortably against the headrest, and my eyes looking out an open window, best alleviated any eye strain. To block the distracting mayhem of the Cassiopeia Milky Way, I used my hands to form a rectangular frame through which I could see only NGC 281 and its immediate surroundings. Still the nebulosity was visible only with averted vision; the nebula quickly vanished with a direct gaze.

it visually back then, but for the fact that it was not plotted in the *Norton's Star Atlas* I used in the early 1960s.

I did not see NGC 281 visually (at least that I recall) until January 7, 2002. That night, out of curiosity, I wanted to know how many nebulae I could see in Cassiopeia with my 4-inch refractor. NGC 281 was first on my list. Imagine my surprise, however, when I saw it first in 7 × 50 binoculars. I was so taken by its sepulchral presence that I returned to it night after night, just to be certain my eyes were not deceiving me. I surprised myself further when, given time, I could *just* detect it with the unaided eye, but only with averted vision. Without flinching, I recorded the object in my logs and marked it for inclusion in my *Deep-Sky Companions: The Caldwell Objects* – as one of the "20 Spectacular Non-Caldwell Objects." Then, on October 7, 2002, I turned my telescope to it again and began my first official hidden treasure observation.

To see it visually under a dark sky, try using various observing positions. I found that

In 7 × 50 binoculars the 4′ cluster (IC 1590) within NGC 281 is immediately obvious as a magnitude 6.3 "star." But with attention, you should be able to see a halo of faint starlight surrounding the cluster; it spreads out uniformly in an east–west direction, but also slightly north. IC 1590 has 63 probable members within its tiny disk, and Brent A. Archinal suspects many more stars in the nebula may be associated with the cluster.

Guillaume Bigourdan's (1899–1917) discovery of IC 1590 was no less of an achievement than Barnard's. It takes a certain visual perspicacity to go beyond what one is expected to see (a nebula) and to find something new within it. Bigourdan was a highly esteemed visual observer. He found about 100 new nebulae and clusters with the 12-inch refractor at Paris Observatory, France, and IC 1590 was one of them. Of it, he writes, "In the region of NGC 281, there is in addition to the nebulosity suspected near BD +55° 191′, a large number of stars forming a very large cluster, without concentration." Hal Corwin (California Institute of Technology) notes that Bigourdan's position of the cluster is about 3′ southeast of the obvious group of stars taken to be IC 1590. "Without better evidence from Bigourdan's published material," Corwin says "there is not much point in trying too hard to find this object. It sits in the middle of a large region of star formation, and any position we take in the area will get us some hot, young stars."

As for the nebula, well, "All I can say is that I have . . . seen it under all atmospheric conditions, at every hour of the . . . night, and I have never failed to be thrilled with its beauty and romance. What romance." A distant relative of mine, Lieutenant-Colonel E. J. O'Meara, penned these words in his 1935 autobiography *I'd Live It Again*; he's not describing NGC 281 but the Taj Mahal around 1912. Yet these same words capture the very wonderment I feel whenever I look at NGC 281, whose ghostly white "edifice" is bejeweled with stellar gems.

At 23× the glow is immediately apparent, even with light from a near first-quarter Moon in the sky. The first thing I notice is a nebula, like breath on a mirror, forming a concentric circular glow around the 9th-magnitude star at the heart of IC 1590. But with a gentle tap of the telescope tube, butterfly wings of ever so slightly denser material can be seen extending east–west of HD 5005. A closer look reveals that the eastern wing is a tad brighter than the western wing, which is also slightly larger. A fainter extension of light, like a comet's dim dust tail, extends to the south-southwest and terminates near a magnitude 9.4 star (*c*).

Take time to relax your gaze and view the nebula with averted vision. It should, over time, transform NGC 281 from a uniform glow, to one with a discrete dappled texture. The nebula is enchanting, and the

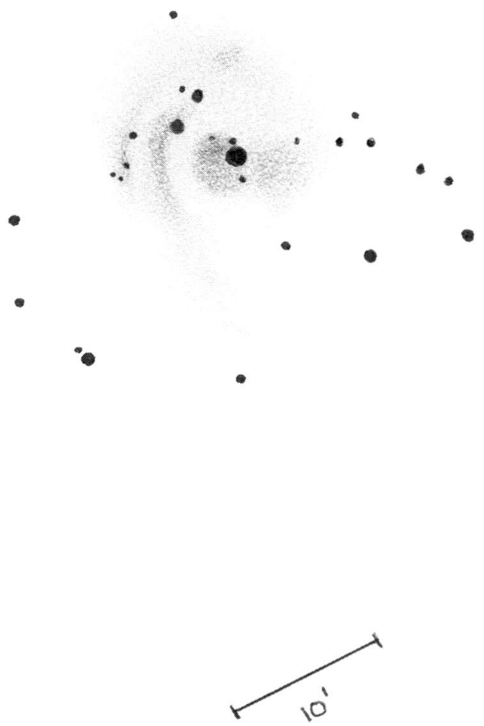

longer you look the more you should see. But be warned, the details are fleeting. For instance, once my eye catches a patch of nebulosity with averted vision, it tries to snatch a glimpse of it with direct vision. Just then, my averted vision picks up a different patch of nebulosity, and the process repeats itself. That's the frustrating consequence of trying to see subtle details separated by equally subtle deviations in brightness. Seeing the nebula in this way is like taking a walk across a mysterious isle of light and shadow, a place where mist and fog weave back and forth like restless spirits. It is reminiscent of how the star detective of Carolyn Keene's popular Nancy Drew mystery stories imagined the Grand Isle – the haunt of the famous pirate Lafitte and his men. "[A]ccording to tradition," Drew says to her friends in *The Ghost of Blackwood*

Hall, "when burying treasure, he always murdered one of his band and left his ghost to guard the hidden loot." With every sweep of the eye, I feel as though I'm being taunted by these bothersome spirits.

With moderate and high powers, the densest regions of the nebula's core remain visible, while the remainder fades away. Look for a channel of darkness 2′ northeast of HD 5005 – between it and two 10th-magnitude suns (*e* and *f*). A curved spit of bright nebulosity slices through two stars. Another dark channel, concentric with the previous one, lies between the spit just mentioned and another, fainter, spit equidistant to the northeast. These ripples of light and dark are reminiscent of those seen at the heart of the M8, the famous Lagoon Nebula, and look like shock waves pushing through the dim gas.

One "buried" treasure appears as a tight "nebulous" knot about 5′ north-northeast of IC 1590 (*d*). This glow is not a nebula but a tight clustering of about a dozen faint suns – too faint to be resolved with any confidence in the 4-inch. In agreement with Corwin, it's difficult to follow or judge the size or extent of IC 1590 because there appear to be many spots (like *d*), namely mini-clusterings – tight knots of dim stars within a larger concentration of brighter suns. Of course, the real treasure is at the heart of IC 1590. When viewed with a variety of powers up to 302×, HD 5005 is a beautiful triple star joined to the southwest by a tight 11.5-magnitude pair of suns. All told, 15 suns can be seen clearly in the 4-inch, and they form an interesting pattern – one of a shopping cart, a poor cousin to NGC 2169 in Orion (Hidden Treasure 36).

The nebula's largest and most prominent dark cloud of gas – a dense wedge of matter covering the southwest quadrant of the nebula in photographs – is conspicuous only in photographs of that region. It is the photographic appearance that gives rise to the nebula's popular nickname: the "Pacman Nebula," honoring a video-game icon first introduced in 1980, more than a quarter-century ago. Although the dark nebula itself is not seen projected against bright nebulosity in the 4-inch, it is indirectly conspicuous by creating the opening that forms Pacman's mouth.

If we accept NGC 281's distance as 9,400 light-years, the nebula measures 96 × 82 light-years; IC 1590 in comparison, spans 17 light-years across. In apparent size (35′ × 30′) it is inferior to M42, the Orion Nebula (1°.5′ diameter), but NGC 281 is six times farther away. If NGC 281 were at the distance of the Orion Nebula, it would be nearly 2.5 times larger than that impressive gallery of glows and be second only to the Great Eta Carinae Nebula (Caldwell 92), the original celestial Taj Mahal, in apparent splendor.

Before you move away from this region, do not overlook its other "hidden treasure," Eta Cas. William Herschel discovered this glorious double star in August 1779. It carries a most delightful nickname: the "Easter Egg Double." The magnitude 3.5 primary (type G0V) and magnitude 7.4 secondary (type *dM0*) are separated by 13″ (2003). The primary has been seen with a pale white, yellow, or topaz hue; the secondary shines with a ruddy purple, garnet, lilac, or lavender sheen. At 167× in the 4-inch, I see the pair as a lemon-yellow sun with a dusty rose companion. The pair is even more dramatic at 23×, because the secondary looks more like a crimson spark. Seeing the pair at this magnification also lends credence to its other nickname: The Balloon and Gondola.

Hidden Treasure 4

NGC 288

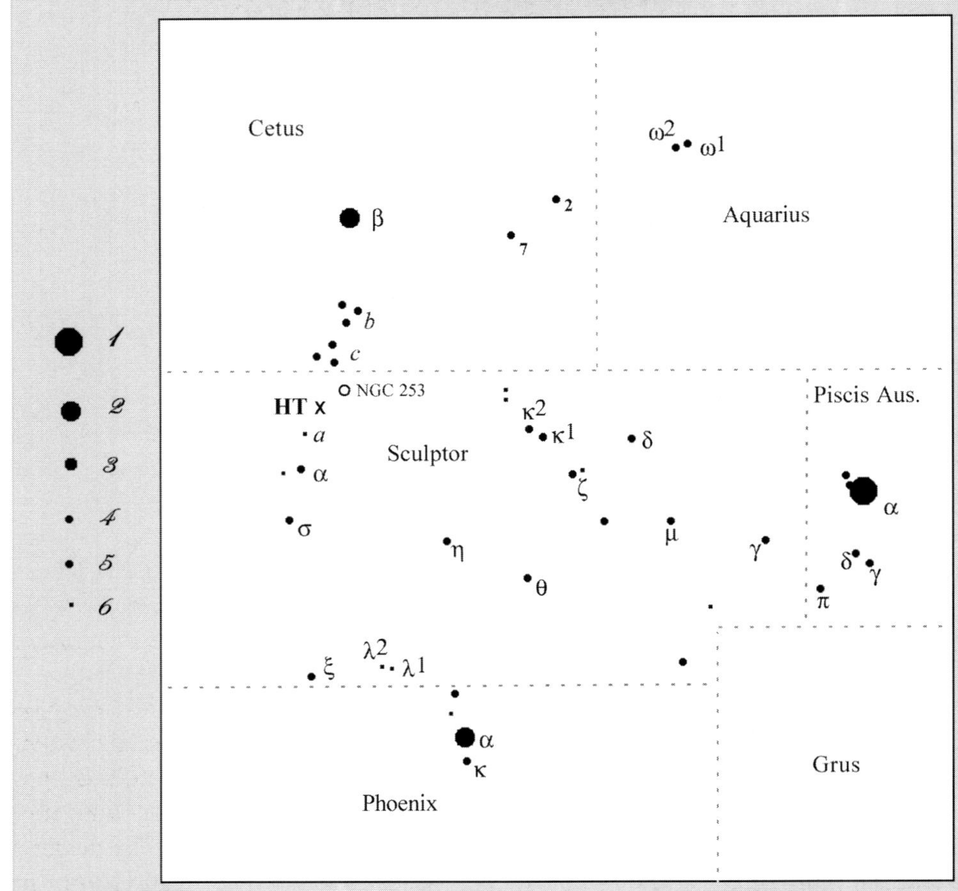

Tirion: Chart 18

Uranometria: Chart 307

4

NGC 288
Type: Globular Cluster
Con: Sculptor

RA: 00^h 52.8^m
Dec: −26° 35′
Mag: 7.9 (O'Meara); 8.1
Diam: 13′
Dist: 27,500 light-years
Disc: William Herschel, 1785

W. HERSCHEL: [Observed October 27, 1785] Considerably bright, irregularly round, 8′ or 9′ in diameter. A great many of the stars are visible so there can remain no doubt but that it is a cluster of very small stars. (H VI-20)

NGC: Globular cluster, bright, large, little extended, stars of magnitude 12 to 16.

SCULPTOR IS A SHADOWLAND constellation – one dim enough and low enough (as seen from mid-northern latitudes) to cause consternation over whether it's a "ship" worth raiding. Well, it is. While it's true that Sculptor's brightest stars shine between 4th and 5th magnitude, they are not as low in the sky as some northern observers might think. As Richard Hinkley Allen notes in *Star Names: Their Lore and Meanings*, "The constellation culminates with the bright star of the Phoenix on the 17th of November, and is visible from the latitude of New York City." In fact, our third hidden treasure, globular cluster NGC 288 in Sculptor, is almost 10° farther north than the most southerly Messier object (open cluster M7, in Scorpius). It is also comparable in brightness to about one-third of all the Messier and Caldwell globular clusters.

As if to support the myth that Sculptor and its treasures are "invisible" from mid-northern latitudes, Adm. William Henry Smyth omitted the constellation from his *Cycle of Celestial Objects*; in that work, Smyth included only those constellations he could see from England. The irony, of course, is that Smyth's fellow countryman, William Herschel, discovered NGC 288 from England.

NGC 288 is a hidden treasure for reasons other than the dimness of its parent constellation. The globular is simply overshadowed by one of the southern sky's greatest attractions – the oblique spiral galaxy NGC 253 (Caldwell 65). Although NGC 288 lies only

$1\frac{3}{4}^\circ$ to the southeast of NGC 253, that's about $1\frac{3}{4}^\circ$ too far to be placed in the same low-power field as NGC 253 in many modern telescopes.

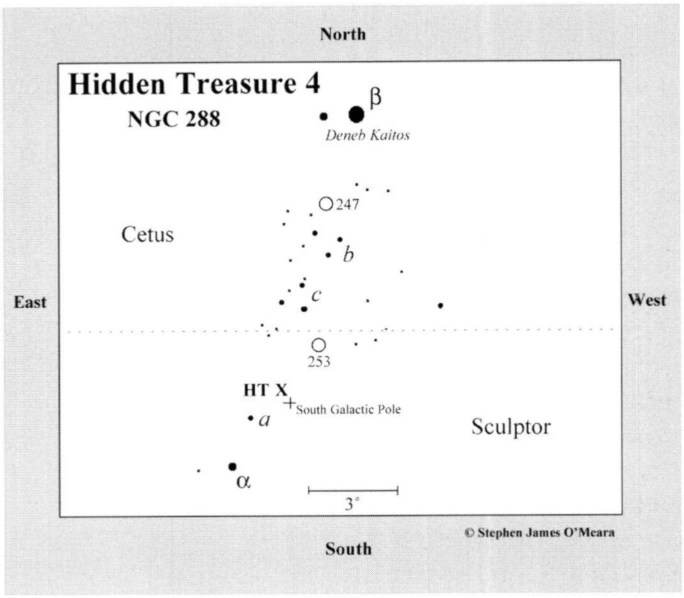

With a field of view of nearly 3°, however, my Tele Vue 4-inch refractor removes NGC 288 from the shadows of obscurity and thrusts it into the "spotlight" with NGC 253. Few sights in the sky are as rewarding as seeing this dramatic pairing of deep-sky splendors. Yet, Patrick Caldwell-Moore did not include NGC 288 in his popular list of deep-sky objects, and Robert Burnham Jr. fails to highlight it in his *Celestial Handbook*, even in passing. The Rev. Thomas W. Webb acknowledges NGC 288 in his *Celestial Object for Common Telescopes*, though he simply calls it "bright." The late Walter Scott Houston, *Sky & Telescope*'s long-cherished Deep-Sky Wonders columnist, was well aware of the cluster's obscurity, writing "This interesting globular cluster is often overlooked by amateurs, though I readily find it [from Connecticut] whenever the sky is especially clear. One August, after a cold Canadian high had swept the air clean, binoculars were sufficient to distinguish NGC 288 . . ."

From Hawaii, NGC 288 is easily spotted in 7 × 50 binoculars and looks like a smooth orb of light shining at 8th magnitude. It resides less than 9° southeast of Beta (β) Ceti (Deneb Kaitos), 3° northwest of Alpha (α) Sculptoris, and just 40′ north and slightly east of the South Galactic Pole. If your telescope is on a polar-aligned equatorial mount, and you know how to find NGC 281 (Hidden Treasure 3), then center NGC 281 and swing the telescope 70° to the south; NGC 288 should be centered, because the two objects share the same right ascension.

Star hopping to the cluster is easy if you can find Alpha Sculptoris first. Simply look for 7th-magnitude Star *a* a little more than $1\frac{1}{2}^\circ$ to the northwest. NGC 288 is a little less than $1\frac{1}{2}^\circ$ further to the northwest. Otherwise, you can star hop from Beta Ceti. Use binoculars to look for a 1°-wide triangle (*b*) of 5th- and 6th-magnitude stars about 4° to the south-southeast; the 9th-magnitude galaxy NGC 247 (Caldwell 62) lies 1° to the north of that triangle. Next, look for an equally sized triangle (*c*) of slightly dimmer suns just 2° to the southeast; its northernmost star is a binocular double. The Great Sculptor Galaxy, NGC 253, lies about $1\frac{1}{2}^\circ$ southwest of Triangle *c*. NGC 288 is $1\frac{3}{4}^\circ$ to the southeast of NGC 253.

In my antique telescope, the cluster is a little ball of "gas" with a slightly condensed

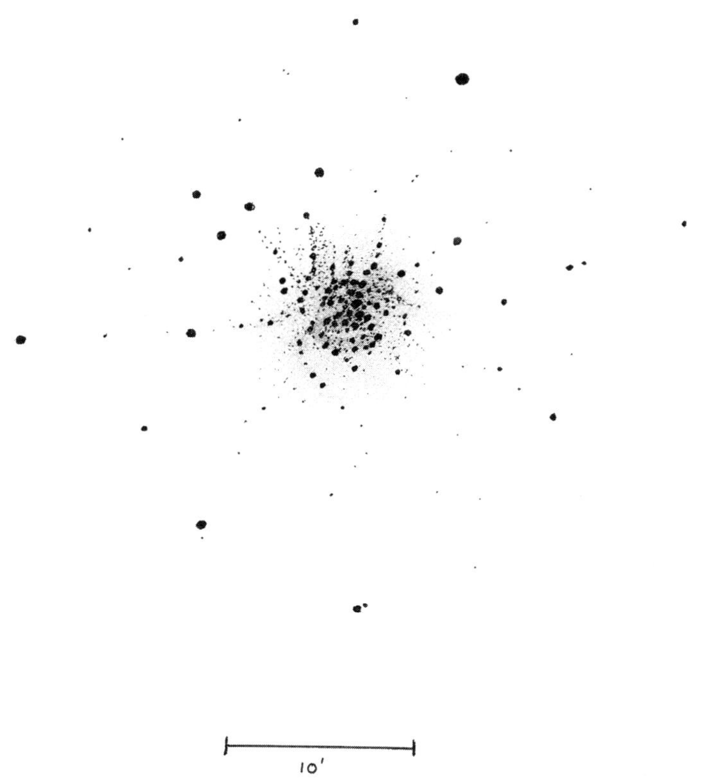

core. Averted vision reveals a slight granular texture. The cluster comes to life in the 4-inch. At 23×, it's a perfect sphere of stellar foam that seems to froth out of a squashed dipper asterism of surrounding starlight. A closer inspection shows the globular to have an irregular border, one that is fractured into clumps of stars, many of which are immediately resolvable. In brightness and form, the cluster reminds me of M71 in Sagitta. Like that more northerly globular, NGC 288 does not have a strong central condensation, which is interesting since William Herschel lists it under his Class VI objects (very compressed and rich clusters of stars). At least in a 4-inch telescope, NGC 288's core appears to be a large mottled region, perhaps 6′ across, whose overall brightness is greater than its equally large halo of dim suns. With averted vision, this bright inner region seems to scintillate with frenetic energy, like bees swarming around a hive.

NGC 288 is well resolved at all magnifications, especially at the fringes, which, as I said, become patchy with clumps of starlight. The cluster's brightest members shine at magnitude 12.6, and its horizontal-branch magnitude is 15.3, so even a modest-sized amateur telescope can penetrate deep into the globular's inner sanctum. This feat is best accomplished at high magnifications but, be warned, even at 182× the relatively low surface brightness of the cluster makes seeing any patterns difficult. Be prepared to spend a good evening plumbing its depth.

What should be immediately apparent at high power is the core's boxy shape. This feature also shows up well in short-exposure photographs of the cluster. The details in this region can be grouped into three categories: (1) a faint blanket of similarly bright suns that forms a soft backdrop to (2); (2) several distinct strings of brighter stars whose overall arrangement gives the cluster its boxy shape; (3) a cluster marred by dark lanes and voids. One prominent dark lane curves around a squashed apostrophe of stars just south of the cluster's vacant core, while two rectangular vacancies can be seen just north of it. With a critical investigation at 182×, I can see a crown of stars superimposed on the apostrophe. The strings of stars rimming the two dark bays to the north form a fine Omega (ω) pattern with north up. In larger telescopes, the cluster is quite remarkable. Christian Luginbuhl and Brian Skiff compare their view of NGC 288 through a 12-inch reflector to that of the great globular M4 in Scorpius, though they admit it is not as rich.

If we accept NGC 288's distance to be 27,500 light-years from the Sun, its true physical diameter then is 104 light-years across. The cluster is in an oddball retrograde orbit traveling at a speed of about 305 kilometers per second some 39,000 light-years from the Milky Way's center. It's a loose halo cluster whose members each contain, on average, anywhere from $\frac{1}{13}$ to $\frac{1}{20}$ as much iron as does the Sun. That's about the same abundance that each cluster member has in globular cluster NGC 1851 in Columba – a galactic veteran of 14 billion years. How old is NGC 288? Well, that has been the matter of debate for many years. Some researchers claimed it was the same age as another aged globular – NGC 362 in Tucana. Others found a significant discrepancy in the two clusters' ages. The most recent findings, however have shown that NGC 288 is some 2 billion years older than NGC 362, making it about 13 to 14 billion years old, which places it among the very oldest globular clusters.

If such bickering seems inconsequential, it's not! Age battles in the galactic halo are of vast importance to astronomers. Knowing the age of globular clusters in the halo sets time constraints for the origin of the halo and the start of the collapse of the galaxy. The age of the oldest globular star clusters is also arguably the best observational lower limit we have to the age of the universe, which, at the moment, is generally accepted to be about 13.7 billion years old.

If you live in southern USA or at similar latitudes, there's a fantastic variable star $9\frac{1}{2}^\circ$ to the southeast of NGC 288 where you'll find the scarlet carbon star R Sculptoris. The American Association of Variable Star Observers (AAVSO) has it ranging between magnitude 9.1 and 12.8 (photographic) every 370 days, while Australia's Canberra Astronomical Society has it ranging between magnitude 5.8 and 7.7 (visual) every 370 days. Regardless, the star is a scarlet wonder and a must see. Interestingly, another red variable, S Sculptoris lies about 10° to the southwest of NGC 288; it varies from magnitude 5.5 to 13.6 every 365 days, making it a perfect year-long observing project. Monitoring long-period variables is the best way for beginners to hone their skills at making magnitude estimates, since their periods and rates of variability are so well known that accurate predictions can be made. If you join a world-renowned organization like AAVSO, you can compare your magnitude estimates each month with those of other members.

Hidden Treasure 5

NGC 404

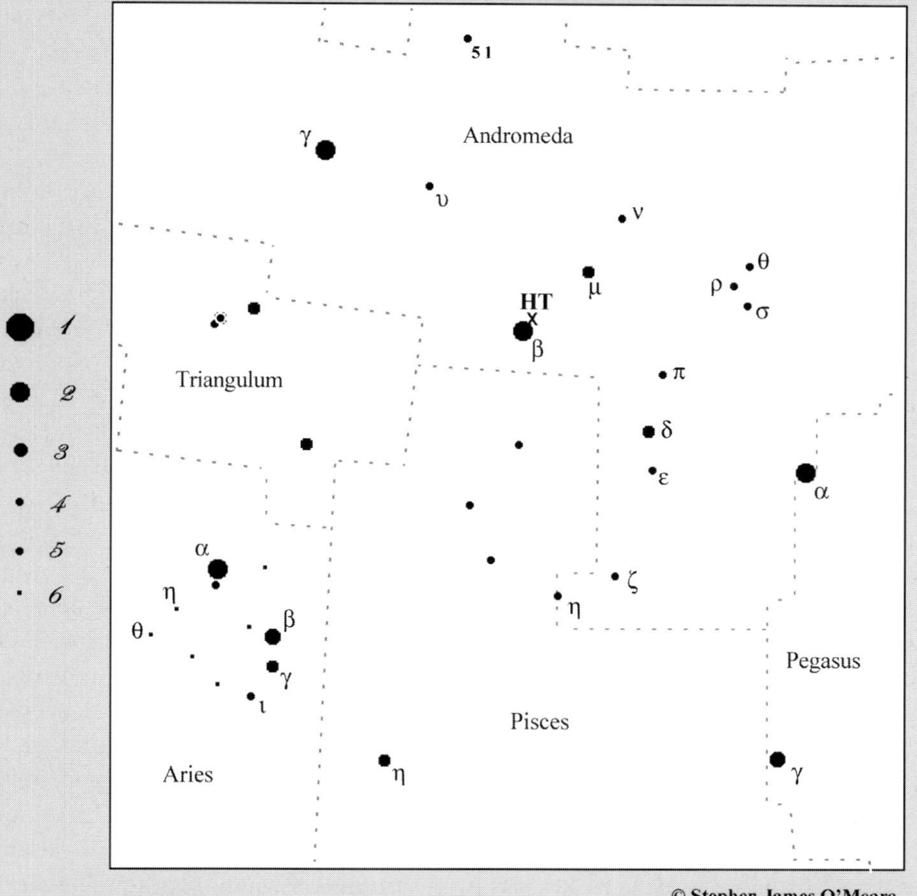

Tirion: Chart 4

Uranometria: Chart 91

5

Mirach's Ghost, Lost Pearl Galaxy
NGC 404
Type: Dwarf Lenticular Galaxy (S0)
Con: Andromeda

RA: 01^h 09.4^m
Dec: $+35°$ $43'$
Mag: 9.8 (O'Meara); 10.3
SB: 14.0
Diam: $6.6' \times 6.6'$
Dist: 8 million light-years
Disc: William Herschel, 1784

W. HERSCHEL: [Observed September 13, 1784] Pretty bright, though Beta Andromedae is in the field, considerably large, round, bright middle. (H II-224)

NGC: Pretty bright, considerably large, round, gradually brighter in the middle, Beta Andromedae south following.

NGC 404 IS TRULY A HIDDEN treasure. This bright lenticular galaxy lies a little more than 6′ northwest of the 2nd-magnitude gold star Beta (β) Andromedae (Mirach). Beta's light interferes with – but does not drown out – this round little gem, whose face shines with the luster of a freshly polished pearl. Unfortunately, the galaxy lies *so* close to Beta that it is not plotted on many star atlases. When NGC 404 is plotted, Beta's symbol overlaps it, so it's easily overlooked. For these reasons, I call NGC 404 the Lost Pearl Galaxy. Imagine a loose pearl rolling across the deck of a pirate ship, until it wedges firmly against the ship's brass.

NGC 404 is the nearest lenticular galaxy to our Milky Way, approaching us at a speed of 35 kilometers per second. At a distance of 8 million light-years, it lies at the edge of the Local Group and belongs to a small group of galaxies in the Coma–Sculptor Cloud. No doubt, the galaxy is quite small. In his 1988 *Nearby Galaxies Catalog*, R. Brent Tully lists NGC 404 as only 16,000 light-years in diameter with a total luminosity of only about 450 million Suns. But in 2003 Russian astronomer N. A. Tikhonov (Special

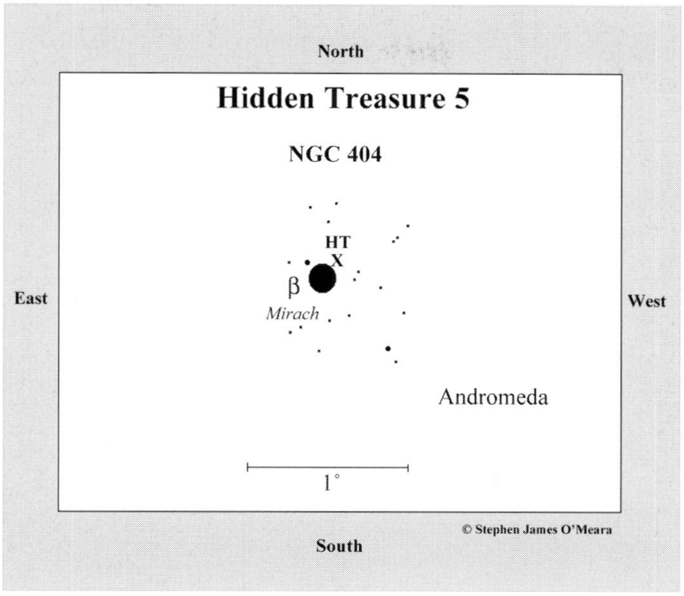

Astrophysical Observatory) and his colleagues reported that observations with the Hubble Space Telescope suggest that the galaxy is about 11 million light-years distant and the size of its disk (which is dominated by red giant stars) exceeds 65,000 light-years. Through large Earth-based telescopes NGC 404 appears to be a roundish spiral. Its small bulge contrasts well with a surrounding dust ring. In the nineteenth century, Ireland's Lord Rosse (William Parsons, Third Earl of Rosse) believed he had resolved NGC 404 into individual stars with his 72-inch reflector – the Leviathan of Parsonstown. Was this an illusion? Perhaps. Only recently has the Hubble Space Telescope, a superior 94-inch mirror, performed such a feat.

Many observers have found NGC 404 by chance, including me. Seeing such an unsuspected glow so close to Beta Andromedae causes immediate alarm. The mind wants to dismiss it as a ghost image – a phantom glow caused by the reflection of bright starlight at an optical surface – of Beta (Mirach's Ghost). But when the phantom glow does not change its position relative to Beta when the telescope is moved, something more tantalizing comes to mind; as James Mullaney reveals in his delightful book *Celestial Harvest,* NGC 404 has yet another nickname: "Comet Komorowski (& many others)." This galaxy, he says, is the "sky's finest example of a false comet."

I tried to see NGC 404 with 7 × 50 binoculars without success. Observers with larger binoculars should definitely try. Although I could not make out the magnitude 9.8 galaxy, I could detect the magnitude 8.8 star 6′ east and ever so slightly south of it. The galaxy is *just* visible in a 60-mm refractor. Rev. Thomas W. Webb found NGC 404 "rather faint," but "easy" in a $9\frac{1}{3}$-inch speculum mirror reflector – his description demonstrates just how poorly these mirrors reflected light.

In my 4-inch, the galaxy has a milky-smooth texture at 23× and can be seen both with a direct gaze and with averted vision – if you know where to look. Before I reached that conclusion, I had some fun experimenting with the galaxy's visibility at low power. It's natural to want to place Beta at the center of the field of view. But resist the temptation. Doing so places not only Beta but also NGC 404 in your eye's cone-rich fovea. All that is visible of the galaxy is its starlike core, so, chances are, you won't notice it. You want NGC 404 to fall on the rod-rich retina of

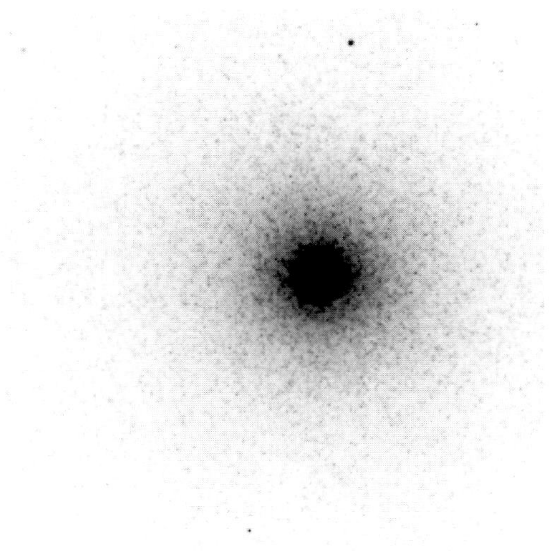

your eye, which means, placing Beta away from the center of vision. When I place Beta anywhere away from the field's center, the galaxy's diffuse form swells into view and is very obvious as a concentrated puff of light 6′ northwest of Beta.

A careful investigation at 23× will reveal its starlike core. The nineteenth-century observer Heinrich d'Arrest called it a "strong nucleus." High-resolution images show it well, and a 1995 Hubble Space Telescope image reveals it to be a bright, unresolved point source (less than 1 parsec across) surrounded by what is presumed to be star-forming regions. The disk of the galaxy is dominated by mostly red giant stars, while the spheroidal bulge consists of red giant stars and the more advanced Asymptotic Giant Branch (AGB) stars. These AGB stars represent the very last phase of normal stellar evolution and are well on their way to becoming planetary nebulae. The galaxy is surrounded by a large face-on donut shaped halo of neutral hydrogen gas, which extends out to an overall diameter of 9′. A study in 2004 by Mexican astronomer Maria Soledad del Rio (Universidad de Guanajuato) and her colleagues found that most or all of this gas is likely the result of a merger with a 15.5-magnitude dwarf irregular galaxy.

When I changed the magnification from 23× to 72× in the 4-inch and moved Beta out of the field of view, the galaxy had an enticingly mottled appearance, as if it were breaking up into individual starlight – a feeling Lord Rosse had shared. But at 101×, I realized that the mottled nature is due, in part, to some superimposed suns. One star lies just beyond the galaxy's north flank; it's enough to make one question whether it is a supernova, so be careful. I noticed two other dimmer concentrations of light at 168× – one at the galaxy's south flank and another at the southeast flank. The galaxy takes magnification surprisingly well, and I was able to survey it quite comfortably at 303×. The core remained stellar, a fantastic pip of intense light, surrounded by a slightly offset halo of light that favors the north. This bright

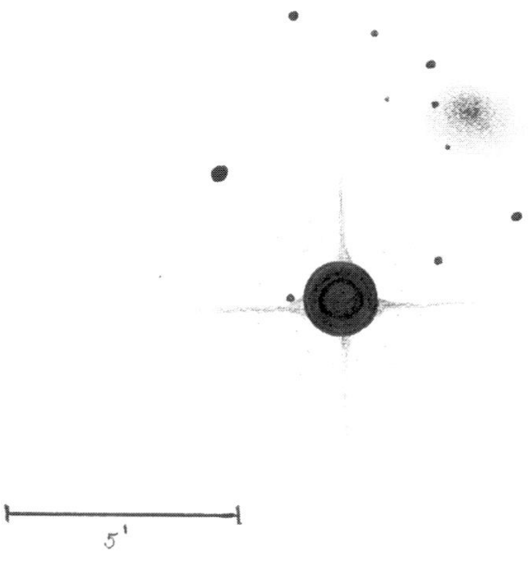

inner halo is surrounded by a larger ellipse of largely uniform light. No spiral structure can be seen with such a small instrument, though large telescopes have found crescents of darkness, like dust lanes in the galaxy's disk. It is up to you now to go out and find this lost pearl and admire its beauty.

Hidden Treasure 6

NGC 584

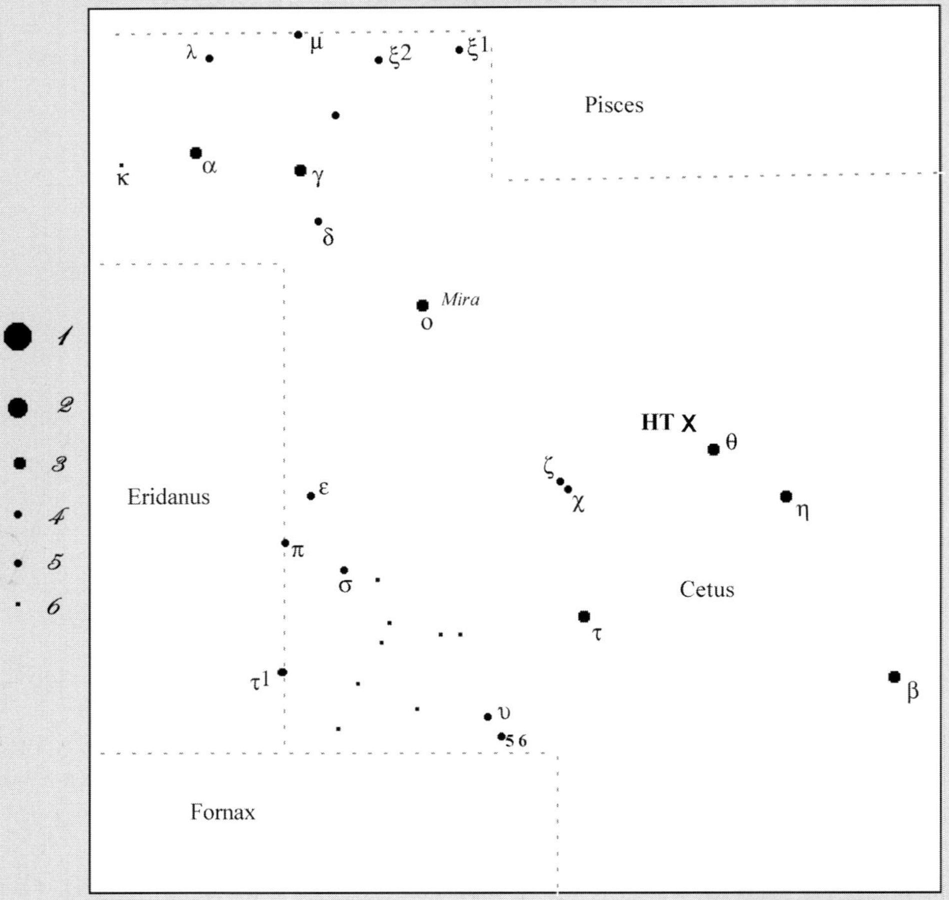

Tirion: Chart 10

Uranometria: Chart 263

6

Little Spindle Galaxy
NGC 584 = IC 1712
Type: Lenticular Galaxy (S0?)
Con: Cetus

RA: 01^h 31.3^m
Dec: –06° 52′
Mag: 10.4 (O'Meara); 10.5
SB: 12.2
Dim: 3.8′ × 2.4′
Dist: 76 million light-years
Disc: William Herschel, 1785

W. HERSCHEL: [Observed September 10, 1785] Considerably bright, pretty small, round with a much brighter middle, 5 or 6 minutes distant from III-431 [NGC 586]. (H I-100)

NGC: Very bright, pretty large, round, much brighter in the middle, preceding of 2.

NGC 584 IS A STUNNINGLY SMALL, and deceivingly bright, lens-shaped galaxy riding on the back of Cetus, the Whale. It is a fine example of how a tiny galaxy with a moderately faint listed magnitude is quite easy to see.

William Herschel, the object's discoverer, cataloged it as a Class I object (bright nebulae). In comparison, he classified the previous hidden treasure, NGC 404 (Hidden Treasure 5) in Andromeda – a lenticular system with a brighter apparent magnitude and larger apparent size – as a Class II object (faint nebulae). The reason for this apparent discrepancy is the intensity of NGC 584's bulge. Note too that Herschel saw it as a round object and not elongated.

A member of the Cetus–Aries Cloud of Galaxies, NGC 584 is quite large, spanning 93,000 light-years of space. It is also distant; at 76 million light-years, it is nearly 10 times further than NGC 404. But while NGC 404 has a total luminosity of only about 450 million Suns, NGC 584 has a total luminosity of 17 *billion* Suns. In Brent Tully's *Nearby Galaxies Catalog*, and NASA's *Extragalactic Database*, NGC 584 is listed as an elliptical galaxy. It has been further subclassed as an E4 galaxy (where E0 is round and E10 is highly elliptical). But evidence is growing that it is a lenticular (S0) system. If so, NGC 584 has one of the most luminous bulges known.

6

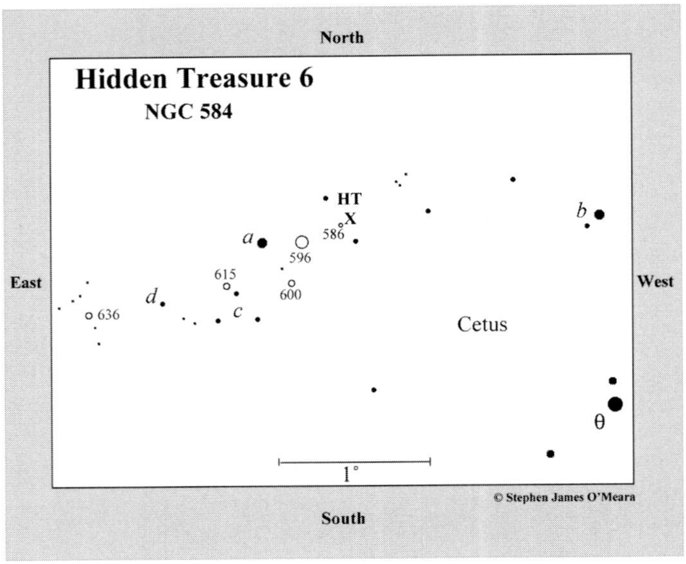

In a classical sense, elliptical galaxies have a bright nucleus and a featureless envelope; the stars are highly reddened and have random orbits; they swarm about the nucleus like comets around the Sun. In general, they rotate slowly and thus are not very flattened but prolate (like an American soccer ball) or triaxle. Roger Davies (Institute of Astronomy, University of Cambridge) and his colleagues say that kinematic studies of NGC 584 support the idea that it is a lenticular galaxy – an intermediate system between the ellipticals and the spirals.

Davies' data also show that NGC 584 is a relatively rapid rotator and has a slightly tilted disk within its bright bulge, which we see inclined 35° from edge-on. The major axes of the bulge and of the outer, diffuse envelope differ in position angle by about 15°. As advances in telescope power and technology progress, astronomers will be able to probe deeper into elliptical galaxies, which are more complex creatures than we once believed. It still remains unclear whether lenticulars evolved from ellipticals or the other way around. One popular theory is that ellipticals evolve primarily due to galactic mergers. The fraction of ellipticals with shells and disks suggests that all large ellipticals have had at least one merger event in their history, which could possibly explain NGC 584's rapid rotation and its arguably lenticular nature.

Beyond M77, the most famous Seyfert galaxy in the heavens, objects in Cetus are easy to overlook, probably because, as the late *Sky & Telescope* columnist Walter Scott Houston wrote, "[Cetus] is a vast stellar desert." But, as Houston reminded his readers, it is also "a rich area for galaxy hunters." NGC 584 is the northernmost object in a roughly 1°-long, meandering chain of five galaxies about $2\frac{1}{4}^{\circ}$ north of magnitude 3.6 Theta (θ) Ceti – a fine golden star 115 light-years from the Sun; look for a 6th-magnitude neighbor, about 10′ due north. NGC 584 is only about 40′ west-northwest of a magnitude 5.7 star (*a*) that's 97 light-years distant. It's nestled between a magnitude 7.4 star and a fine telescopic double star, which the nineteenth-century observer Adm. William Henry Smyth took a special liking to. Here's how he described the scene: "A tolerably bright round nebula, of a pearly tint, just above the Whale's back . . . The field is very interesting, for nearly south of the little nebula is a neat double star, the components of which are of the 9th and 11th magnitudes."

If your telescope is on an equatorial mount, you can try locating NGC 584 this way: move the scope $1\frac{1}{4}^{\circ}$ north of Theta, where you will encounter a pair of 6th- and 7th-magnitude stars (*b*), oriented northwest and southeast and separated by about 8′. NGC 584 is about $1\frac{3}{4}^\circ$ east and slightly north of that pair. Houston notes that the galaxy has been seen with an aperture of $2\frac{1}{2}$ inches. With extreme patience and time, I have glimpsed it at the limit of my antique $1\frac{1}{4}$-inch telescope. Regardless, it is a cinch in anything larger.

If you're a beginner and are used to seeing galaxies as large and amorphous glows, this one might throw you off base. You're truly looking for something very small and bright. In the 4-inch at 23×, NGC 584 is almost stellar – a fuzzy navel within a smear of light. The galaxy is so small and its core so bright and stellar that the system is easy to sweep over. The galaxy's fuzzy nature shows up better at 72×, but I find 101× provides the most comfortable view in the 4-inch.

At 101×, the galaxy displays a tack-sharp core, an egg-shaped inner lens and a strong and equally sharp needle of light that slices through the major axis of the outer lens, which fades to insignificance under higher powers. I call NGC 584 the Little Spindle because it looks like a smaller and dimmer version of NGC 3115 in Sextens (Caldwell 53). As Smyth wrote, the galaxy has "a pearly tint" and, like a pearl, its beauty is magnified by its simplicity.

Most surprising is that a companion galaxy, NGC 586, is quite easily visible at 101× in the 4-inch just 4.3′ southeast of NGC 584. This non-interacting galaxy has a listed magnitude of 13.2, but that must be in error by a magnitude or more. Like NGC 584, NGC 586 has a *very* bright nucleus within a shell of faint light.

Once you finish admiring these galaxies at high power, change back to low power and relax your gaze. If you avert your vision 25′ to the southeast, you should see the nearly 3′-wide galaxy NGC 596, another elliptical system that shines at magnitude 10.9. It makes a wonderful companion to NGC 584, though NGC 596 is almost hidden in the glare of Star *a*, which is only about 12′ to its east. Much more difficult to see is the magnitude 12.4 barred spiral galaxy NGC 600. It is but a breath of light barely visible in the 4-inch under a very dark sky – and that's

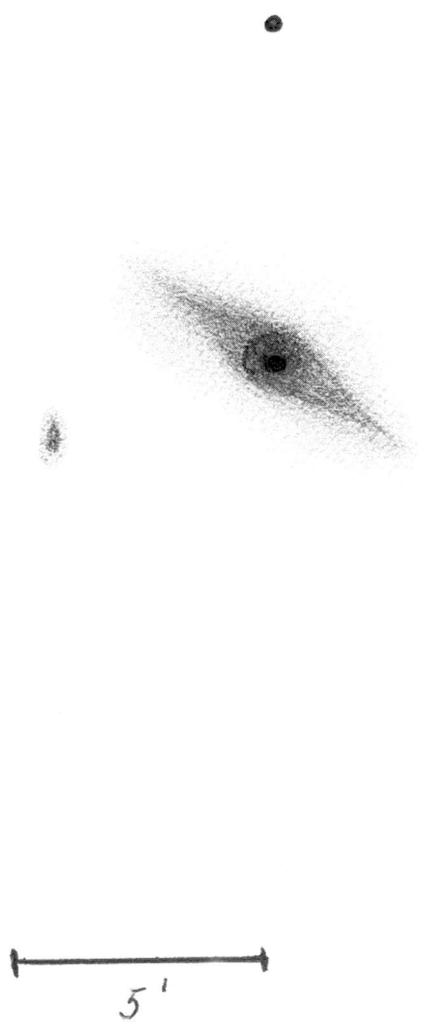

after *much* painstaking search and attention with averted vision and heavy breathing. Almost 30′ southeast of Star *a*, and just 6′ east-northeast of a magnitude 8.5 star, is the nearly edge-on spiral galaxy NGC 615, the easternmost galaxy in the chain of five described above. Although NGC 615 shines at magnitude 11.6, Brian Skiff and Christian Luginbuhl say it's easily seen in a 6-inch telescope. It also shows well in the 4-inch, again because the galaxy has a bright nucleus in a cocoon of light.

Finally, if you want a further challenge, go after NGC 636, a magnitude 11.5 elliptical galaxy nearly $1\frac{1}{2}^{\circ}$ southeast of Star *a*. With an apparent diameter of only 2.7′, the galaxy is easy to sweep over at 23×. So use at least 72× and star hop from NGC 615, which is just northeast of a 12′-wide equilateral triangle of 9th-magnitude stars (*c*). Just 30′ due east of the triangle's center is a solitary 8.5-magnitude star (*d*). NGC 636 lies 35′ east and slightly south of that solitary sun. When Houston saw the galaxy in a 10-inch telescope, he thought it was a planetary nebula at first. Indeed, the galaxy has a bright pip at the center, surrounded by a circular ring of mottled light with two slightly elongated tips. The field is busy with dim suns but the galaxy stands out quite nicely from them.

By the way in the 1888 *New General Catalogue* and the 1908 *Second Index Catalogue*, J. L. E. Dreyer uses the same coordinates for NGC 584 as he does for IC 1712, so the two objects are the same. E. E. Barnard must have overlooked this Herschel discovery when he independently found it and relayed its position to Dreyer.

Hidden Treasure 7

NGC 659

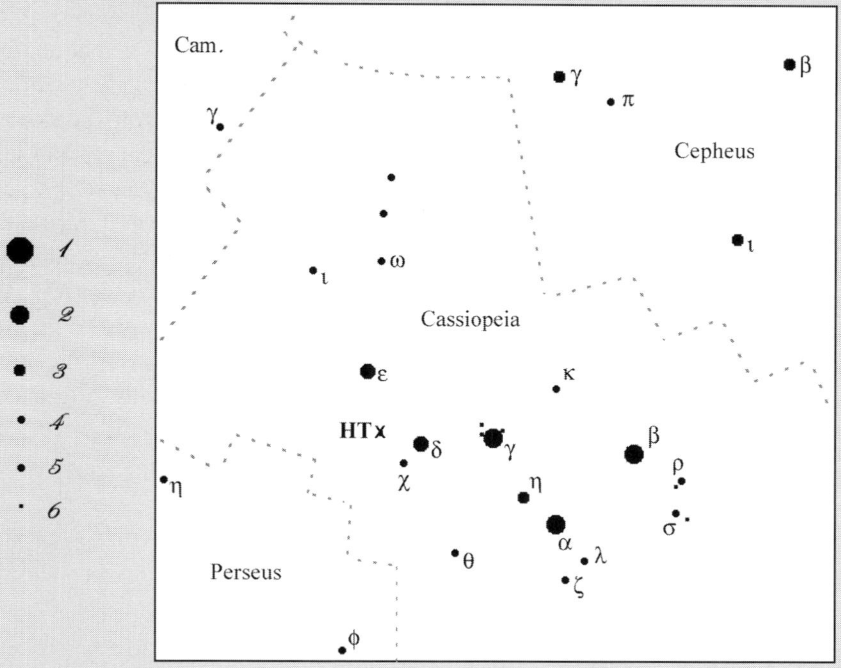

Tirion: Chart 1

Uranometria: Charts 16 & 37

7

Yin-Yang Cluster
NGC 659
Type: Open Cluster
Con: Cassiopeia

RA: $01^h\ 44.4^m$
Dec: $+60^\circ\ 40'$
Mag: 8.2 (O'Meara); 7.9
Diam: 6.0′
Dist: 8,200 light-years
Disc: Caroline Herschel, 1783

W. HERSCHEL: [Observed November 3, 1787] A small cluster of small stars, not very rich, C.H. 1783. (H VIII-65)

NGC: Cluster, little rich in stars, stars bright.

ONLY $2\frac{1}{2}^\circ$ EAST-NORTHEAST OF 3RD-magnitude Delta (δ) Cassiopeia in the constellation's famous W asterism, you will find five open star clusters forming an incomplete 2°-wide circle around a diamond of roughly 7th-magnitude stars. They are in order of equinox 2000.0 right ascension: M103, Trumpler 1, NGC 654, NGC 659, and NGC 663 (Caldwell 10). Our interest, is in the star cluster NGC 659, discovered by the great Caroline Herschel (see Appendix A).

NGC 659 is not a sight that will take your breath away. Shining at 8th magnitude, it is a gasp of dim light in my antique telescope. The cluster can also be spotted in 7 × 50 binoculars, though only with difficulty; there is barely enough contrast to distinguish it from the Milky Way background. Although I succeeded with averted vision under dark Hawaiian skies, I believe the task might actually be simpler from a suburban location, where the Milky Way background is subdued.

While NGC 659 does not inspire visual greatness, it is on a par with the celebrated M103 in almost all of its characteristics: both NGC 659 and M103 measure 6′ in apparent diameter, lie about 8,000 light-years from Earth, have true physical diameters of about 15 light-years, contain around 180 members, and have similar ages; NGC 659 is 20 million years young, while M103 is 5 million years older. When NGC 659 was born, the Pleiades were just emerging from their dusty cocoon, grasses were making their first appearances on Earth, and the mighty mastodon was beginning to roam the North American continent.

One significant difference in the two clusters is that M103 shines about $\frac{1}{2}$-magnitude brighter than NGC 659; it also has brighter stars at its core. If we can dare to be bold, we

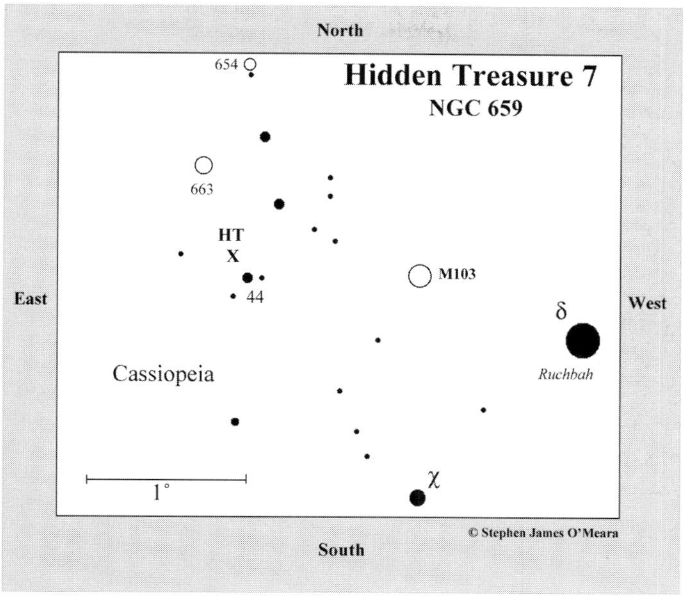

could call NGC 659 and M103 – which are separated by only about $1\frac{1}{4}^{\circ}$ of sky – Cassiopeia's Dim Double Cluster. Of course, it's hard to neglect the other open star clusters in the region, all of which have ages ranging from 15 to 26 million years. To honor the bloom of so many clusters in so close a region in so close a time, I call this rich region of Cassiopeia The Queen's Garden.

Finding NGC 659 is a cinch. From Delta Cas, use your naked eyes or binoculars to locate 5th-magnitude Chi (χ) Cas, which is nearly $1\frac{1}{2}^{\circ}$ to the southeast. The 6th-magnitude star 44 Cas forms the northeast apex of a near-equilateral triangle with Delta and Chi Cas. NGC 659 blooms only about 10′ north-northeast of 44 Cas. You could also first locate M103, which is 1° northeast of Delta Cas. A simple push of the tube $1\frac{1}{4}^{\circ}$ due east will take you to 44 Cas and NGC 659.

At 23× in the 4-inch, NGC 659 is a subtle glow that seems to hide in the shadow of 44 Cas and its nearby stellar companions. It has a 3′-wide core that, with averted vision, is a tight, almost circular, collection of 10th- and 11th-magnitude suns; it is the main jewel in a necklace of four comparably bright suns that arc from the south to the northwest. The entire necklace is surrounded by a larger and fainter halo of suns, which glint in and out of view. Sticking with a garden theme, NGC 659 looks like a pale violet among roses, shining through mist and moonlight.

With 101× the cluster breaks down into individual packets of stars. The core, however, remains distinct. It is a rosette of a half dozen suns, one of which is a close 11th-magnitude double star that resolves further into a triple at higher magnifications. If you add the four roughly

12th-magnitude suns to the west, the core has an apostrophe shape. A separate, though fainter, apostophe-shaped congregation of stars lies southwest of the core; it is a mirror image of the first, so together, with a stretch of the imagination, they can be seen as the well-known Yin-Yang symbol, which, in Chinese philosophy, represents the idealized balance of the two opposing but equal forces in nature. It is a law of unity, which seems appropriate for open clusters, whose gravitationally bound members must travel together around the galaxy in relative harmony.

As with many of these sparse open clusters in the Milky Way the question becomes, how far does the cluster really extend? On one particularly fine night, I experimented and found that the cluster remained a coarse scattering of suns at 504×. Many of the stars were seen in pairs or tiny groupings, which were symmetrically placed about the central rosette and connected by strings of stars. If you drifted away from the center, it was easy to get lost in the surroundings. It was as if the cluster was trying to make a point about keeping one's center. If you lose it, you will lose your balance.

Hidden Treasure 8

NGC 772

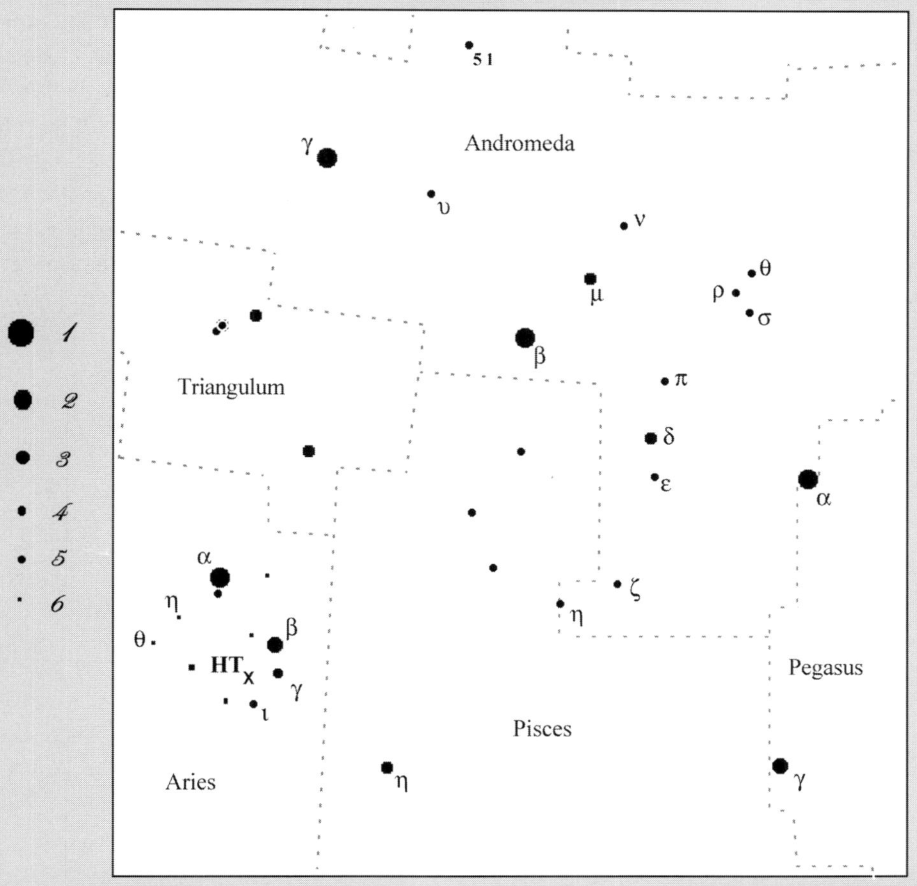

© Stephen James O'Meara

Tirion: Charts 4 & 10

Uranometria: Chart 129

8

Fiddlehead Galaxy
NGC 772
Type: Spiral Galaxy (Sb)
Con: Aries

RA: 01^{h} 59.3^{m}
Dec: $+19°$ $00'$
Mag: 9.9 (O'Meara); 10.3
Dim: $7.1' \times 4.7'$
SB: 14.0
Dist: 106 million light-years
Disc: William Herschel, 1785

W. HERSCHEL: [Observed November 29, 1785] Very bright, large, round, with a much brighter middle, not easily resolvable, 4′ in diameter. (H I-112)

NGC: Bright, considerably large, round, gradually brighter in the middle, [mottled].

IF THERE IS ONE CONSTELLATION admired by astronomical historians and naked-eye observers, but neglected by deep-sky observers, Aries, the Ram, is it. Messier lists no deep-sky objects in this small but visually quaint constellation. Nor do we find any in the Caldwell list. Yet Aries has a hidden treasure that is just as bright as M91, M98, and M99 in Coma Berenices, as well as M108 and M109 in Ursa Major. The question is how many have seen it?

NGC 772 is easy to overlook because it lies within an acute triangle of three extremely popular Messier wonders: M33 (the Great Triangulum Spiral), M74 (the famous Phantom Galaxy) in Pisces, and M45 (the wondrous Pleiades in Taurus). And while NGC 772 is nearly $3\frac{1}{2}$ times more distant than M74, it is only $\frac{1}{2}$-magnitude fainter. If NGC 772 were at the same distance as that galaxy, it would rival M51, the Whirlpool Galaxy in Ursa Major, in visual grandeur.

In high-resolution photographs, NGC 772 is a stunning early spiral system seen 37° from edge-on. It has several tightly coiled arms and one strong, asymmetric arm crossing the disk westwards. That one bright arm coils around the nucleus like a fiddlehead unfolding, which is why I call it the Fiddlehead Galaxy. The galaxy's arms are well formed and relatively smooth, indicating only a small rate of star formation in the relatively recent past – except in the single high-surface-brightness arm which is lightly peppered with faint H-II regions. The relatively smooth multiple arms on the opposite side

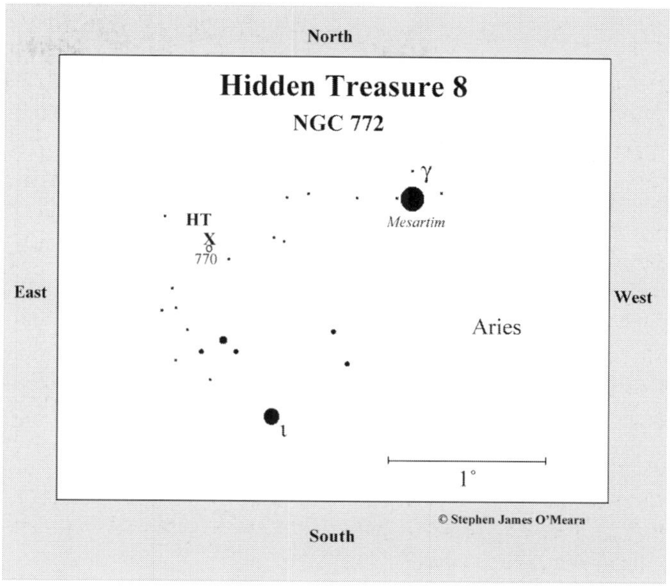

are defined by fragments of spiral dust lanes. These fragments have a delicate, almost feathery texture, like a whirlpool of cirrus cloud known as the Mare's Tail.

A faint plume of matter can also be seen fanning toward two dwarf satellites. It would seem that the dwarf companions have gravitationally disturbed NGC 772, and are perhaps responsible for the prominence of the western spiral arm (like the one in M51 that connects to its companion). But the two dwarf satellites have very anomalous redshifts – roughly 10 times that of NGC 772. Thus, Halton Arp includes NGC 772 in his *Atlas of Peculiar Galaxies* as Arp 78. "The latter two [dwarf companions]," Arp argues, "are further examples of small companion galaxies which have redshifts that do not correspond to the standard cosmological-velocity redshifts."

NGC 770, on the other hand, appears to be at about the same distance as NGC 772, so it is most likely a true companion to the larger spiral system. It has a separation of only 160,000 light-years from its larger companion, and its apparent size indicates an actual diameter of only about 40,000 light-years. NGC 772 is comparatively large indeed, having a true linear diameter of 208,000 light-years, making it roughly twice the diameter of our Milky Way Galaxy with a total luminosity of 80 billion Suns. The galaxy is whisking away from us at 2,473 kilometers per second.

Finding NGC 772 is a cinch. The starting point is Gamma (γ) Arietis (Mesartim), the First Star in Aries, which was at one time nearest to the equinoctial point. It is also one of the most attractive doubles in the region. Robert Hooke (1635–1702), a founding member of the Royal Society in England and, who, in 1664, made the earliest sighting of a dark spot on Jupiter (one that revealed the planet's true rotation), also discovered Gamma's duplicity while following the comet of that year: "I took notice that it consisted of two small stars very near together; a like instance to which I have not else met with in all the heavens." Here are two magnitude 4.8 beacons separated by 7.8″, looking like a car's headlights blazing forth on a dark highway. Gamma Arietis is a close neighbor to our Sun, being 160 light-years distant and shining with a combined luminosity of about 50 Suns. NGC 772 is a clean sweep about $1\frac{1}{2}^{\circ}$ to the east-southeast of Gamma. In my nearly 3° field of view, if I place Gamma at the west edge, NGC 772 is virtually centered in the field.

At 23× in the 4-inch the galaxy's appearance is, in fact, reminiscent of M1, the

Crab Nebula in Taurus. The ghostly glow seems extremely lonesome. It has an ever so slightly condensed core surrounded by a weak halo. It's more a star with a very dim sheen around it. In his *Cycle of Celestial Objects*, Adm. William Henry Smyth all but concurs, calling it a "round nebula, closely following γ on the neck of the Ram . . . It is large and pale, but brightens in the centre." But this pale ghost glows more strongly with time and averted vision. Certainly, as with its phantom neighbor, M74, which is only about $6\frac{1}{2}^\circ$ to the southwest, light pollution will affect the visibility of NGC 772.

At 72× the galaxy's core is a white egg inside a mottled nest of light, which tapers into a whisper of light toward the west; here is the "bright" (in photographs), solitary arm of the fiddlehead, still wrapped in its fibrous web. When I switch to 101×, the galaxy's elliptical companion, NGC 770, just *snaps* into view; like NGC 586, the dim companion galaxy to NGC 584 (Hidden Treasure 6), NGC 770 must be brighter than its listed magnitude (12.9). The "apostrophe" shape of NGC 772 is most apparent at 101×. The southeast flank is remarkably rounded and smoothed, while the long western arm is but a slim arc of light with a slight mottled texture. Other condensations can be seen nearer to the nucleus to the northeast and south-southwest. These are very subtle details, however, and require time to see in a small telescope. Try alternating between different magnifications and gently tapping the tube while looking at the galaxy with averted vision. Don't forget to breath rhythmically and strongly as you look. Plan to spend a few hours on this one.

NGC 772 was the host of two supernovae that could be seen simultaneously in 2003: supernovae 2003hl and 2003iq. M. Moore and W. Li discovered supernova 2003hl at magnitude 16.5 on August 20, 2003, on images taken with the University of California at Berkeley's Katzman Automatic Imaging Telescope; it was 24″ east and 13″ south of the nucleus. Supernova 2003iq was discovered on October 8, 2003, by French amateur astronomer Jean-Marie Llapasset; the star was magnitude 16.4 at discovery, 5″ east and 46″ south of NGC 772's nucleus, and about 37″ southwest of SN 2003hl. Both were Type II supernovae.

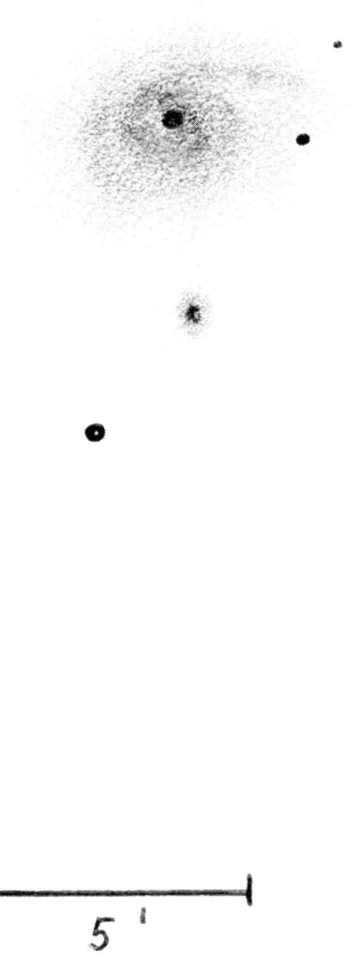

Hidden Treasure 9

NGC 908

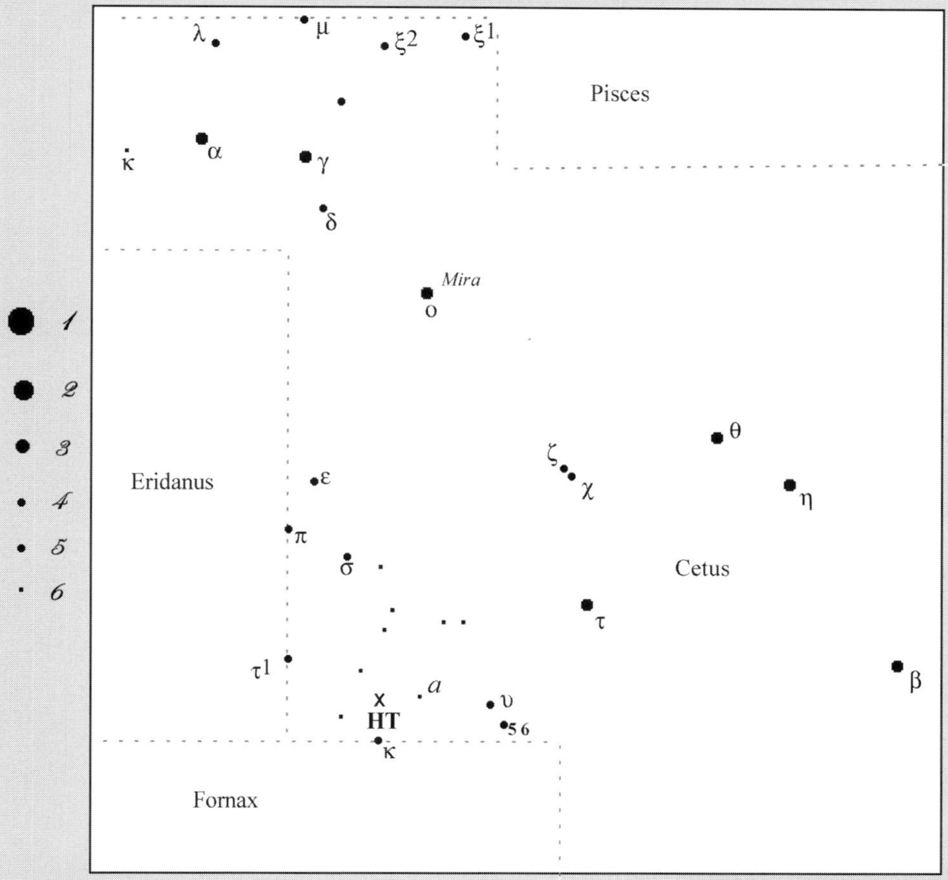

Tirion: Chart 10

Uranometria: Chart 263

9

NGC 908
Type: Spiral Galaxy (Sc)
Con: Cetus

RA: 02^h 23.1^m
Dec: −21° 14′
Mag: 10.4 (O'Meara); 10.4
Dim: 6.1′ × 2.7′
SB: 13.1
Dist: 58 million light-years
Disc: William Herschel, 1786

W. HERSCHEL: [Observed September 20, 1786] Considerably bright, very large, extended in a direction south preceding-north following, above 15′ in length. (H I-153)

NGC: Considerably bright, very large, extended.

TO THE GREEKS OF OLD, CETUS WAS not so much a Whale as it was a Sea Monster, a serpentine, Nessie-like creature that Poseidon sent to devour Andromeda for her mother's constant vanity. Cetus was turned to stone when Perseus, the hero, held the severed head of gorgan-haired Medusa before the monster's eyes. The legend of Cetus slithered its way into natural history when Pliny the Elder (AD 23–79) wrote that in 58 BC, the Roman politician Marcus Aemilius Scaurus (163–88 BC) had found the "spine of the sea serpent killed by Perseus at Joppa" and displayed it in Rome. The great skeleton measured 40 feet in length and the vertebrae measured six feet in circumference. Some scholars theorize that the bones might have belonged to a sperm whale that had beached itself near modern-day Jaffa (Joppa).

But Pliny truly believed in dragons. In faraway India, he said, these creatures were large enough to prey on elephants by dropping out of trees and strangling them; the crests on their heads were so large that they could sail to Arabia to hunt. (It's possible that Pliny loved his wine, either that, or he had heard magnified tales of python behavior.) The legend of Cetus and its hideous looks lived on through the ages. Indeed, in his 1603 *Uranometria*, Johann Bayer, depicts Cetus as a sea serpent with the head and teeth of a greyhound, an armored crest, the body of a dolphin, and a posterior that resembles the coiled and forked tail of some demonic snake. The depiction only reflects the

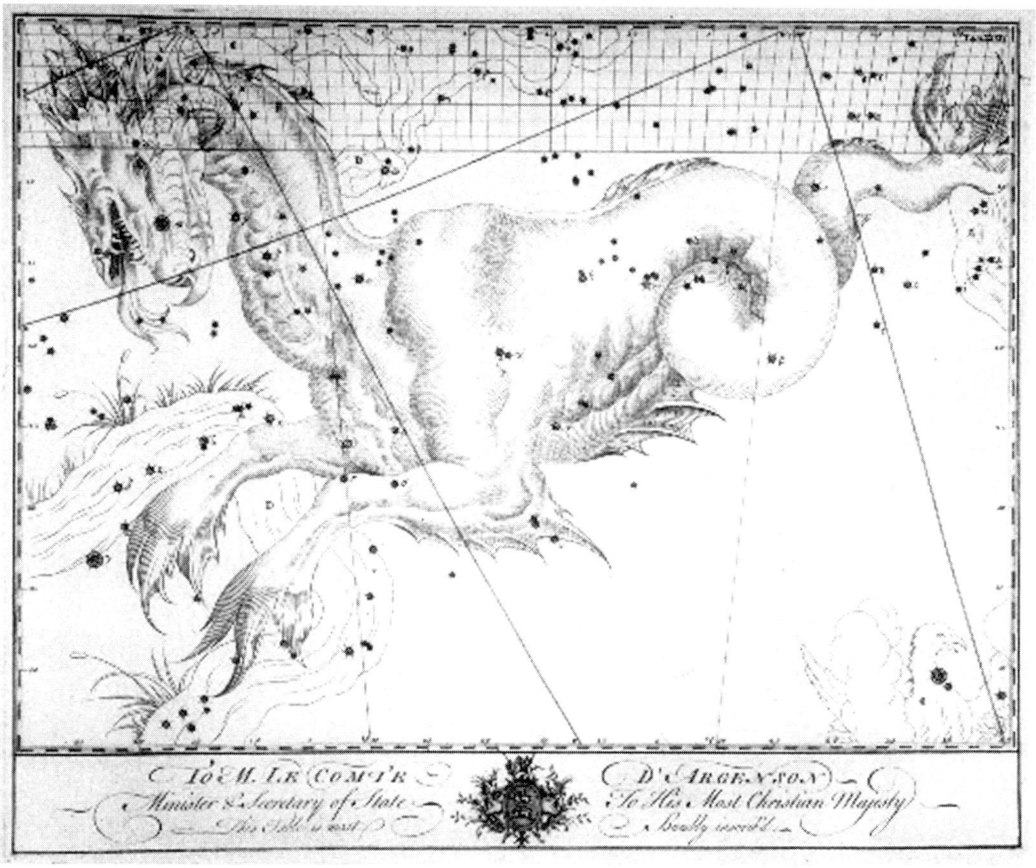

ignorance of the times, when men had little knowledge of the nature and appearance of creatures in the sea. In his 1995 book *Dragons, A Natural History,* Karl Shuker theorizes that the Greek legend of Cetus grew out of rare sightings of oarfish, whose snakelike bodies can be 30 feet in length and have a red head crest.

Mythology aside, the fact is that the classical depiction of Cetus looks nothing like the humpback whales of today. That is why so many skywatchers who grew up reading H. A. Rey's *The Stars: A New Way to See Them,* have adopted Rey's more modern interpretation of the constellation's stars. Whereby Beta (β) Ceti (Deneb Kaitus, "the Tail of Cetus") becomes the tip of the Whale's snout, and the famous circlet – including Alpha (α) Ceti (Menkar, the "Nose") – is the Whale's tail. In Rey's depiction of the Whale, 4th-magnitude Upsilon (υ) Ceti is the bottom of the Whale's belly. It is at Upsilon you need to start your journey to find the next hidden treasure, NGC 908, a magnificent starburst galaxy.

NGC 908 belongs to the Cetus–Aries Cloud of Galaxies. It is another respectably large system, spanning 86,400 light-years in true physical extent. It has a total mass of 100 billion Suns and a total luminosity of 30 billion Suns. It is receding from the Sun at 1,508 kilometers per second. In photographs the galaxy is a beautiful pinwheel seen 20° from edge-on. It looks as if M51 has been steamrolled. The spiral arms are very massive and bright. Three dominant ones spiral outward from the central region, and detailed images reveal four crossings of the major axis by arms on both ends. The galaxy's central lens is elongated, which may indicate the presence of a small bar, although its official classification is that of a normal Sc-type spiral.

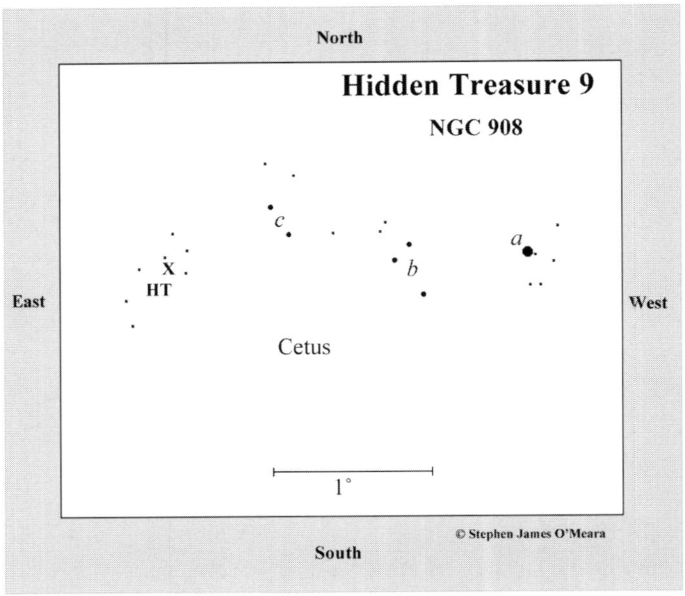

The spiral structure is also asymmetrical: on the west side, there is one thick prominent arm, while on the eastern side, the spiral arm is split in two. The forked or double arm on the eastern side is very interesting and it appears to have been pulled out of its normal symmetry, Indeed, this frayed appearance is probably due to a gravitational interaction with an unseen companion – one that may lie behind the galaxy as seen from our vantage in space. Although this object has not been identified, if we extend the curve of the forked arms out 16.6′ to the northeast, we come to a 15th-magnitude galaxy, MCG-4-6-36. Unfortunately, its redshift is considerably larger than NGC 908's; conventional science says that these objects cannot be related, but the geometry of the two is interesting.

Look closely at the photograph on p. 56 and you'll see that the galaxy's arms are freckled with bright clumps of starlight, as if the galaxy has hives. That's because NGC 908 is a starburst galaxy, one that is undergoing a period of intense star formation – another indication that NGC 908 had a recent encounter with another object. Usually starburst galaxies can manufacture solar-mass stars at the phenomenal rate of hundreds of millions per year. At such a rate, the galaxy would have used up all its star-forming material on the order of 100 million years. Naturally, then, these starburst episodes must have occurred in the relatively recent past. Generally these episodes are confined to very luminous clusters around the galaxy's nucleus. They are the most dense and intense star-forming regions known. Other starburst galaxies include M82 in Ursa Major and the Ringtail galaxies NGC 4038–9 (Caldwell 60 & 61) in Corvus.

Unless you're using a "go-to" telescope, you may have to work to find this treasure. Again, you should start at Upsilon Ceti, which may be a difficult naked-eye catch from a suburban sky; if so, use binoculars to locate it. It lies in the far southeastern quadrant of the constellation and is the brightest star around for many degrees. You'll know when you have it, because 5th-magnitude 56 Ceti will be about $1\frac{1}{2}°$ to the southwest and, closer in, 5th-magnitude 57 Ceti will be 15′ to the north-northwest. Once you're certain you have Upsilon, center it in your telescope and then move a little more than 3° due east where you'll find a solitary 6th-magnitude star (*a*) bordered by a 10′-wide trapezoid of 8th- and 9th-magnitude suns 7′ to the southwest. If you hop an additional 45′ due east, you'll encounter a 20′-long acute triangle of 8th- to 9th-magnitude stars (*b*). Center the northernmost star (the faintest of the three) and move 50′ east-northeast. Here, two roughly 9th-magnitude stars (*c*) will be oriented northeast–southwest and separated by about 12′. The galaxy is about 50′ southeast of this stellar pair. Look for a 10th-magnitude cigar-shaped glow (6′ long) with a stellar core. The core is so bright that it is *barely* visible in 7 × 50 binoculars under a dark sky, just southeast of a little triangle of roughly 10th-magnitude suns.

At 23× in the 4-inch the galaxy is immediately obvious as a slightly elongated haze surrounded by a trapezoid of suns. With averted vision, the galaxy has a stellar nucleus within a dense core of light. An outer envelope tapers off in an east–west direction. At 72× and 101×, the galaxy breaks down into finer details. It is symmetrical along the east–west axis. The dense core is no longer entirely stellar. Instead, it appears to consist of a tack-sharp stellar nucleus surrounded by a fuzzy elliptical core, which is at the center of a broad lens; that lens is slightly tilted with respect to the galaxy's major axis. A dense lash of material lies on the western side of this inner lens; in fact, the western side of the outer envelope appears to be bulging out from the inner lens.

Although I cannot see it, this marks the location of one of the galaxy's dense spiral arms. A dense clump of matter lies at the extreme eastern end of the galaxy, where I can see a faint swirl of material – a spiral – sweep to the north. Christian B. Luginbuhl and Brian A. Skiff say NGC 908 is a beautiful object in a 12-inch telescope. And through an 11-inch Celestron at Lake San Antonio, California, amateur Jamie Dillon referred to NGC 908 as "a gratifying galaxy, with a big fat fancy spiral and twirling dust lanes, no clear core, but otherwise like a junior version of [NGC] 253."

On December 20, 1994, Universal Time, A. Williams and R. Martin discovered a 17th-magnitude supernova about 24″ east and 4″ north of the center of NGC 908. The new star was near maximum light when found on images taken at the Perth Observatory as part of the Perth Astronomy Research Group's automated supernova search. Called Supernova 1994ai, the new star was a Type Ic supernova (one that fades slowly). Type Ic supernovae presumably occur when massive stars self-destruct after ejecting their outer envelopes of hydrogen and helium. While this particular supernova was too faint for a visual discovery, observers should definitely be on the lookout for others. One must be careful, though, not to mistake a star clump for a supernova candidate.

Hidden Treasure 10

NGC 1023

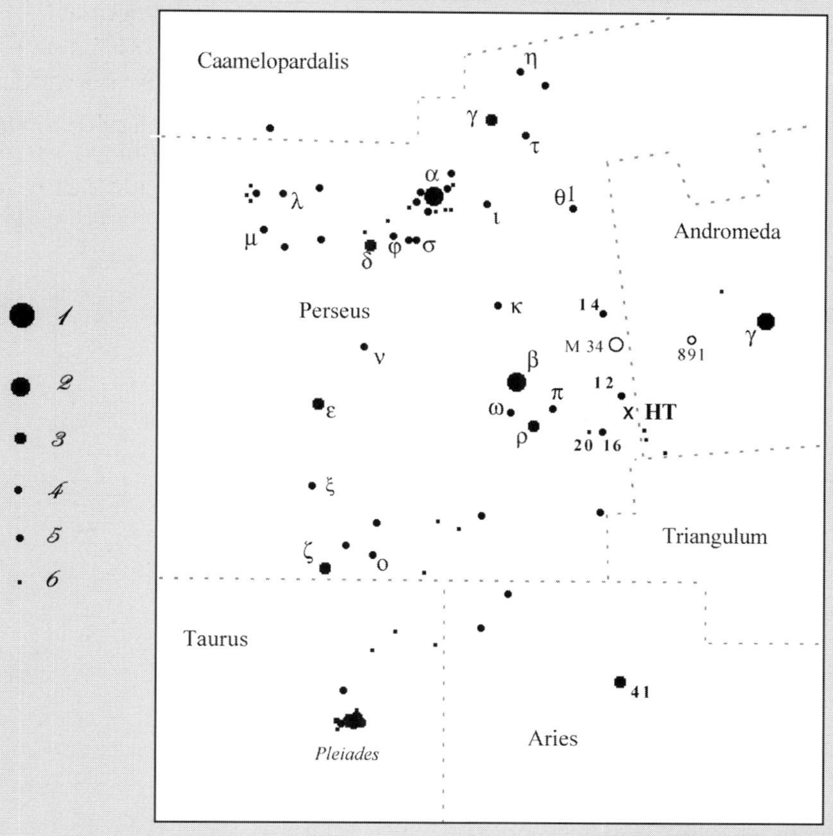

Tirion: Chart 4

Uranometria: Charts 62 & 93

10

Perseus Lenticular
NGC 1023
Type: Barred Lenticular Galaxy (SB0)
Con: Perseus

RA: 02^h 40.4^m
Dec: $+39^\circ$ $04'$
Mag: 9.3 (O'Meara); 9.3
Dim: $7.5' \times 3.0'$
SB: 13.1
Dist: 34 million light-years
Disc: William Herschel, 1786

W. HERSCHEL: [Observed October 18, 1786] Considerably bright, much elongated, 12° in a direction south preceding-north following, very bright nucleus, near 10′ in length. (H I-156)

NGC: Very bright, very large, very much extended, *very* much brighter in the middle.

AT A DISTANCE OF 34 MILLION light-years, NGC 1023 in Perseus is the nearest lenticular galaxy to the Milky Way. It is also the brightest of 40 galaxies belonging to the Triangulum Spur of galaxies, of which NGC 891 (Caldwell 23) in Andromeda is a member. But being the brightest does not mean being the largest. While NGC 1023 is moderately sized, spanning 65,000 light-years in length, it is 40 percent smaller than NGC 891. The reason NGC 891 appears dimmer (albeit only slightly) than NGC 1023 is because it is tilted a mere 6° from edge-on. As a result, much of its luminous matter is obscured by dust, which runs perpendicular to the galaxy's mid plane. NGC 1023, by comparison, is inclined 18° from edge-on, so we see more of its star-studded face, which has a total luminosity of 20 billion Suns. NGC 1023's recessional velocity (624 kilometers per second) is about 100 kilometers faster than NGC 891's.

As its discoverer William Herschel first noted in 1786, NGC 1023 sports an extremely bright nucleus, which is fairly common for a lenticular galaxy. Lenticulars are noted for their prominent central bulges, which are surrounded by a disk of luminous matter that lacks coherent spiral structure. In high-resolution images, the nucleus of NGC 1023 appears to be in the center of a diffuse bar near the minor axis of the projected lens. Radio studies of the ionized hydrogen gas outside NGC 1023 suggest that the galaxy suffered a tidal interaction, sometime in the past, with one of its dwarf companions, the most obvious of which is an irregular system (NGC 1023A) to the east and slightly south of the major axis. The optical appearance of NGC 1023, however, is quite relaxed and shows no sign of that event.

Spectroscopic observations of the central parts of NGC 1023 with the 6-meter telescope of the Special Astrophysical Observatory, have revealed that the galaxy's stellar-like nucleus is substantially younger than its surrounding bulge: the mean age of the nuclear stellar populations is 7 billion years. As with NGC 891 and other spiral galaxies, including our own Milky Way, NGC 1023's rotational velocity is far greater than one would expect if the galaxy had no matter other than stars, gas, and dust – the implication being that the galaxy is embedded within a halo of dark matter that exerts gravitational forces but emits no measurable electromagnetic radiation. Hubble Space Telescope (HST) observations have shown the galaxy's nuclear region (out to a distance of about 16 light-years from the nucleus) to be rotating at about 70 kilometers per second. One explanation for such rapid core rotation is the presence of a supermassive black hole with an estimated mass of around 40 million solar masses.

A total of 11 globular clusters have been found in NGC 1023. Most appear similar to those of our own Milky Way; NGC 1023 globulars are at least older than 5 billion years. NGC 1023's brightest globular cluster, recently studied by the HST, is much more compact than Omega Centauri (Caldwell 80) and has a linear diameter of 24 light-years – which is huge, especially if you consider that the nearest star to our Sun is only a little over 4 light-years away. It also contains an enhanced amount of cyanogen, which makes it similar to some of the globulars found within M31. In fact, it is most similar to G1 (Mayall II), the brightest globular cluster in M31. Unlike G1, NGC 1023's brightest globular appears to be very highly flattened with an ellipticity of about 0.37, which is similar to the most flattened globular clusters found in the Large Magellanic Cloud.

A new population of faint (23–24-magnitude) star clusters was also recently discovered in NGC 1023. They extend from 22 to 49 light-years from the galaxy's center and are thought to be superclusters formed by the merging of various clusters. Such superclusters often form in interacting galaxies and have very large diameters exceeding 45 light-years. These objects are suffering from strong tidal forces, yet they are still able to survive.

This visually simple but astrophysically enigmatic galaxy can be relatively easy to find. It lies $3\frac{3}{4}^\circ$ almost due south of M34. If you have an equatorial mount with setting circles, just center that bright Messier open cluster with low power and swing your scope $3\frac{3}{4}^\circ$ to the south. If you live under a dark sky, all you need to do is find 5th-magnitude 12 Persei, which is $2\frac{1}{2}^\circ$ due south of M34, or just south of the midpoint between the 2nd-magnitude stars Beta (β) Persei (Algol), the famous Demon Star, and Gamma (γ) Andromedae (Almaak). If you live under suburban skies, use binoculars and star hop from Beta Per, which is the northernmost star in a bright trapezoid with Omega (ω), Rho (ρ), and Pi (Π) Persei. Next look for 4th-magnitude 16 Persei, which is the west-southwest tip of an acute triangle with Rho and Pi Per. You'll find 12 Per about $2\frac{1}{2}^\circ$ to the northwest. There is no mistaking 12 Per, since it is the brightest star in the region.

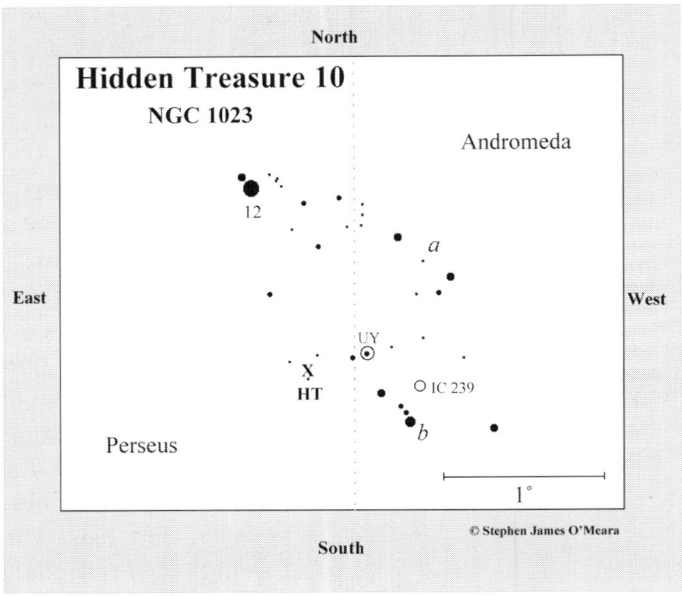

NGC 1023 is immediately north of that pair of stars.

At 23× in the 4-inch, the galaxy is a tiny ellipse of light nestled among a splash of field stars. The elliptical shape is quite easy to see in a 4-inch under a dark sky. Under suburban skies, the object most likely would appear like a planetary nebula with a bright central star. Note, however, that Christian B. Luginbuhl and Brain A. Skiff say the galaxy is "easily visible" in a 6-inch. Admiral Smyth saw this "curious body . . . [t]hough pale . . . very distinct" in his 6-inch refractor. He also notes that, though it is classified as lenticular in form, the object appears elliptical: "an appearance owing, perhaps, to its being a vast ring lying obliquely to our line of vision." Smyth's words reflect the Herschelian view of the universe. William Herschel believed that the Milky Way was a lens-shaped agglomeration of stars. Extensive nebulae, he argued, were "telescopic milky-ways" – other "island universes," as Immanuel Kant had previously referred to them. We simply see these telescopic "milky-ways," he said, at various orientations. How prophetic!

In the 4-inch, the galaxy looks best at 72× and higher powers. Roger Clark, in his *Visual Astronomy of the Deep Sky,* suggests using 243× in a 4-inch telescope, 148× in a 6-inch telescope, and 104× in an 8-inch. The reason being, NGC 1023's core is quite brilliant and takes magnification well. The surrounding lens fades gradually away from this center, however, so you must sacrifice

Through a telescope 12 Per is a close pair of stars, the fainter companion is itself a double.

NGC 1023 is a little more than 1° south-southwest of 12 Per. You can try sweeping directly to it, but you might want to star hop, because hunting for this galaxy is more like hunting for a planetary nebula. First, move $1\frac{1}{4}^{\circ}$ to the southwest, where you'll encounter a pair of 7th-magnitude suns oriented northeast–southwest (*a*). Center the southwestern member of that pair (which has an 8th-magnitude companion about 10′ to the southeast. Now sweep 1° south-southeast to a magnitude 6.5 star (*b*), which marks the southern end of a 15′-long arc of slightly fainter suns. NGC 1023 is about 45′ northeast of Star *b;* it is a tiny sliver of light nestled inside a triangle of roughly 9th-magnitude stars. The easternmost star of that triangle itself is the westernmost star in a smaller acute triangle of dimmer suns. The southernmost star in the larger triangle is double.

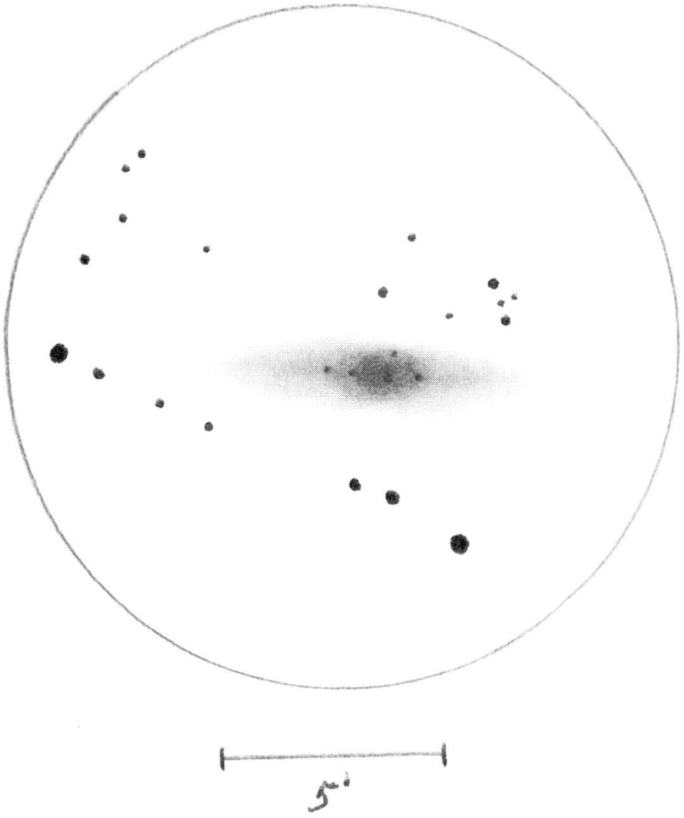

the galaxy's major axis, which the eye extends into a line when seen at the limit of visibility. In reality, NGC 1023's disk is quite featureless.

As for the galaxy's dim dwarf companion (NGC 1023A), I could not see it. It is located on the eastern side of NGC 1023 slightly south of the center-line and appears superimposed on the image. Radial velocity plots, however, indicate it is actually a background object some 6 million light-years in the distance. Two young blue star clusters have been found in NGC 1023A with estimated ages between 125 and 500 million years, making these objects similar to the young clusters found in the Large Magellanic Clouds. They most probably formed during a period of enhanced star formation, which was stimulated by a close encounter with NGC 1023. California amateur Jamie Dillon detected NGC 1023A in an 11-inch f/4.5 Dobsonian at 420×; in an 18-inch, he says it looks like a little "splotch" on the galaxy's eastern flank.

Before leaving this intriguing treasure, note that NGC 891 lies only $4\frac{3}{4}^{\circ}$ northwest of NGC 1023. Now would be a good time to search for that Caldwell object and compare the two views. If you look at an atlas that has the two galaxies plotted, you can use your imagination and see them as two pirate ships waiting for unsuspecting prey off the coast of the most frequented island in the area – M34.

one to see the other well. But it is through such studies at various powers that you can see the most detail.

At high power, the galaxy becomes a very puzzling sight. The sharp center is surrounded by a bright lens of light, punctuated on the west, northwest, and east by nebulous knots or stars. Together these appear to form a broken ring around the nucleus. A long "bar" of light extends east along the galaxy's major axis. This too appears beaded, as if it were comprised of faint starlight. A similar "bar" or lane of brightness is equally prominent extending along the galaxy's major axis to the west. But these bars are obviously illusions created by a chance alignment of a few field stars along

Hidden Treasure 11

NGC 1232

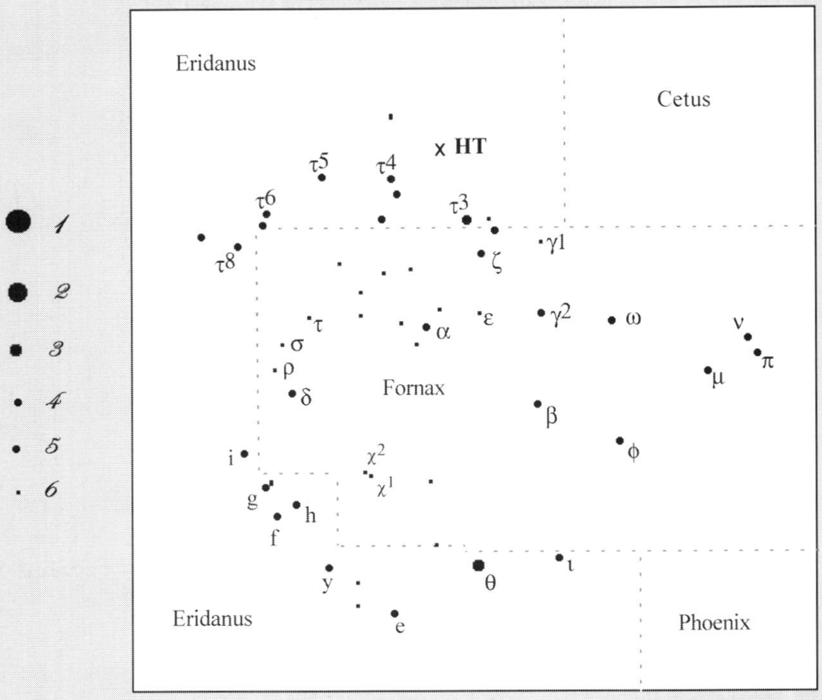

Tirion: Chart 18

Uranometria: Chart 311

11

The Eye of God Galaxy
NGC 1232
Type: Mixed Spiral Galaxy (SABc)
Con: Eridanus

RA: 03^h 09.8^m
Dec: $-20°$ $35'$
Mag: 9.5 (O'Meara); 10.0
Dim: $7.2' \times 5.7'$
SB: 13.8
Dist: 65 million light-years
Disc: William Herschel, 1784

W. HERSCHEL: [Observed October 20, 1784] Faint nebula, very large, a little brighter middle, round, 7 or 8′ in diameter. (H II-258)

NGC: Pretty bright, considerably large, round, gradually brighter in the middle, [mottled].

NGC 1232 IN ERIDANUS IS ANOTHER enigma. The galaxy is about as bright as M100 in Coma Berenices, at a declination slightly further north than M41 in Canis Major, and is nearly 15° higher in the sky than M7 in Scorpius, but few observers turn their telescopes to it. The object is not in the Herschel 400 list. Admiral Smyth and Rev. T. W. Webb pay it no heed. Ernst Hartung does not mention it in his *Astronomical Objects for Southern Telescopes,* Patrick Caldwell-Moore did not include it in his Caldwell list, and it is not mentioned in the *Deep-Sky Observer's Handbook* (Vol. 4 *Galaxies*). Yet I have spied NGC 1232 in my $1\frac{3}{4}$-inch antique telescope as a dim glow (with time, breathing, and a bit of patience) under dark skies. And though I could not detect it in my old 7 × 50 binoculars under a dark sky, I believe someone with younger eyes could with 10 × 50 or larger binoculars.

Regardless of what you will ultimately think of the galaxy's visual appearance, it is a cosmic wonder in photographs, a veritable maelstrom of starlight, Nature's attempt to mimic what can only be imagined as the Eye of God. To peer into the lens of light at the core of NGC 1232's swirling arms is to peer into a well of eternal questions: "The ways of God in Nature, as in Providence," writes the Cambridge Platonist Joseph Glanville (1636–1680), "are not as *our* ways; nor are the models that we frame any way commensurate to the vastness, profundity, and unsearchableness of His works, *which have a depth in them greater than the well of Democritus.*"

NGC 1232 is a fantastic system, belonging to the equally fantastic Fornax Cluster and Eridanus Cloud of galaxies. In his *Galaxies and the Universe,* Gerard de Vaucouleurs notes that the Eridanus Cloud also includes the NGC 1209 and NGC 1332 groups. The five brightest members of the Eridanus Group are NGC 1232, NGC 1398 (Hidden Treasure 19), NGC 1187, NGC 1300, and NGC 1407. Our target has a linear diameter of 131,000 light-years, a total mass of 120 billion Suns, and a total luminosity of some 40 billion Suns. We see the system nearly face on, so its arms appear open and loose. Here we have the prototype of a multiple-armed spiral galaxy. Look at the photograph on p. 67. Note that the arms do not spiral out from the edge of a high-surface-brightness disk. Only two major arms begin at this rim. After about one-quarter revolution, each branches into two fragments, then, further on, branch again. The system is so loose that these branching fragments do not interact. Seen as a whole, almost every spiral form is crooked. The pattern resembles that of M101 but is slightly more regular.

Note also the long row of suns, to the west, which is apparently formed by H-II regions. Indeed, all the arms are filled with H-II regions. Halton Arp lists NGC 1232 as the 41st object in his 1966 *Atlas of Peculiar Galaxies.* Of it, he remarks: "Companion spiral wound in same sense as parent. Note split of companion's arm further into centre." A barred spiral galaxy (NGC 1232A) lies 4′ to the east of NGC 1232; whether this is a true companion or a chance alignment is a matter of debate. Certainly, it too is highly resolved and displays distorted knotty arms. If it does lie at the same distance, then NGC 1232A is separated from its larger companion by 134,000 light-years.

NGC 1232 and NGC 1232A are actually a central and pivotal part of Arp's redshift argument, which challenges the almost universally held belief that the large redshifts of quasars and other active galaxies are due to cosmic expansion. Arp's *Atlas of Peculiar Galaxies* contains photographs of 338 peculiar galaxies that appear to be interacting or linked by bridges of luminous matter. If the frequency of discordant galaxies is greater than the statistics of chance projection allow, or if the bridges of matter are real, then there may be a need for new redshift models. Arp argues there is.

In the case of NGC 1232, Arp notes that the companion galaxy, which is near the end of one of NGC 1232's spiral arms, shows the same resolution of knots and features as the main galaxy. Inward from this companion, the arm splits, strangely into a channel about the width of the companion. Arp says this is evidence that the companion originated within the main spiral and traveled outward along the arm. The problem for conventional theorists is that the redshifts of NGC 1232 and NGC 1232A are discordant. The conventional theory therefore places the companion, NGC 1232A, at nearly four times the distance of the primary. How then, Arp argues, can the resolution of the two galaxies be the same? And, he says, is it *really* a coincidence that the companion is located near the *end* of a disturbed spiral arm?

Another interesting anomaly is that almost opposite the companion, the arm of NGC 1232 exhibits a strange deformity or flattening. When Arp took a spectrum of the region he discovered a highly redshifted object of over 28,000 kilometers per second (almost one-tenth the velocity of light) which far exceeds that of the main galaxy. This highly redshifted galaxy which Arp calls Galaxy B is extremely blue and cannot be any kind of normal galaxy. Is it really a mere coincidence that this highly redshifted (background?) object is located precisely at the point where the arm is most highly disturbed?

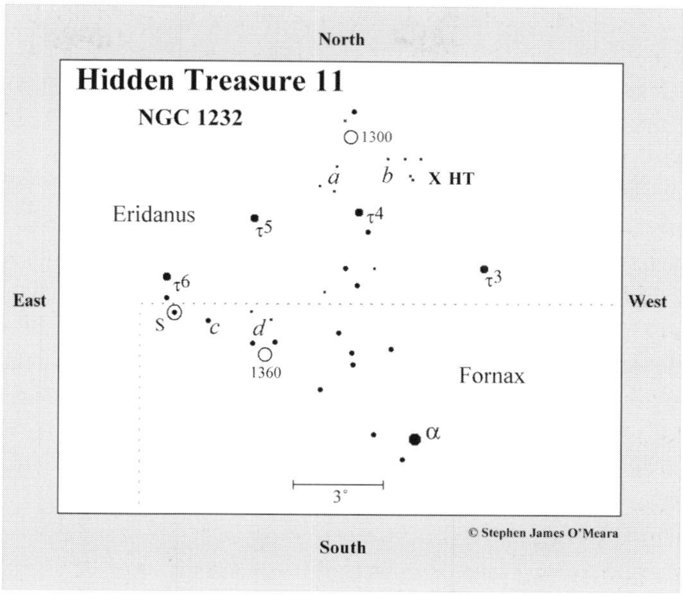

Arp also argues that at the distance of NGC 1232 one does not see background galaxies through the disk of a spiral galaxy. The dust and obscuration in the disk of such spirals, particularly near the arms simply forms an impenetrable screen. If, however, by some miracle we could see a background object, it would have to be heavily reddened – yet this is a highly redshifted *blue* object. All of this, he says, is strong evidence that all of the objects described are at or near the same distance from us, and the conventional redshift as a distance indicator is misleading.

Alas, evidence is mounting to support the conventional theorists. So, at least for now, there is no need for new physical redshift models, though the debate continues.

To find this controversial object, start at 3rd-magnitude Theta (θ) Eridani (Acamar). A fist (~10°) to the northeast lies 4th-magnitude Alpha (α) Fornacis. A little more than 7° to the northeast is 4th-magnitude Tau^4 (τ^4) Eridanus. It is the westernmost of three 4th-magnitude stars forming an arc $6\frac{1}{2}^\circ$ wide and oriented north-northwest–south-southeast; the other two stars are Tau^5 (τ^5) and Tau^6 (τ^6) Eri. Use binoculars to confirm these stars, which are part of a longer chain of similarly bright suns. You want to center Tau^4. NGC 1232 is $2\frac{1}{2}^\circ$ to the northwest, immediately southwest of a 1°-wide, sideways Y of 7th- to 8th-magnitude suns (*b*).

You could also move 1° northeast of Tau^4 Eri. That's where you will find a 1°-wide isosceles triangle of 6th- to 7th-magnitude suns (*a*). If you center the triangle, then sweep $2\frac{1}{2}^\circ$ due west, you should have NGC 1232 near the center of your field. These hops are large, but Eridanus is not in the plane of the Milky Way, so the star fields are not as rich as those, say, in Cygnus. The star patterns mentioned above stand out prominently against the background sky.

In the 4-inch at 23×, NGC 1232 appears as a round, comet-like glow about 7′ west-southwest of a magnitude 9.2 star – a comet without a strong central condensation. Anyone sweeping past this object unexpectedly under dark skies would certainly get a start. I fear, however, that from a sky with any significant light pollution, the galaxy may be a wash. I consider it the M74 of

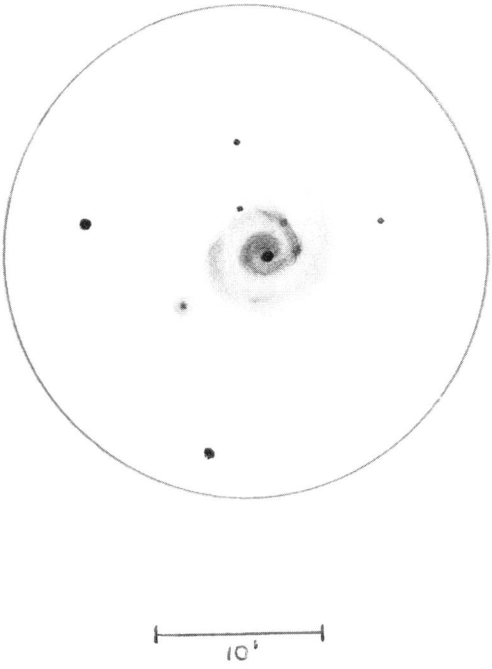

the *Hidden Treasures* catalog, because, as I recount in *Deep-Sky Companions: The Messier Objects*, M74 remained elusive to me under bright city skies with a telescope; not until I moved to a dark-sky site did I finally see the galaxy . . . in binoculars.

Under a dark sky, the galaxy is more than an amorphous haze, if you give it time. *Sky & Telescope*'s late Deep-Sky Wonders columnist Walter Scott Houston, called it a "gem." Interestingly, he had not heard of anyone seeing detailed structure in it with telescopes any smaller than 12 inches. I found his comment surprising because even at 23×, if I relax and wait for moments of pure clarity, I can see a dim pip at the core of the galaxy and some mottling that suggests a broken ring.

But the key to success in seeing detail in this visually challenging galaxy is to observe it in layers, using different magnifications. I agree, you will not see much in this galaxy with a quick look. You have to want to study it, and that requires time. I began my journey with 72×, which seemed to show best the galaxy's sharp, starlike core. I also got suggestions of spiral structure close to the central lens, curving to the southwest. Once you suspect some detail, take a rest, breath in some fresh air, then return to the telescope. Now use your averted vision and see if the "amorphous" haze does not start to splinter into all manner of enhancements across the galaxy's face. Don't worry about trying to pinpoint any of them. Just let them pop in and out of view.

When I put in 101×, a dim arm arcing to the southwest immediately snapped into view. Then, after a few minutes, this arm split into two distinct and parallel features. With averted vision and 168×, a few distinct stars or knots could be seen superimposed on these arms. (Houston notes that his best view of this galaxy was with 150×.) Finally, I zoomed out, using 41× (a 22-mm Nagler and a 1.8× Barlow) to survey the area surrounding the galaxy's inner lens. That's when I got my best suggestion of a fainter spiral arm beyond the two already mentioned.

When John Herschel observed NGC 1232 from the Cape of Good Hope with his 18-inch f/13 speculum telescope, he noticed something peculiar: "first," he said, the galaxy was "very gradually then pretty suddenly brighter towards the middle." That was with the view with his right eye. Then he switched to his left eye: "With the left eye I see it mottled. (N.B. This is no doubt a distant globular cluster.)" Steve Coe had a similar impression through his 13-inch f/5.6 reflector, saying "Going to 165× makes the arms appear very mottled . . ." He also spotted two H-II regions.

Once you have finished straining your eyeballs on NGC 1232, take a break, but be ready to return to the field for yet another time-intensive challenge: the search for 1232A. It's important to forget the listed magnitude of faint galaxies, they can be in error of up to three magnitudes. I spent a half hour looking for NGC 1232A and had complete success. It came in and out of view several times as a pulsating and fairly sizable glow. It's easy to pinpoint, all you have to do is use your averted vision and look midway between, and a little south of, a line between the magnitude 9.2 star and the nucleus of NGC 1232. I was shocked to read that the galaxy is listed as magnitude 14.7; it is more like magnitude 12. Good luck!

Hidden Treasure 12

NGC 1291

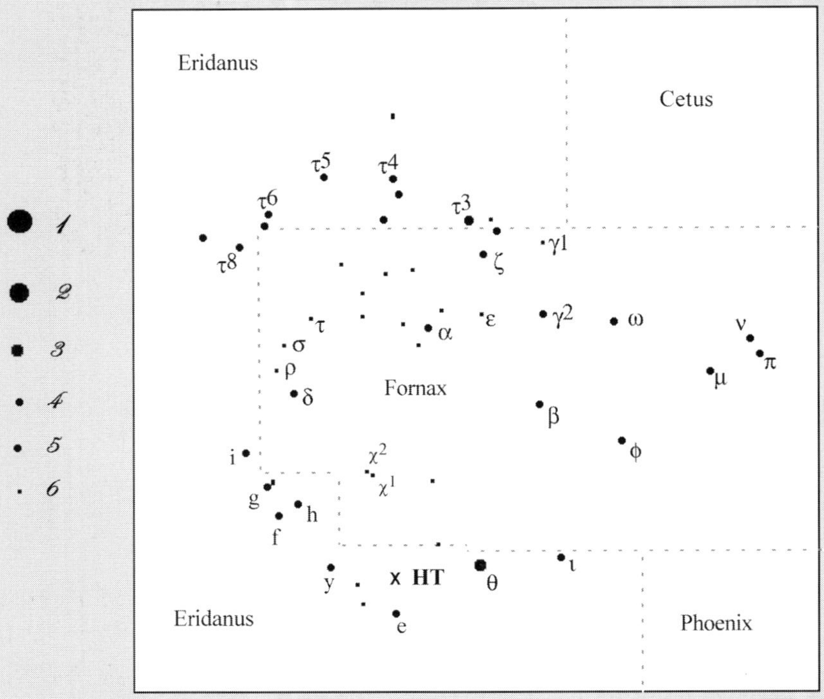

Tirion: Chart 18

Uranometria: Chart 311

12

Snow Collar Galaxy
NGC 1291
Type: Barred Lenticular Galaxy (R)SB(s)0/a
Con: Eridanus

RA: 03^h 17.3^m
Dec: $-41°$ $06'$
Mag: 8.6 (O'Meara); 8.5
Dim: $11.1' \times 10.1'$
SB: 13.4
Dist: 28 million light-years
Disc: James Dunlop, included in his 1827 catalog

J. HERSCHEL: Globular cluster, very bright, pretty large with a much brighter middle. (h 2521 = h 2518 = NGC 2169)

NGC: Globular cluster, very bright, pretty large, round, much brighter in the middle, extremely [mottled].

NGC 1291 IS THE 487TH OBJECT listed in James Dunlop's 1827 *A Catalogue of Nebulae and Clusters of Stars.* Of its appearance through a 9-inch reflector, he wrote: "A pretty bright round nebula, about $1\frac{1}{2}'$ diameter, very bright and condensed to the centre, and very faint at the margin; with a very [faint] star about 1′ north, but not involved." Note, however, that the NGC description in the table above calls it a globular cluster. It reveals how little we understood these misty glows at the turn of the twentieth century.

Interestingly, when John Herschel observed this object from the Cape of Good Hope, he also referred to NGC 1291 as a "globular cluster." Confusing the matter is the object he listed as object No. 2518, now known as NGC 1269, which is only $2\frac{1}{4}'$ west of NGC 1291. Either Herschel had unknowingly discovered a comet – his description of the object certainly sounds like one, "very bright, round with a gradual much brighter middle" – or he had made a second observation of NGC 1291, since there is nothing at that position.

Today we know NGC 1291 to be a special type of galaxy – one whose morphology is intermediate between a "pure" lenticular system, which does not show any spiral structure, and an early type spiral with a large central region. It belongs to the Dorado Cloud of galaxies and has a recessional velocity of 829 kilometers per second. In most long-exposure photographs, NGC 1291 displays a fairly large nucleus on a weak bar that marks the major axis of a little, elongated, bright lens; a faint outer ring emerges from two very weak spiral arms or arcs. But a more detailed view reveals it to have an outer ring, lens and bar, a secondary lens and bar, and a nucleus.

It is a fairly large system, measuring 88,000 light-years in linear extent, and shines with a total luminosity of some 20 billion Suns. In his *Nearby Galaxies Catalog,* R. Brent Tully lists the galaxy's inclination as 28° from face-on, but it may be as little as 17°; observations of the galaxy's molecular-hydrogen clouds (hotbeds of star formation), however, seem to indicate an even smaller inclination of possibly 6°. Indeed, a few regions of star formation appear in the galaxy's thin-armed fragments. Weak dust lanes can also be seen in the bar, and they are organized in a clockwise spiral pattern – the same sense as in the outer spiral arms. The pattern of the dust is controlled by rotation. Interestingly, observations of the galaxy's molecular-hydrogen also reveal a large central hole, about twice as large as the central bar. But such holes are common to these systems, whose rotating bars can channel gas and dust into the center of the galaxy and trigger bursts of star formation. Winds driven by supernovae explosions or massive star formation could also clear the inner regions of molecular-hydrogen gas.

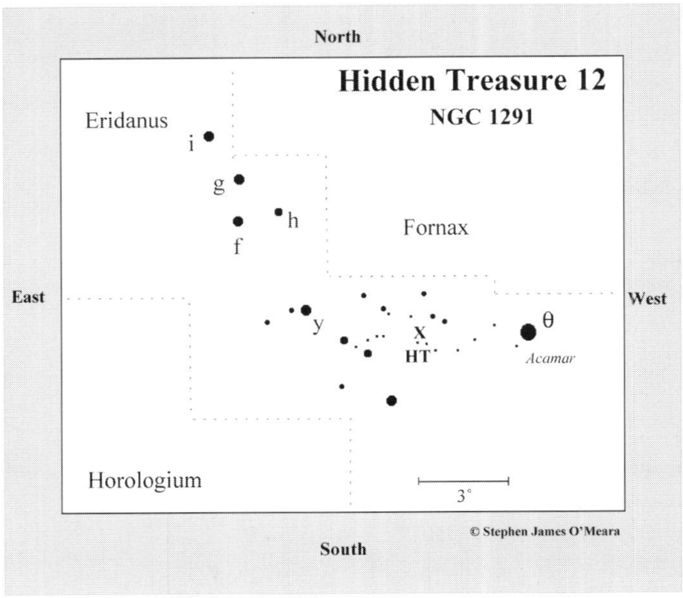

Recent X-ray data confirm that the interstellar medium in the bulge of NGC 1291 is dominated by hot gas. It seems that within the morphological class Sa, there can be significant differences in the gas content of the bulge, with the more massive bulges being likely to contain the hot gas. Since NGC 1291 is a bulge-dominated Sa galaxy, this scenario fits nicely. The source of X-ray emission appears to be the galaxy's central point-like nucleus and some 50 additional sources, which may be low mass X-ray binaries. NGC 1291 is a galaxy that is obviously still forming and therefore typical of morphological Sa-type galaxies.

With a declination of −41°, NGC 1291 is a poor target for mid-northern observers. But it is a fine object for those living at more southerly latitudes in the USA and beyond. So some observers will have to take a road trip to seek out this treasure. Anyone who does will be, literally, doubly rewarded; the galaxy lies near a fantastic guide star, 3rd-magnitude Theta (θ) Eridani (Acamar), one of the most visually delicious double stars in the heavens. G. Piazzi (1746–1826) discovered Acamar's duplicity, which is easily split in a 2-inch glass. The magnitude 4.1 secondary is separated from its magnitude 3.2 primary by 8.4″. The stars lie 161 light-years distant and are hot *A*-type suns. Ernst G. Hartung (1893–1979) describes this brilliant white pair as "one of the gems" of the southern sky, which "completely dominates the field of a few scattered stars."

To find NGC 1291, just center Acamar, then sweep your telescope $3\frac{1}{2}^{\circ}$ due east. At 23× in the 4-inch, the galaxy shows a sharp inner core surrounded by a bright elliptical lens, which is itself wrapped in a circular halo of faint light. When I place my 22-mm eyepiece into a 1.8× Barlow (41×), NGC 1291 looks more like a planetary nebula than a galaxy, which is interesting if you consider that the NGC lists it as a globular cluster. But the reason for the NGC description becomes obvious when the galaxy is studied at higher powers. That's when several dim stars or knots can be seen superimposed on this globular glow. Hartung concurs, writing that this "extra-galactic system looks not unlike a distant unresolved globular cluster; the edges diffuse away gradually to about 2′ diameter with a small very bright central region which shows a strong prism band. It is easy for small apertures." Thus, the fact that James Dunlop and John Herschel both recorded this object as a globular is no mystery.

At 101× in the 4-inch, the starlike core sits inside a circular annulus of bright light, which is surrounded by a large diffuse lens. Both the inner ring and lens appear mottled. Now, try studying this region with the highest reasonable power; the galaxy's high surface brightness and intense core is very

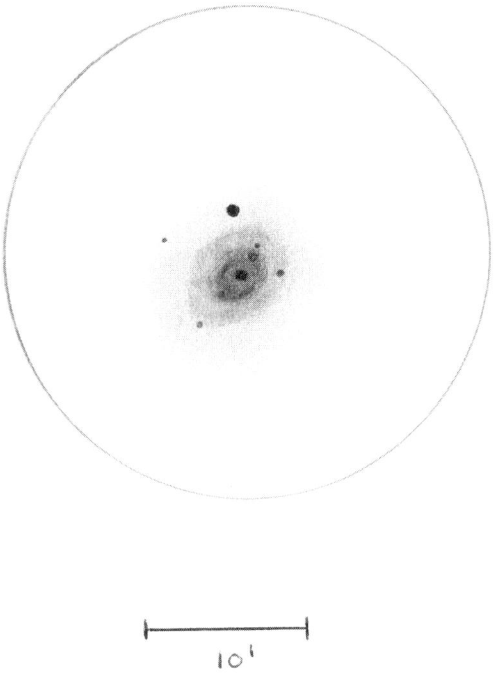

amenable to the use of high magnification. The core is so intensely bright that I followed it into astronomical twilight, losing sight of it only 45 minutes before sunrise.

If you want a challenge, try looking for the two broad arcs of light, which line the ansae of the inner lens. Finer enhancements appear and disappear all around the circumference; what I see depends on where I focus my attention with averted vision. The great Italian astronomer Angelo Secchi (1818–1878) described a similar phenomenon while observing M57, the Ring Nebula in Lyra. Secchi believed he resolved the ring into minute stars "glittering like stardust." In the case of NGC 1291, however, the phenomenon is most likely due to the fact that dim field stars pepper the face of the galaxy. It's more like looking at snowflakes falling on a fur collar; as soon as one snowflake lands, it quickly dissolves, only to be replaced with yet another snowflake.

Before moving on, return your thoughts to Theta Eri. It's not widely known, but the nineteenth-century Scottish clergyman and amateur astronomer Thomas David Anderson – who discovered a nova in Auriga in 1892, and one in Perseus in 1901 – argued that Theta Eri had dimmed over the centuries, since it was listed by Ptolemy and Al-Sufi as being of the first magnitude; Anderson's arguments appear in the 1893 issue of *Knowledge.*

Hidden Treasure 13

NGC 1316

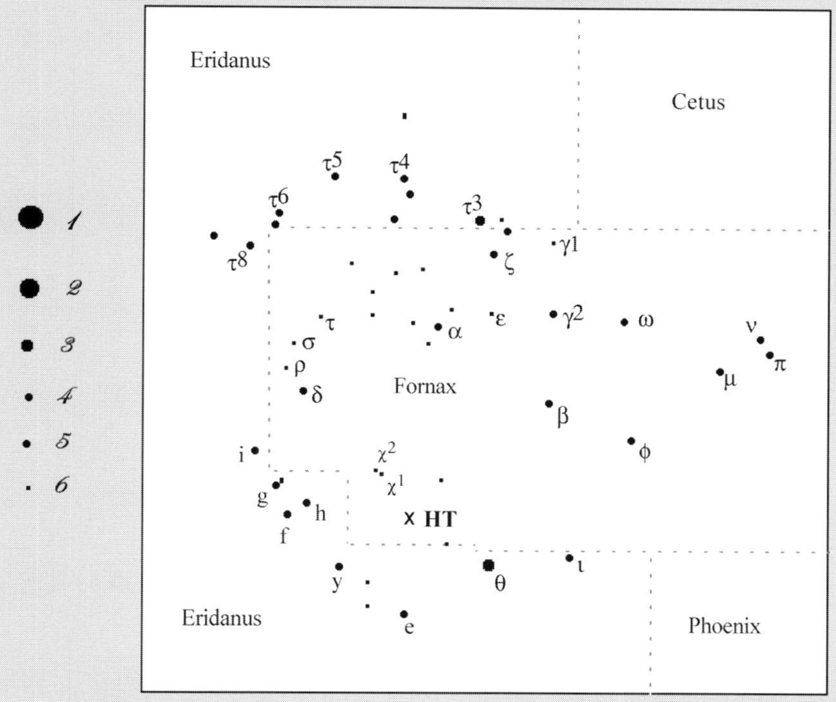

Tirion: Chart 18

Uranometria: Chart 355

13

Fornax A
NGC 1316
Type: Peculiar Mixed Lenticular Galaxy (SAB0p)
Con: Fornax

RA: 03^h 22.7^m
Dec: $-37°$ $12'$
Mag: 8.2 (O'Meara); 8.2
Dim: $13.5' \times 9.3'$
SB: 13.3
Dist: 55 million light-years
Disc: James Dunlop, included in his 1827 catalog

J. HERSCHEL: Very bright, pretty large, little extended, very suddenly very much brighter in the middle to a nucleus 2 in diameter. (h 2527)

NGC: Globular cluster, very bright, pretty large, round, much brighter in the middle, extremely [mottled].

NGC 1316 IS THE BRIGHTEST AND most peculiar member of the Fornax Cluster of galaxies. This peculiar elliptical or lenticular galaxy is the well-known radio source Fornax A, one of the closest and most famous sources in the Southern Hemisphere. The morphology of NGC 1316 resembles NGC 5128 in Centaurus (Caldwell 77), which is also identified with a strong radio source: Centaurus A. Indeed, NGC 1316 has radio lobes extending over several degrees of sky. And like NGC 5128, NGC 1316 has a large bulge, a system of dust filaments across the minor axis with spiral segments, ripples, and loops in its outer shell – all tidal signatures of a merger.

Indeed, NASA's Hubble Space Telescope (HST) has found evidence for such a merger, or mergers. In a stunningly clear image of the galaxy's nuclear region, smoggy wisps of dust – "Like dust bunnies that lurk in corners and under beds," as one press release put it – are seen projected against the galaxy's bright nucleus. These wisps have the eerie shape of a hideous pendulum: "a huge pendulum such as we see on antique clocks," writes Edgar Allan Poe in his famous work *The Pit and the Pendulum.* "There was something, however, in the appearance of this machine which caused me to regard it more attentively." For the astronomers studying the dusty pendulum in the core of NGC 1316,

the message seemed clear: sometime during the last 100 million years, NGC 1316 collided with and swallowed a gas-rich companion galaxy.

Wide-field imagery from Cerro Tololo Interamerican Observatory in Chile corroborates this theory. The images show a bewildering variety of ripples, loops, and plumes immersed in the galaxy's outer envelope – again, all tidal signatures of a merger. In fact, the narrowest loops are believed to be the stellar remains of other spiral galaxies that merged with NGC 1316 some time during the last few billion years.

And there's more evidence. In 1999 Carl Grillmair (California Institute of Technology) and his colleagues, used HST to study NGC 1316 and found surprisingly few globulars around it. Instead, they discovered a large population of old and relatively dim star clusters, which contain only a few thousand stars each. The results took Grillmair and his team by surprise. Usually, when galaxies merge, the resulting clusters are very bright, very blue, and very young. But the star clusters detected in NGC 1316 are too old to have been born during the collision that produced the dusty debris seen in the HST image, Grillmair says. "[T]hey have not been around long enough to have been torn apart by galactic tidal forces." It's possible that the dim, old clusters are the remains of an earlier collision.

But in 2003, a US team of scientists, led by Dr. Paul Goudfrooij of the Space Telescope Science Institute in Baltimore, Maryland, did a follow-up study with a more sensitive camera. For the first time, his team saw evidence of the gradual disruption of star clusters created during a past merger. They found that the relative number of low-mass clusters is significantly lower in the inner regions than in the outer regions, by an amount consistent with theoretical predictions. The team's results have improved our understanding of how elliptical galaxies and their star clusters may have formed during galaxy mergers and then evolve to resemble "normal" elliptical galaxies after several billions of years.

As for the source of NGC 1316's radio emissions, they could be generated by a massive black hole lying at the center of this massive galaxy, which spans 84,000 light-years of space and has a total luminosity of some 60 billion Suns. The opposing jets in the radio data extend out about 16,000 light-years before abruptly terminating. The inner 1,000 light-years of the jets contain discreet knots while the outer parts are continuous, which indicates strong interactions with the surrounding medium. The lobes' radio emitting electrons have a lifetime of perhaps 100 million years. If so, then the nucleus was particularly active at least 100 million years ago. At present, however, the nucleus is very faint in X-ray emissions, which is a good indicator that activity has either decreased markedly or become

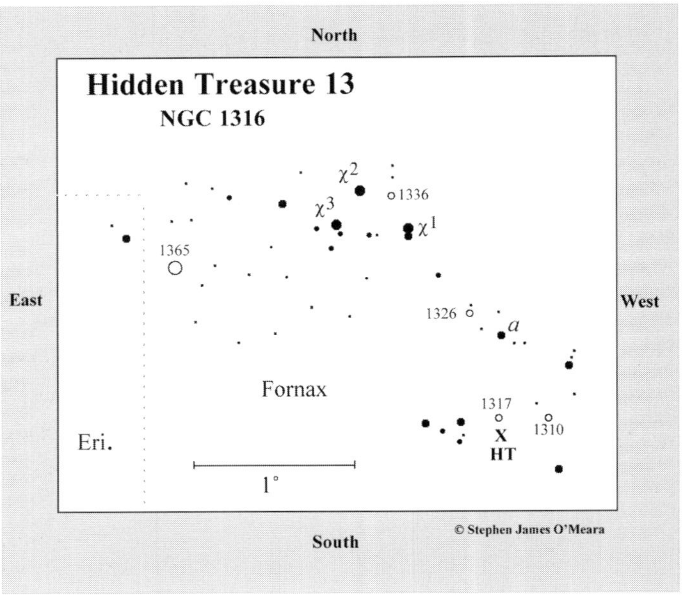

dormant. This also suggests the possible abundance of dormant quasars in nearby galaxies, if quasars are thought to be the central hot spots of very active galactic nuclei.

James Dunlop discovered this fantastic system on September 2, 1826. He, of course, was totally unaware of the drama unfolding before his eyes. Through his 9-inch reflector, which he had set up at his house in Paramatta, New South Wales, he simply saw "a rather bright round nebula, about $1\frac{1}{2}'$ diameter, gradually condensed to the center." He included it as the 548th object in his 1827 *A Catalogue of Nebulae and Clusters of Stars.* Dunlop also discovered the galaxy's "unswallowed," magnitude 10.8 companion, NGC 1317 (a tight spiral with a very bright core), which lies about 6′ to the north. Like NGC 1232 in Eridanus, John Herschel believed it to be a globular cluster. And, once again, it is no wonder. For all intents and purposes, the object looks like some distant globular cluster with weak arms.

To find this dynamic duo, first locate 3rd-magnitude Theta (θ) Eridani (Acamar). One fist (10°) to the east-northeast is an obvious 3°-wide, Y-shaped asterism of 4th- and 5th-magnitude stars (f, g, h, and i Eridani). About $3\frac{1}{2}^\circ$ northwest of the southwestern-most star in the Y (h Eri), is a 30′-wide acute triangle of 6th-magnitude suns: Chi^1 (χ^1), Chi^2 (χ^2), and Chi^3 (χ^3) Fornacis. Confirm these with your binoculars. NGC 1316 lies $1\frac{1}{2}^\circ$ south-southwest of Chi^1 (a nice double star). From Chi^1, hop 45′ to the southwest where you'll find a roughly magnitude 8.5 star (*a*). NGC 1316 is less than 40′ south of Star *a*. The galaxy is bright, as bright as M32 in Andromeda, and it is easily seen from a dark sky in 7 × 50 binoculars. And like M32, its light is very condensed. It's an even finer sight in my antique telescope, because it shows the galaxy's bright core surrounded by a uniform haze of light.

At 23× in the 4-inch, NGC 1316 is a beautiful milky glow with a starlike nucleus at the center of a tight inner core. At a glance, the galaxy looks round, but if you continue to look hard at the core, you should get a nagging sensation that the galaxy has a nub immediately to the north, and it does. It's the tiny glow of NGC 1317, which, by the way, looks brighter than the listed magnitude.

With higher powers, the galaxy loses its circular shape. While NGC 1316's inner core remains round, the galaxy's outer extensions form an ellipse oriented northeast–southwest. In images, the galaxy is truly elliptical in form, being 39° from edge-on,

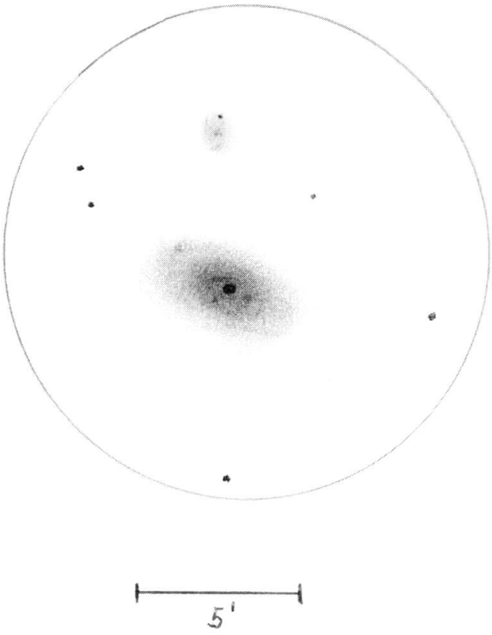

and slightly flattened – a consequence of NGC 1316 rotating so rapidly around it minoraxis.

At 101× and 168×, the nucleus of the galaxy is a brilliant bead of light. It sits at the center of a dappled or mottled inner lens, which seems to have a bar aligned with the major axis, though this could be illusory. I see no further definition in the outer envelope. At these high powers, I also suspect some spiral structure to NGC 1317.

Other observers have failed to see much more detail in larger scopes. Most note a starlike nucleus to NGC 1316. Luginbuhl and Skiff, who note that the galaxy is visible in a $2\frac{1}{2}$-inch telescope, see a "bright, non stellar core" in a 12-inch telescope. The question is, can you see more?

By the way, two supernovae erupted in this galaxy within two months: Supernova 1980N occurred on December 12, 1980, 220″ east and 20″ south of the nucleus; Supernova 1981D was found on March 3, 1981, 20″ west and 100″ south of the nucleus. Both shined around magnitude 12.5 at the time of discovery.

Hidden Treasure 14

Melotte 20

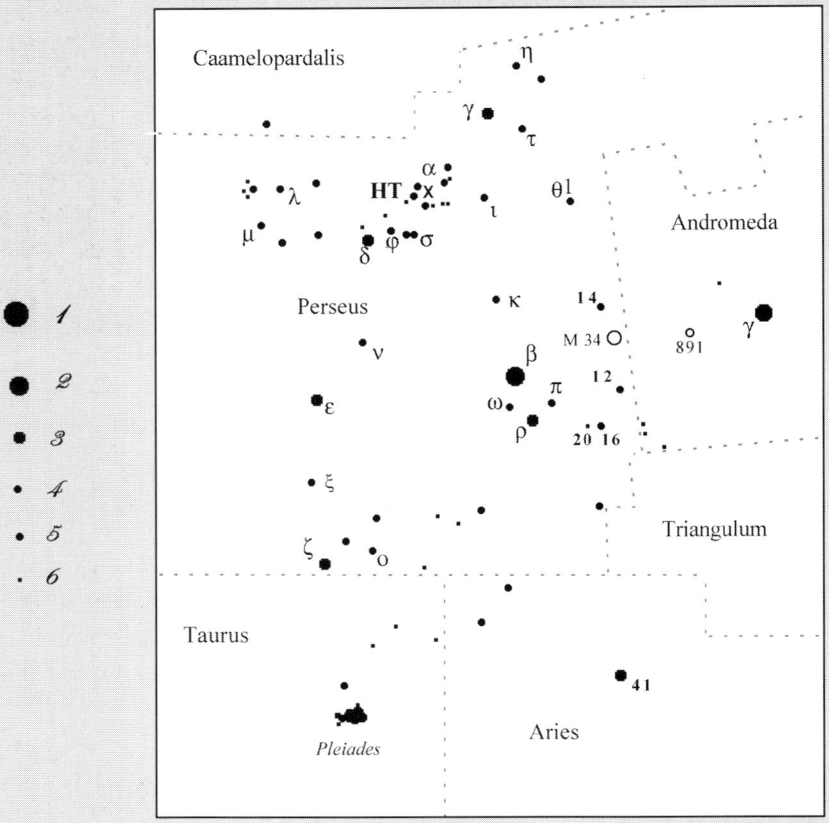

Tirion: Chart 4

Uranometria: Chart 63

14

Alpha Persei Moving Cluster, Little Cloud of Pirates
Melotte 20
Type: Open Cluster
Con: Perseus

RA: 03^h 24.3^m
Dec: +49° 52′
Mag: 2.3 (O'Meara); 2.3
Diam: 300′
Dist: 535 light-years
Disc: Giovanni Batista Hodierna, before 1654; arguably known throughout antiquity

HERSCHEL: None.

NGC: None.

OF ALL THE LEGENDS IN THE NIGHT sky, none is as great or as dramatic as that of Perseus, the hero, and Andromeda, the woman chained. It is a stirring tale of adventure, betrayal, and lust, one that revolves around a proverbial victim of circumstances and an alert opportunist.

Perseus was the son of the comely mortal Danae and great Zeus (*Per Zeus,* "fathered" by Zeus). Beautiful Andromeda, an Ethiopian princess, was the daughter of King Cepheus and Queen Cassiopeia. The two lovers owe their union to Andromeda's mother, who had a serious problem: *vanity.* Cassiopeia loved to boast that her daughter was fairer than the Nerieds, who were the sea's most exquisitely beautiful nymphs. When the Nerieds heard the flaunts, they complained to their father, Poseidon, who became enraged. To punish the queen, he sent a flood and a sea monster (Cetus) to drown and devour the inhabitants of Ethiopia. As the monster terrorized these people, Cepheus consulted the Oracle of Ammon. To his horror, the oracle told him to sacrifice his only daughter to the hideous beast.

What's a king to do? Give up his own flesh and blood to save the lives of many? Or save the life of his treasured daughter? With great reluctance the king gave in to the prophesy. "Gods," Robert Burnham, Jr., comments in his *Celestial Handbook,* "even the best of them, seem to lack the most elementary sense of justice; they dispense punishments at random and appear to be quite satisfied as long as *someone* suffers sufficiently in order to 'atone' for the original offense."

Stripped naked, bound by the arms and waist to coastal stone, poor Andromeda awaited her fate. But the tide turned when youthful Perseus flew down to her on winged sandals. Perseus nearly mistook her pure white form for marble, except that he noticed her flaxen hair blowing in the gentle breeze, and her brilliant eyes welling with tears.

"O virgin, undeserving of those chains, but rather of such as bind fond lovers together," Perseus said, "tell me, I beseech you, your name, and the name of your country, and why you are thus bound."

As the hero listened to Andromeda's woeful tale, his muscles tensed. The youth was already feeling quite the champion, because he had just slain the gorgon Medusa, whose gruesome head (with those I'll-turn-you-to-stone eyes) he had in a leather sack. Perseus whispered some words of hope to Andromeda. The enterprising young man also made her a bargain: if he slayed this beast, she would be his bride. Andromeda agreed and sealed the promise with a kiss.

In his 1856 book, *The Heroes,* Charles Kingsley continues the story in a most illustrious fashion:

> Perseus laughed for joy, and flew upward, while Andromeda crouched trembling on the rock, waiting for what might befall.
>
> On came the great sea-monter, coasting along like a huge black galley, lazily breasting the ripple . . . His great sides were fringed with clustering shells and sea-weeds, and the water gurgled in and out of his wide jaws, as he rolled along, dripping and glistening in the beams of the morning sun.
>
> At last he saw Andromeda, and shot forward to take his prey, while the waves foamed white behind him, and before him the fish fled leaping.
>
> Then down from the height of the air fell Perseus like a shooting star; down to the crests of the waves, while Andromeda hid her face as he shouted; and then there was silence for a while.
>
> At last she looked up trembling, and saw Perseus springing toward her; and, instead of the monster a long, black rock, with the sea rippling quietly about it.

Perseus had held the blood-freezing head of Medusa before the monster and turned it to stone. The story has a rare happy ending: the gorgon and the sea monster are slain, the beautiful maiden is saved by a handsome hero, and the lovers marry. They become the great grandparents of yet another hero – the mighty Hercules.

Perseus is unmistakable in the night sky. The core of the constellation is a prominent spine of bright suns known as the Segment of Perseus. This graceful arc follows "a very brilliant part of the Via Lactea [(Milky Way)]," as Adm. Smyth wrote. The brightest part of the Segment measures about 12° in length and consists of the stars Eta (η), Gamma (γ), Alpha (α), Sigma (σ), Phi (ϕ), and Delta (δ) Persei. A fainter extension curves to the northeast from Delta and includes the stars Mu (μ), Lambda (λ), and 48 Persei. So the overall shape of the spine is a J or fishhook.

The brightest star in the Segment, and in the constellation, is Alpha Per. Its classical name, Mirfak, Mirphak, or Marfak, means Elbow. It derives from the Arabic *Marfik al Thurayya* (Elbow Nearest the Many Little Ones). While *al Thurayya* is the descriptive Arabic word for the Pleiades, it makes little sense. Even if we concede that the Arabs could have depicted Perseus differently in their day, it is hard to imagine Alpha as an elbow *near* the Pleiades, which is some 25° to the south-southeast. It is instead my belief that the Arabian word *al Thurayya* was, at least in this case, referring to the "little ones" near Alpha Persei. That early Arabian skywatchers, at times, gave different stars the same name is a known fact. The name Meissa (Lambda [λ] Orionis), for instance derives from the Arabic *Al-Maisan* (The Proudly Marching One, or The Shining One). But the Arabs first gave that name to Gamma (γ) Geminorum (see Hidden Treasure 29). Indeed, the fourth-century Arabian lexicographer Al Firuzabadi believed the title *Al-Maisan* could be applied to any bright star. It is more believable, then, and perhaps more illuminating, to think that the Arab skywatchers of old were arguably the first to see the glittering gems associated with the Alpha Persei Moving Cluster – a large scattered cluster of stars whose brightest members lie between Alpha and Delta Persei, the very heart of the famous Segment.

Like the majority of bright stars in Taurus, Orion, and Ursa Major, those in Perseus, are part of a true physical system. In 1910 Arthur Eddington first described it as a widespread moving cluster and P. J. Melotte cataloged it in 1915. In 1930 Harvard astronomer Harlow Shapley wrote that while the group might be placed in the class of "very loose and irregular clusters" typified by the Hyades, he thought it better placed with the Orion Nebula cluster in the class of "star associations," or widespread moving clusters, such as the Ursa Major group.

Today we know the cluster consists of some 50 stars splashed across 47 light-years of space. About a dozen of these suns are within 2° of Alpha Per and shine at magnitude 6 or brighter. So, as with the Pleiades, the region appears luminous with direct vision and resolved with averted vision. The Italian astronomer Giovanni Batista Hodierna (1597–1660) first logged these stars as a "nebulous" object in his *De Admirandi Coeli Caracteribus* of 1654 (which is mainly why I include it in the *Hidden Treasures* list.) He described its appearance through a simple Galilean refractor at 20×: "[It] shines on the right side of Perseus near the Milky Way, where, next to a star of high luminosity (which is also shining from the same side), there is also a certain nebulosity."

A look at this region today in even the smallest of binoculars is one of abject wonder. In his *Astronomy with an Opera-Glass* (1923), Garrett Serviss called the field one of the finest in the sky, with "[s]tars conspicuously ranged in curving lines and streams. A host follows Alpha from the east and south." In 7 × 50 binoculars from Hawaii, the cluster has an exquisite serpentine body, one befitting of the sea monster that mythical Perseus had turned to stone. Seen with southwest "up," the beast's hideous head is a conspicuous 1°-wide circlet of 6th-magnitude suns with Alpha Per marking its gaping jaws.

"Many Little Ones" of Arabia, which are in Hodierna's "nebulosity" (a testament to the poor quality of his lenses).

The cluster's stars are principally luminous *O*- and *B*-type supergiants and comprise what is known as the Alpha Persei *OB*3 Association. The group is moving about 10 miles per second in the general direction of Beta Tauri. But don't start plotting its motion just yet, because it will take the group about 90,000 years to change its present position by 1°. Mirfak, the cluster's brightest member is a spectacular supergiant star of spectral type *F*. It is also a borderline Cepheid variable with a titillating color that has been described as brilliant lilac and ashy. ROSAT X-ray observations covering an area of about 10 square degrees detected about 160 X-ray sources, 88 of which have an optical counterpart. And in 2003, John R. Stauffer (California Institute of Technology, Pasadena, CA) and his colleagues identified 27 brown dwarf candidates. If these stars are cluster members they have masses less than 0.1 Sun. A 1992 study by Charles Prosser (Lick Observatory) of 132 low-mass members, suggests an age of about 80 million years, making it about the same age as the Pleiades.

I spent days observing the cluster, which is a spectacle at 23×. Countless hundreds of dim suns form all manner of

Sigma Per marks the rising hump of the monster's charging body, while Delta Per is the tip of its tail.

Through the 4-inch at 23×, the near 3° field is one of stellar plenty, a dazzling array of diamond studs set in the sequined fabric of the Milky Way. It can also be seen as a *cheval-defrise* – a jagged obstacle set atop a fortress to prevent pirates from trespassing or escape. Or, as I enjoy most, a "little cloud of pirates" – a phrase I borrowed from Robert Louis Stevenson's *Treasure Island*: "Suddenly, with a loud huzza, a little cloud of pirates leaped from the woods on the north side . . ." The name is most fitting, especially if these stars are indeed the geometrical patterns: loops, arcs, hooks, lines, angles, and other miniature asterisms. What impressed me most, however, was the virtual lack of any stunning doubles – just patterns. The cluster is one of the true circumpolar clusters in the northern skies, never setting from the latitude of New York City. So unless you live south of this line, the hero never rides into the sunset.

Hidden Treasure 15

NGC 1333

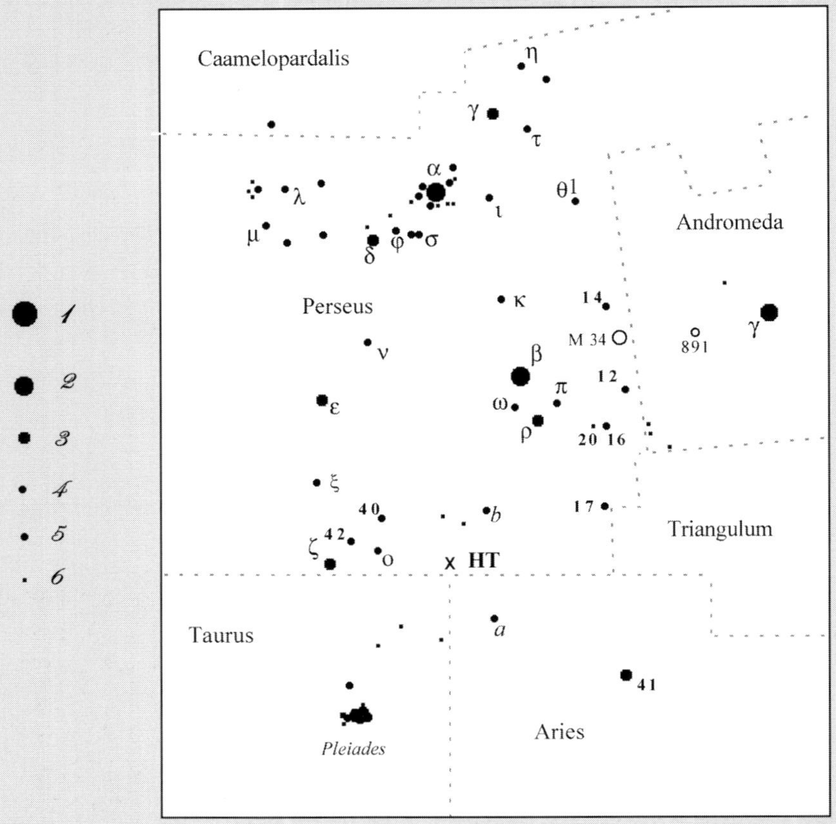

© Stephen James O'Meara

Tirion: Chart 4

Uranometria: Chart 94

15

Embryo Nebula, Phantom Tiara
NGC 1333
Type: Reflection Nebula
Con: Perseus

RA: 03^h 29.3^m
Dec: $+31^\circ$ 25′
Mag: 5.7
Dim: 6′ × 3′
Dist: ~1,100 light-years
Disc: Eduard Schönfeld, 1858; independently discovered by Horace P. Tuttle, 1859

HERSCHEL: None.

NGC: Faint, large, magnitude 10 star north following.

NGC 1333 IS A SIMPLE BUT beautiful reflection nebula, and a highly active star-formation region, within the Perseus molecular cloud complex. It lies $3\frac{1}{4}^\circ$ southwest of Omicron (o) Persei and almost nicks the spot where Perseus, Taurus, and Aries meet. At magnitude 5.7 the nebula is quite bright, but it's also very tiny. It belongs to a class of nebulae, which I call "dwarf nebulae" – reflection, emission, or dark, nebulae that have apparent diameters 10′ or smaller.

On most star charts, the object is far from eye-catching. In the second edition of Tirion's *Sky Atlas 2000.0*, for instance, the symbol for NGC 1333 is a tiny box no bigger than the disk representing a 4th-magnitude star. The box looks like an afterthought, perhaps a move to flesh out that region of sky with pretty symbols. Certainly observers moving their fingers across their star charts while selecting objects for a night's work would not stop there. NGC 1333 is literally dwarfed by the Pleiades star cluster just $8\frac{1}{2}^\circ$ to the southeast, and by the challenging but famous California Nebula (NGC 1499) a similar distance to the northeast. Yet, surprisingly, at the eyepiece, NGC 1333 is easier to see than the Merope Nebula (IC 349).

If you find it hard to believe that this tiny nebula can be seen in a small telescope, consider this: Eduard Schönfeld discovered NGC 1333 in 1858 with the 3.1-inch Fraunhofer comet seeker at Bonn Observatory in Germany. He was observing the positions of stars for the *Bonner Dürchmusterung* – a set of 37 large star charts and a three-volume catalog giving the approximate positions of 324,198 stars, down to magnitude 9 and fainter, lying between the declinations $+90^\circ$ and -2°, when he came across this "large" and "faint" object. I find these adjectives pretty remarkable considering how tiny and bright the object truly is.

Horace P. Tuttle independently discovered the nebula on February 5, 1859; he *was* hunting for comets with the 4-inch f/8 Merz comet seeker at Harvard College Observatory when he chanced upon it.

Independent discoveries continue to this day. For instance, when Paul Clark (North Wales, UK) came across the object with his $5\frac{1}{2}$-inch Orion Maksutov telescope, he said, that at first, "I thought I'd found a comet. A bright oval coma with dense nebulosity hiding a star as the nucleus." I am no exception. Several times I have swept up NGC 1333 during my comet hunts (even in moonlight) with the 4-inch refractor. And Christian Luginbuhl and Brian Skiff say it's not difficult to see in a 60-mm refractor. Tomm Lorenzin in his *2000+ Catalog* adds that NGC 1333 can be picked up in 14 × 70 binoculars on a fine night. Ironically, in Volume 2 of the Webb Society's *Deep-Sky Observer's Handbook,* D. A. Allen describes NGC 1333 as "[n]ot an easy object unless the sky is very dark"; he was using the great 60-inch reflector at Mount Wilson Observatory in California. Allen's comment only demonstrates an awesome truth, namely that some objects are just not well suited for extreme apertures and high magnifications. This fact might also explain why William Herschel missed NGC 1333 in his great sky survey. (Remember this whenever you hear some observer shout, "You couldn't possibly have seen such and such, because *I* couldn't see it in my monster telescope.")

Allen's comment reminds me of some observations William Sheehan and I made of M57, the famous Ring Nebula in Lyra, with two different 40-inch telescopes. In July 1992, we were using the 40-inch reflector atop Pic du Midi in the French Pyrenees to study Mars. One night, for a change, we turned the telescope to M57. While the Ring was fantastically obvious in the finderscope, we were surprised to see only two stars centered in the eyepiece of the main telescope. It turned out that, at 1,200×, we were looking at the area of sky *inside* the Ring; the two stars we saw were the planetary's central star and its similarly bright companion to the northwest. When we slewed the telescope to look at the nebula, we were again surprised at how pale and ill-defined it looked. This observation stood in stark contrast to a low-power view we had of it with the 40-inch reflector at Lick Observatory several years later, when the Ring was an intensely bright vortex with a pearly luster. We could not see the central star nor its companion at such a low magnification because the Ring's central region was filled with luminous gas. Large telescopes – especially the older models that, because of their long focal lengths, cannot offer decent low-power views of deep-sky objects – cannot perform well on "large" and diffuse nebulae, especially when the atmosphere is not stable.

To find this interesting object, first locate 3rd-magnitude Zeta (ζ) Persei about 8° north-northwest of the Pleiades. The star is easily visible even under city lights. Zeta Per is the easternmost star in a roughly 2°-wide triangle with 4th-magnitude Omicron (o) Per and 5th-magnitude 42 Per. Omicron itself is the easternmost star in a 5°-wide triangle with magnitude 4.5 Star *a* and 5th-magnitude Star *b*. NGC 1333 lies just south of the center of that larger triangle, nestled between two $\frac{1}{2}$°-wide triangles (*c* and *d*) comprised of 7th- and 8th-magnitude stars. I find it's best to locate the larger triangle first with either the naked eye or binoculars, then center your finder at the triangle's heart. You will be within sweeping distance

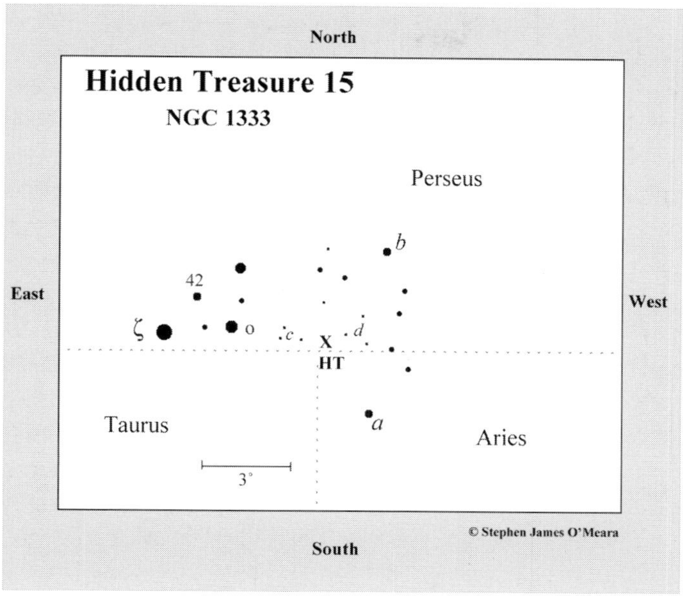

surrounding this star. A dark finger immediately to its southwest separates a fissure of light. A wider bay of darkness separates that fissure from a very curious dappled cloudlet of dimly glowing gas. I could not detect any of these more southerly features, but those with larger telescopes should definitely try.

In the 4-inch, the bright nebula around Star *b* is connected to Star *a* by a thin film of glowing gas, like a comet's tail, through which shine two dimmer suns: the brightest lies roughly midway between Stars *a* and *b* and slightly east; the faintest is a very dim (roughly 13th-magnitude) sun just southeast of Star *a*. A hyperfine pair of roughly 12th-magnitude stars lies northeast of Star *a;* use high power on a very steady night to resolve the pair.

If you're having trouble eyeing these details, try studying the region first with high power, so you can record only the brightest regions. Then zoom out gradually, using progressively lower powers. At high power, center Star *a* and use your averted vision. Do you see two dim knots of nebulosity – one to the north, the other to the northwest very close to Star *a*? At low power, study the region northwest of the midpoint between Stars *a* and *b*. Do you see two fine filaments of glowing gas oriented northwest–southeast? One actually extends into that region from Star b. The other is a lone filament separated from the other by a dark lane. When all these details are seen together, NGC 1333 looks like an embryo. Star *a* is the embryo's eye, The round bulbous cloud around Star *a*

of the nebula. Again, this is a dwarf nebula, so it is *extremely* small at low powers. Just look for a moderately bright star with a shimmering halo of light.

The very simplicity of NGC 1333 is its attraction. It's the face of a shy specter masked in a white veil. A magnitude 10.5 star (labeled *a* in the accompanying graph) illuminates the nebula and adorns it like the crown jewel in a phantom diamond tiara. At first glance, the nebula has a round form with a uniform luster, but closer inspection reveals it to be a mottled kidney bean of light, oriented north-northeast–south-southwest. After making several observations of the nebula over the years, I find it much more complex than I had first imagined. Three fainter "diamonds" mark the tiara's head band. The brightest of these, a roughly 12th-magnitude star (*b*), lies 3′ southwest of Star *a* and illuminates the nebula's southwestern corner. In photographs, thin black clouds of obscuring matter slash the face of the bright nebulosity

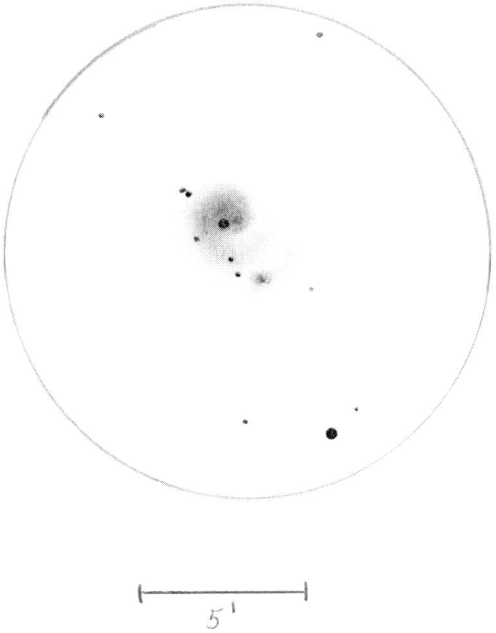

is the embryo's head. Star *b* marks the embryo's *okole* (Hawaiian for buttocks) and the two filaments are the embryo's arms and umbilical cord. I've nicknamed NGC 1333 the Embryo Nebula.

Hal Corwin (California Technical Institute) notes that early observations of NGC 1333 led some to believe that the nebula varied in brightness. "[The] observations with small telescopes were contradictory enough to lead to suggestions that it might be variable," Corwin explains. This perception was exacerbated by Tuttle's discovery observation, in which he inverted the field directions. Interestingly, part of the object seems to be a collapsing protostar. If the density and position of dust clouds surrounding the protostar in NGC 1333 are changing, and if that protostar is changing in intensity over time, then it's possible that the nebula surrounding it could reflect these changes.

Recent near-infrared images also reveal some 80 pre-main-sequence stars embedded in this region of active star formation. The dark nebula Barnard 205 lies just to the south of NGC 1333, and its two fingers of darkness seem to be pinching the reflection nebula's southern tip. This dark nebula stretches for nearly $\frac{1}{2}^{\circ}$ and is a wonderfully complex agglomeration of shadows. Exploring this hidden corner of Perseus through the 4-inch at low power is like tiptoeing through an old, forgotten house dusted with cobwebs and littered with keepsakes.

Hidden Treasure 16

NGC 1360

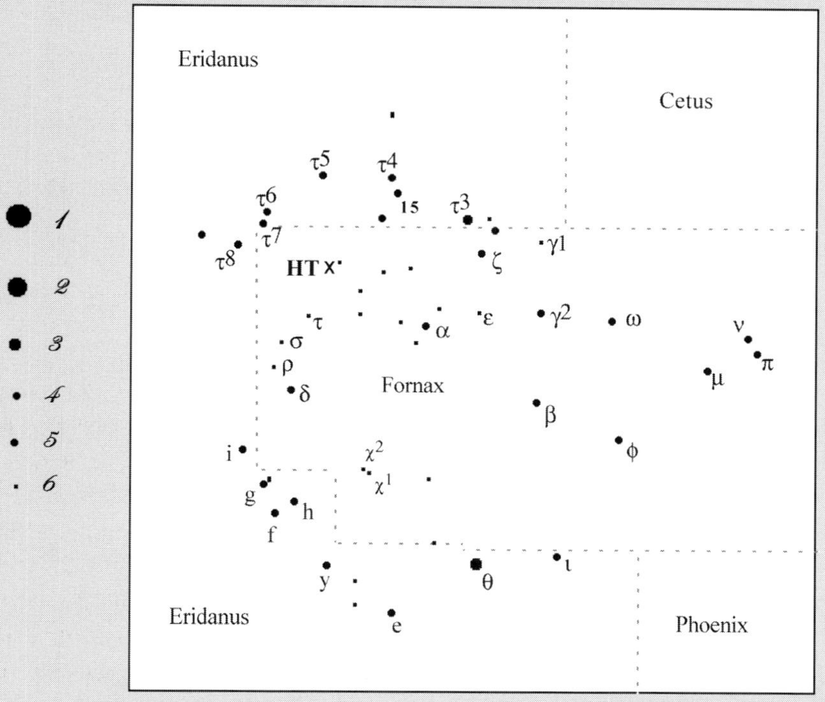

© Stephen James O'Meara

Tirion: Chart 18

Uranometria: Chart 312

16

Comet Planetary
NGC 1360
Type: Planetary Nebula
Con: Fornax

RA: $03^h\ 33^m\ 14.6^s$
Dec: $-25^\circ\ 52'\ 19''$
Mag: 9.1 (O'Meara); 9.4
Dim: $9' \times 5'$
Dist: ~848 light-years
Disc: Lewis Swift, 1857

HERSCHEL: None.

NGC: Magnitude 8 star in bright, large nebula, extended from north to south.

CONCEALED AMONG THE DIM SUNS of Fornax, mixed within the wide fanfare of galaxies that populate the region, lies one of the sky's brightest planetary nebulae, one which richly rewards small telescope users: NGC 1360. This often-overlooked gem has an immediate wow-what-is-that appeal, especially when stumbled upon at the eyepiece. NGC 1360 was long described as a peculiar object until Rudolph Minkowski identified it as a planetary nebula in 1946, though it took a while for the planetary to lose its "peculiar" status. In 1968 Lubos Kohoutek called it an "uncommon planetary." Robert Burnham Jr. refers to NGC 1360 as a "peculiar nebula . . . usually classified as a planetary." Steven J. Hynes lists NGC 1360 in his *Planetary Nebulae* as a Type III planetary (a planetary with an irregular disk).

Today we know NGC 1360 is typical of a large evolved planetary nebula. It is also one of the few known examples of a large, high-excitation planetary. Italian astronomer Mario Perinotto (Observatory of Florence), who studied the central star's spectrum in the early 1980s, placed it in the Wolf-Rayet category of planetary nebulae, meaning the central star appears to be in a state of flux and is violently ejecting matter into the surrounding shell.

The action, of course, is centered at NGC 1360's bright (magnitude 11.4) central star, which has a temperature of at least 85,000 kelvin – that's some 14 times hotter than the Sun and at least 540 times more luminous. The central star is a hot subdwarf with only about one-tenth the Sun's diameter and 0.55 its mass. It is a suspected variable star, and, while some early studies revealed it to be a spectroscopic binary, no modern evidence supports that claim.

The central star does, however, have a claim to fame. As reported in the 1996 *Monthly Notices* of the Royal Astronomical Society, Melven G. Hoare

(Max-Planck-Institute for Astronomy) and his colleagues used the *Extreme Ultraviolet Explorer* (EUVE) satellite to obtain the first extreme ultraviolet spectrum of a planetary nebula's central star – that of NGC 1360. Since most of a central star's emission is emitted in the extreme ultraviolet, the EUVE satellite observations will help astronomers better understand the excitation of winds and nebulae around all hot stars. In 1978, the *International Ultraviolet Explorer* (IUE) found that fast winds arise from planetary nebulae. These winds have remarkable speeds of 2,000 to 4,000 kilometers per second, making them among the fastest moving phenomena in the universe. The IUE detection of fast winds is a nice confirmation of the prediction of the interacting winds theory first proposed in 1977 by Chris Purton, Pim Fitzgerald (University of Waterloo, Canada), and Sun Kwok (University of Calgary, Canada).

The interacting wind theory is a simple, one-dimensional model of wind-blown planetary nebulae. In this model, "fast winds" from the nucleus of the planetary nebula blow into the star's former envelope from the inside. A shock that moves outward into the old red-giant envelope and a second interior radiationless shock close to the star, where the radial component of the wind speed drops from its 600 miles per second to Mach 1 in the hot bubble interior. The actual observed nebula consists of the slowly expanding red-giant-envelope halo, which is often ionized and heated by stellar radiation, and the outward moving shock on the inside edge of the halo. Mario Perinotto (Observatory of Florence) confirmed that such fast winds from central stars are common. We now know that NGC 1360 has two obvious gas shells. The inner one is expanding at 27 kilometers per second, while the outer shell is crawling along at 7 kilometers per second. The planetary itself is moving toward us at about 48 kilometers per second.

Not much is heard about NGC 1360's visual appearance from amateurs living at mid-northern latitudes, which is surprising. Then again, as seen from the north, Fornax, the Furnace, is not among the most glamorous constellations, as its 110 stars, from 4th- to 7th-magnitude, all lie near the southern horizon in mid December. Still, NGC 1360 is at nearly the same declination as Antares and the globular cluster M4. In fact, Lewis Swift discovered the nebula in 1857 from his home in Cortland (Monroe County) in western New York.

At the time Swift was new to astronomy (he would, however, become one of America's leading astronomers of his day). Inspired by the epochal Leonid meteor shower of 1833, the Great Comet of 1843, and later the astronomical books of Thomas Dick, Swift purchased in 1855 a damaged 3-inch objective from a peddler for five dollars. After making a brass mounting and an eyepiece for the telescope, he began a survey of the night sky. Two years later he discovered the extremely comet-like glow of NGC 1360.

News of astronomical discovery traveled slowly in those days, especially across the great seas. And amateurs sweeping the skies worked largely in isolation, which is perhaps why Friedrich Winnecke is mentioned as the object's co-discoverer; he found NGC 1360 in January 1868 with a 3.5-inch comet seeker at Karlsruhe. The Latvian native Eugen Block, who studied astronomy at Dorpat and became an assistant there in 1868, also independently discovered NGC 1360

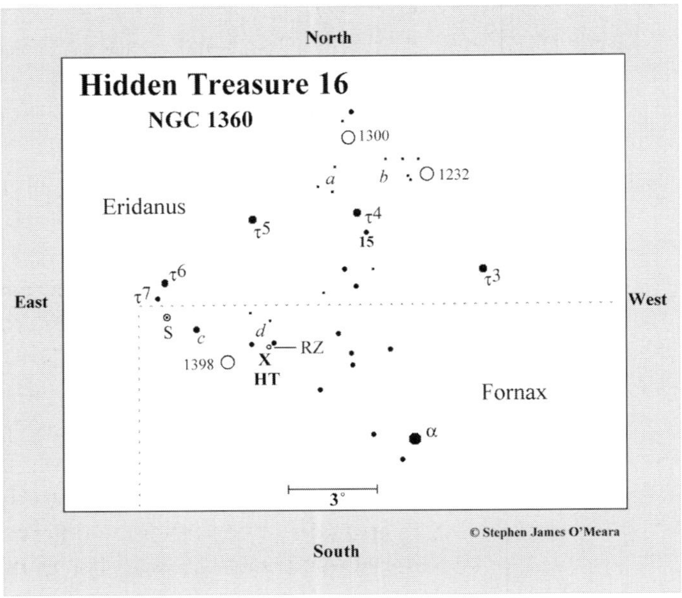

on December 17 that same year. The astronomical heritage of this southern object then belongs to the mid-northern latitudes. Not surprisingly, in *The Australian Guide to Stargazing* (2001 edition), Gregg D. Thompson calls this object the "Comet Planetary."

You will find this "peculiar planetary" about $5\frac{1}{2}^\circ$ northeast of magnitude 3.8 Alpha (α) Fornacis, less than 15′ southeast of a magnitude 6.4 star. (Note that the second edition of *Sky Atlas 2000.0* curiously fails to show NGC 1360.) Observers at mid-northern latitudes might want first to find 3rd-magnitude Gamma (γ) Eridani (Zaurak), which is two fists (about 20°) to the west-southwest of Beta (β) Orionis (Rigel). One fist (10°) to the southwest of Beta Ori is a $6\frac{1}{2}^\circ$-long arc of three 4th-magnitude stars – Tau^4 (τ^4), Tau^5 (τ^5), and Tau^6 (τ^6) Eri – oriented roughly east to west. The arc is easy to confirm in binoculars because both ends have a 5th-magnitude companion: 15 Eri is about $\frac{3}{4}^\circ$ southwest of Tau^4, and Tau^7 (τ^7) lies about $\frac{3}{4}^\circ$ south-southeast of Tau^6.

Now center Tau^7 in your telescope. Just 35′ southwest is the peculiar and uncertain variable S Fornacis. (The star shines at magnitude 8.6 today, but in 1899 three observers (Hartwig, Holetschek, and Abetti) claimed that it was as bright as magnitude 5.5; no variability has been detected since.) Next, hop 1° west-southwest to a magnitude 6.7 star (*c*). If you sweep 2° due west of it, you will find yourself at the center of an $1\frac{1}{2}^\circ$-long trapezoid (*d*) comprised of two 7th-magnitude stars to the north and two 6th-magnitude stars to the south. Center the westernmost of those two southern stars. Note that the red semi-regular variable RZ For lies only about 7′ to the east-southeast; the star varies between magnitude 9.1 and 10.0. NGC 1360 is only 15′ southeast of the variable. You cannot miss it. NGC 1360 is not only bright as planetaries go, it's huge; while many planetaries measure mere arcseconds across, NGC 1360 spans a respectable $6\frac{1}{2}'$, making it about five times larger than M57, the Ring Nebula, on the plane of the sky. If we accept NGC 1360's distance as 848 light-years, its true linear extent is 2.2×1.2 light-years. The Ring, by comparison, is nearly 300 light-years further away and only 0.4 light-years wide.

In their *Revue des constellations,* R. Sagot and J. Texereau state that NGC 1360 has been detected from France in a 2.2-inch refractor. From more southerly latitudes, it is visible in 7×50 binoculars, and it is a beautiful oval glow in my antique scope under dark skies.

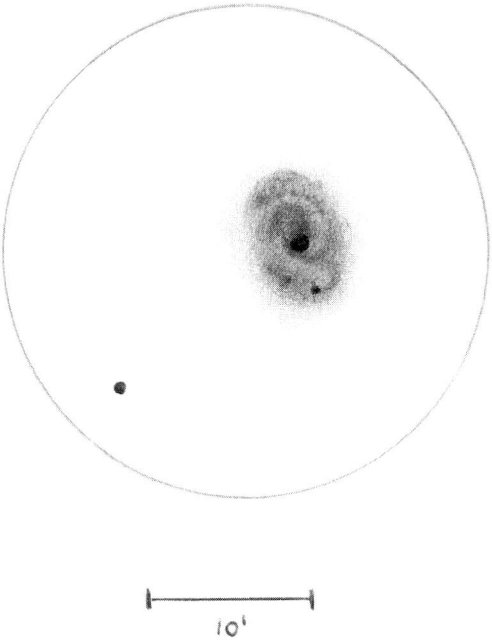

Through the 4-inch, NGC 1360 is a simple yet spectacular sight – a conspicuous central star surrounded by a fantastic oval shell that's evident at 23×. With concentration, two diffuse arcs of light can be seen surrounding the blazing central star. These arcs are not symmetrical. With imagination they look like the imprint of a "lipstick kiss" on a mirror.

With higher power it's still a glorious object, especially if you can appreciate an incredible amount of extremely subtle detail. There's so much to see at the limit of detection. I could spend endless hours on this nebula. But what's most obvious is a bushy brow of nebulosity immediately northeast of the central star. This is part of the nebula's inner shell of gas, which appears to be broken. Two other segments of that shattered shell can be seen: a mottled arc immediately southeast of the central star, and a mottled triangular patch of light immediately southwest of the central star. A look at the opening photograph reveals that this fuzzy wedge of light is probably due to the presence of two close field stars superimposed on the nebula. The outer shell (or crown) has sharp boundaries on the northeastern and southwestern sides. A tiny knot of material (or a very dim star) lies at the outer shell's northeastern edge just beyond the inner brow. And a 13th-magnitude star lies within the outer shell's sharp southern rim and is quite difficult to see. The star seems to separate that sharp rim into two segments.

Seen together, all these features make the nebula look like a spiral galaxy. Aesthetically, the planetary looks best at 72×. The overall shell appears delicately mottled, like slightly curdled milk, or an egg made of white marble. If you have any problems with light pollution from your observing site, try using either an O-III or UHC filter to boost the contrast between the nebula and the background sky. Good luck!

Hidden Treasure 17

NGC 1365

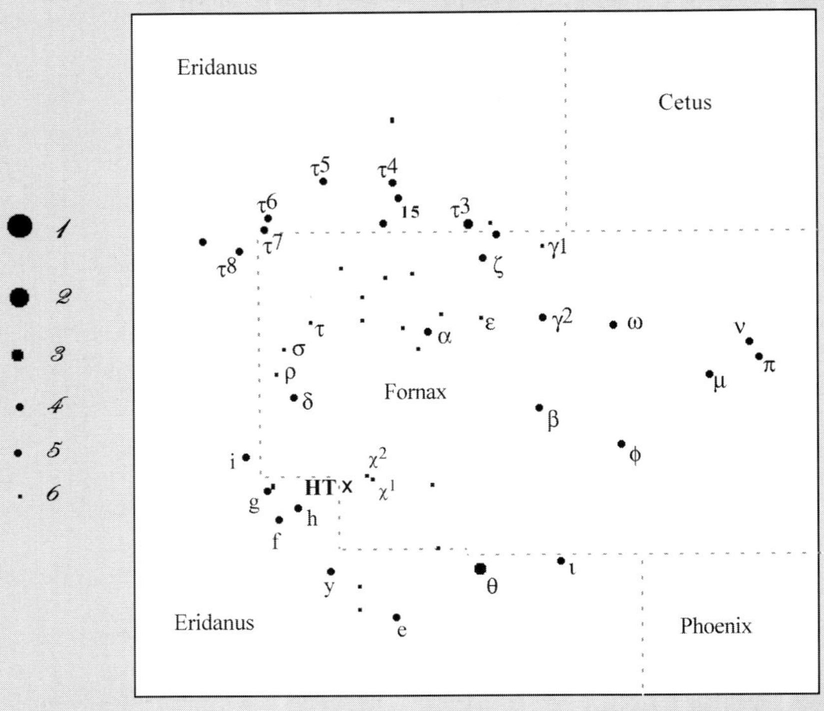

© Stephen James O'Meara

Tirion: Chart 18

Uranometria: Chart 355

17

NGC 1365
Type: Barred Spiral Galaxy (SBb)
Con: Fornax

RA: 03^h 33.6^m
Dec: −36° 08′
Mag: 8.3 (O'Meara); 9.3
Dim: 11.2′ × 5.9′
SB: 13.6
Dist: 60 million light-years
Disc: James Dunlop, included in his 1827 catalog

J. HERSCHEL: A very remarkable nebula. A decided link between nebulae M51 and M27. Centre very bright; somewhat extended; gradually very much brighter to the middle; a 13th-magnitude star near the edge of the halo involved. The area of the halo very faint; general position of the longer axis 20.8° whole breadth = 3′. (h 2552)

NGC: *Remarkable*, very bright, very large, much extended, resolvable to a nucleus.

NGC 1365 IS ONE OF THE MOST stunning barred spiral galaxies in the night sky. This impressive system has a linear diameter of 160,000 light-years and a luminosity of 200 billion Suns, so it is a match for our Milky Way. In photographs, NGC 1365 displays near-perfect symmetry. Two long, arabesque arms spring from a central bar centered on a dazzlingly bright nucleus. That bar measures an awesome 94,000 light-years in length – about three times the distance of the Sun from the Milky Way's center.

Careful inspection of the structure in NGC 1365, however, reveals that the galaxy is a less-perfect system than NGC 1300 in Eridanus: the prototype, and most glorified, barred spiral galaxy in the heavens. (Because of its faintness, low surface brightness, and southerly location, NGC 1300 is not a *Hidden Treasure* galaxy, though it is in my list of 20 additional objects; see Appendix C.) In high-resolution images, NGC 1365's bar is ever so slightly warped and less smooth than that in NGC 1300; its arms are also more open, making it more of an intermediate class of barred spiral – between an SBb and an SBc.

Since NGC 1365 is oriented 27° from edge-on, its spiral nature is comfortably revealed. It has four well-developed arms – two major ones and two rudimentary ones. As with other barred spirals, dust lanes appear on opposite sides of its bar. These lanes lie on the bar's leading edge relative to the galaxy's direction of rotation. Interestingly, the galaxy's outer regions show a decline in rotational rate, which indicates that NGC 1365 lacks a very massive dark halo. This may also be the reason why NGC 1365 has such a very strong bar.

Less prominent are wisps of spiral structure winding around the core between the galaxy's two major arms. In color photographs, the bar and its immediate surroundings are populated by cool, yellow suns. In contrast, the long graceful arms at the end of the bar are peppered with hot blue stars and pink star-forming regions; these mark the locations where interstellar matter is being compressed at the crests of density waves, triggering star formation. Maximum star formation in these arms occurs at the ends of the bar.

NGC 1365 is both a Seyfert and a starburst galaxy. Being a Seyfert galaxy means that its small and intensely bright nucleus behaves like a low-energy quasar, ejecting matter episodically into space. Indeed, radio images of NGC 1365's core show jet-like structures originating from the nucleus and extending about 5″ (~1 million light-years) in position angle 125°; these jets are aligned with the minor axis of the galaxy. The production of the energy feeding the jet probably comes from a supermassive black hole at the galaxy's heart.

The Seyfert nucleus, which is a weak radio source, is surrounded by a 8″ × 20″ ring of radio emission. This radio ring contains a number of hot spots, which may indicate the presence of supernovae and supernova remnants. One bright compact radio source (NGC 1365:A) has been classified as a "radio supernova" and is some 100 times brighter than the bright supernova remnant, Cas A, in our Milky Way.

NGC 1365:A is located within one of the superstar clusters that have been discovered in the nuclear region by the Hubble Space Telescope (HST). Indeed, the HST images reveal that the Seyfert nucleus is surrounded by an intense starburst region. The complex patchiness of this region suggests that the galaxy's ancient red stars are probably being overwhelmed by the light of new red giants and supergiants formed in the starburst, which may be young globular clusters. At visible wavelengths, the starburst region is a maelstrom of hot young star clusters and dark dust lanes, which, in infrared light, contain still more newborn clusters. Thus a black hole is probably not the dominant contributor to the energies of the central region.

NGC 1365 was also part of the HST's Key Project on the Extragalactic Distance Scale, which sought to measure the Hubble constant to an accuracy of 10 percent. To achieve that goal, the team used HST to observe 18 galaxies out to 65 million light-years. They discovered almost 800 Cepheid variable stars, a special class of pulsating star used for accurate distance measurement. The team measured the distance to NGC 1365 at 60 million light-years and the Hubble constant at 70 kilometers per second per megaparsec, with an uncertainty of 10 percent. This means that a galaxy appears

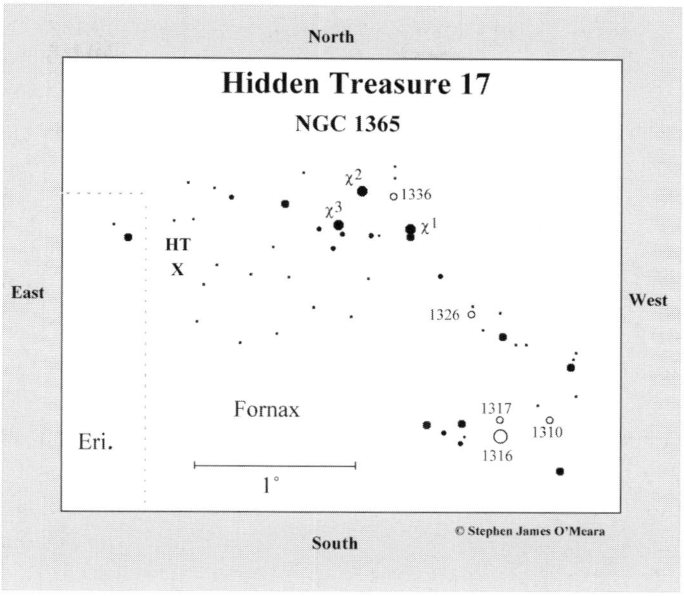

to be moving 160 thousand miles per hour faster for every 3.3 light-years away from Earth.

While NGC 1365 is commonly classified as the largest member of the Fornax Cluster of galaxies, its redshift velocity does place NGC 1365 at the same distance as the Fornax system, but its large angular size makes its membership questionable. NGC 1365's size is comparable to that of the largest spirals in the Virgo Galaxy Cluster which are also at nearly the same distance as NGC 1365. One recent discovery might help to solve the dilemma. On August 24, 2001 (Universal Time), the Rev. Robert Evans of Hazelbrook, New South Wales – history's greatest visual supernova hunter – discovered a 14th-magnitude Type II supernova 90″ west and 10″ south of the galaxy's nucleus. Supernovae are among the most luminous phenomena in the universe and have been used as light beacons to measure cosmological distances. Two other supernovae have been discovered in this galaxy: a magnitude 13.5 object in 1983, which was 57″ west and 30″ south of the nucleus; and a magnitude 16.5 object in 1957; it was 54″ west and 75″ north of the nucleus.

To find this remarkable object, first locate 3rd-magnitude Theta (θ) Eridani (Acamar). One fist (10°) to the east-northeast is an obvious 3°-wide Y-shaped asterism of 4th- and 5th-magnitude stars, including f, g, h, and i Eridani. About $3\frac{1}{2}^\circ$ northwest of the southwesternmost star in the Y (h Eri), is a 30′-wide acute triangle of 6th-magnitude suns: Chi^1 (χ^1), Chi^2 (χ^2), and Chi^3 (χ^3) Fornacis. Confirm these with your binoculars. NGC 1365 lies a little more than 1° east-southeast of Chi^3 (χ^3). While most sources list the galaxy as magnitude 9.3, I estimated its magnitude and found it to be a full magnitude brighter. It is visible with averted vision in 7 × 50 binoculars and is a dim round nebula in my antique telescope.

At 23× in the 4-inch, NGC 1365 is quite a beauty, an attention grabber. But the galaxy's low surface brightness might make it a challenge under suburban skies. It is well worth the effort to try and find it, though. As Walter Scott Houston wrote, "its obviously lenticular form and bright central part . . . reminds one of binoculars views of the Andromeda Galaxy, M31." But that's with a bit of magnification. At low power, under a dark sky, NGC 1365 appears as a sizable, uniform glow. But with just a little bit of patience, the galaxy's bright Seyfert nucleus pops into view; then, after the passage of a little more time, some finer details – the

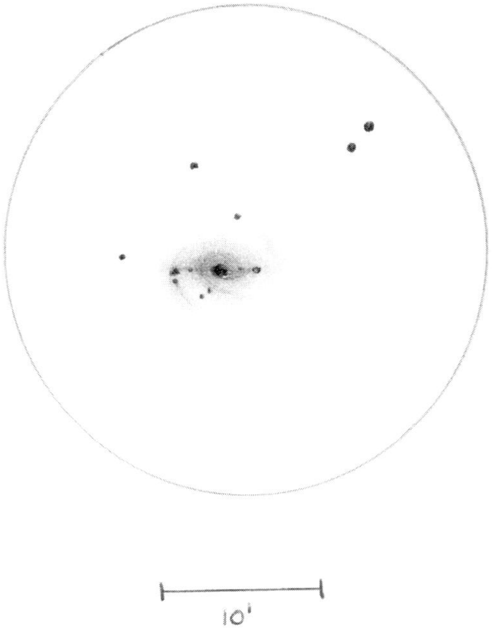

bar, knots, and arcs – can be seen. You need higher magnifications to differentiate clearly any of these details.

The details at 72× to 168× are extremely titillating, especially since the view is complicated by a few moderately bright field stars superimposed on the galaxy's face. You need enough magnification to separate these stars from the galaxy's bar and sweeping spiral arms. The nuclear region is rather small, about 3′, and it's best to try to examine the galaxy with as much power as possible at first, then move progressively to lower and lower powers. The exact power you use will depend on the night and the comfort level provided by your field of view. What's most difficult and most time consuming, I find, is trying to separate the weak spiral region surrounding the nucleus from the bar. Again, try using high magnification first, especially since this region also includes a 13.5-magnitude star 1.3′ northwest of the nucleus; this star could be mistaken for a supernova. An equally bright star or knot lies on the tip of the western bar. A slightly brighter star appears just off the tip of the eastern bar, which also appears mottled. These stars probably gave Sir John Herschel the illusion that he was resolving this nebula, as he describes in his second observation of the object: "very bright, extended, resolvable nucleus; or has 2 or 3 stars involved."

In many respects, NGC 1365 looks like a planetary nebula, like NGC 7009 (Caldwell 55), the Saturn Nebula, in Aquarius. Both have a bright central nucleus, tight elliptical core, large iris of light, and a bar that glows like the rings of Saturn seen edge-on. The difference, in visual appearance, of course, is that NGC 1365 has two long arms sweeping off the edgewise "ring."

Curiously, I found the most conspicuous of the two long spiral arms to curve southward from the tip of the eastern bar. Herschel found the "preceding Arc" to be brighter. Herschel's view matches what is seen in photographs, taken with large telescopes; see if you agree.

Hidden Treasures 18 & 20

NGC 1399 & 1404

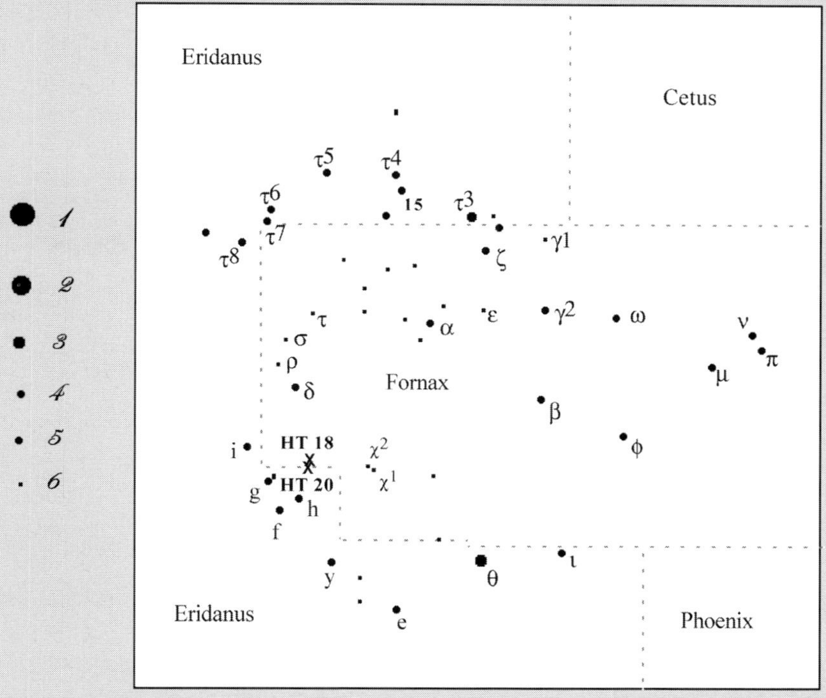

Tirion: Chart 18

Uranometria: Charts 355 & 356

18

NGC 1399
Type: Elliptical Galaxy (E)
Con: Fornax

RA: $03^h\ 38.5^m$
Dec: $-35°\ 27'$
Mag: 8.8
Dim: $8.1' \times 7.6'$
SB: 13.1
Dist: 55 million light-years
Disc: John Herschel, 1835

J. HERSCHEL: Globular cluster, very bright, pretty large, pretty suddenly brighter in the middle, resolvable or resolved, 2′. (h 2569)

NGC: Globular cluster, very bright, pretty large, pretty suddenly brighter in the middle, partially resolved, some stars seen.

20

NGC 1404
Type: Elliptical Galaxy (E)
Con: Fornax

RA: $03^h\ 38.9^m$
Dec: $-35°\ 36'$
Mag: 9.7
Dim: $4.8' \times 3.9'$
SB: 12.7
Dist: 55 million light-years
Disc: John Herschel, 1837

J. HERSCHEL: Very bright, round, pretty suddenly much brighter in the middle, 40′, has a star north following. (h 2571)

NGC: Very bright, pretty large, round, pretty suddenly much brighter in the middle.

---

18 & 20

NGC 1399 AND NGC 1404 ARE TWO giant elliptical galaxies lying 8′ apart in the heart of the Fornax Cluster and Eridanus Cloud of Galaxies. They are the brightest members of a gaggle of 17 moderately bright galaxies lying within a 2° field of view centered on NGC 1399 and 1404. Only two of the 17 galaxies (NGC 1373 and 1392) do not belong to the Fornax Cluster. Those that do are NGC 1374, 1375, 1379, 1380, 1380A, 1381, 1386, 1387, 1389, 1399, 1404, 1427, 1427A, 1428, and 1437; one galaxy, NGC 1382, is a questionable member. Two other bright members of the Fornax Cluster, NGC 1316 (Hidden Treasure 13) and NGC 1365 (Hidden Treasure 17), lie outside this field to the west.

The 2° field of view surrounding NGC 1399 and NGC 1404 is a veritable jackpot of extragalactic wonder equal to, if not more impressive than, that centered on M84 and M86 in the Virgo Cluster of Galaxies. While both views contain about the same number of galaxies to about magnitude 13.5, the Fornax Cluster is more compact and geometrically simple (most of the galaxies lie in a north–south trending line), making it easier to navigate. Celestial pirates should be able to find the riches of the Fornax Cluster more swiftly and confidently than they would in the Virgo Cluster, which can amount to a fuzzy ball of confusion.

Next to the Virgo Cluster, the Fornax Cluster is the richest galaxy cluster within 100 million light-years of the Milky Way. The entire Fornax Cluster (not just the bright core system described above) contains 340 members and many more possible members. The problem is that until more distance measurements are made, astronomers cannot tell whether a galaxy that appears small is a dwarf member of the cluster or is a more distant and unrelated galaxy. The ongoing Fornax Cluster Survey, however, is addressing this problem by using wide-field spectrographs to measure the distances to some 14,000 objects in the direction of the cluster.

Like most of the Fornax Cluster's members, NGC 1399 and 1404 are elliptical galaxies, which is not surprising since ellipticals are not only the most common type of galaxy found in clusters but in the nearby universe. NGC 1399 and NGC 1404 are moderately large systems, measuring 53,000 and 43,000 light-years in true physical extent; they have luminosities of 20 billion and 15 billion Suns, respectively. Both are nearly spherical orbs with classically smooth appearances. This explains why John Herschel believed NGC 1399 to be a globular cluster. Of course, he had no idea that the fuzz balls we call galaxies were distant island universes. Ironically, as with other ellipticals, NGC 1399 and NGC 1404 have shells populated by globular cluster systems (GCS).

The globular cluster systems of galaxies are invaluable probes for measuring the formation and evolutionary history of their hosts. Each cluster has a specific age and metallicity and hence the spectrum and color of an individual cluster is much easier to interpret than the very complex populations of stars seen in their overall integrated light. Their light is dominated by stars with temperatures less than 25,000 K. Star-formation models show this temperature puts the main-sequence turnoff in a region where star formation ended in NGC 1399 about 200 million years ago. Interestingly, NGC 1399 has an extraordinarily rich GCS while NGC 1404 has a poor one. NGC 1399 has some 6,500 globular clusters (±700), while NGC 1404 has only about 725.

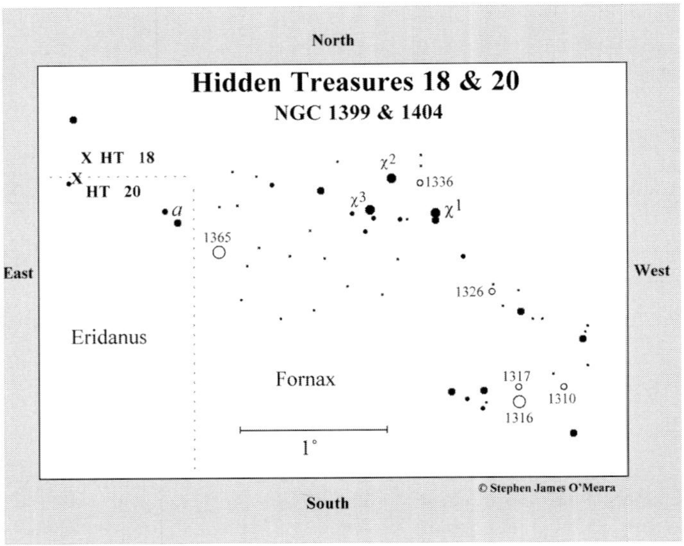

In a 2003 *Monthly Notices* of the Royal Astronomical Society, Kenji Bekki (University of New South Wales, Sydney) and his colleagues proposed that NGC 1404's poor GCS and NGC 1399's exceedingly rich GCS might be explained by tidal stripping. Bekki notes that a tidal stream of intracluster globular clusters orbit the center of the Fornax Cluster. Interestingly, NGC 1399, which is located in the dynamical center of the Fornax cluster, possesses a large extended halo, which could be formed by the accretion of dwarf galaxies. The accretion of a large number of dwarf galaxies, Bekki theorizes, might also help to explain NGC 1399's extraordinarily rich GCS. During the Fornax Cluster Survey, a group of astronomers led by Michael Drinkwater (University of Melbourne, Australia) discovered that indeed, 16 or more galaxies are falling into the main Fornax Cluster at a speed of several hundred kilometers per second. This massive collision of galaxies came as a surprise, because astronomers had previously believed that the Fornax Cluster was quite relaxed.

Support of Bekki's theory also comes from X-ray observations by the Earth-orbiting Chandra telescope, which recorded more than 600 X-ray point sources – over 70 percent of which are associated with known globular clusters surrounding NGC 1399. It also found that NGC 1404 has a comet-like tail, which is consistent with other findings that NGC 1404 is falling toward NGC 1399 through the cluster gas.

To find NGC 1399 and NGC 1404, first locate 3rd-magnitude Theta (θ) Eridani (Acamar), an excellent double star with a 4th-magnitude companion resolvable in a 2-inch telescope; Theta Eri is one of the finer doubles of the southern skies; these clear sapphire gems have been called "radiant" and "dazzling." Early Arabian astronomers knew Acamar as Achernar, because, from their latitude, it used to be the End of the River Eridanus. Today, as Richard Hinckley Allen writes in his *Star Names*, Theta is the "solitary star visible from the latitude of New York City in early winter evenings, low down in the south, on the meridian with Menkar of the Whale." One fist (10°) to the east-northeast is an obvious 3°-wide Y-shaped asterism of 4th- and 5th-magnitude stars, including f, g, h, and i Eridani. About $3\frac{1}{2}^{\circ}$ northwest of the southwesternmost star in the Y (h Eri), is a 30′-wide acute triangle of 6th-magnitude suns: Chi^1 (χ^1), Chi^2 (χ^2), and Chi^3 (χ^3) Fornacis. Confirm these with your binoculars.

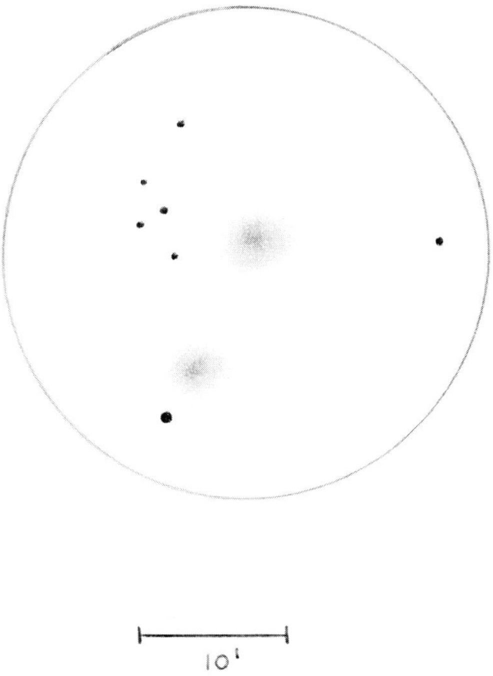

The next step is to locate the 8th-magnitude galaxy NGC 1365 (Hidden Treasure 17), which lies a little more than 1° east-southeast of Chi³ (χ^3). A pair of 8th-magnitude stars (*a*), oriented northeast–southwest, lies about 25′ northeast of NGC 1365; these stars are just over the border in Eridanus. NGC 1399 and NGC 1404 are only about 45′ farther to the northeast. There will be no mistaking this dynamic duo, for they lie only 8′ apart. Note that NGC 1404 rides the border between Fornax and Eridanus, though its celestial address is Fornax.

Both NGC 1399 and NGC 1404 look as smooth as baby skin. These are two galaxies that seemingly defy all reason – at least in small telescopes. No matter what magnification I use, no matter what combination of eyepieces and Barlow lens I try, I cannot make these galaxies appear like anything else but circular glows that gradually get brighter in the middle. It is as if a comet has split in two and the fuzzy fragments are slowly drifting apart. NGC 1399 appears only slightly brighter than NGC 1404. With averted vision and low power, the outer envelopes may have an ever so slightly out-of-round shape, but one has to work hard to imagine it.

The view of these two galaxies seen together in such purity of form is tantalizingly haunting. It is like seeing some ghostly apparition, ill of form, or, the rising mist at sunrise, which, to paraphrase Joseph Conrad in *The Lagoon*, breaks into drifting patches before vanishing into thin flying wreaths.

If you are using a large telescope, though, Larry Mitchell says beware. "The main thing I remember about observing NGC 1399," he says, "is the faint star located about 20″ from the center, that always tricks me into thinking 'supernova.'"

Hidden Treasure 19

NGC 1398

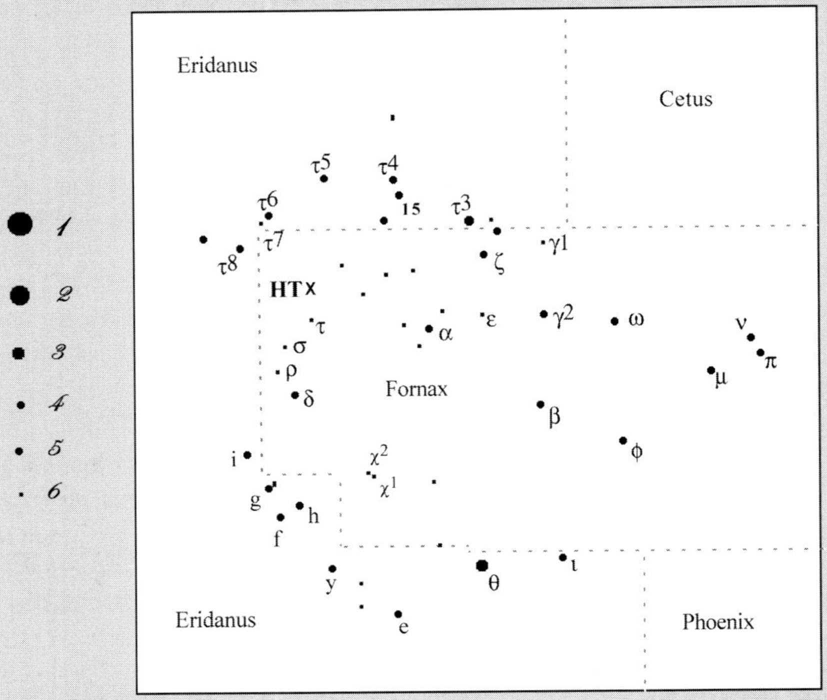

© Stephen James O'Meara

Tirion: Chart 18

Uranometria: Chart 312

19

NGC 1398
Type: Barred Spiral Galaxy (SBab)
Con: Fornax

RA: $03^h\ 38.9^m$
Dec: $-26°\ 20'$
Mag: 9.5
Dim: $7.6' \times 5.3'$
SB: 13.3
Dist: 52 million light-years
Disc: Friedrich August Theodor Winnecke, 1868

HERSCHEL: None.

NGC: Considerably bright, considerably large, round, very much brighter in the middle.

A LITTLE MORE THAN $1\frac{1}{4}°$ EAST-southeast of the bright planetary nebula NGC 1360 in Fornax is the near-equal glow of NGC 1398 – a large, isolated, barred spiral galaxy, a hidden gem far from the central vein of galaxies in the Fornax Cluster.

NGC 1398 is a respectable 9° due north of NGC 1399 (Hidden Treasure 18) – the second brightest galaxy in the Fornax Cluster. Anyone, then, with an equatorial mount can find one or the other quite easily by moving the scope 9° north or south in declination. In fact, if you live at a mid-northern latitude, I suggest you locate NGC 1398 first, then swing the telescope 9° south to find NGC 1399, which would place you at the heart of the Fornax Cluster.

Observers at mid-northern latitudes should also follow the same directions for locating NGC 1360 (Hidden Treasure 16). First find 3rd-magnitude Gamma (γ) Eridani (Zaurak), which is two fists (about 20°) to the west-southwest of Beta (β) Orionis (Rigel). One fist (10°) to the southwest of Beta Ori is a $6\frac{1}{2}°$-long arc of three 4th-magnitude stars – Tau^4 (τ^4), Tau^5 (τ^5), and Tau^6 (τ^6) Eri – oriented roughly east to west. The arc is easy to confirm in binoculars because both ends have a 5th-magnitude companion: 15 Eri is about $\frac{3}{4}°$ southwest of Tau^4, and Tau^7 (τ^7) lies about $\frac{3}{4}°$ south-southeast of Tau^6. Now center Tau^7 in your telescope. Just 35′ southwest is the peculiar and uncertain variable S Fornacis. Next hop 1° west-southwest to a magnitude 6.7 star (*c*). If you sweep 2° due west of it, you will find yourself at the center of an $1\frac{1}{2}°$-long trapezoid (*d*) comprised of two 7th-magnitude stars to the north and two 6th-magnitude stars to the south. Center the westernmost of those two southern stars. The red variable star RZ Fornacis lies only about 7′ to the east-southeast and the planetary nebula NGC 1360 lies only 15′

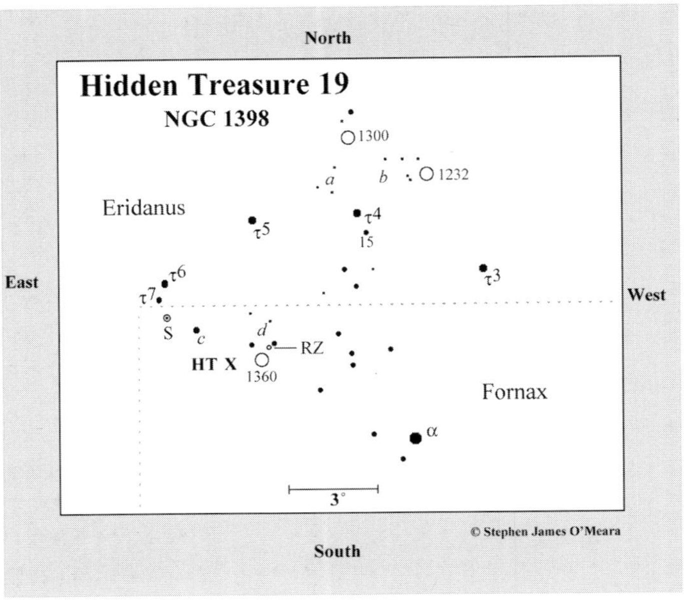

southeast of the variable. NGC 1398 is a roughly $1\frac{1}{4}^{\circ}$ sweep farther to the southeast.

The object is quite conspicuous under a dark sky. Indeed, on the evening of December 17, 1868, Friedrich Winnecke discovered NGC 1398 while sweeping the sky for comets from Karlsruhe, Germany. He was using a modest 4.5-inch refractor built by Reinfelder & Hertel. Winnecke also independently discovered NGC 1360 with a 3.5-inch comet seeker in January of that year. The following autumn, Latvian astronomer Eugen Block independently discovered both NGC 1360 and NGC 1398 on October 18, 1879, with a 4-inch comet seeker. At the time of these discoveries, this region of Fornax belonged to Eridanus.

In high-resolution images, NGC 1398 is an impressive, seemingly symmetrical system. It has a dense inner ring (with a rope-like texture) that surrounds a bright central bulge and bar. The galaxy is inclined 40° from edge-on, so the inner ring is elongated with its major axis parallel to the bar. Ron Buta (University of Alabama) says this alignment is typical for the inner rings of barred spiral galaxies. He also notes that the inner ring is actually a "pseudoring," meaning it is not continuous but comprised of several spiral sections. The brightest segment starts in the northeast just beyond the bar's minor axis and extends through three-quarters of a revolution. The corresponding symmetric component, which begins in the southwest, is not as prominent. Further out, the galaxy has long, narrow blue spiral arms, each of which completes almost a full revolution, possibly forming another pseudoring structure. The arms are dappled with discrete bright areas but these are quite faint when compared to the nuclear region. In the outermost galaxy, the arms appear as dim, flocculent structures. Buta points out that when the inner ring of a galaxy is well defined and conspicuous, the outer ring is usually quite faint, and vice versa.

At a glance, the inner regions of NGC 1398 look strikingly similar to those of M95, a barred spiral galaxy in Leo. These galaxies have nearly the same apparent magnitude and dimensions. But these similarities are an illusion: M95 is twice as close as NGC 1398, and NGC 1398 is twice as large – with a physical diameter of 107,000 light-years and a total mass of 300 billion Suns. Here is another illusion at work. Visually, NGC 1398 looks smaller than the apparent diameter listed in the table, that's because those dimensions include the galaxy's dim

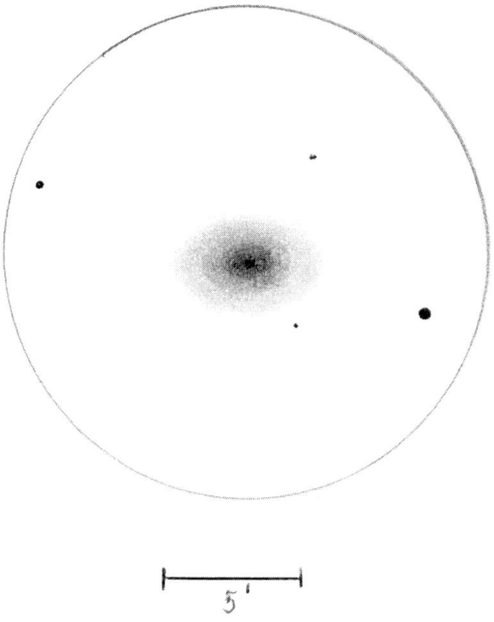

outer arms, which cannot be seen in small telescopes. Christian Luginbuhl and Brian Skiff note that the visual dimensions of this galaxy are only on the order of 1°.5′ × 1′.

Still, the galaxy is quite easy to see. Luginbuhl and Skiff have spied it in a $2\frac{1}{2}$-inch refractor, describing it as a patch 7′ east and a little north of a magnitude 9.5 star. Otherwise, the galaxy shows little detail, though it is not as angelically pure as NGC 1399 and NGC 1404 (Hidden Treasures 18 and 20, respectively). The principal difference is that while NGC 1399 and NGC 1404 get gradually brighter toward the middle, the fuzzy nature of NGC 1398 gets gradually brighter, then much suddenly brighter to a starlike nucleus – like a Russian puzzle box, only this is a jewel within a fuzzy egg, within a fuzzy egg, within a fuzzy egg. Supernova hunters should be aware that NGC 1398 has had one supernova eruption to our knowledge: Supernova 1996N, which was 16th magnitude when it erupted 46″ east and 12″ north of the nucleus.

Hidden Treasures 21 & 23

Kemble's Cascade & NGC 1502

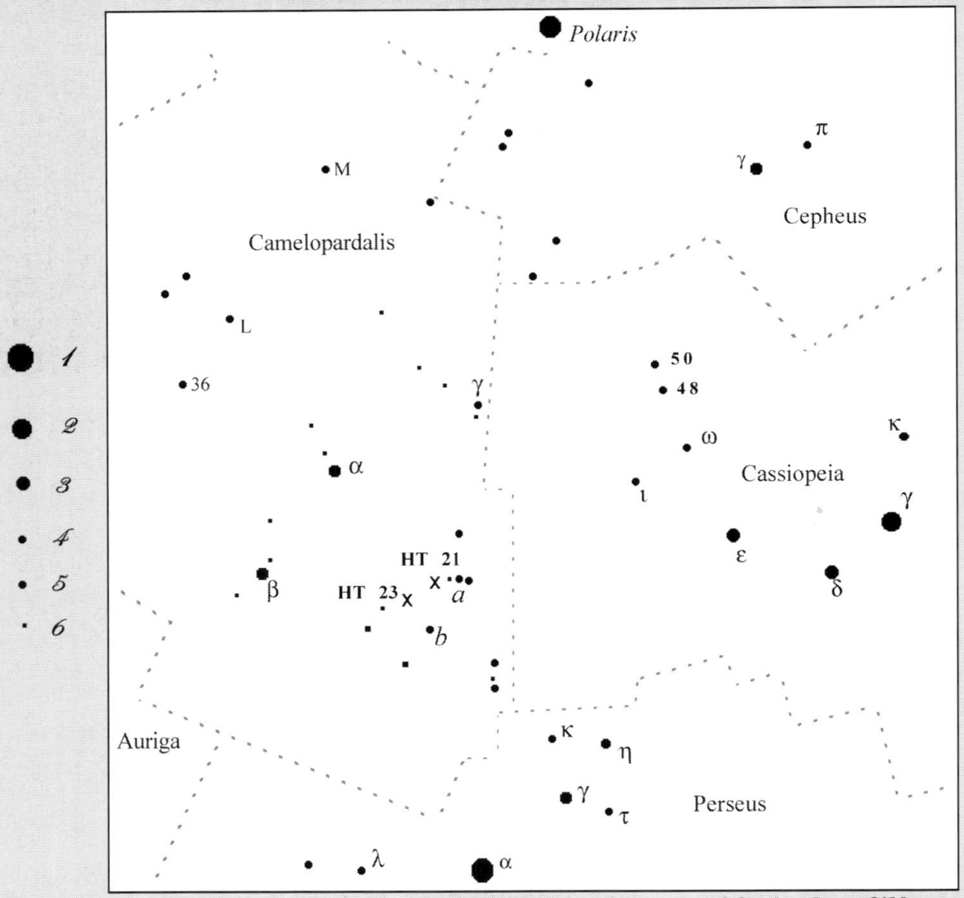

Tirion: Chart 1

Uranometria: Chart 18

21

Kemble's Cascade, Kidd's Creek
Kemble 1
Type: Asterism
Con: Camelopardalis

RA: 03^h 57.4^m
Dec: +63° 04′
Mag: –
Diam: 180.0′
Dist: –
Disc: Lucian J. Kemble, 1980

HERSCHEL: None.

NGC: None.

23

The Jolly Roger
NGC 1502
Type: Open Cluster
Con: Camelopardalis

RA: 04^h 07.8^m
Dec: +62° 20′
Mag: 6.0 (O'Meara); 5.7
Diam: 20′
Dist: 2,650 light-years
Disc: William Herschel, 1787

W. HERSCHEL: [Observed November 3, 1787] A cluster of stars, pretty rich and considerably compressed, a little elongated, 3′ or 4′ diameter, irregularly faint. (H VII-47)

NGC: Cluster, pretty rich, considerably compressed, irregular figure.

21 & 23

Whenever we venture into the night, our eyes unconsciously try to distinguish familiar shapes and forms in the nocturnal landscape. It's a primitive response that stems from the days when our distant ancestors needed to know friend from foe. It must have been the same when early man first looked into the night sky with wondering eyes. Here were myriad sparks in an unfathomable void. How their senses must have reeled. The stars were untouchable. Nor could they be heard or smelt. They could only be seen. They were more mysterious than any night creature whose watchful eyes reflected light just beyond the glow of early human's protective fires.

Unlike the animals that threatened attack, though, the stars remained fixed to the firmament. They moved together as one, like some distant migrating herd. The longer humans gazed into the peering eyes of night, the more comfortable and understanding they became with the stars. Fear waned.

Soon the brightest stars became familiar. Some were used in navigation. Others were organized into recognizable patterns – ones that reflected the simplicity of the times. Our early ancestors saw bears, dogs, and hunters; later, strong men and remarkable women. They placed personal treasures among the stars (a crown, a harp) and other objects they adored (a woman's hair). They removed from Earth the things they feared (a scorpion, a serpent, and a dragon). Fertile imaginations began to weave myths and legends that united these new patterns among the stars, to personalize them. Their myths reflected the struggles of their times and the morals they wished to preserve. The world of lights in the night sky became a safe record of their lives and beliefs. The sky was an open book for all to read. The stories could not be destroyed as long as oral tradition remained alive.

To this day we still look up into the night sky and see these age-old patterns, and we still enjoy the myths associated with them. But we have also penetrated the imaginary sphere of the naked-eye sky. Armed with binoculars and telescopes we have explored and mapped new vistas. And though we no longer fear the stars, we have yet to lose that time-honored desire to make order out of chaos.

We forever seek patterns in the sky, because that is how we get acquainted with the night. Once we apply binoculars or a telescope, the number of patterns one can create are limited only by the imagination. Still, only a few microconstellations have found their way into both the literature and our hearts. By far, one of the most popular micropatterns of the modern sky is a long chain of stars in Camelopardalis, which rides high in the north on cold February evenings.

The late Franciscan and amateur astronomer Lucien J. Kemble (1922–1999) discovered the asterism in 1980 with a pair of 7 × 35 binoculars. At the time, he was sweeping the skies above his home in Saskatchewan (Alberta, Canada), when he noticed "a beautiful cascade of faint stars tumbling from the northeast down to the open cluster NGC 1502 [(Hidden Treasure 23)]." Kemble immediately wrote a letter of discovery to *Sky & Telescope* columnist Walter Scott Houston, who, in turn, popularized the find in his December 1980 Deep-Sky Wonders column. To honor the discoverer, Houston later suggested the asterism be called Kemble's Cascade. He was "shocked" when he first saw it in binoculars because, he said, "there was a totally unexpected gem . . . a celestial waterfall of dozens of 9th- and 10th-magnitude stars. Down it went $2\frac{1}{2}^\circ$ before splashing into NGC 1502."

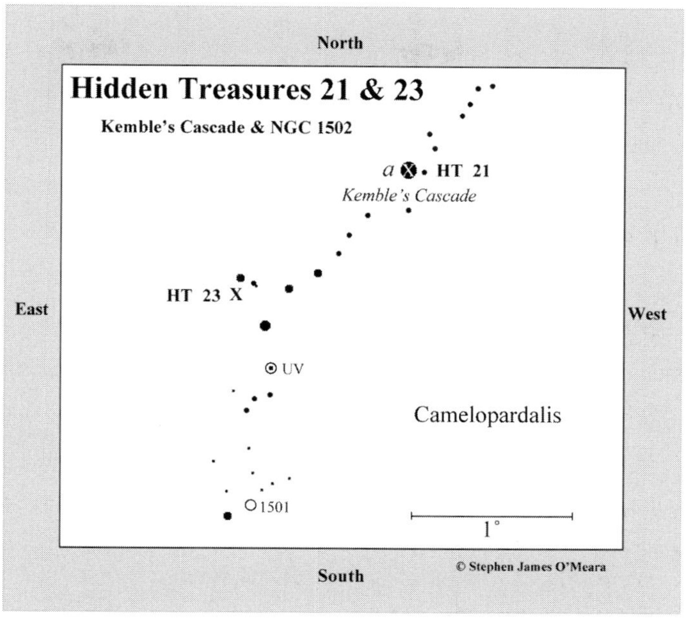

How did such a beautiful sight go unnoticed for so long? There are several reasons. First, Camelopardalis is one of the most obscure constellations in the night sky; while its borders cover a significant 757 square degrees of sky, Camelopardalis has no stars brighter than fourth magnitude. Second, for some reason, pallid regions of the naked-eye sky, especially those near the celestial poles, seem unattractive to backyard astronomers. And third, and perhaps most important, only three stars in the Cascade are bright enough to have been included in the most popular atlas of the day: *Norton's Star Atlas.*

While many sources say that Kemble's Cascade has 20 "members," that is highly subjective. In Kemble's original drawing, which appears in Houston's original article, twice that many stars can be counted. What you see depends on what aperture you use and what magnitude limit you impose. At a glance with binoculars at least 20 stars can be seen in the Cascade from a dark sky. These are the obvious members of the group. The brightest shines at magnitude 5.0, and it is your guiding light.

To find Kemble's Cascade, first locate 5th-magnitude Star *a.* It forms the eastern apex of a near-equilateral triangle with 4.5-magnitude Iota (ι) Cassiopeiae and 4th-magnitude Eta (η) Persei. Your fist held at arm's length (10°) will fill the triangle. Even if you cannot see Star *a* with your naked eyes, just imagine the equilateral triangle in your mind and sweep your binoculars across its location. There will be no mistaking the correct 5th-magnitude star, because Kemble's Cascade flows right by it – from the northwest to the southeast.

If you live under a bright suburban or city sky, Eta Per should not be difficult to see. It's about 8° north-northwest of Alpha (α) Persei, at the end of a straight line with Alpha and Gamma (γ) Persei; it's also the same star we use (along with Epsilon (ε) Cas) to find the famous Double Cluster in Perseus. Although Iota Cas is only slightly fainter than Eta Per, it is in a more isolated region of sky. The best way to find Iota is to imagine a straight line passing through Delta (δ) and Epsilon Cas – the easternmost stars in the famous W asterism – and continue it for another 5° to the northeast. Iota will be just north of that end point.

Star *a* is part of a binocular asterism that I call the Frigate Bird – a mini-constellation

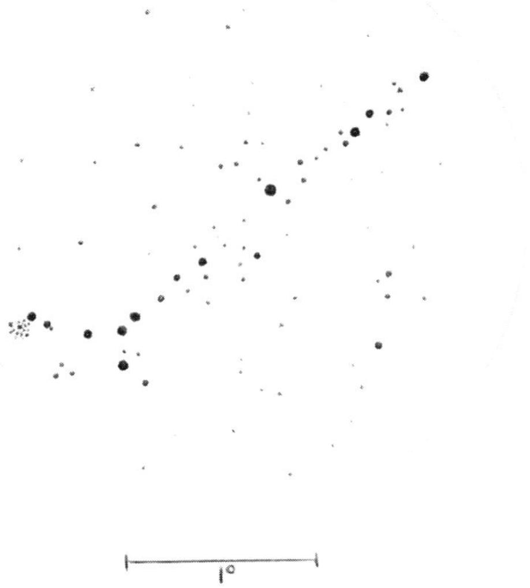

comprised of six 5th- and 6th-magnitude stars. Star *a* marks the bird's eye, while two 5th-magnitude suns – one about $2\frac{1}{2}°$ to the north-northwest and one about 2° to the south are the wings. Three 5th- to 6th-magnitude suns comprise the bird's tail, which extends about 1° west of Star *a*. The Cascade begins at an 8th-magnitude star 1° northwest of Star *a* and 25′ north-northeast of the easternmost star in the bird's tail. It flows southeast until it collides with a bright "boulder" (Star *a*), where it makes a slight jog to the west, as if taking the path of least resistance. It resumes its southeastward fall for another 1° before it widens into an inverted Y and splashes down onto open cluster NGC 1502 at the southeastern arm of the fork. While the main action appears to end here, a rivulet of suns follows the southwestern arm of the inverted Y and gathers in a faint pool of starlight near the 8th-magnitude semiregular variable star UV Cam just $\frac{1}{2}°$ further on. Few asterisms have been so aptly named as Kemble's Cascade, and fewer asterisms still have its added dimension of action.

While binoculars and rich-field telescopes give the best views of the Cascade, do explore the chain with higher powers because it includes many interesting pairings and a few doubles. When sweeping the entire chain at 72×, I noticed several matching pairs of stars aligned parallel in places like the rails of a train track. But the entire scene looks like a railroad track that's been badly twisted during an earthquake of substantial magnitude. The train itself can be seen "wrecked" at the southeast terminus – the crumpled pile of suns known as open cluster NGC 1502.

If you are under a dark sky, remove your eye from the telescope and look up at Star *a* in Kemble's Cascade. Now turn your attention to a point about $1\frac{1}{2}°$ to the southeast and concentrate on it with averted vision. Make peripheral eye sweeps of the area and breath through your nose slowly. Inhale and

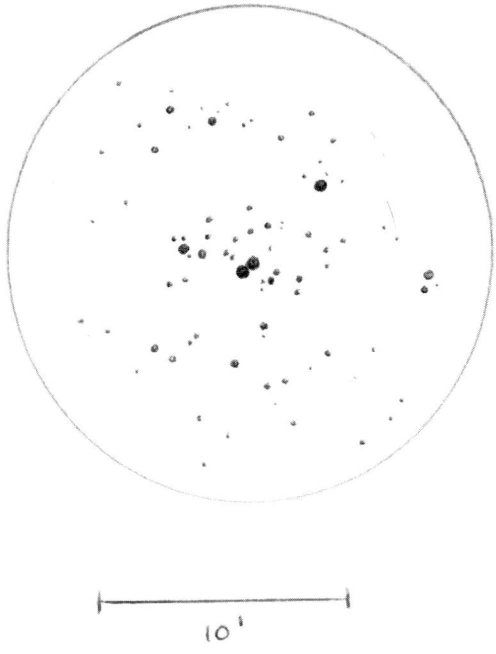

exhale to counts of 10 as you look. Do you see a dim concentration of light? If you do, you've just seen open cluster NGC 1502 with the unaided eye. Magnitude estimates of the cluster vary. In their book *Star Clusters,* Archinal and Hynes list it as magnitude 6.9. It's magnitude 5.7 in both the *Deep-Sky Field Guide* and Luginbuhl and Skiff's *Observing Handbook and Catalogue of Deep-Sky Objects.* My visual magnitude estimate of 6.0 was made with 7 × 50 binoculars and is a happy medium.

Although the cluster measures 20′ across, its core is very tight and bright. Four of the cluster's stars are immediately obvious in small binoculars. Through the 4-inch at 23×, a brilliant double star (ADS 2984AB) shines like a finely cut topaz near the cluster's center. Both components shine at magnitude 6.9 and are simply mesmerizing. Admiral William Henry Smyth called it a "neat double star," whose components are "decidedly red." But under higher powers, this double transforms into what Houston called a "celestial fireworks display involving nine components, seven of which are between 7th and 13th magnitude and are within reach of a good 4-inch telescope."

The nineteenth-century astronomer Wilhelm Struve (STF) discovered the main 7th-magnitude pair in 1830 when they were separated by 18″ in position angle 303°. They haven't changed position much: in 2002 they were only 17.7″ apart in position angle 305°. While both are spectroscopic binaries, the northern component (HD 25638) is the variable star SZ Camelopardalis, which dips 0.3 magnitudes every 2.7 days. It is the brightest star (by a few tenths of a magnitude) in the cluster when the variable is at maximum. The estimated mass of the eclipsing body is 18.6 Suns. It lies 0.07″ from its primary and has been resolved by speckle interferometry – a method of "beating the seeing" by taking hundreds of extremely fast exposures to freeze the blurring effect of the atmosphere, then combining the images. This tertiary body is also believed to be a close binary. Luginbuhl and Skiff note that the A component may also be variable. These highly luminous (spectral type *B*0) stars appear to be the only members of 5-million-year-young NGC 1502 not on the main sequence. In age, NGC 1502 is comparable to NGC 663 (Caldwell 10) in Cassiopeia, and NGC 1893 in Auriga, which recent estimates place at 9 and 4 million years young, respectively. NGC 1502 is dimmed by about 0.5 magnitude by intervening interstellar dust between 815 light-years and 2,800 light-years. This dust cloud keeps NGC 1502 from having a more prominent position in the naked-eye sky.

Sharing the center-stage spotlight is yet another double, actually a triple star, of equal fascination. Wilhelm Struve also found these stars in 1830. The brightest members shine, once again, at equal magnitude (10.0) and are separated by 22.5″ at position angle 334°. The more southerly of the pair has a magnitude 10.5 companion only 5.6″ away at position angle 132°. Together Struve 484 and 485 look like a miniature version of the famous Trapezium in M42, the Great Orion Nebula.

In the 4-inch more than a dozen suns, many in pairs, surround NGC 1502's tight core of double stars. Archinal and Hynes list, however, about 63 members, so those with larger telescopes should be prepared for an even more outstanding display. The brightest members form a lazy X shape centered on the bright double star Struve 485.

One lane of stars crosses Struve 485 roughly from west to east, while the other crosses it from the southwest – but the northeastern arm curves abruptly to the north and opens into a Y of stars. It takes little imagination to see these stars and this pattern as the skull-and-cross-bones symbol of the pirate's Jolly Roger. With that in mind, we can now return to low power and, borrowing from the pages of J. M. Barrie's play *Peter Pan,* see NGC 1502 as Captain Hook's ship, the *Jolly Roger,* moored in Kidd's Creek (Kemble's Cascade) under a thick fog. And what of the cluster's stars? They are Tinkerbell (Struve 485) and her luminous fairy friends dancing unnoticed above the clouds of Neverland.

What makes an object a hidden treasure is not so much what it means to the astrophysical community but what it means to you. The fact that Kemble's Cascade is not a true cluster does little to change its wide appeal. The fact that NGC 1502 is a true cluster does little to stifle the imagination. We look at celestial objects because they entertain us. Who would argue that Kemble's Cascade is not a near-perfect alignment of stars? Who would argue that NGC 1502, with its magnificent doubles, does not have the power to grab our attention and not let go? No, even on the coldest of winter nights, the opportunity to see a little line of stars splashing down onto a galactic cluster spanning more than 15 light-years across space gives us the peace and serenity we seek each time we step outside with our telescopes, even if it only means we go off to bed with sigh and a smile.

Hidden Treasure 22

NGC 1501

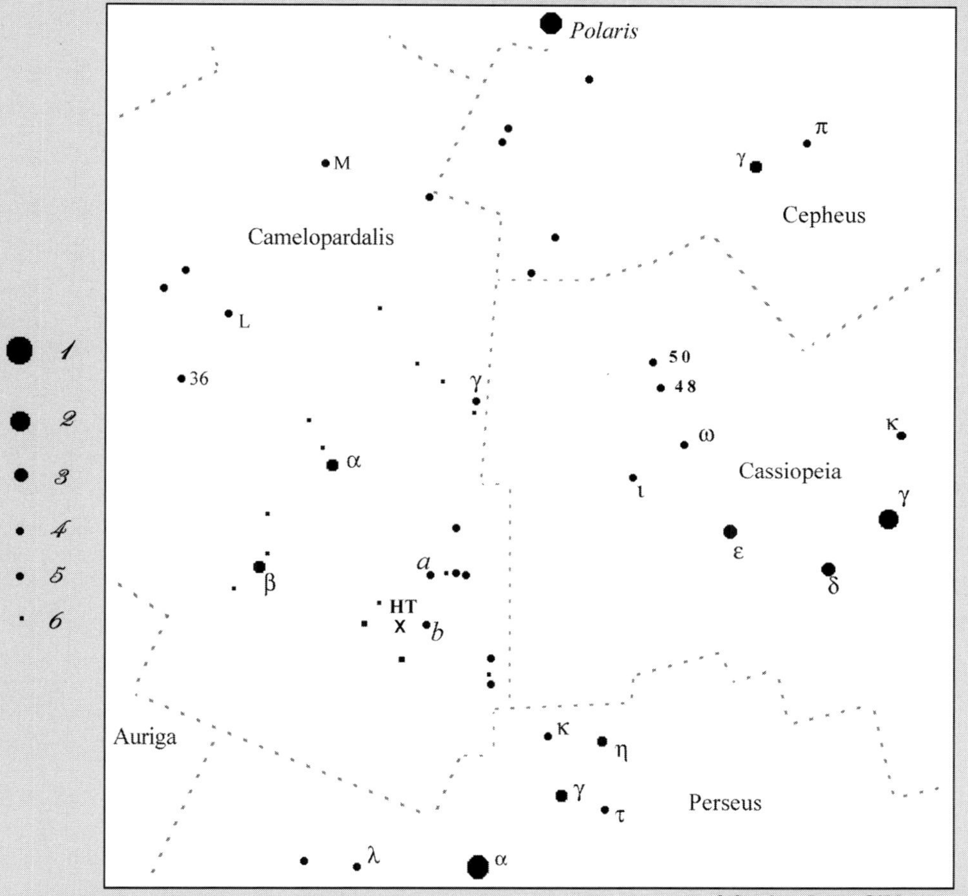

Tirion: Chart 1

Uranometria: Chart 18

22

Oyster Nebula*, *Blue Oyster Nebula
NGC 1501
Type: Planetary Nebula
Con: Camelopardalis

RA: $04^h\ 06^m\ 59.4^s$
Dec: $+60°\ 55'\ 14''$
Mag: 10.6 (O'Meara); 11.5
Dim: $56'' \times 48''$
Dist: ~5,000 light-years
Disc: William Herschel, 1787

W. HERSCHEL: [Observed November 3, 1787] A pretty bright planetary nebula near 1′ in diameter, round of uniform light and pretty well defined. With 360 magnified in proportion, but still the borders are pretty abruptly defined and a little elliptical. (H IV-53)

NGC: Planetary nebula, pretty bright, pretty small, very little extended, 1′ in diameter.

FOR SUCH AN OBSCURE CONSTELLATION, Camelopardalis certainly has its wealth of celestial treasures: the galaxies IC 342 (Caldwell 5), NGC 2403 (Caldwell 7), and NGC 2655 (Hidden Treasure 48); Kemble's Cascade (Hidden Treasure 21); open cluster NGC 1502 (Hidden Treasure 23); and the bright, though woefully neglected, planetary nebulae IC 3568 (Hidden Treasure 64) and NGC 1501.

Popularly known as the Oyster Nebula, NGC 1501 has a peculiar texture. In images taken with large telescopes, NGC 1501 displays a conspicuous magnitude 14.5 central star surrounded by a highly dimpled shell of pale blue gas. Tug the imagination, and you can see that star as a perfect pearl inside an oyster's rough and irregular shell. NGC 1501 is 5,000 light-years distant, and its expanding shell of gas measures 1.4 light-years wide in true physical extent. By comparison, it's $4\frac{1}{2}$ times farther away than the Ring Nebula (M57) in Lyra and $3\frac{1}{2}$ times larger. In distance and size, NGC 1501 is more comparable to NGC 3195 (Caldwell 109) in Chamaeleon.

Judging by the size of its shell, NGC 1501 appears moderately young. Consider a smoke ring produced by someone puffing on a cigarette. The smoke ring is densest

Observations of the shell's spectrum show that NGC 1501 is a thin ellipsoid with a pair of large lobes and a multitude of smaller bumps spread across the nebula's surface (the rough and irregular texture of the oyster shell). In a 2001 *Astronomy and Astrophysics Abstract*, R. Ragazzoni (Astronomical Observatory of Padova) and colleagues revealed NGC 1501's three-dimensional structure to be a boiling complex of bubbles. The main nebula consists of an almost oblate ellipsoid with a bright equatorial belt. The shell's surface is deformed by several bumps reminiscent of boiling convection cells. The complexity of these expanding shells is a result of an interaction between the intense and fast wind from the extremely hot central star that contains about 55 percent of our Sun's mass.

and smallest when first formed close to the lips. As time passes, the smoke ring expands and becomes less dense. Ultimately the ring becomes so tenuous that it can no longer be seen. In the natural world, volcanoes also produce smoke rings that behave in similar fashion; the photograph here shows one being expelled by Italy's Stromboli Volcano in the South Tyrrhenian Sea. Likewise, small and bright planetary nebulae are younger than their larger and fainter counterparts.

In his book *Cosmic Butterflies: The Colorful Mysteries of Planetary Nebulae,* Sun Kwok explains the simplest way to estimate a planetary's dynamical age; it is a model that has been used for a century or so: "If planetary nebulae are expanding at a constant rate, then one can calculate the age of the nebula by dividing its size by the expansion rate." The largest planetary nebula known, has a diameter of 4 light-years. Given that a typical expansion velocity of a planetary's shell is 20 kilometers per second, the life span of a planetary nebula works out to be 30,000 years. The 1.4-light-year-wide shell surrounding NGC 1501's central star, then, is nearing middle age.

As with the central star in NGC 1360 (Hidden Treasure 16), NGC 1501's central star is a Wolf-Rayet star – a very luminous, very hot star whose outer shell was expelled during its red-giant phase. There are two types of Wolf-Rayet stars: Type WN, whose spectrum shows prominent emission lines of helium and hydrogen, and type WC, whose spectrum is hydrogen deficient and dominated by carbon, oxygen, and helium lines. NGC 1501's central star is a WC-type Wolf-Rayet star. It is also a variable, showing erratic brightness variations as large as 0.1 magnitude but with 10 independent pulsation periods ranging from 19 to 87 minutes. These periods change dramatically over month to year time scales. These period changes are related to the way vast streams

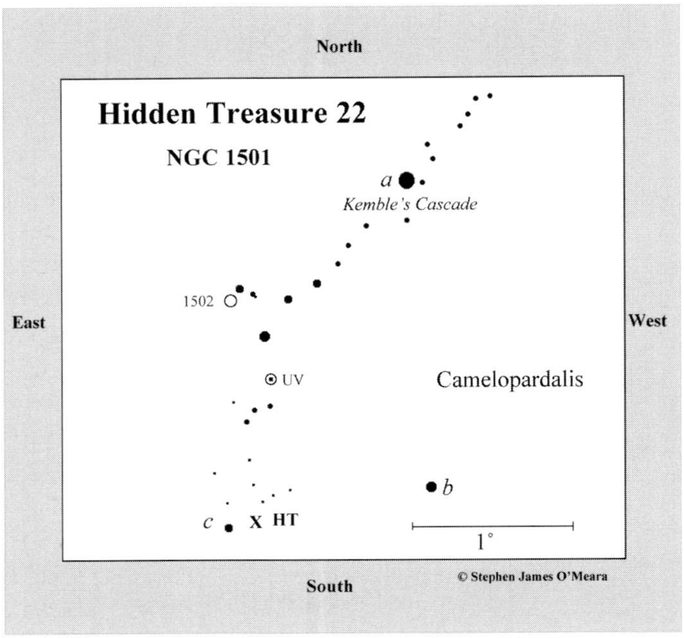

Heinrich d'Arrest thought it conspicuous in a $4\frac{1}{2}$-inch refractor. Using his massive 72-inch reflector at Birr Castle in Ireland, Lord Rosse (William Parsons, Third Earl of Rosse) spied the central star.

To find this peculiar object, locate 5th-magnitude Star *a* in Kemble's Cascade. It forms the eastern apex of a near-equilateral triangle with 4.5-magnitude Iota (ι) Cassiopeiae and 4th-magnitude Eta (η) Persei. Star *a* is part of a binocular asterism that I call the Frigate Bird – a mini-constellation comprised of six 5th- and 6th-magnitude stars. Star *a* marks the bird's eye, while three 5th- to 6th-magnitude suns comprise the bird's tail about 1° west of Star *a*. Two 5th-magnitude suns – one about $2\frac{1}{2}°$ to the north-northwest and one (labeled *b*) about 2° to the south are the wings. You want to train your telescope on Star *b*. NGC 1501 is about $1\frac{1}{4}°$ east-southeast of that star, just 10′ northwest of a magnitude 7.5 star. You can also use the detailed finder chart to star hop to it from Kemble's Cascade.

Under a very dark sky, NGC 1501 can be seen with 7 × 50 binoculars if you know exactly where to look and brace the binoculars well. It is a conspicuous 11th-magnitude "star" in my antique scope. At 23× in the 4-inch, the planetary is in a rich field and, even with direct vision, looks like an 11th-magnitude star. So if you are just sweeping the area at low power, it's doubtful that you'd take notice of it. But if you know where it is, even a short or brief perusal of the area

of hot gas flow beneath the star's surface. And they are evidence that the hot central star is evolving rapidly.

One of the reasons NGC 1501 may have been neglected is that some sources list its photographic magnitude (13.3) rather than its visual magnitude. And I find its listed visual magnitude (11.5) to be one magnitude too faint. The object is actually quite bright and easy to see in small apertures, being slightly fainter and a little smaller than M76 in Perseus. The difference, of course, is that we see M76 as dumbbell shaped and NGC 1501 as an ellipse.

In his *Cycle of Celestial Objects,* Adm. William Henry Smyth called it "A bright planetary nebula, of a bluish white tint . . . It is a curious body; and was watched under the total lunar eclipse of the 13th of October, 1837, being well seen during that shadowy obscurity which an Italian would call *un tenebroso orrore.*" Interestingly, the Rev. T. W. Webb calls it "small and dim." While

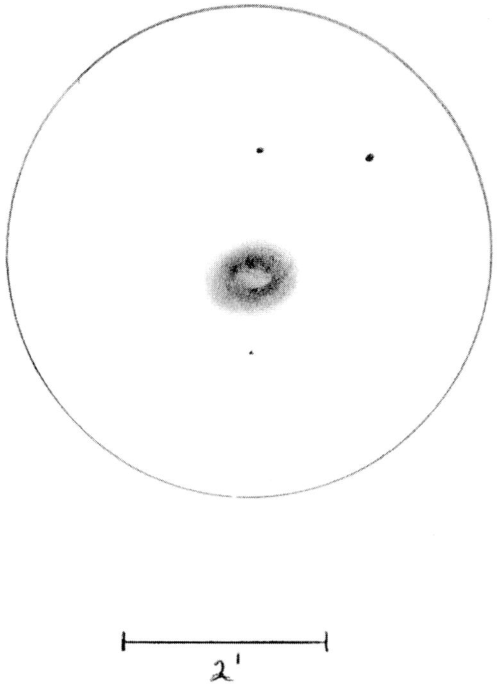

with peripheral vision will reveal that it's not stellar. It's a tad more tenuous. And with time, averted vision should show its disk well, but not dramatically well.

With 101×, the planetary immediately reveals itself as a glorious orb glowing densely even with direct vision. Fainter stars can be seen around the nebula. With averted vision a faint annulus can be seen quite readily and it looks like the Ring Nebula at low power. There also appear to be some enhancements in the ring that pop into view with averted vision, though it's hard to pinpoint the exact location of these lights. This phenomenon is quite common to planetaries. Also, although the planetary has a high degree of concentration and surface brightness at low power, at high power I find this apparent "denseness" fades dramatically. In small apertures, the planetary should look best at 227×, though you should find your own comfort zone. I thought 302× was a bit much, but that might also have been due to eye strain after 3 hours of intense observing, so do try the highest power possible that makes you feel comfortable.

At high power the ellipse is quite evident, as is the "hole" with averted vision. The north and south sides of the inner ellipse are brighter than its outer extensions and the northern arc appears to be comprised of two bright beads. Try as I might, I could not see the central star, which has been reported by several amateurs to be visible in an 8-inch telescope, even if the ring itself is blown to near invisibility. Observing from the dark skies of Mount Pinos in California, Kent Wallace found NGC 1501 to be a moderate-sized circular disk in his 8-inch f/10 reflector at 65×. He too found the disk mottled at 200×. He also notes that the nebula responds well to O-III and UHC filters. Alas, he too could not see the central star.

If you're looking for a true treasure hunt, then, your challenge is to see NGC 1501's elusive 14.5-magnitude pearl. The key to success, I suspect, is a combination of aperture and magnification. Use as high a power as the evening will take to open up the lips of the blue oyster and see if you can't visually grab your prize.

Hidden Treasure 24

NGC 1535

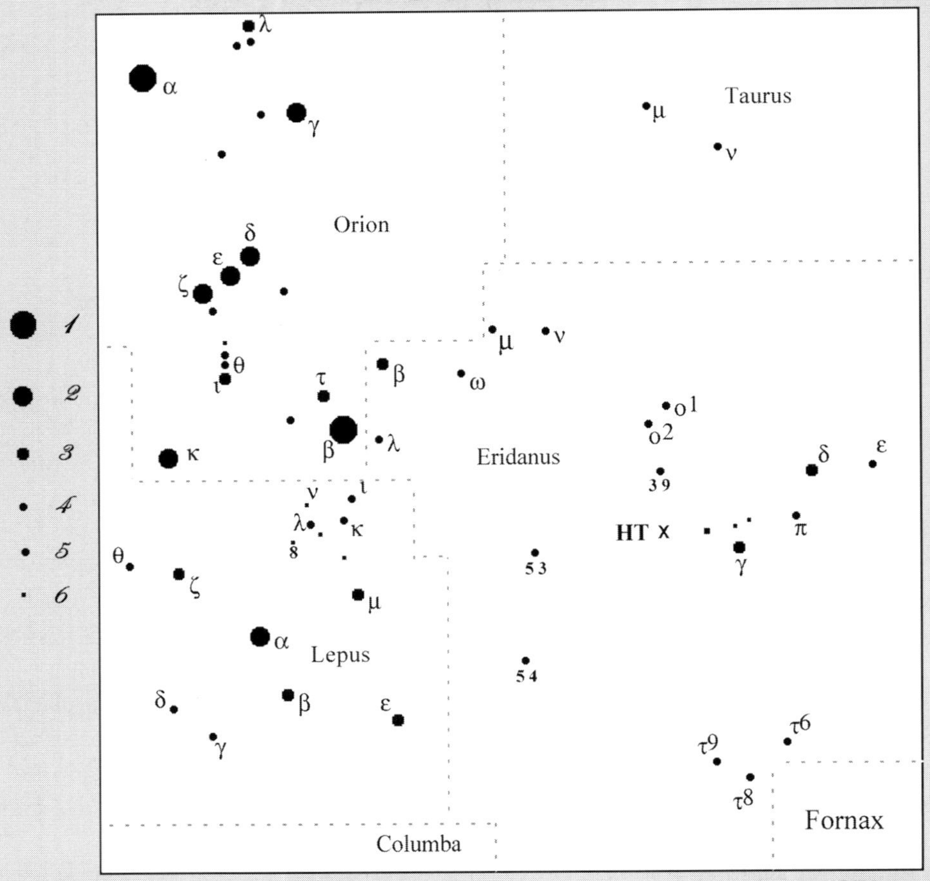

Tirion: Chart 11

Uranometria: Chart 268

24

Cleopatra's Eye, Celestial Jellyfish, Ghost of Neptune Nebula
NGC 1535
Type: Planetary Nebula
Con: Eridanus

RA: $04^h\ 14^m\ 15.8^s$
Dec: $-12°\ 44'\ 22''$
Mag: 9.1 (O'Meara); 9.6
Dim: $48'' \times 42''$
Dist: ~5,200 light-years
Disc: William Herschel, 1785

W. HERSCHEL: [Observed February 1, 1785] Very bright, perfectly round or a very little elliptical planetary but ill-defined disk. (H IV-26)

NGC: Planetary nebula, very bright, small, round, pretty suddenly, then very suddenly brighter in the middle, [mottled].

PLANETARY NEBULA NGC 1535 IN Eridanus is one of the sky's brightest and most satisfying objects of its class. In color photographs it shines like an aquamarine gem locked in ice. Others have likened it to an eye. Indeed, its piercing glow has earned it the nickname Cleopatra's Eye, in deference to the last great queen of Egypt, whose seductive gaze was enhanced by her use of blue eye shadow. The name Cleopatra's Eye, in turn, was inspired by the Eye of Horus, a powerful symbol used in ancient Egyptian funerary rites and decoration to protect one from evil.

If NGC 1535 is hypnotic, it's because its physical structure draws you in. This multiple shelled planetary nebula consists of a bright inner annulus with a much fainter, though uniformly bright, crown. In images taken with large telescopes the inner annulus displays a classically sharp, though irregularly round, outer rim, while its inner edge is greatly serrated. Seen together, these features look characteristically like the iris of a human eye. A magnitude 11.6 central star burns at the center of the shell. And it is this bright spark of light that draws our attention, because it seems to mirror the light of the viewer's soul.

NGC 1535 is a moderately high-excitation planetary nebula with features very similar to the Eskimo Nebula (Caldwell 39). Its shells appear slightly elliptical, with the major axis of the inner ring oriented approximately at position angle 35°. From our vantage point on Earth, however, we are looking straight down the major axis of an inverted funnel of gas. Recent models of NGC 1535's structure describe a bipolar inner shell, which is being constricted by a ring of dense material in the equatorial plane of the central star. We see one lobe expanding toward us at a rate of 20 kilometers per second; the largely spherical outer shell is expanding at about half that rate. Ultraviolet observations, particularly those taken with the Hopkins Ultraviolet Telescope aboard the *Astro-1* Space Shuttle mission in December 1990, confirm the presence of a fast wind streaming out from the planetary's 0.67 solar-mass central star, whose surface temperature is near 70,000 K. The measured speed of this wind is fantastic – about 2,000 kilometers per second. But such speeds are common to these objects. In fact, they are theoretically mandatory to produce the observed density and expansion velocities of planetary nebulae.

That NGC 1535 is not a more popular object among amateurs is amazing. William Lassell described it as the most interesting and extraordinary object of the kind he had ever seen. And Christian Luginbuhl and Brian Skiff rate it as "one of the best planetaries for amateur viewing." While NGC 1535 is in the southern constellation Eridanus, its declination of nearly -13° places it at essentially the same height above your horizon as M72 and M73 in Aquarius. And it shines roughly at the same magnitude as the famous Ring Nebula (M57) in Lyra. But now consider that NGC 1535 is nearly five times more distant than M57, thus a physically larger system; it measures 1.2 light-years in true extent, while the Ring Nebula spans only 0.4-light-years across.

One reason NGC 1535 is neglected is its seemingly dreary surroundings. It lies in a field devoid of bright stars. To find this "celestial jellyfish," as the late *Sky & Telescope* columnist Walter Scott Houston called it, start with brilliant Beta (β) Orionis (Rigel). Nearly 20° to the west-southwest is the obvious magnitude 3.0 star Gamma (γ) Eridanus (Zaurak). A fist held at arm's length will cover about 10° of sky, so you want to look for the Gamma Eri "two fists" west-southwest of Rigel.

Zaurak has an interesting history, in that its name is derived from the Arabic *Al Nair al Zaurak,* meaning "the Bright Sar of the Boat." But this name wandered there from the constellation Phoenix, whose Alpha (α), Kappa (κ), Mu (μ), Beta (β), Nu (ν), and Gamma (γ) stars form an elegant curving line like that of a primitive boat. And though I have not seen it mentioned elsewhere, Gamma Eri is also part of a similar curving line consisting of Gamma, Pi (π), Epsilon (ε), and Delta (δ) Eri; and the bright star in this more northerly boat is indeed Gamma.

There is no mistaking Gamma Eri, because it is the brightest star in the region. Train your binoculars on it, and you will see another elegant curve of three 6th-magnitude stars just north of it. If you center the easternmost star (*a*) in that curve in your telescope, you will be within striking distance of NGC 1535, which is almost $2\frac{1}{2}^\circ$ due east. Start by moving 45′ east-northeast to a solitary magnitude 8.4 star (*b*). Then hop 35′ east-northeast to a magnitude 7.7 star (*c*). A greater 55′ hop to the north-east will bring you to a magnitude 7.9 star (*d*). You'll find the 9th-magnitude planetary just $\frac{1}{2}^\circ$ east-southeast of that sun. If you observe under a dark sky, the nebula should be visible as a dim star in 7 × 50 binoculars. It is a cinch in my antique scope.

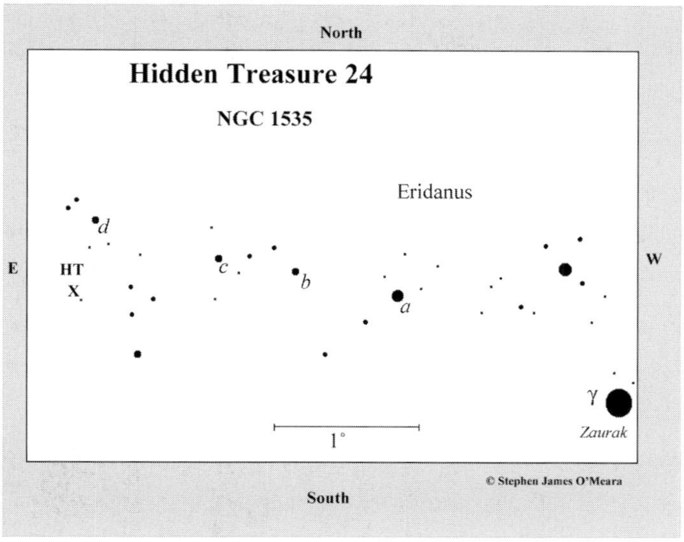

At 23× in the 4-inch, NGC 1535 is a tiny blue-gray disk, which is so nearly stellar that I would have passed over it in a sweep. With averted vision, it looks slightly fuzzy. The nebula's round appearance and pale blue color at 72× are reminiscent of Neptune's disk seen through backyard telescopes, which is why I call it the Ghost of Neptune Nebula.

At high power the structure in the annulus is complex. The problem with eking out details, though, appears to be one of separation. In other words, it's difficult to separate the bright central star from the equally bright inner annulus, which extends to about 20″ in diameter. Do not be afraid to use very high power, even though the nineteenth-century English observer Thomas W. Webb said that with a telescope of $3\frac{7}{10}$ aperture, the nebula was "bright and round, with low powers . . . but not bearing magnifying."

The reason that many amateurs have failed to see the planetary's central star in telescopes 6-inches and smaller is because they failed to separate the central star from the inner annulus. The key then to seeing the central star is to use *very* high power. I also recommend using your *foveal* vision at high power, which will help you to pick out the central star while suppressing the surrounding nebulosity. How high a power should you use? On one night of marginal seeing, I used up to 504× – but that was just to study the central star and the inner annulus; the outer envelope, or crown, being tenuous and of uniform brightness, is best observed at low power. Still, on that marginal night I saw the central star and at least four discrete knots asymmetrically positioned along the inner ring. The problem was that, due to the variable seeing, the details sailed in and out of view.

On another extremely clear and transparent morning with much steadier seeing, I cleanly separated the central star from its inner ring, which was dappled with knots. At low power, the outer shell, or crown, was sharper and more circular than that of any planetary I've seen. In photographs, a faint star appears in the northern section of the nebula's crown, but I could not detect it. If it is there, it blends in well with the nebulosity.

Seeing a dappled inner ring probably inspired many of our astronomical heroes of yesteryear to see this object as more of a star cluster than as a nebula. For instance, in his second observation of the object, William Herschel said that the object is "round on the borders and is probably a very compressed cluster of stars at an immense

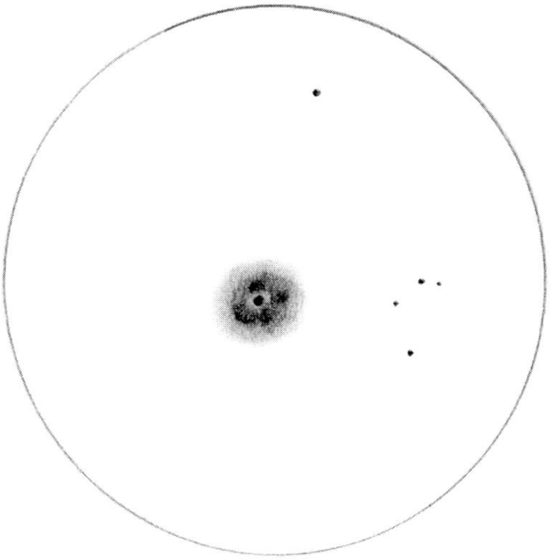

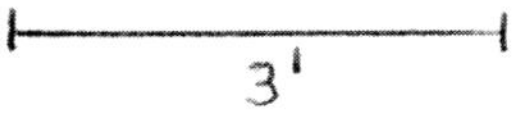

distance." Heinrich d'Arrest could not see the nebula's crown, but he did believe he could resolve the nebula's edges; he also saw the nebula shine with a "light blue" color. And Lord Rosse, peering through his magnificent 72-inch Leviathan at Parsonstown, said he could make out the nucleus as "granular."

How fantastic a time it must have been to be looking through such a behemoth, peering deep into the heavens and trying to fathom the riddles of the universe. How Herschel's theories of the origin and nature of the nebulae must have influenced the wide-open minds of his contemporaries and those immediately following. How limiting it proves the eye and mind to be when it comes to interpreting the cosmos.

I failed to see any *intense* blue coloration; overall the nebula was a pale blue-gray, if not mostly gray. Perhaps the color will show well only in larger instruments. Observing from the Southern Hemisphere, Ernst J. Hartung described the nebula as "bright pale blue" in a 12-inch reflector. Pennsylvania amateur James Mullaney describes the color as "pale blue-green." What color(s) do you see?

Hidden Treasures 25 & 26

NGC 1528 & 1545

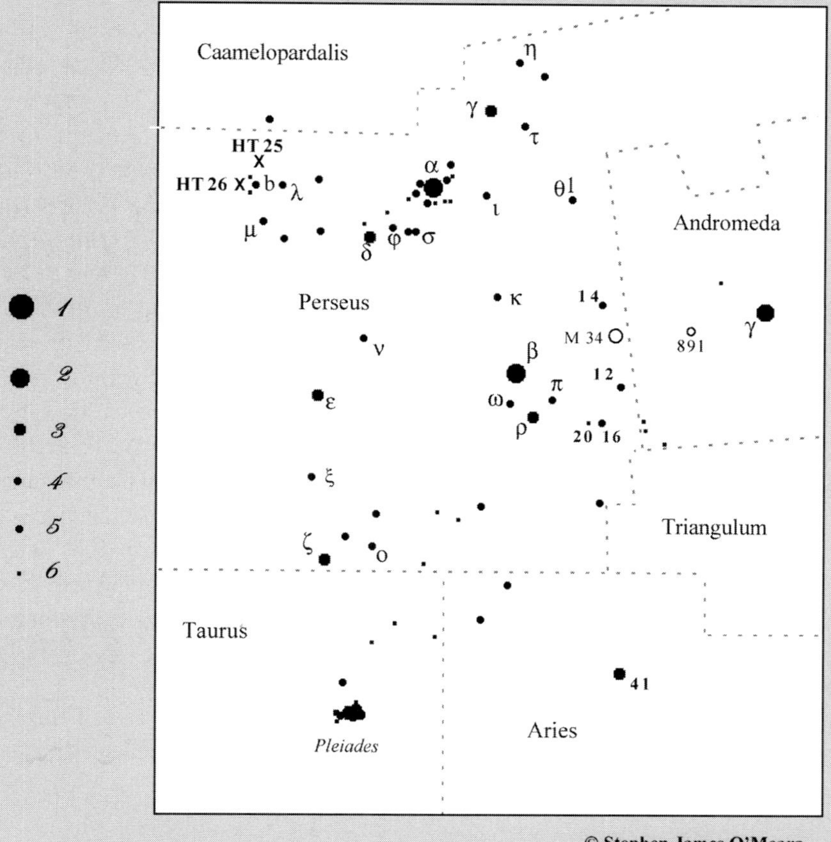

Tirion: Charts 1 & 4

Uranometria: Chart 39

25

m & m Double Cluster
NGC 1528
Type: Open Cluster
Con: Perseus

RA: $04^h\ 15.3^m$
Dec: $+51°\ 13'$
Mag: 6.2 (O'Meara); 6.4
Diam: 18.0′
Dist: 2,430 light-years
Disc: William Herschel, 1790

W. HERSCHEL: [Observed December 28, 1790] A beautiful cluster of large stars, very rich and considerably compressed. About 15′ in diameter. (H VII-61)

NGC: Cluster, bright, very rich in stars, considerably compressed.

26

m & m Double Cluster, Running Man Cluster, Magic Pentagram Cluster
NGC 1545
Type: Open Cluster
Con: Perseus

RA: $04^h\ 20.9^m$
Dec: $+50°\ 15'$
Mag: 6.6 (O'Meara); 6.2
Diam: 12.0′
Dist: 2,470 light-years
Disc: William Herschel, 1790

W. HERSCHEL: [Observed December 28, 1790] A coarsely scattered cluster of [bright] stars, pretty rich. (H VIII-85)

NGC: Cluster, pretty rich in stars, little compressed, stars [bright].

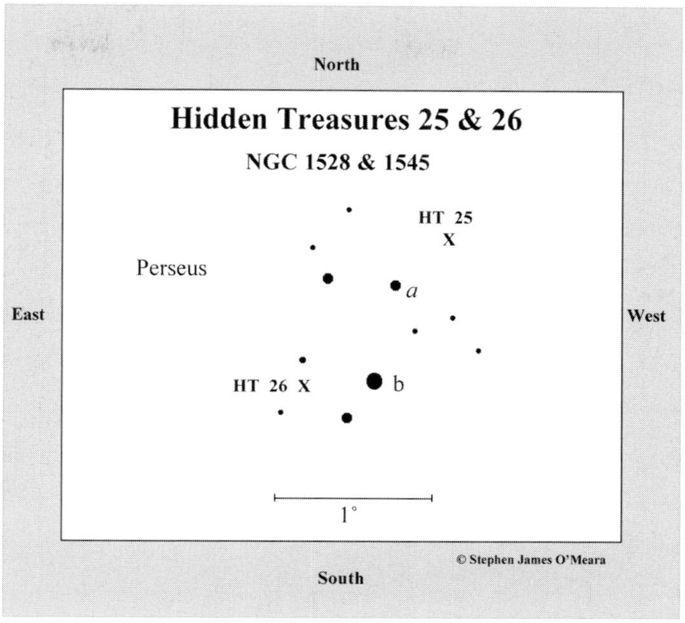

DURING THE COOL WINTER nights of the Northern Hemisphere, our galaxy's outer arms arc gracefully over the North Celestial Pole like a garden trestle covered in snow. Night after night, this rich stellar region sails up and over the pole until, by February, the Perseus Arm of our spiral galaxy dominates the view after sunset. And it is here that many amateurs will train their telescopes and discover some of the sky's most prominent star clusters and stellar associations, including the famous Double Cluster (Caldwell 14), M34, and the Alpha Persei Moving Cluster (Hidden Treasure 14). But there is a rather distinctive cluster, NGC 1528, which is casually overlooked in favor of its more popular neighbors.

William Herschel, the object's discoverer, classified it as one of his "Pretty Much Compressed Clusters of [bright] or [faint] Stars." The Rev. Thomas W. Webb called it a "[b]right" and a "good low-power object." And the Third Earl of Rosse observed two red stars in the cluster through the great Leviathan telescope at Birr Castle in Ireland.

How observers prior to William Herschel had missed this jewel of the night borders on the mysterious. You'll find NGC 1528 1° north-northwest of 5th-magnitude b Persei in the Champion's knee, and just 30′ northwest of a magnitude 7.5 star (*a*). NGC 1528 shines at magnitude 6.2 and makes a decent naked-eye challenge for anyone under dark skies. Only on one night, though, did I suspect seeing its core of brightly scattered suns. It is a fine object for binoculars, and its brightest members will start to appear to the dedicated observer even in the smallest pair.

In 7 × 50 binoculars the cluster is round and well formed. Its brightest stars appear unmistakably in my antique telescope against the cluster's unresolved background, which shines with a cometlike luster. With concentration, stars clump together to form a jumble of mismatched suns in a milky haze – like harbor lights seen from a ship in a fog. Yet few guide books pay much attention to this hidden treasure. In *Celestial Harvest*, however, James Mullaney sings praises to the cluster, sharing these comments by various observers: "Gorgeous . . . glows like a gem," and "Radiant cluster – entire zone truly inspiring." Another observer suggests using low power to appreciate this cluster fully. But while NGC 1528 is aesthetically appealing at low

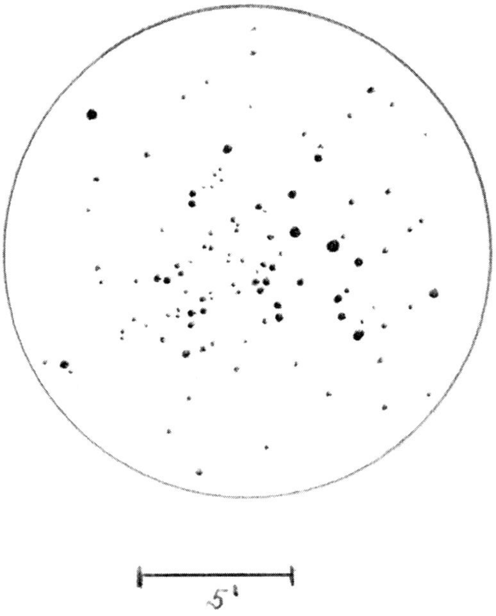

powers, high power should not be neglected, which I will soon explain.

The cluster is indeed large and bright at 23× in the 4-inch. It reveals a veil of well-resolved suns scintillating in front of a fuzzy backdrop of fainter stars. The brightest ones are curved in rows, like waves. Brent Archinal and Steven Hynes catalog some 165 stars in NGC 1528, with the brightest shining at magnitude 8.7, splashed across 13 light-years of space. So NGC 1528 has the same linear extent as M34 in Perseus. NGC 1528, however, is about 1,000 light-years more distant than that more famous cluster. And, with an estimated age of nearly 370 million years, NGC 1528 is nearly three times older than M34; it's closer in age to M39 in Cygnus. But these are modestly young ages. The oldest known open cluster in our galaxy is NGC 6791 in Lyra, which is some 7 to 8 *billion* years old.

If you relax your gaze at low power, NGC 1528's brightest stars form a teardrop-shaped asterism that is elongated along a northwest–southeast axis; its narrow tip points to the southeast. In *Star Hopping for Backyard Astronomers,* Alan M. MacRobert notes that through a 6-inch reflector under suburban skies, a dark U-shaped bay can be seen at the cluster's western end, where it is outlined by seven bright stars. That U can also be extended a bit further to the southwest, making the asterism look like the Sickle of Leo. Now relax your gaze again as you slightly defocus the telescope; do you see how the U forms the brighter half of a broad and lazy "m"-shaped asterism, the southeasternmost star of which is a fine double star. Return to sharp focus and scan the quadrant of sky immediately to the northeast of the "m." Here you will find a coarse scattering of dimmer suns, which, if you concentrate, form a near mirror-image of the "m." That's one reason why I call this cluster, the "m & m" cluster. But there's another reason.

The Russian astronomer, A. V. Loktin (Ural University Astronomical Observatory), has cataloged NGC 1528 and NGC 1545, which is less than $1\frac{1}{2}°$ to the southeast, as a possible binary cluster – one that probably formed through tidal capture. The two clusters share similar galactic coordinates, ages, distances, and mean radial velocities. If they are a true pair, then Perseus has more than one Double Cluster: NGC 869 and NGC 884 (Caldwell 14), and NGC 1528 and NGC 1545, what I call the "m & m" Double Cluster. Paradoxically, by the way, double clusters seem to be quite rare in the Milky Way, and yet they are quite common in the Large Magellanic Cloud.

But don't move on to NGC 1545 just yet; there's more to see in NGC 1528. If you slightly defocus the telescope again and sweep your vision across NGC 1528's face, you may notice that MacRobert's dark bay is but one of many wisps of darkness that meanders through the region. Under dark skies, the cluster appears to be composed of shards of starlight separated by dark cracks, as if a crystal teardrop had accidentally shattered on a sheet of black ice. Under high power and sharp focus, NGC 1528 is comprised of a healthy dose of mini-asterisms, all of which look hazy at low power but resolve into distinct geometrical patters under 101×. A beautiful Y-shaped mini-asterism (like M72 in Aquarius) can be found a few arcminutes northeast of the northeasternmost star in the U (near the center of the cluster). With high powers, many of the stars form lovely pairs, triangles, arcs, and squiggly lines. The visual fabric of this cluster is extremely patchy.

Finally, if you use your imagination to see this cluster as an open bag of "m & m's"; you can go hunting for colored ones – specifically, four intensely red carbon stars have been found in the cluster, which might, or might not, be members. Try as I might, when I visually sifted through my "bag," I did not notice any pronounced color differences. Seeing these intensely red carbon stars might require larger apertures than a 4-inch. Regardless, looking for fine details in NGC 1528 is like using a geologist's loop to look for the fine and varied crystalline structures locked in a shard of granite. So look for your own patterns, use your imagination, and set out on your own treasure hunt.

NGC 1545 is a most fascinating object for small telescope users. First, it's a difficult object to see with the naked eye for two reasons: at 6th magnitude, the cluster borders on the limit of naked-eye visibility for an extended object; and it lies less than $\frac{1}{2}°$ east-northeast of a pair of 5th-magnitude stars,

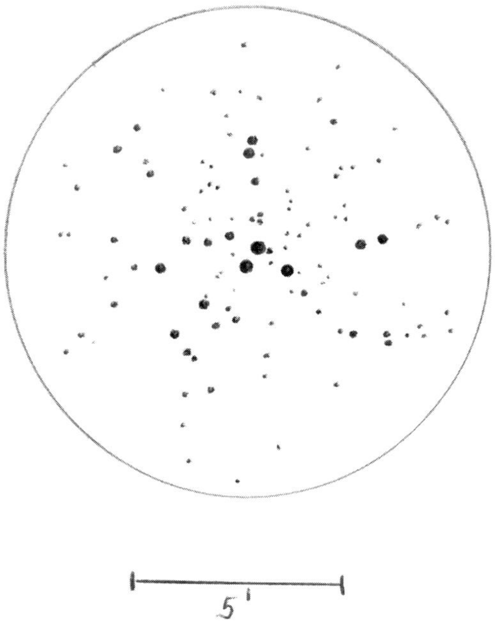

the brightest of which is b Persei. Under dark skies, all I see is a fuzzy wedge of light, which is only the combined glow of the two 5th-magnitude suns. So don't be fooled.

In fact, in 7 × 50 binoculars, the idea of a cluster in the listed position seems ludicrous. All that is visible are two stars (magnitude 7.7 and 7.1) oriented north–south, respectively. If you look very hard with averted vision, you might also detect a magnitude 8.1 star immediately north-northwest of the brighter (southern) star in the pair. So, at first blush, it appears that NGC 1545 is indeed like M40. But it isn't. In my antique telescope, for instance, I can see some dimmer suns swimming around these three suns and, at 23× in the 4-inch, a clear pattern of about a dozen suns emerges. The close southern pair of stars is actually a tight triangle of suns, which lies at the center of a five-pointed-star-shaped pattern like a pentagram, the ancient Gnostic "blazing star," which was a symbol relating to the magic and mystery of the night-time sky. The Pentagram in NGC 1545 is comprised of about a dozen 10th- to 11th-magnitude suns, many of which are in pairs. These stars are only the first tier of bright stars that can be seen in small telescopes with direct vision. It's amazing how many stars suddenly appear with averted vision. The Pentagram appears to be surrounded by a ghostly mist that wafts in and out of view with averted and direct vision, respectively.

With 72× and 101×, nearly three dozen additional dim suns can be counted inside an imaginary pentacle surrounding the Pentagram. By alternating between averted and direct gazes, the cluster's crisp background stars seem to appear and disappear like lights blinking on a Christmas tree. I find the affect most pronounced when I avert my gaze away from the cluster toward the 2 : 00 position angle and then sweep in an arc between the 2 : 00 and 4 : 00 position angles. Seen another way, the stars look like the stick figure of a man running.

While trying to detect faint stars can become an obsession, don't forget to scrutinize the cluster's brighter suns. Most striking are the color differences in an acute triangle at the cluster's core. The brightest star shines with a lemon-orange light, while the companion to the northeast is straw-colored; it is also an easily resolved double. And the sun marking the west-southwest corner of the triangle has an icy blue hue. The bright star marking the northern point of the Pentagram is another fine double star, STF 519 (Wilhelm Struve 519): a yellow magnitude 7.7 primary and a white magnitude 9.4 secondary 17.9″ to the north-northwest. Finally, look at the southeastern arm of the Pentagram. It curves southward, then north, forming a fine ellipse of starlight, in the shape of a head. This seems appropriate considering that the pentagram once symbolized the relationship of the head to the four limbs, signifying, in essence, that every man and woman is a star.

Hidden Treasure 27

NGC 1647

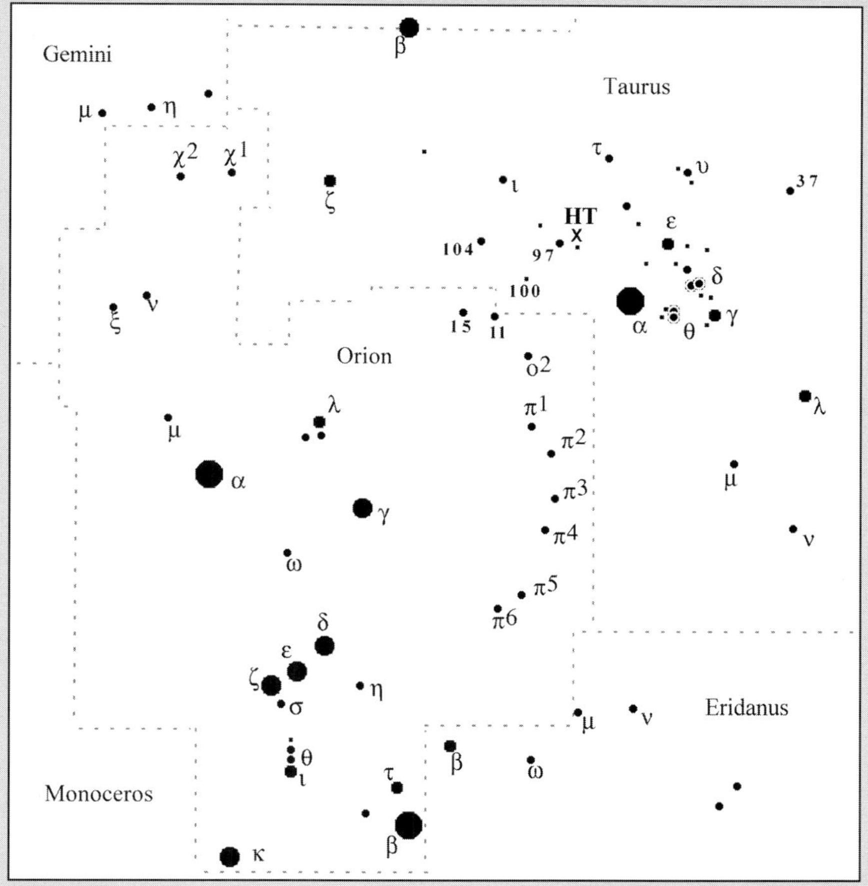

Tirion: Charts 5 & 11

Uranometria: Chart 134

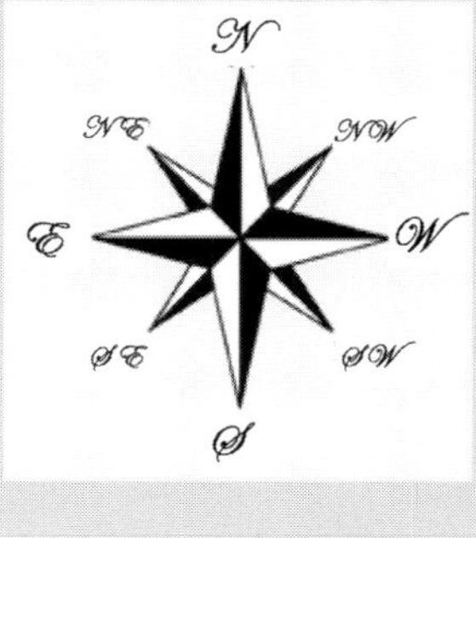

27

Pirate Moon Cluster
NGC 1647
Type: Open Cluster
Con: Taurus

RA: 04^h 45.7^m
Dec: +19° 07′
Mag: 6.2 (O'Meara); 6.4
Diam: 40.0′
Dist: 1,700 light-years
Disc: William Herschel, 1784

W. HERSCHEL: [Observed February 15, 1784] A cluster of considerably large, very coarsely scattered stars, perhaps a projecting point of the Milky Way. (H VIII-8)

NGC: Cluster, very large, stars [bright], scattered.

Cast adrift on tempestuous seas
with his black hearted scoundrel,
Katelin swore she'd have her revenge . . .
only to be undone by the searing rapture
she found in the embrace of a loving rogue
beneath a glowing Pirate Moon.

Pirate Moon, Kathleen Drymon

I THANK MY WIFE, DONNA, FOR introducing me to NGC 1647, a wonderful but somewhat neglected open cluster near the V-shaped face of Taurus, the Bull. One night Donna stood in the driveway of our home in Volcano, Hawaii, and pointed her 7 × 35 binoculars toward the Hyades cluster (Caldwell 41), then in the eastern sky. Suddenly she exclaimed that she had discovered a comet and pointed a finger to a spot about $3\frac{1}{2}°$ northeast of Aldebaran. There, to my surprise, I saw a beautiful open cluster that had, for all these years, escaped my gaze. Later, while studying the Hyades for *Deep-Sky Companions: The Caldwell Objects*, I learned that the magnitude 6.4 glow of NGC 1647 could be seen with the naked eye from a dark sky. Indeed, in *Star Clusters*, Brent Archinal and Steven Hynes write that NGC 1647 is "just visible to the unaided eye, an easy binocular object, and a fine sight in a telescope, but rather overshadowed by its near neighbor, the Hyades."

With the unaided eye the cluster displays a mottled core surround by a vaporous halo of uniform light. Through 7 × 50 binoculars NGC 1647 is a lovely cluster, appearing as a round ghostly glow with an apparent size larger than that of the full Moon, which is why I call it the Pirate Moon Cluster; the cluster's stars are congregated into disparate bunches, which mimic the dark and bright

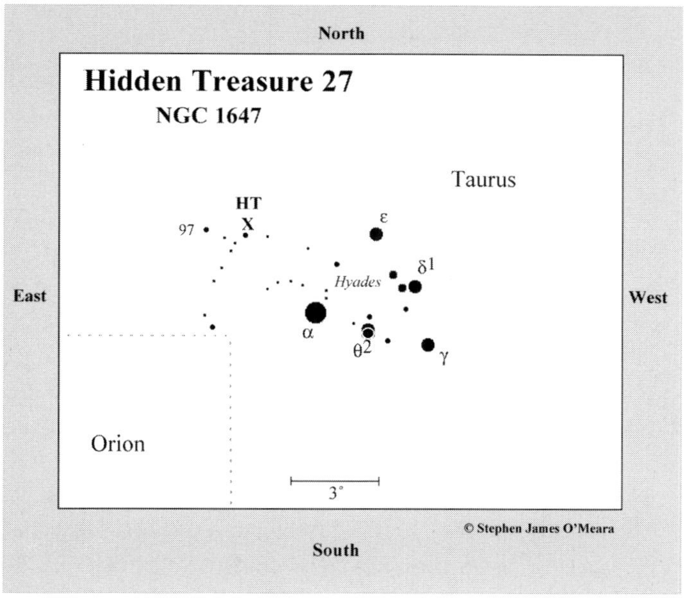

Treasure 93), which has the same apparent diameter. But the reason is obvious. First NGC 1647 is not projected against a dense part of the Milky Way. And though parts of Taurus are seen against the outer reaches of the Milky Way, NGC 1647 lies in a bay between two luminous folds of starlight. As Sigfrid Raab writes in his 1922 book, *A Research on Open Clusters*, "[NGC 1647 is a] large, very thin [cluster], distinctly separated from the environs and probably containing exclusively bright stars." So the cluster is not projected against a rich star field. Second, though the cluster does not have a strong central condensation, most of its brighter stars are found in the central 15′, which matches the field of view used by William Herschel in his sweeps. IC 4756, on the other hand, *is* seen against a rich Milky Way star cloud and it has no apparent central condensation of starlight. Indeed, William Herschel's own words that he thought NGC 1647 was "perhaps a projecting point of the Milky Way," suggests that this coarse scattering of stars might actually not be a cluster but a portion of Milky Way.

features we see on the naked-eye Moon. (Besides, what's a pirate story without a ghostly looking Moon?) Undoubtedly, since I had lived most of my life in the Boston area, light pollution must have drowned out the cluster's pale glow, making it easy to sweep past without notice.

It might seem odd that William Herschel would have discovered NGC 1647 but not IC 4756 in Serpens Cauda (Hidden

The Rev. T. W. Webb's description of the cluster in his *Celestial Objects for Common Telescopes* is rather perplexing, for he writes: "Stars 8.5 to 10, round wide pair" (I believe

cluster's face. A careful study of the brightest stars will reveal that many of them have single or multiple companions. In fact, the cluster is alive with doubles. Some mini-clusters can also be found amid the chaos. In fact, the two 9th-magnitude doubles at the cluster's core each have a dim companion, making the central triangle comprised of two mini-triangles and a solitary sun. Other asterisms include a tiny cross (just southeast of the southeasternmost star in the central triangle) and a faint clone of M73, which lies to the west of the westernmost double in the central triangle. Since NGC 1647 is not densely packed, the view remains best at low power; in fact, in his *Visual Astronomy of the Deep Sky,* Roger Clark suggests 20× for a 6-inch and 27× for an 8-inch.

there should be a comma between "round" and "wide pair").

In the 4-inch at 23×, the cluster is quite round, and about a third of its 200 stars can be found within $\frac{1}{2}^{\circ}$ of its core, where a fine pair of 9th-magnitude double stars dominate the view. These stars and a similarly bright companion equidistant to the southeast give the cluster a distinct, trianglar-shaped heart. Other bright stars spiral out from this triangular core in crooked or disjointed arms. The overall impression is one of a face-on spiral galaxy whose dashing arms are resolved into streams of individual suns.

At higher powers, the cluster does not have the snap it does at lower powers. But do spend some time sweeping across the

Despite its prominence in the sky, NGC 1647 is a poorly studied cluster. Aside its proximity to the Hyades as seen projected on the sky, NGC 1647 has nothing to do with it. At 190 million years old, NGC 1647 is about 400 million years younger than the Hyades and 11 times more distant. In true physical extent, NGC 1647 spans 20 light-years of space, while the Hyades spans only 14.5 light-years. NGC 1647 is more similar to M7 in Scorpius in both age and true physical extent. Had NGC 1647 been about 1,000 light-years closer, it would have been just as apparent as M7 but seen much higher in the northern sky. In 1996 Geffert and his colleagues visually inspected NGC 1647 on a *National Geographic Society–Palomar Observatory Sky Survey* print and determined the position of the cluster's center; they also determined that the Cepheid variable SZ Tauri (whose apparent visual magnitude ranges from 6.50 to 8.86 every 3.15 days) is probably not a member of this cluster.

Hidden Treasure 28

IC 418

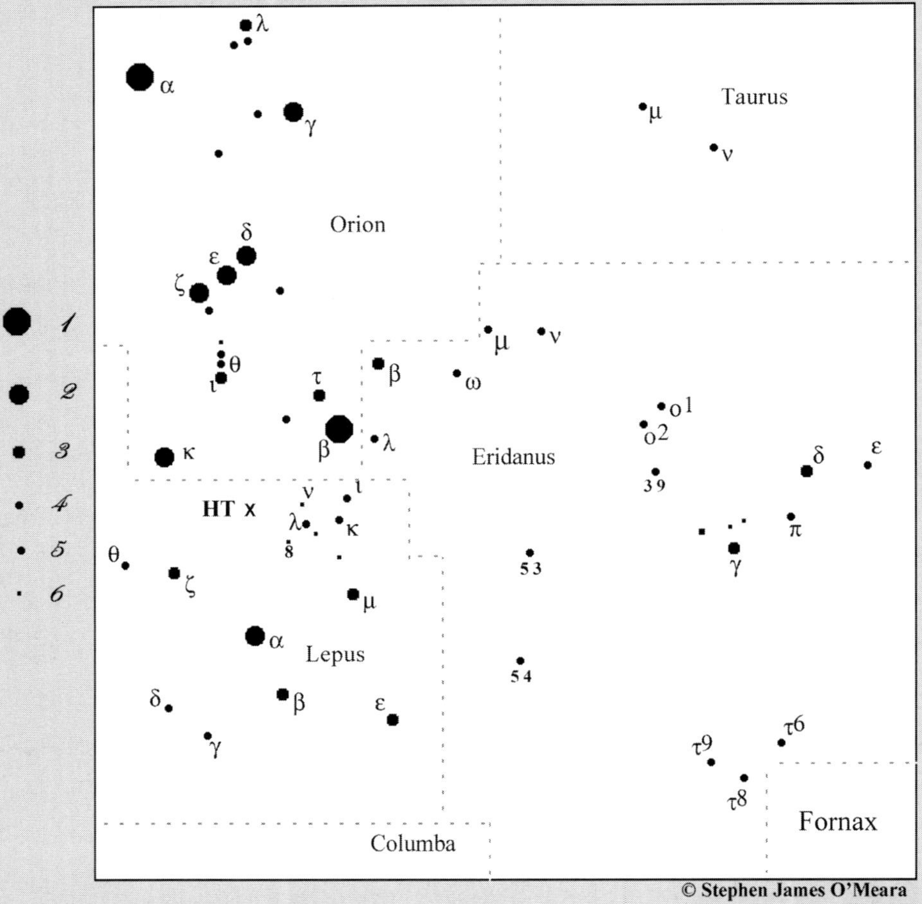

Tirion: Chart 11

Uranometria: Chart 270

28

Spirograph Nebula, Raspberry Nebula, Colored Contacts Nebula
IC 418
Type: Planetary Nebula
Con: Lepus

RA: $05^h\ 27^m\ 28.2^s$
Dec: $-12°\ 41'\ 50''$
Mag: 9.1 (O'Meara); 9.3
Dim: $14'' \times 11''$
Dist: ~2,000 light-years
Disc: Discovered on photographic plates taken at Harvard College Observatory under the direction of E. C. Pickering.

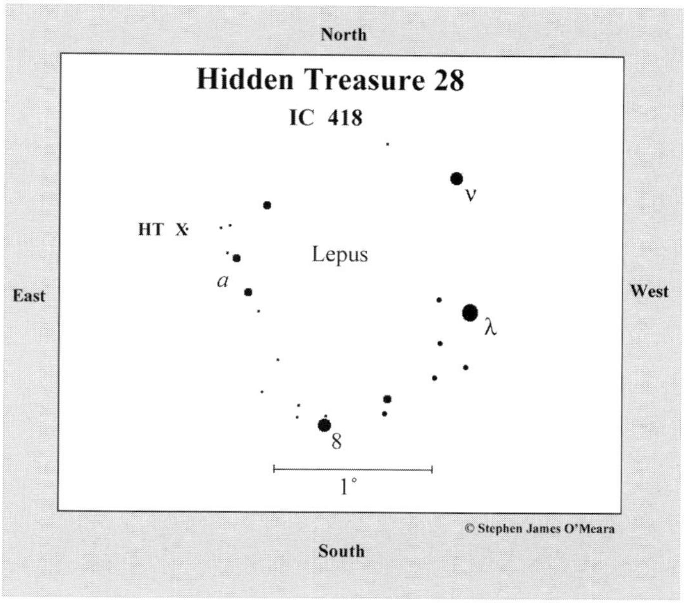

PLANETARY NEBULA IC 418 is a tiny tick of light in northern Lepus, just beneath Orion's feet. It is visually crushed by the stellar luminaries surrounding it. Brilliant Beta (β) Orionis (Rigel) lies only $5\frac{1}{2}°$ to the northwest, Kappa (κ) Orionis (Saiph) is about the same distance to the northeast, and Alpha (α) Leporis (Arneb) is roughly the same distance to the south-southeast. In fact, IC 418 is a little northeast of the midpoint between Rigel and Arneb. It is part of a 4°-wide, Little Lepus asterism – two joined trapezoids of binocular stars that mimic the shape of the main body of the Hare; it is comprised of Iota (ι), Kappa (κ), Lambda (λ), Nu (ν), 8 Leporis, and IC 418. The planetary's light is so concentrated that it appears as a 9th-magnitude "star" in hand-held binoculars. IC 418 also marks the northeast arm of a $\frac{1}{2}°$-long, Y-shaped asterism (*a*) of binocular stars (oriented north–south) at the northeast corner of Little Lepus; I call this other asterism the Martini Glass, and IC 418 is the "olive."

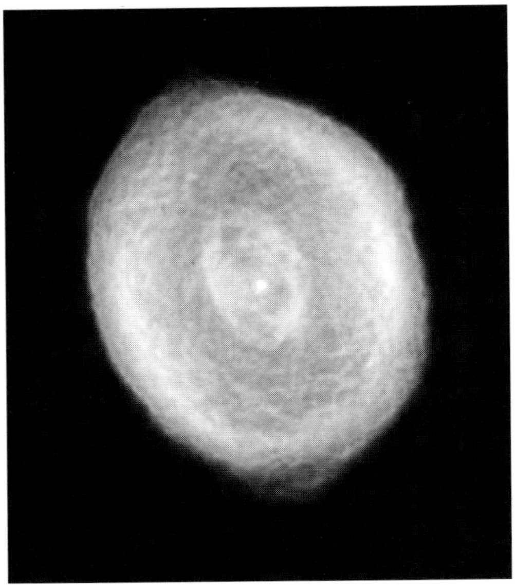

IC 418 dodged detection until its slightly swollen form was detected on photographic plates taken at Harvard Observatory in Cambridge, Massachusetts. I call it the Chameleon Nebula for two reasons: its tiny size makes it blend into the background starlight, and the color one sees (if any) seems to depend on the aperture and magnification used. For instance, in his 12-inch reflector, the late southern sky explorer Ernst Hartung saw a "bright bluish disk . . . with a bluish central star." I get mixed impressions: at 72× in the 4-inch the nebula alternates between bluish and yellowish tints, while at 101×, the star looks blue and the ring yellow. Steve Gottlieb (El Cerrito, California), using a 18-inch reflector at 82×, saw the nebula with a "slight reddish tinge at its edge." Others have described the nebula as being pink, orange, and brown.

Many of these colors can be seen in a stunning Hubble Space Telescope (HST) image (shown here in black and white), which shows the planetary's magnitude 10.5 central star burning within a blue inner ellipse surrounded by a pale lavender crown with a fiery yellow rim. All is "[g]lowing like a multi-faceted jewel." The HST image also shows the nebula patterned with an intricate weave – one that astronomers have likened to a drawing made by a spirograph. With imagination, the entire scene looks as if someone has dropped a ruby into a well of purple mist, creating a complex arrangement of ripples. The strange texture of IC 418 may be related to the chaotic winds flowing from the central star (HD 35914) – a confirmed variable star that shows irregular brightness changes over a period of days and cyclic variations every 6.5 hours. Of course, it is the ultraviolet radiation streaming from this hot white dwarf star that excites the surrounding atoms in the nebula causing it to fluoresce. Interestingly, mid-infrared and radio images have found the nebula to be carbon rich. The intensity of the infrared emission spills over to the visual end of the red spectrum. So, while the planetary shines with a typical blue-green light, some red coloration warms its visual hue. What colors an observer sees in IC 418, then, may be determined by the color sensitivity of his or her eyes. Observers should try observing IC 418 with an H-Beta filter. When Gottlieb tried, he said "the halo brightened while the central star dramatically faded, leaving a more noticeable disk."

IC 418 lies about 2,000 light-years away and spans 0.1 light-years across. But this is the diameter of the main shell and crown. Astronomers have also detected a more extensive halo out to at least 30″ in apparent diameter, which swells the planetary's true physical extent to 0.3 light-years. The tightness of IC's shell suggests that it is very

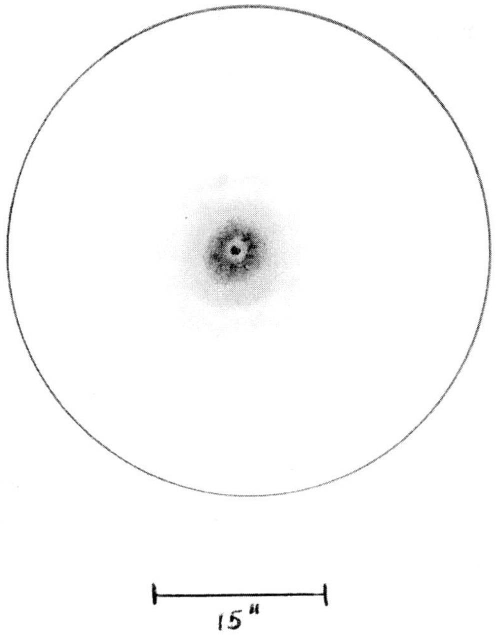

young, perhaps only a few thousand years old.

Visually, IC 418 is a fine object for small telescope users, though it requires high magnifications to see its finer details. In the 4-inch, the planetary is a pale blue star at 23×. Only on nights when the atmosphere is extremely calm and steady can I convince myself that at low power IC 418 looks ever so slightly larger than the surrounding stars – and that's with averted vision and much time trying. At 72× the planetary still appears stellar with a glance, but now the use of averted vision will begin to show its slightly swollen character. Now take the time to study what you see really. In fact, try placing IC 418 in and out of your foveal vision. When I do this I can clearly separate the planetary's 10th-magnitude central star from the surrounding annulus, which, as I mentioned earlier, appears to be a mix of blue and yellow light. This impression is consistent with reality, since the nebula's outer rim is "yellow-orange" (like smog) and its core is blue. By moving my eye around, I am probably catching quick glimpses of both parts of the nebula. Overall, the nebula is slightly elongated north–south.

Now increase the power. When I use 101×, the central star is clearly separated from a yellow ring of light. With keen averted vision I also suspected a crown of extremely dim light. Increasing the power to 151× reveals the yellow ring of light surrounding the central star to have a bright inner edge that appears symmetrically beaded, like a pearl necklace. It's hard not to imagine a complete string of miniature pearls lining the ring's inner edge. Adding a 1.7× Barlow (302×) helps to separate the central star further from the ring, which is still extremely radiant. In fact, on one night, when I was observing from the 7,000-feet elevation of Mauna Loa volcano, I placed the 3-mm eyepiece in a 1.8× Barlow, then placed this combination in a 3× Barlow (907×) – and the inner ring and central star were still manageable. This was the most power I have ever used on a deep-sky object with this telescope. Despite all my efforts, I could bring out the object's blue coloration more than the yellow, which is odd, since my eyes tend to be blue sensitive.

That there is a red planetary nebula in Lepus with a carbon-rich shell is interesting because 7° to its west-southwest is R Leporis, Hind's Crimson Star. This Mira-type variable is a prime example of a carbon star – the reddest stars known. It has been called a "gleaming crimson jewel." Julius Schmidt (1823–1884) described it as "of the most intense crimson, resembling a blood-drop on the background of the sky; as regards depth of colour, no other star visible in these latitudes could be compared with it." If you want to follow this ruby star through its light changes, R Lep varies between magnitude 7.3 and 9.8 every 420 days. To see this glistening drop of blood as red as a Vampire's kiss, look for it when the star is at minimum light. As with the waxing and waning of the red planet Mars, the intensity of the color diminishes with the greater brilliancy of the star's light.

Hidden Treasure 29

Collinder 69

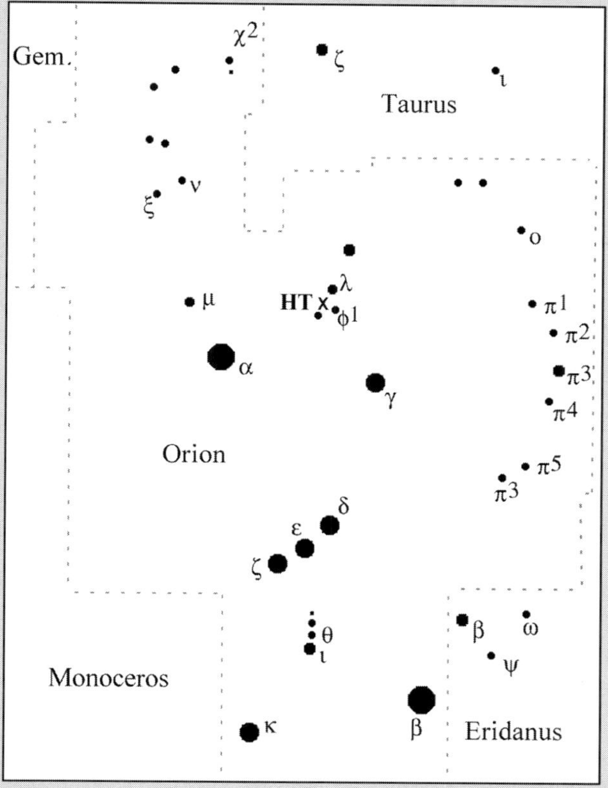

© Stephen James O'Meara

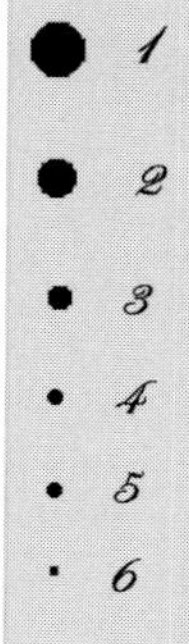

Tirion: Chart 11

Uranometria: Chart 181

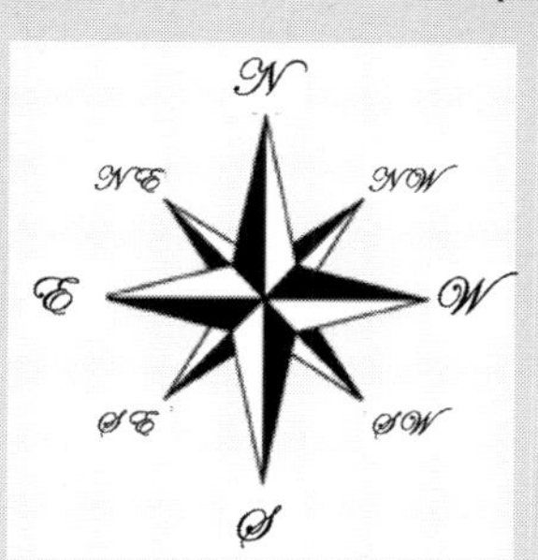

29

Lambda Orionis Cluster, Aunt Margaret's Mirror
Collinder 69
Type: Open Cluster
Con: Orion

RA: 05^h 35.0^m
Dec: $+09°$ $56'$
Mag: 3.0 (O'Meara); 2.8
Diam: $70.0'$
Dist: 1,600 light-years
Disc: Since historical times; Ptolemy mentioned it in his *Almagest*

HERSCHEL: None.

NGC: None.

ORION, THE HUNTER, HAS HAD HIS head in the clouds ever since humans first began to look upon it with wonder. Orion's Head is marked by 3rd-magnitude Lambda (λ) Orionis and its two, 4th-magnitude attendants: Phi^1 (ϕ^1), and Phi^2 (ϕ^2) Ori. These stars form a distinct isosceles triangle midway between, and a little north of the Hunter's shoulders: Alpha (α) Ori (Betelgeuse) and Gamma (γ) Ori (Bellatrix). In his 1892 *Celestial Handbook,* Jules Colas writes, "In looking at this triangle nobody would think that the moon could be inserted in it; but as the distance from λ to ϕ^1 is $27'$, and the distance from (ϕ^1) to (ϕ^2) is $33'$, it is a positive fact."

While Orion's Head does not lie along the ecliptic today, so the Moon cannot be directly compared with its three brightest stars, the vernal equinox lay there in 4,500 BC. Indeed, Lambda constituted the Euphratean lunar station *Mas-tab-ba-tur-tur,* meaning The Little Twins. That name is a bit confusing, since it also applies to Gamma and Eta (η) Geminorum. Equally confusing is Lambda's popular name today – Meissa. It derives from the Arabic *Al-Maisan* (The Proudly Marching One or The Shining One). But the Arabs first gave that name to Gamma (γ) Geminorum – a star in the sixth Lunar Mansion, which was known as *Al-Han'ah,* The Brand-Mark (as on a camel's or horse's neck).

The confusion lies with the Arabic names for the fifth and sixth Lunar Mansions – *Al-Haq'ah* (The Circle of Hairs or The White Spot) and *Al-Han'ah,* respectively. As George A. Davis, Jr., explains in his paper "Pronunciations, derivations, and meanings of a selected list of star names," published in the January 1944 *Popular Astronomy*: "[s]ome early commentators confused the two because of the similarity of names and because the name *Jauza'* was frequently used to designate the stars in Gem[ini]." The fourth-century Arabian lexicographer Al Firuzabadi was most likely the perpetrator, since he believed the title *Al-Maisan* could be applied to any bright star. Perhaps to abolish the confusion, Gamma Gem is known today as Alhena.

It was not uncommon for the Arabs to have many names for their beloved stars, whose presence breathed life into the barren desert at night: "Thou canst not know how much we Arabs depend on the stars," Shaykh Ilderim told Ben Hur at the Orchid of Palms. "We borrow their names in gratitude, and give them in love."

The original Arabic name for Lambda Ori and its attendant stars was *Al-Hak'ah* or *Al-Haq'ah* (and has descended to us as *Heka* and *Hika*). These words refer to the fuzzy appearance of Orion's Head as "a White Spot." Brent Archinal finds this description of Orion's head certainly relevant as, he says, "it describes the whole group of three stars and perhaps the cluster Collinder 69 itself." Archinal's contention is supported by the Roman astronomical poet Gaius (or Marcus) Manilius who wrote in the first-century AD:

> On each shoulder a bright star displayed
> And three obliquely grace his hanging blade.
> In his vast head, immersed in boundless spheres,
> Three stars less bright, but yet as great he bears . . .

Ptolemy (AD 83–161) included these "boundless spheres" in his *Almagest,* the most authoritative astronomical manual throughout the whole of Europe and the Near East until at least the middle of the fifteenth century. He described them collectively as "The Nebulous One," or as Kenneth Glyn Jones has in *The Search for the Nebulae,* a "Mistiness in Orion's Head." The Persian astronomer Al-Sufi (AD 903–986) refined the object's appearance in *The Book of the Fixed Stars,* writing that "[t]he first star of the Giant [Orion] is the cloud which is in the head and which is composed of three close stars." And Ulugh Begh (1394–1449), the celebrated Tartar prince and astronomer, likewise realized that the "nebula" in Orion's head was caused by the proximity of three close stars: "No.1 Orionis. The northernmost of those in the head of the Giant: however, they are three close together" Al-Sufi then is arguably the first to imply that the fuzziness is an illusion.

It is no surprise that dim stars close to one another can have a fuzzy, cometlike appearance. In fact, even the best comet observers, when working near the limits of their vision, have been fooled by such close pairings of stars. This phenomenon may have been known, at least indirectly, to Pliny the Elder, who in his *Natural History* says, "Some comets move, like planets, while others remain fixed." The most famous fuzzy double of them all, of course, is M40, in Ursa Major, a simple duet of telescopic suns that the great Polish astronomer Johannes Hevelius (1611–1687) apparently mistook as a nebula in his "old, imperfect instrument."

But there is more here than meets the eye. It is a true cluster of some 20 stars spanning an area of sky of about 70′ in apparent diameter, which Per Collinder first recognized in his 1931 paper "On structural properties of open galactic clusters and their spatial distribution." Raise a pair of binoculars to Lambda and you will immediately see that Lambda and Phi^1 are part of a pretty chain

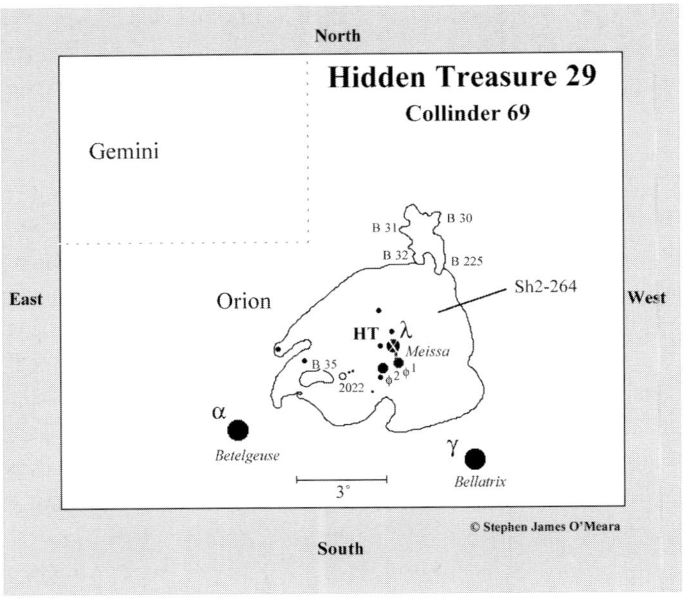

of a half-dozen 5th- to 6th-magnitude suns aligned roughly north–south. Rays of fainter stars stream eastward from this line like the fins of a robinfish. So the fuzzy nature of the Head of Orion, from under a dark sky, is truly the accumulated haze teeming with suns. "Although there is no nebula here," writes Garrett P. Serviss in his 1923 *Astronomy with an Opera Glass*, "yet these stars, as seen with the naked eye, have a remarkably nebulous look." But unbeknown to Serviss, there is; Orion's Head is immersed in a roughly 4°-wide cloud of gas and dust – the emission nebula Sharpless 2–264 and a dappling of dark nebulae discovered by E. E. Barnard during his great photographic surveys of the heavens.

Galileo was probably the first to see these stars as a cluster: "A diagram of the nebula called that of Orion's Head in which I have counted twenty-one stars." Seeing so many stars and no nebulosity in this region was testament to his growing belief – that all nebulae would be resolved into stars under the power of the telescope. The fact that he did not include the nebula component of Orion's Sword in another drawing is not a comment about his ability as an observer but of his resolute state of mind; as far as Galileo was concerned, nebulosity did not exist, so why record it. If true, it demonstrates the power of belief over truth.

Robert D. Mathieu (University of Wisconsin, Madison) and his colleagues studied the Lambda Orionis region with the 3.5-meter WIYN telescope atop Kitt Peak in Arizona – with a spectrograph that can deploy optical fibers to analyze the light from 100 celestial objects simultaneously. His data revealed an additional 266 low-mass, pre-main-sequence stars among 40,000 field stars. After using these data to recreate the star-formation history of the region, Mathieu and his collaborators concluded that the region was "profoundly affected by a supernova 1–2 million years ago." The entire region, then, presents a snapshot of the molecular cloud 1–2 million years after newly formed, Type-*OB* stars ravaged it with their radiation. A separate proper-motion study of stars in the region suggests that the star formation was concentrated in an elongated cloud extending from the dark nebula Barnard 35 (about $2\frac{1}{2}$° southeast of Lambda) to the Barnard 30 dark cloud (about 3° northwest of Lambda).

It may be coincidence, but the view of this region in the 4-inch at 23× is one of a pyramid of bright stars surrounded by an ellipse of dim suns. The brightest star, of course, is

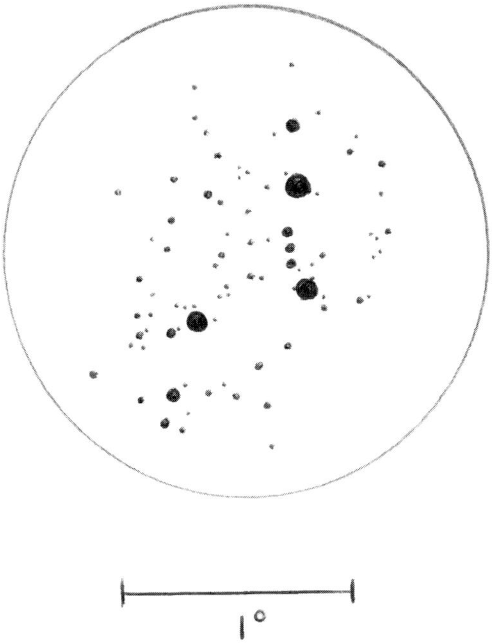

Lambda, a Type-*O* beacon with a luminosity of about 9,000 times that of the Sun. Lambda is also a fine double star. The lemon-yellow primary has a 5th-magnitude, lemon-lime companion to the northeast. William Herschel discovered the pair in 1779. At the time, the stars were separated by 5.8″. By 1822 the gap had narrowed to 5.5″, then to 4.2″ when Friedrich Georg Wilhelm von Struve measured them in 1830. Popularly known today as Struve 738, the stars are now separated by 4.3″, making them a delightful sight in small telescopes. Lambda has two fainter companions as well – an 11th-magnitude sun 28″ to the south and a 10th-magnitude sun 78″ to the west.

In his *Celestial Handbook,* Robert Burnham, Jr., writes, "Many and curious are the reported discrepancies in the colors of this pair, the more strange since both stars are Type *O* and should appear simply white." The Adm. William Henry Smyth called them "a neat double" with "pale white" and "violet." The Rev. T. W. Webb saw them as "yellowish" and "purple." William T. Olcott found them as "yellow" and "red"; Burnham found them "sparkling white with a hint of light amber tint."

While long-exposure photographs show the entire region covered by an extensive but faint diffuse nebulosity, the question is, can this be seen with the naked eye. In other words, is there more to the mistiness that cloud's Orion's Head than the illusion of clustered starlight? I believe the nebula can be seen but only from a very dark-sky site, with averted vision, only when the air is *very* dry. To see it, you must sweep your averted gaze back and forth between Orion's shoulders and look for an ever so slight increase in brightness in the area. But I would like to leave the verdict up to you. Perhaps the use of a nebula filter will resolve the issue.

Seeing the nebula with the unaided eye is one of those observations that will perhaps forever reside in that fuzzy razor-thin world that lies between belief and disbelief. And this leads me to my final point. You may be wondering why I called this cluster Aunt Margaret's Mirror. It has nothing to do with me. (I don't have an Aunt Margaret, at least to my knowledge.) No, it refers to Sir Walter Scott's tale, "My Aunt Margaret's Mirror," from his 1828 *The Keepsake:* "And amidst such shadowy and doubtful light . . . imagination frames her enchanted and enchanting visions, and sometimes passes them upon the senses for reality." These words, I find, clearly mirror the difficulty of seeing the emission nebula clouding Orion's Head.

Actually, the story's narrator was commenting on his Aunt Margaret's superstitious tendencies. As the reader learns, she avoids looking into a mirror, when she is alone in her chamber for the evening.

29

> I will tell you that I myself, like many other honest folk, do not like to see the blank black front of a large mirror in a room dimly lighted, and where the reflection of the candle seems rather to lose itself in the deep obscurity of the glass, than to be reflected back again into the apartment. That space of inky darkness seems to be a field for fancy to play her revels in. She may call up other features to meet us, instead of the reflection of our own; or, as in the spells of Hallowe'en, which we learned in childhood, some unknown form may be seen peeping over our shoulder.

The oval shape of the dim stars that form an ellipse around the bright stars of Orion's Head is eerily like seeing the profile of a face in an oval mirror. Do you dare to look?

Hidden Treasures 30-32

NGC 1981, Collinder 72, NGC 1977

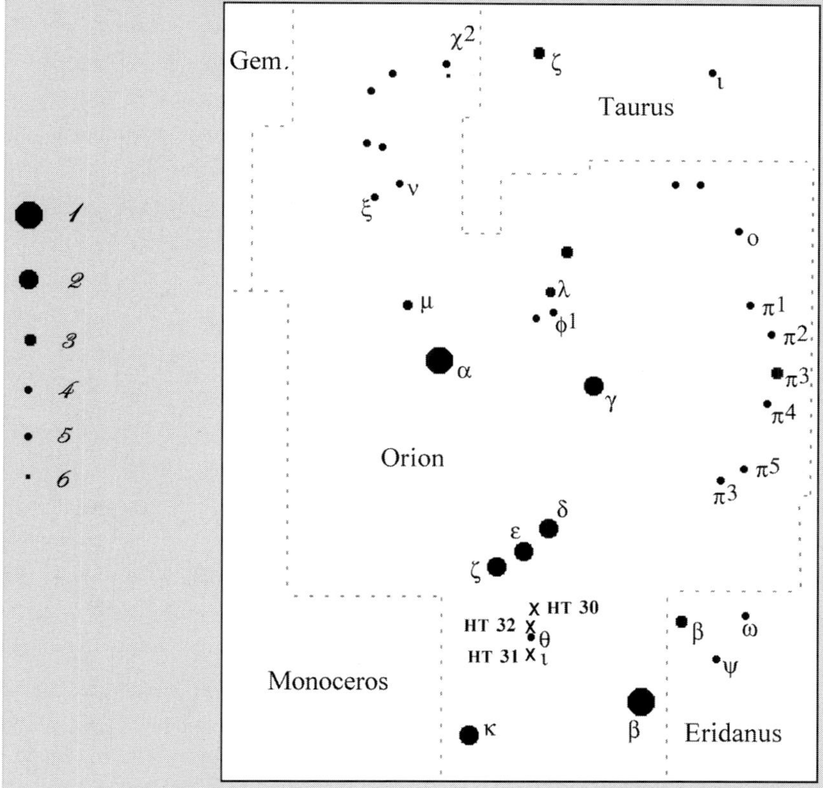

© Stephen James O'Meara

Tirion: Chart 11

Uranometria: Chart 181

30

Coal Car Cluster
NGC 1981
Type: Open Cluster
Con: Orion

RA: 05^h 35.2^m
Dec: −04° 26′
Mag: 3.7 (O'Meara); 4.2
Diam: 28.0′
Dist: 1,300 light-years
Disc: John Herschel, 1827

J. HERSCHEL: Very bright cluster, a little irregularly round, scattered [bright] stars. (h 362)

NGC: Cluster, very bright, little rich, stars [bright], scattered.

31

Lost Jewel of Orion
Collinder 72
Type: Open Cluster
Con: Orion

RA: 05^h 35.4^m
Dec: −05° 55′
Mag: 3.0 (O'Meara); 2.5
Diam: 15.0′
Dist: ~1,500 light-years
Disc: Per Collinder, listed in his 1931 paper "On structural properties of open galactic clusters and their spatial distribution"

HERSCHEL: None.

NGC: None.

32

Mermaid's Purse, Running Man Nebula
NGC 1977
Type: Emission and Reflection Nebula
Con: Orion

RA: 05^h 35.5^m
Dec: $-04°$ $52'$
Mag: 6.3 (O'Meara)
Dim: $20' \times 10'$
Dist: ~1,500 light-years
Disc: William Herschel, 1786

W. HERSCHEL: [Observed January 30, 1786] The 1st and 2nd c Orionis, and the stars about them, are involved in an extremely faint, unequally bright, milky nebulosity. (H V-30)

NGC: *Remarkable*, c^1 42 Orionis and nebula (V 30).

WHEN YOU LOOK AT ORION'S SWORD with the naked eye, what do you see? Most observers would agree that the Sword is comprised of a short row of three naked-eye stars. Of these stars Manilius, a Roman, wrote: "[Orion] marches on and measures a vast space. / On each shoulder a bright star displayed / And three obliquely grace his hanging blade." And with little effort, most observers today can also see that the middle star in the Sword, Theta (θ) Orionis, looks fuzzy. Indeed, the middle star is M42, the Great Nebula in Orion. Few sights in the night sky excite beginners more than being able to detect this, the most glorious of all nebulae, without optical aid. M42's beauty is only magnified in binoculars and telescopes, which also reveal its little companion nebula, M43. The Great Nebula in Orion was the first object that Sir William Herschel observed on March 4, 1774, with his first workable homemade telescope (a 5.5-foot focal-length reflector); and it was the last object he viewed on January 19, 1811, with his 40-foot reflector. The view of M42–M43 is so comely that it has the power to draw our attention away from other bright nebulae and clusters in the immediate vicinity. Any of these lesser known objects would be stellar showpieces had they not been so close to this "single misty star, / Which is the second in a line of stars / That seem a sword beneath a belt of three," to quote from "Merlin and Vivien," the sixth poem in Lord Alfred Tennyson's 1857 *Idylls of the King*.

While we are fast to say that the Orion Nebula is the middle of three naked-eye stars in Orion's Sword, it's not true. It is the third of *four* deep-sky objects that make up Orion's Sword; actually, in the Bayer depiction of Orion, the Sword is fully sheathed, so we should say that the chain of four objects marks the Sword's scabbard.

OPEN CLUSTER NGC 1981

The northernmost star in Orion's Sword is not a star at all but the 4th-magnitude open star cluster NGC 1981 (Hidden Treasure 30) – a clip of jewels set into Orion's black-leather scabbard. This loose gathering of 20 or so suns is spread across an area of sky comparable to that of the full Moon. Eight of these scintillating stars can be easily seen with 7 × 50 binoculars: look for a tight arc of three 6th-magnitude suns (oriented north–south) set between a solitary 7th-magnitude star to the east and four roughly 7th-magnitude suns, in a Y-shaped pattern, to the west. Seen with north up in the antique telescope, these stars join with others to form what looks like an old coal car. The northernmost arm of the "Y" is itself a little arc of three suns; here is the heap of coal sticking up out of the basket.

NGC 1981 is the northernmost object in the Orion *OB*1c Association and measures 11 light-years across in true physical extent. Alan MacRobert in his popular book *Star Hopping*, calls NGC 1981 a "bevy of stellar beauties." He continues, "Most telescopists miss the beauty of this little-known cluster because of its relative sparseness and large size. However, discerning binocular users will immediately recognize the charm of NGC 1981, especially amid such grand surroundings." Indeed, in the early twentieth century, Harvard astronomer E. C. Pickering regarded it as one of his top 60 objects in the heavens, perhaps because it contains two stunning double stars: Struve 750 and Struve 743. They are the brightest stars in the cluster and form a nice binocular pair.

Through the 4-inch, Struve 750 is a wonderful magnitude 6.4 and 8.4 pair, separated by 4.2″; in his *Cycle of Celestial Objects*, Adm. William Henry Smyth called it a "delicate double star in the wide-spread cluster on Orion's sword." He saw the components as "lucid white" and "pale blue." I see them as glacial blue and pleasantly pink. Struve 743, on the other hand, is a very tight magnitude 7.7 and 8.2 pair, separated by 1.8″. Curiously, the closer pair was discovered first, in 1830, and the wider pair a year later. Struve 743 is a fine test for a 3-inch lens, because even the slightest bit of atmospheric turbulence will cause it to blend into a single orb.

Of the cluster as a whole Smyth penned, "The principal members of this group of stars, are of the 6th and 7th magnitudes, with some smaller; and from their brightness and disposition form a capital test for the light of a telescope. It was examined by [John Herschel], and entered in his Catalogue of 1830; whence it may lay claim to being an aggregated and connected assemblage, and, comparatively speaking, not very remote from us."

These words are extremely revealing, because they tell us that William Herschel failed to notice the group as a cluster, which supports the argument that the stars of NGC 1981 would have been too separated in his telescope to be easily discerned as a stellar aggregation. At 72× and 101× in the 4-inch, the cluster divides into three distinct

nebulosity that looks like a stain in the Orion starfield, like wet blotter. The glow is evident only by the absence of starlight, especially when the field is compared to other starfields around it. Such a sight can only be appreciated with a wide field of view under a dark sky. So I encourage everyone who has wide-field capability to hunt down this ghostly glow – "a lovely baleful star / Veiled in gray vapor"; to quote again from "Merlin and Vivien."

groupings, with the number of faint stars increasing dramatically toward the west. In fact, the Y-shaped asterism appears to be the focus of a vast and irregular display of some two dozen dim background suns.

The region is not as free from nebulosity as it might look. Long-exposure photographs and charge coupled device (CCD) images show a dim, sepulchral veil of emission nebulosity pervading the area of the cluster. The question is, can we detect any of it visually? Sweeping the entire region with the nearly 3° field of view of the 4-inch, I noticed something spectacular. The cluster NGC 1981 and the nebula NGC 1977 appear to be surrounded by a grayish "fog," which is best visible if I sweep the scope in large arcs around these objects. It's a dimly lit

THE LOST JEWEL OF ORION

There's a beautiful object in Orion that's literally been in a haze. It lies just $\frac{1}{2}^\circ$ due south of M42, the Orion Nebula, and is one of the brightest objects in its class. The problem is that many observers do not know it exists.

Readers who know Orion well may be wrinkling their foreheads wondering what I'm talking about, since any decent star atlas will show the beautiful emission nebula, NGC 1980, $\frac{1}{2}^\circ$ due south of M42. The nebula is associated with Iota (ι) Orionis, the brightest star in Orion's Sword, but it's not the object I'm thinking of, even though many observers don't recognize it as a separate object from the Orion Nebula. NGC 1980 does, however, have its own history. It was discovered on January 31, 1786, by William Herschel, who wrote, "ι Orionis with its neighboring stars are involved in an extremely faint milky nebulosity to a great extent." He cataloged it as the 31st object in his list of very large nebulae (his Class V objects).

Anyone who has seen the Orion nebula in a wide-field telescope from a dark-sky site knows that Iota Orionis is connected to the Orion Nebula by a long, faint loop of nebulosity. Of that loop and its dark interior, nineteenth-century English observer, Thomas W. Webb wrote, "A considerable aperture will show how beautifully one [bright] star, nearly opposite the great dark opening, is encompassed by a spiral mass of haze."

Robert Burnham, Jr., amplifies Webb's words in his *Celestial Handbook,* writing, "On the east side [of the Orion Nebula] there are a few thicker and brighter filaments which curve southward in the direction of the bright star Iota Orionis; the most prominent of these is very clear even in a 6-inch glass. A huge arc of faint nebulosity continues outward from this area, turning back to the westward as it passes Iota, then curving north to join the main nebula again on the west side." Iota Orionis, then, is near the apex of this vast loop, which measures some 30 light-years wide.

NGC 1980 is in a selected list of Herschel's best 400 objects compiled by the Astronomical League, and some observers have seen it with binoculars: "The final binocular 'sweep' of the night," writes Massachusetts amateur Lew Gramer, who observed NGC 1980 in February 1999 with 16 × 70 mounted binoculars, "turned up a clear blob of unstructured haze, surrounding the beautiful bluish 3rd-magnitude star Iota Orionis. This otherwise featureless nebulosity was well separated from the swirling confusion of M42 but was slightly more noticeable north of bright Iota."

Anyone who has seen NGC 1980 through a telescope may have also noticed that Iota Ori, otherwise known as Struve 752, is a wonderful triple star. The white sapphire primary is attended by a 7th-magnitude aquamarine companion 11″ to its southeast, and a 10th-magnitude star burns with a pale mango hue 49″ to the east and slightly south of Iota Ori. And there's more.

The region also sports two additional double stars: Struve 747, a pair of 5th-magnitude diamonds separated by 36″; and Struve 745, a pair of 9th-magnitude gems 29″ apart. Iota and Struve 747 make a charming naked-eye double separated by 8′. If the coincidence of three multiple stars in a small region of space seems extraordinary, it's not, at least not in this case.

If you look at Iota Orionis and its surroundings carefully, you'll see that Iota is the brightest member of a 15′-wide cluster of 30 suns embedded in the milky haze of NGC 1980. Indeed, both the nebula and the cluster are part of the Orion Molecular Cloud Complex – a giant swath of interstellar gas and dust some 1,500 light-years distant, centered roughly on the Orion Nebula.

Since Herschel saw stars "involved" in the nebula, some claim he also discovered the cluster. Again, as with NGC 1977, that's a matter for interpretation. Consider this passage from Adm. William Henry Smyth's *Cycle of Celestial Objects*:

> There is a glow about this object, when viewed under favouring circumstances; yet I cannot assert that the nebulosity in which it is enveloped is clearly seen. But under proper means it is well worth scrutiny; for nebulous stars are certainly among the most remarkable objects in the heavens and perhaps should be distinguished from stellar nebulae in being

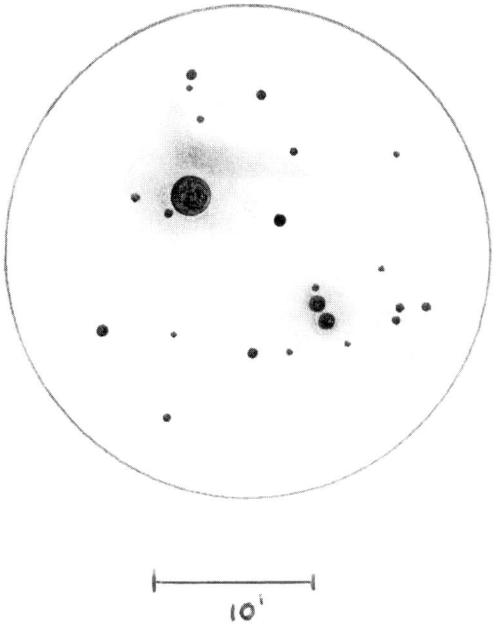

of a less doubtful character, as to the state of condensation, the central matter in such being suddenly vivid and sharply defined.

Smyth's description of Iota being a nebulous star rather than a stellar nebula is most telling. The cluster embedded in NGC 1980 went unnoticed until 1931, when Per Collinder published his seminal paper, "On structural properties of open galactic clusters and their spatial distribution." Today the cluster carries the name Collinder 72 (Cr 72), but don't bother looking for it on any popular star atlas. It's simply not plotted. For that reason, I call Cr 72 the Lost Jewel of Orion. Besides, the cluster's brightest stars form a pattern reminiscent of a large, pear-shaped diamond set in a pendant that dangles from a gossamer necklace. Iota is the blinding glint of sunlight reflecting off one of the diamond's cleanly cut facets.

In a more piratical sense, Cr 72's distinct star pattern bears an uncanny resemblance to a hangman's noose tied to the crossbar of a gibbet at Execution Dock. For more than four centuries, pirates – such as Captain Kidd, John Gow, and other notorious felons – were hanged at Execution Dock, which was on the north bank of the Thames, a mile downstream from the Tower of London. "The gallows was set up on the shore near the low tide mark," David Cordingly writes in *Under the Black Flag*. "After the pirates had been hanged, their bodies were slowly submerged by the swirling waters of the incoming tide." With imagination, NGC 1980, Webb's "spiral mass of haze," can be seen as the swirling waters of the Thames lapping against the nefarious gallows (Cr 72).

THE NGC 1977 COMPLEX

If you doubt M42 has the ability to draw attention away from other objects in the surrounding region, consider NGC 1977 (Hidden Treasure 32). This 6th-magnitude nebula lies only $\frac{1}{2}^\circ$ north of M42. And though NGC 1977 is one of the brightest objects in its class, the nebula was not noticed until 1786, when William Herschel swept it up with his 18.7-inch f/13 speculum reflector. The keen-eyed observer cataloged it as a Class V object, meaning it is a very large nebula. (An explanation of the Roman numeral classes in William Herschel's system can be found on p. 5.

Herschel also mentions that stars are involved with the nebula. In fact, a sparse cluster of about a dozen stars does exist here, and some catalogs prefer to list NGC 1977 as a cluster not a nebula. Since Herschel saw stars in the nebula, does that mean he also discovered the cluster?

Arguably, no. First, the stars would have been too sparse and spread out in Herschel's

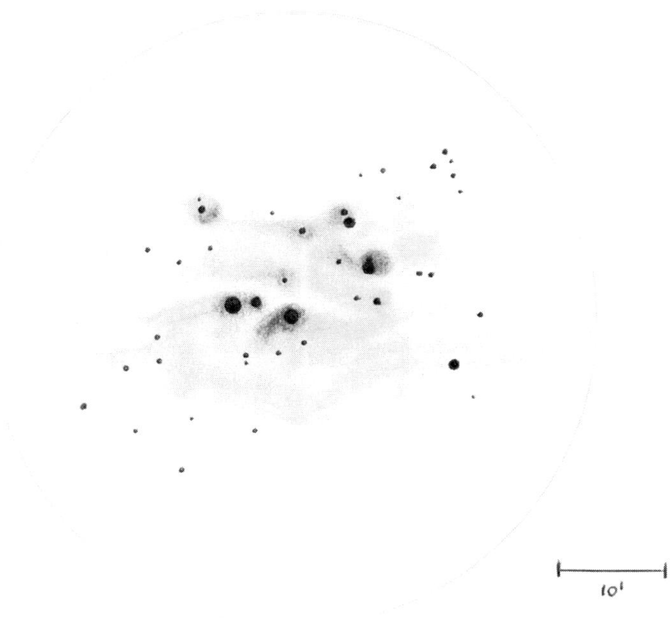

see the stars that we now know belong to a galactic cluster, it is unclear whether he viewed them as a gravitationally bound system, especially since he does not mention which stars are involved in the nebula or how many there are. Consider Herschel's discovery observation of NGC 7076 in Cepheus. "A few [faint] stars with suspected nebulosity, with 300 many very [faint] stars intermixed with the former, so as to make a Cluster." He listed this new object as a Class VII object (pretty much compressed clusters of large or small stars), not a nebula. Interestingly, it is not a cluster at all but a planetary nebula. Since Herschel does not mention the word "cluster" in his observation of NGC 1977, he must have perceived the nebula as being in the process of forming individual, scattered stars, which had not yet gathered under the influence of gravity into a cluster.

Herschel was a skilled observer, but he did not understand, the way we do today, the true nature of the objects he surveyed. Given the historical evidence, William Herschel discovered a nebula, not a cluster, and he classified it as such. Not until 1958 were the stars associated with NGC 1977 officially classified as a cluster (OCL 525.1). This poorly studied group is considered a cluster only because it is part of the Orion Molecular Cloud Complex – a giant cloud of interstellar gas and dust some 1,500 light-years distant. This giant cloud is centered roughly on M42–M43 and includes

tiny 15′ field of view for him to perceive a collected gathering of suns. Second, the idea that stars would be involved with nebulosity was not extraordinary. Initially, Herschel believed that all nebulae resolved into stars, provided they were seen with a telescope of sufficient aperture and magnification. Later, after careful observation of numerous deep-sky objects, he reasoned that all of space must be filled with luminous matter totally unknown to us. This luminous matter (nebulae), he believed, would condense over time into smaller networks of shining clouds, which would then condense into *individual*, scattered stars. Under the influence of gravity, these individual stars would, again, over time, be attracted into *clusters*. Finally, the light given off from these clusters would eventually collect into diffuse clouds. And the cycle would repeat.

Given this knowledge, how does NGC 1977 best fit into Herschel's evolutionary scheme? While it is likely that Herschel did

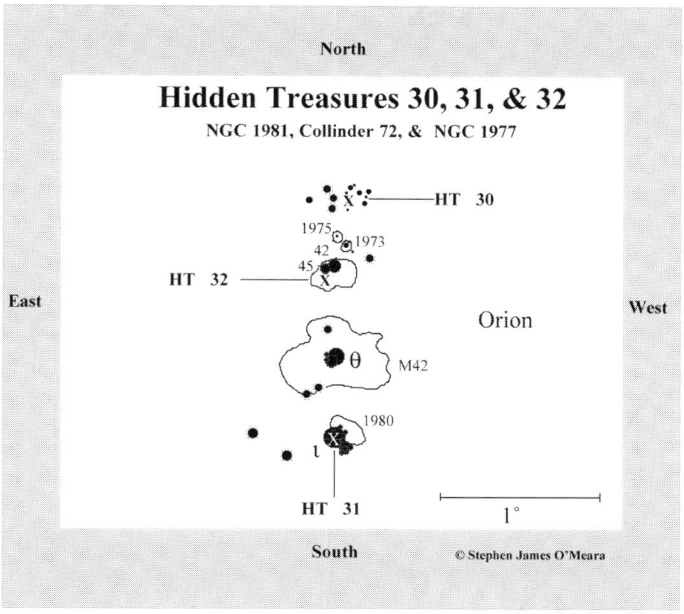

NGC 1977, M78, Barnard's Loop, and the numerous bright and dark nebulae near Zeta Orionis (see Hidden Treasure 34). The cloud, which formed when a density wave triggered the collapse of gas and dust into hot young stars, measures several hundred light-years across. Collectively, this swath of new suns is known as the Orion *OB*1 Association. The association is further divided into subgroups: *OB*1a, which includes the stars in and around Orion's Belt; *OB*1b, which covers the region northwest of Orion's Belt; and *OB*1c, which contains Orion's Sword. The very youngest of these stars, those found in M42–M43 are sometimes considered a separate subgroup, *OB*1d. The stars involved in NGC 1977, then belong to the Orion *OB*1c group of this spectacular realm of telescopic riches.

NGC 1977 is fainter than the Orion Nebula and not as extensive. Still, it is a marvelous sight. A close pair of 5th-magnitude stars (42 and 45 Ori) reside in the nebula's brightest section. Johann Bayer (1572–1625) saw only one star here with the naked eye, which he labeled "c" on his famous *Uranometria* (1603) atlas. Galileo and other telescopic observers resolved the star into two components (c^1 and c^2), and John Flamsteed (1646–1719) relabeled them 42 and 45 Ori, respectively. This explains the curious start to Herschel's discovery description of the object, "The 1st and 2nd c Orionis, and the stars about them," and the 1888 *NGC* description, "*Remarkable,* c^1 42 Orionis and nebula (V 30)."

Since 42 and 45 Ori are aligned with the nebula's major axis, it is difficult to differentiate between true nebulosity and the "fuzzy" nature of two dim stars seen close together with the unaided eye. Then again, I have seen several reports of observers seeing the nebula with the unaided eye, with some saying it is "easy." No matter, the nebula is easy to detect in 7×50 binoculars; through my antique telescope, the two 5th-magnitude stars look like car headlights emerging from a fog.

Through the 4-inch at 23×, which gives a nearly 3° field of view, NGC 1977 is a glorious network of nebulous fabric. It looks elegant, like wrinkled silk. At first the glow has a roughly uniform texture, but that illusion breaks down after spending any time with it. Discerning eyes should see NGC 1977 as three distinct, and near parallel, banks of bright nebulosity, sliced and separated by dark streams of obscuring matter. Pockets of glowing gas, like white foamed reefs, surround a half dozen or so of the region's brightest stars. Each has a tapered tail that, with imagination, appears to have been shaped by wind. Here we see the rugged terrain of a celestial starscape, one sculpted by the turbulent action of luminescent gas, which has been heated by a torrent of energetic ultraviolet light streaming from young hot stars for several million years. In color images, much of the northern and southern tracts of nebulosity shine with a pale sapphire sheen, evidence that starlight is being scattered by minute dust particles within the cloud. The central dust cloud wears a negligee of red light, indicating that strands of hydrogen gas have been excited by starlight to emit a rosy glow. Seen in another way, the bright patches of glowing gas, especially those around the brightest suns, look like spits of sand whose shoulders have been caressed by the constant flow of Orion's dark waters – the bleak and murky dark nebulae. These impressions are much more apparent when seen through a telescope. Photographs give a harsher impression, as if someone had attacked the "soft earth" of this nebula with a rake.

Two of the light islands mentioned above have separate NGC numbers: 1973 and 1975 (see photo). Heinrich Ludwig d'Arrest (1822–1875) discovered them both while at Copenhagen Observatory. He found NGC 1973 on December 16, 1862, and NGC 1975 on October 3, 1864. No doubt these objects stood out as separate objects from NGC 1977 because of the telescope he used – the 11-inch f/17.5 Merz refractor. That telescope is a near twin of the 15-inch Merz refractor at Harvard College Observatory in Cambridge, Massachusetts, which I have used to survey the Orion Nebula and its environs. So I can say with confidence that large-aperture refractors with long focal lengths are notoriously blind to dim and extensive nebulae. Even at the lowest powers attainable, which hover around 180×, these fantastic instruments can magnify any tenuous sheen of a nebula to near invisibility. The eyepieces used with these scopes were the very antithesis of those wide-field ones popular today. Indeed, nineteenth-century eyepieces had infinitely small fields of view, so one could not fully appreciate the grandeur of a large nebula unless it was carefully studied, bit by bit, over the course of hours or days. Only the brightest sections of NGC 1977, then, would have stood out in d'Arrest's eyepiece as he swept the sky for nebulae. Considering that a thick dark lane also separates the brightest region of NGC 1977 (that to the south) from its dimmer northern half, it is not surprising that d'Arrest believed he had stumbled upon two new objects. Of course he had simply picked out two bright regions within NGC 1977.

This fact is reflected in the 1888 *NGC* descriptions of NGC 1973 and NGC 1975: "Magnitude 8–9 star involved in a nebula (V 30)," and "bright double star involved in nebulosity (V 30)." The "(V 30)" stands for William Herschel's H V-30, or NGC 1977. So, though these nebulae have separate NGC numbers, they are part of the same nebulous complex originally discovered by William Herschel. Note that many sources abbreviate the name of this complex by calling it either "1973+," or "1973,75,77." But such designations are deceptive, because they make diminutive NGC 1973 the principal nebula instead of NGC 1977.

Through the 4-inch, the main body of NGC 1973 appears as a circular patch, 5′ wide. It is associated with the variable star KX Orionis and looks like the head of a comet just beginning to shine; with averted vision a stubby tail can be seen flowing to the east. NGC 1975 is a similar but much dimmer "comet" 5′ northeast of NGC 1973. In a fanciful way it looks like the depiction of Halley's Comet in the famous Bayeaux Tapestry, having a round head, narrow neck, and broomlike appendage for a tail.

The entire expanse of the NGC 1977 complex (including NGC 1973 and 1975) covers an area $20' \times 10'$. Look for a dim, wedge-shaped cloud of glowing gas between the bright southern section of NGC 1977 (the part surrounding 42 and 45 Ori) and NGC 1975. It is an extremely subtle expanse of milky light. This section of the nebula was probably too tenuous for d'Arrest to see. If you own a telescope that offers a wide field of view, be sure to use averted vision and sweep the telescope gently back and forth from east to west, then from north to south. The longer you do this, the better your chances are of seeing the extremely long filaments that stretch out to the west and east of the southern section of NGC 1977; the one to the east is bifurcated. The one to the west has a feathery texture. The northern section of the nebula also has long extensions, but only to the east. The west side, beyond NGC 1973,75, looks more rounded and ill defined. Seen together, these long tendrils give the nebula an uncanny resemblance to a skate egg case, otherwise known to beachcombers as the Mermaid's Purse, which is the name I have given to NGC 1977. The nebula's most common nickname, however, is The Running Man. It is based on the photographic impression of the dark nebulosity silhouetted against the glowing gas. When seen with north up, the dark regions look like a man running in full stride.

Now go out on some clear night, when Orion stands prominently in the south, and look at his Sword Scabbard with your unaided eye. And behold, for the first time, the Great Nebula of Orion blazing at the heart of a bright and scintillating row of hidden treasures – the dazzling jewels of the Orion Molecular Cloud Complex.

Hidden Treasure 33

NGC 1999

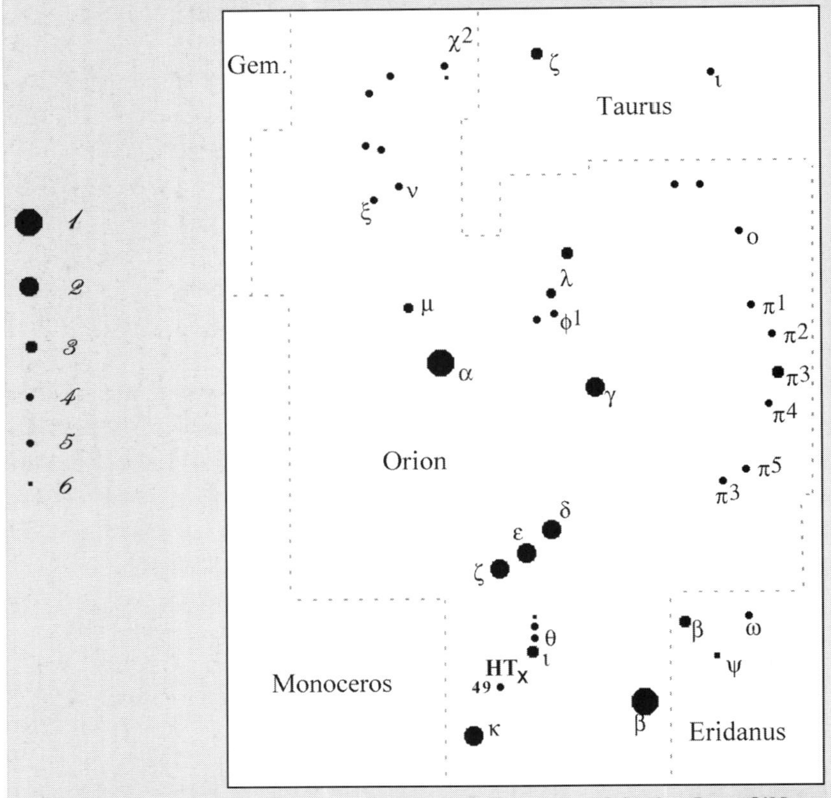

Tirion: Chart 11

Uranometria: Charts 270 & 271

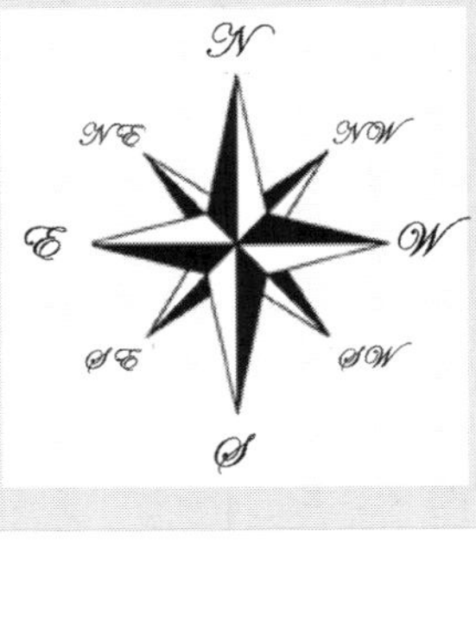

33

The 13th Pearl, Rubber Stamp Nebula
NGC 1999
Type: Reflection Nebula
Con: Orion

RA: 05^h 36.5^m
Dec: $-06°$ $42'$
Mag: 9.5
Dim: $2' \times 2'$
Dist: ~1,500 light-years
Disc: William Herschel, 1785

W. HERSCHEL: [Observed October 5, 1785] A star with milky chevelure or a very bright nucleus with milky nebulosity. (H IV-33)

NGC: Stars of magnitude 10 and 11 involved in nebulosity.

OUR 33RD HIDDEN TREASURE IS THE 33rd object that William Herschel classified as a planetary nebula (Class IV). But NGC 1999 is not a planetary nebula as we know it today – the visible breath of a dying star – but a reflection nebula – the swaddling clothes of a young, hot star. Of course, Herschel only meant that, through the telescope, the visual appearance of the object somewhat resembles the planet Uranus. He described his Class IV objects as being stars with burs, with milky chevelure, with short rays, and remarkable shapes.

NGC 1999 is also a member of my own class of nebulae, which I call Dwarf Nebulae – bright nebulae measuring 10′ or less in diameter; on some colored star atlases, such as Tirion's *Sky Atlas 2000.0,* these tiny nebulae are displayed as thin green boxes about the size of a 4th-magnitude star. I call NGC 1999 the 13th Pearl, not only for its particularly round and polished appearance through a telescope, but because it is one of 13 Dwarf Nebulae in the area surrounding Orion's Belt and Sword Scabbard: IC 423, IC 426, IC 430, IC 431, IC 432, IC 435, M78, NGC 1973, 1975, 1999, 2064, 2067, 2071.

The Deep-Sky Field Guide identifies NGC 1999 as an emission nebula. It is, in fact, a bright but tiny reflection nebula about $1\frac{1}{2}°$ south-southeast of the Orion Nebula. So NGC 1999 does not emit visible light. It shines by reflected, or scattered, starlight. Imagine a pirate ship sailing through a fog at night by lanternlight – wherever the pirates roam, the surrounding fog will shine with a ghostly glow, leaving only darkness in their wake. In the case of NGC 1999, the nebula is being illuminated by the variable star V380 Orionis. This hot, young star has a surface temperature of about 10,000 °C (nearly twice that of our own Sun) and a mass of about 3.5 Suns. The star is so young that the nebula we see is the dust and gas left over from the star's formation.

NGC 1999 has a near claim to fame in the annals of astronomical history: in 1946–1947, the first three Herbig–Haro objects were discovered in images of NGC 1999. Herbig–Haro objects – named after independent discoverers George Herbig (University of California, Berkeley) and Mexican astronomer Guillermo – are small bright nebulae created in rich star-forming regions, where fast-moving jets of material ejected from very young stars collide with the interstellar medium. The collision of these jets with the surrounding gas generates strong shock waves, which move at speeds topping 100,000 miles an hour, exciting atoms along the way and causing the nebula to glow.

Indeed, NGC 1999 lies in the great star-forming region known as the Orion dark cloud, or Orion A. It is a small but beautiful part of the much larger Orion Complex – a series of enormous clouds of interstellar gas and dust centered roughly on the Orion Nebula. In wide-field images taken with large telescopes and hydrogen-alpha light, tiny NGC 1999 (which measures about 1 light-year in physical extent) is surrounded by a much more extensive, spade-shaped array of nebulous patches and filaments, all of which appear to be in a state of ordered chaos. The entire region, which could span 10 million light-years, is rife with bowshocks, marking the locations where vigorous stellar winds from young, hot stars slam into the surrounding interstellar medium.

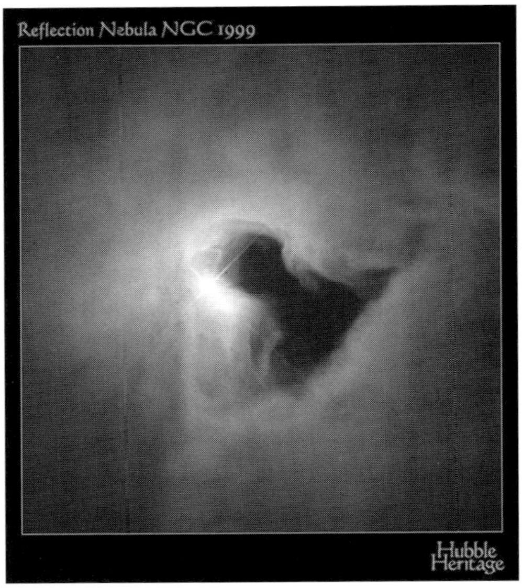

With imagination, NGC 1999, in these special photographs, looks like a hot pearl burning in the palm of a calloused hand whose outstretched fingers radiate with pain. These "fingers" may, in fact, be visible indicators of fast winds ripping through the surrounding dust and gas as it blows outward from NGC 1999's core.

That core, as imaged by the Hubble Space Telescope in January 2000, shows a devilishly dark cloud, resembling a rubber stamp, silhouetted against the brilliant reflection nebula. The dust comprising that dark mass, called Parsamian 34, is so dense that it blocks the light behind it just as a dense cloud here on Earth will hide a full Moon from view. While Parsamian 34 dominates the view in the HST image, it actually obscures only a small portion of the visible nebula.

NGC 1999 is not difficult to find. It is only 50′ south-southeast of Iota (ι) Orionis, the brightest star in Orion's Sword Scabbard. But remember, this is a dwarf nebula, so it is tiny (2′). Fortunately all you have to do is use

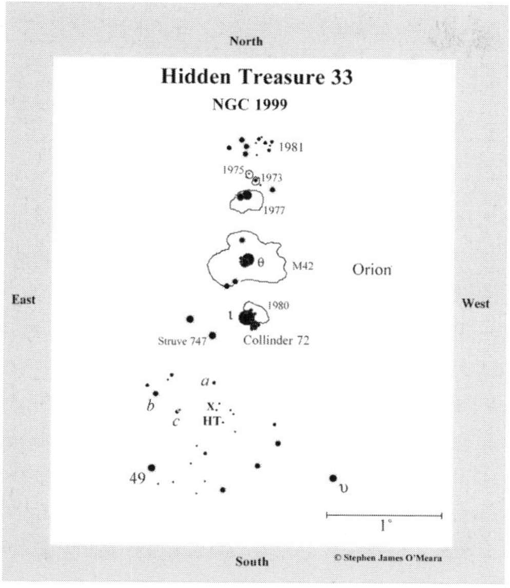

low power to look for the magnitude 10.4 star illuminating the nebula, then you can concentrate on seeing the phantom glow with higher powers. If you live under a suburban sky, start by centering Iota Ori. Next, move 20′ to the southeast, where you'll find Struve 747, a pair of 5th-magnitude stars separated by 36″. Another 20′-hop south of Struve 747 is a magnitude 7.4 star (*a*). NGC 1999 lies only 15′ due south of that star. Observers under dark skies can start by locating 5th-magnitude 49 Orionis, about $1\frac{1}{2}°$ southwest of Iota Ori. Center that star, then move 40′ to the north-northwest, to a magnitude 6 star (*b*). A close pair of stars (magnitudes 8 and 9) lie 15′ to the southwest (*c*). NGC 1999 is 20′ due west of that close pair. As with a small planetary nebula, once you locate the field, use averted vision and wait for a "star" to swell a full magnitude in brightness.

At 23× in the 4-inch, NGC 1999 looks simply like a faint fuzzy star. Still, I was somewhat surprised at how easy it was to see. The few sources that mention NGC 1999 as a target for amateur observation mention that a 6-inch is required to see it. I believe the problem is one of proper magnification. Dwarf nebulae, especially those like NGC 1999, which have a relatively bright star associated with them, require substantial magnification. Fortunately, NGC 1999 is bright enough, and condensed enough, to handle high power.

Under magnifications ranging from 101× to 303×, NGC 1999 is a perfect example of a Class IV Herschelian object, being a star surrounded by a round milky glow, with burs, short rays, and other remarkable shapes. With time, that roundish glow becomes a broken annulus, one that appears more elliptical than round. The ellipse is tilted northwest to southeast, with the northeastern segment being the brightest part. Look for a knot in the northeastern "brow" of nebulosity due north of V380 Ori. If you use averted vision, a fine ray streaks northward from it like a comet's ion tail.

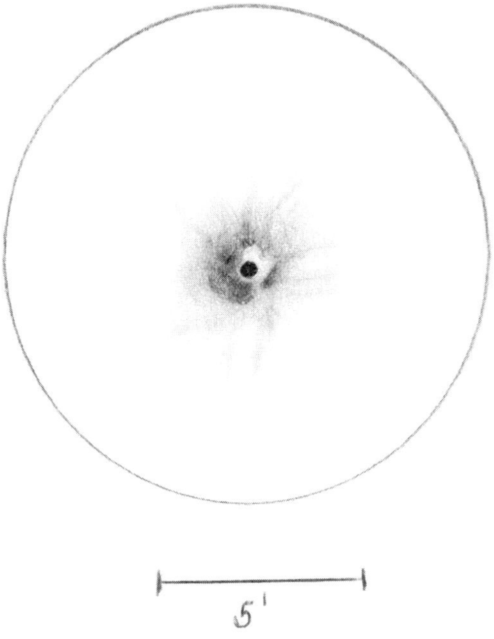

The brow's east-northeastern side appears beaded or mottled. Though fainter, the details here are a near mirror image of those in the northeastern brow. A tiny knot lies due south of V380 Ori and a dimmer one lies on the west-southwest flank of the brow. The principal difference in the two brows is that the northern one is broad and uniform, while the southwestern one is highly serrated, with tiny rays shooting off to the west and southwest. Also, a broader and fainter ray can be seen extending from the southern knot to the southeast. All these details are hyper fine and require a lot of patience to see. But they are apparent under different magnifications. Also, you need good atmospheric stability. If you do not see the rays on one night, try again on another.

As for the dark obscuring cloud, Parsamian 34, it lies about 15″ west-southwest of V380 Ori. I could not see it directly; it's arguable though, that one could infer its presence from the absence of nebulosity in that region in my drawing. Christian Luginbuhl and Brian Skiff definitely saw the dark bay in a 10-inch, and recorded the full "T" or "rubber stamp" in a drawing based on views through the 42-inch reflector at Lowell Observatory.

By the way, Parsamian 34 is an example of a Bok globule – a small cloud of cold gas, molecules, and cosmic dust, named after the late Harvard College Observatory astronomer Bart J. Bok who proposed in the late 1940s that such clouds may be stellar nurseries. Bok argued that these diminutive nebulae, which span less than 10,000 astronomical units, were once attached to umbilical filaments of neutral hydrogen gas. Radiation from nearby stars separated, then compressed, them, and the orphaned dark clouds continued to contract under their own gravity until they formed the tiny globules we see today. Keith Noll, an astronomer at the Space Telescope Science Institute in Maryland, notes that Parsamian 34 probably has enough mass to form one or two stars.

The story behind the creation of each Bok globule is like that of the legendary black pearl. According to Tahitian mythology, the black pearl illuminated the heavens and was used by gods to create the firmament of bright stars. Today these precious gems are found only in black-lipped oysters. And as with Bok globules, the luster of a Tahitian black pearl can reveal Nature's deepest mysteries.

Hidden Treasure 34

NGC 2024

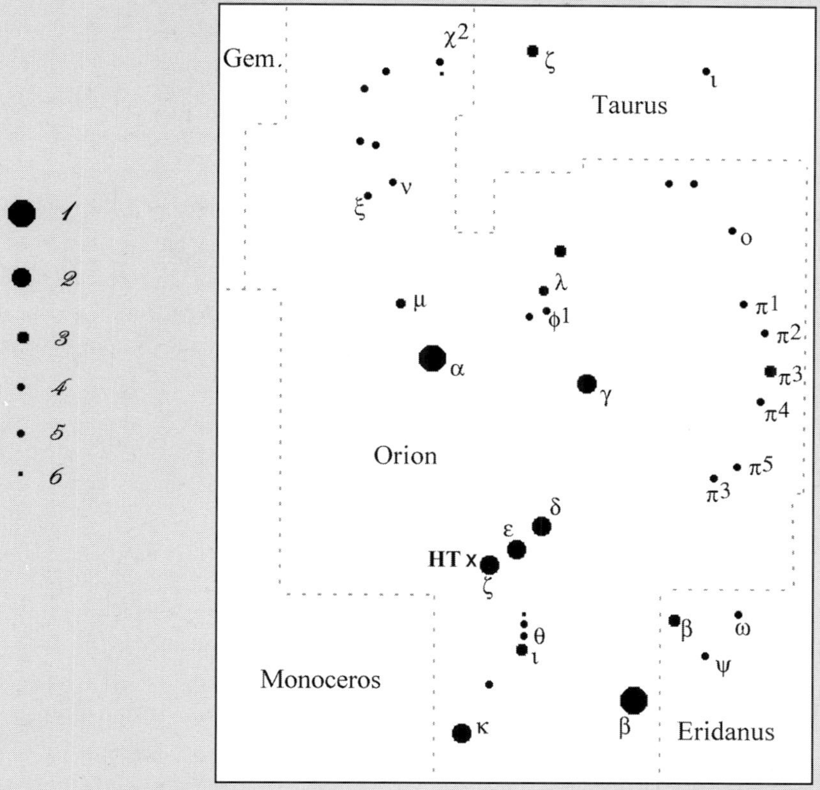

© Stephen James O'Meara

Tirion: Chart 11

Uranometria: Chart 226

34

Lips Nebula, Flame Nebula, Burning Bush Nebula, Tank Tracks Nebula, The Ghost of Alnitak, One Piece
NGC 2024
Type: Emission Nebula
Con: Orion

RA: 05^h 41.9^m
Dec: −01° 51′
Mag: 7.2
Dim: 30′ × 30′
Dist: ~1,300 light-years
Disc: William Herschel, 1786

W. HERSCHEL: [Observed January 1, 1786] Wonderful black space included in a remarkable milky nebulosity, divided in 3 or 4 large patches; cannot take up less than ½ degree, but I suppose it to be much more extensive. (H V-28)

NGC: Remarkable, irregular, bright. *Very* large, black south preceding included.

JUST 15′ NORTHEAST OF 2ND-magnitude Zeta (ζ) Orionis (Alnitak), the eastern star of the three forming Orion's Belt, is the bright but elusive emission nebula NGC 2024, a perfect example of a hidden treasure. In color photographs, NGC 2024 is a blazing crown of flamingo-pink light girdled by a frayed, black ribbon of dust. Visually it is a pale, sepulchral glow that could easily be mistaken for an optical reflection – the ghost of Alnitak.

A spectacular H-II region, NGC 2024 is part of the Orion B Molecular Cloud Complex, one of the closest sites of recent and massive star formation. As with all H-II regions, energetic photons from young stars strip the surrounding hydrogen atoms of their electrons, causing the gas in the visible nebula to glow. Indeed, NGC 2024 contains an embedded cluster of young stars, which is being partially obscured by the dark dust lane. The dust is so dense that even the Hubble Space Telescope (HST) could not detect stars in visible light toward the cluster's core. Infrared observations, however, have detected at least 94 cluster members in an area centered on the cluster, which spans 4.8 square arcminutes. These stars are only some 300,000 years young – about as young as those stars hiding behind dusty cloudscapes in the Trapezium region of the Great Orion Nebula.

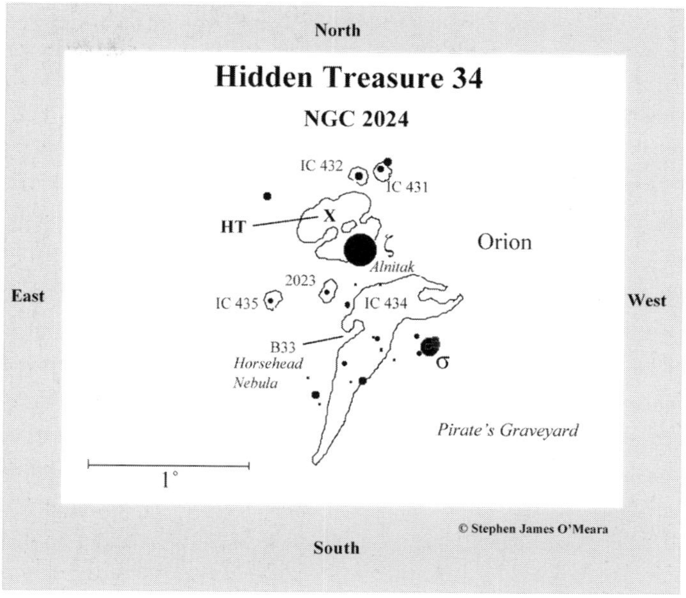

While NGC 2024 has been well studied at all wavelengths, the main star responsible for ionizing the whole nebula, which measures 11.4 light-years across, has yet to be identified – though, Peter J. Barnes (University of Illinois) and his colleagues suggest that several *B*-type stars could be responsible. NGC 2024, in fact, harbors young stars at every stage of evolution: hot *B* stars that ionize the H-II region, the infrared cluster, protostars, and possibly some condensations that may be on their way to becoming protostars. HST and very large array observations of the embedded infrared cluster have also revealed more than 20 possible "proplyds" – disks of dust and gas surrounding newly formed stars – with sizes comparable to those expected for infant solar systems.

If we had telescopes large enough, and eyes powerful enough to see beyond the realm of the visible spectrum, what wonders we could see as we pushed aside the veil of dust and gas that hides the mystical treasures of NGC 2024 from view. Sitting comfortably behind the eyepieces of our telescopes, we could behold in a region of sky no larger than the full Moon all the secrets of stellar evolution. As we focused our attention on one, and then another dusty starscape, we could watch creation unfold before our eyes. We would, in essence, be having silent communion with God.

To find this remarkable nebula just point your telescope to Alnitak, which is from the Arabic word, *Al Nitak*, meaning "The Girdle." Before you go any further, stop, and pay tribute to one of the sky's finest triple stars in the heavens. Alnitak, though it might appear singular in appearance, is a close double star – a magnitude 1.9 primary with a magnitude 3.4 secondary 2.6″ to the southeast. A third companion, shining at magnitude 9.5, lies 57″ northeast of the primary. Thomas W. Webb said the primary pair were "of some nondescript hue, about which observers do not agree." But this is not totally the case. For instance, Adm. William Henry Smyth saw the primary as "topaz yellow" and the secondary as "light purple," William Tyler Olcott saw them as "yellow and blue," Richard H. Allen records them as "topaz yellow and purple," and I see them as topaz and rose. Then again, Ernst J. Hartung (1893–1979) calls them a "brilliant white pair," and F. G. Wilhelm Struve calls the secondary *"olivacea subrubicunda,"* which means "slightly reddish-olive."

Although the primary can be resolved in a 2-inch telescope, William Herschel did not note them. My guess is that either his discovery of NGC 2024 and its "[w]onderful black space" whisked his attention away from that bright star, or the seeing was terrible on that discovery night. Regardless, Alnitak's duplicity went unnoticed until 1819, when the German amateur astronomer Georg Karl Friedrich Kunowsky (1786–1846) first resolved the star in a 4.5-inch Fraunhofer refractor.

Without question, the only reason NGC 2024 is a hidden treasure is because of its close proximity to a bright star. Had the nebula been farther away from its bright companion, it would have not only been discovered sooner but also would have become an object as popular as M42–M43. But don't be deceived by NGC 2024's listed magnitude (7.2), which I determined under dark Hawaiian skies. You have to imagine the light of a 7th-magnitude star spread out over an area the size of the full Moon. Yet, at least under a dark sky, I have seen it clearly with 7 × 50 binoculars, especially if I occult Alnitak with a distant rooftop or some other sharp-edged structure – natural or otherwise.

Telescopically Webb calls NGC 2024 a "[f]aint but extensive" nebula; he also notes that Heinrich d'Arrest resolved it into four large patches. Other sources claim that the nebula is difficult to observe because of its proximity to Zeta Ori. But such phrases underestimate how easy and wonderful this object can appear under the right conditions. Yes, if you are an observer who simply wants to see and move on, NGC 2024 may be nothing more than a splash of dim nebulosity whose light is overpowered by a very bright star. But I have yet to be in the field with amateur astronomers who did not gasp at the glory of NGC 2024 – especially those using nebula and H-beta filters.

In a rich-field telescope, the nebula's core is, actually, quite bright, and its dark lane quite distinguishable. In fact, the view through the 4-inch f/5 Genesis is outstanding. One clear winter's evening in 2003, I logged that I "[n]ever knew so much detail could be seen." In fact, when studied at all magnifications, the nebula and its most intricate details can be detected with just a little patience. Through larger scopes, hardly any effort is required at all to see the black girdle ribs ripping through the glowing gas, separating it into five distinct sections. Each section I have also seen in the 4-inch, though it takes some concentration and careful study.

The brightest segments of the cloud lie on either side of the north–south-oriented dark lane. The slightly brighter eastern side is further segmented by at least three thin dark lanes, which extend radially from the main girdle like ribs from a breast bone. Another bit of tenuous nebulosity wafts ever so slighty to the east – it being separated from the chest of light by yet another north–south-oriented lane of dust. A line of three telescopic stars lie along the eastern edge of the black sternum. The northernmost star shines the brightest and the southernmost one looks the faintest.

The bright western half of NGC 2024 is triangular shaped, with the apex pointing toward Alnitak. Three telescopic stars frame this triangle. And though it may be illusory, Alnitak appears to be surrounded by a dim

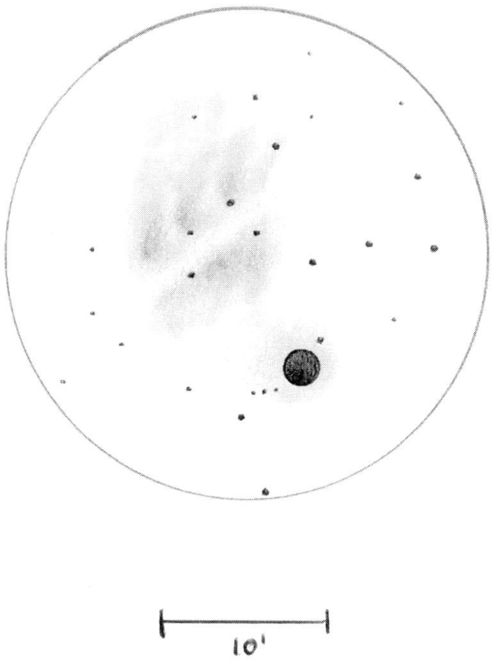

haze of light. If you look at the drawing with northeast up, the nebula looks like a flaming torch, like the one in the hand of the Statue of Liberty.

Seeing these details in a small telescope takes time and patience. But under a dark sky with a large telescope, NGC 2024 comes alive. One memorable view came in January 1990, during my first visit to the Winter Star Party, which is held each year in the Florida Keyes. One night Jack Newton, the renowned Canadian astro-imager, took command of an 18-inch Dobsonian reflector owned by WSP founder Tippy d'Auria, and said he wanted to "tease me" with some sights. He then placed the scope on NGC 2024, which he called the Tank Tracks Nebula. But I saw something different. There in full glory was the imprint of someone's lips on the crystalline sphere of the heavens. A casual scan of the nebula showed all manner of delicate wisps and knots of bright nebulosity interspersed with the intricate folds of obscuring dust. It was as if someone had spilled oil in water and stirred it with a stick – a perfect introduction to what was to become a week of celestial wonder.

PIRATES' GRAVEYARD

As you might imagine, NGC 2024 has many names: Tank Tracks Nebula, Lips Nebula, Flame Nebula, Burning Bush Nebula, and more. But in preparation for this book, I decided to call it One Piece – for the legendary treasure left behind by the Pirate King (Gold Roger) in the Japanese *Manga*[1] by that name. In "One Piece," Gold Roger dies. But just before he does, he reveals that he has buried all his treasure in "one piece" somewhere on the Grand Line – a legendary ocean said to be so dangerous that it is sometimes referred to as the Pirates' Graveyard. In the spirit of the One Piece, then, you can sail across the Pirates' Graveyard (the region around Alnitak) and be on the lookout for Five Ghosts – NGC 2023, IC 435, IC 431, IC 432, and IC 434 – and the Grim Reaper's Horse – Barnard 33. Two things will help your search, averted vision and magnification once the sites are located.[1]

Start with NGC 2023, the brightest of the lot. It lies about 20′ southeast of Alnitak. Look for a 10′-wide glow circling an 8th-magnitude star. While the glow might appear uniformly round, there's some fine wispy structure to it, especially to the

[1] A *Manga* is a professional Japanese comic or anime based on a popular show. The word literally means "irresponsible pictures."

southeast. You'll find IC 435 about 20′ east of NGC 2023. This bright reflection nebula hugs a 9th-magnitude star. Although it is only 4′ in diameter, it too should be easy to spot with magnification. Now return to Alnitak. If you sail about 25′ north you'll find a 7th-magnitude star surrounded by a 5′ glow; that's IC 432. IC 431 is a much more challenging glow, 8′ in diameter, about 12′ to the west-northwest. Look for a close pair of stars (magnitudes 7.0 and 7.7); IC 431 surrounds the dimmer, southeasternmost star in the pair.

Before discussing our fifth ghost, IC 434, swing your telescope over to 4th-magnitude Sigma (σ) Orionis, just 50′ southwest of Alnitak. Sigma Ori is a multiple star of unfathomable depths. Admiral Smyth called it a "fine group of 10 members" that includes a double-triple star. The 4th-magnitude primary has a 5th-magnitude companion 0.3″ away to the east-southeast, so you will need a large telescope to split it. I see 11 stars in the immediate area in the 4-inch. The brightest ones form a linear quadruple (the Sigma system) and a slender triangle of 8th-magnitude stars (Struve 761) just 210″ to the northwest. Since Struve 761 shares the same motion through space as the Sigma system, all must be a little moving cluster. Do you suspect a slight fog around Sigma?

The fifth ghost, IC 434, is a long reef of dim nebulosity in our legendary ocean. The reef begins at Alnitak and runs for more than 1° south-southeastward. It includes what is arguably the most famous, if not the most telescopically desired, dark nebulae of all – Barnard 33, the Horsehead Nebula, our hooded Grim Reaper. Seeing IC 434 is no challenge. Larry Krumenaker in New Jersey observed it in a 2.4-inch refractor in the early 1980s. But seeing the Horsehead is another matter. It may be the most challenging popular object to see in the heavens. Still, the late, famous comet discoverer and variable star observer Leslie Peltier (Delphos, Ohio) saw the Horsehead in a 6-inch refractor; Walter Scott Houston glimpsed it with a 4-inch Clark refractor; and a group of four amateurs sighted it with Snow's 4-inch off-axis reflector.

Some observers have mistaken the dark girdle in NGC 2024 for the Horsehead, which is easy to imagine, especially if you get your mental compass turned around. Some observers, enamored with the Horsehead's magnified appearance in images, go searching for something much larger than it really is and expect to see the Horse's mane and pointed face. The fact is, the Horsehead is a tiny notch of darkness halfway down the sharp and straight eastern fringe of IC 434. The Horsehead Nebula spans only about 5′, and its appearance in small telescopes is one that looks more like a faded thumbprint than a horse's head.

Knowing exactly where to look and using averted vision are the keys to seeing it. If you have located NGC 2023 and the 7th-magnitude sun it surrounds, you are within striking distance of the Horsehead. Just look for another 7th-magnitude star about 10′ to the west-southwest. The Horsehead is almost 10′ due south of it. The sharp eastern wall of IC 434 runs right past that star. The Horsehead is simply a west-facing notch in the wall.

The irony is that I have had more difficulty seeing the Horsehead Nebula in an 18-inch reflector without an H-beta filter than I have in seeing it in a 4-inch rich-field telescope at low power without a filter. The fact is, low

MythologyWeb mythology.com

power concentrates the light of IC 434, making the silhouette of B33 more obvious than at higher powers, which magnifies the dim light and thereby reduces the apparent contrast. Of course, if you have a nebular filter, especially an H-beta filter, which transmits nebular light from the hydrogen-beta (Hβ) emission line, the Horsehead can be seen in hand-held binoculars.

At first I liked to call the Horsehead Nebula the Grim Reaper because I've heard amateurs often say that, "Looking for the Horsehead Nebula will be the death of me." Besides, with imagination, the dark nebula could be seen as the head and shoulders of the Death figure with its ominous hood. But I've since changed my mind. You see, one of the earliest mythological renderings of the Grim Reaper appears in Geoffroy Tory's 1525 *Horae*, which shows a skeleton holding a scythe and an hourglass while riding a *black horse*. So now I see B33 as the head of the Grim Reaper's black horse. Besides, who wants to see the Grim Reaper, anyway? As legend has it, whoever eyes that shadowy figure will . . . well, let's just leave it at that!

Hidden Treasure 35

NGC 2163

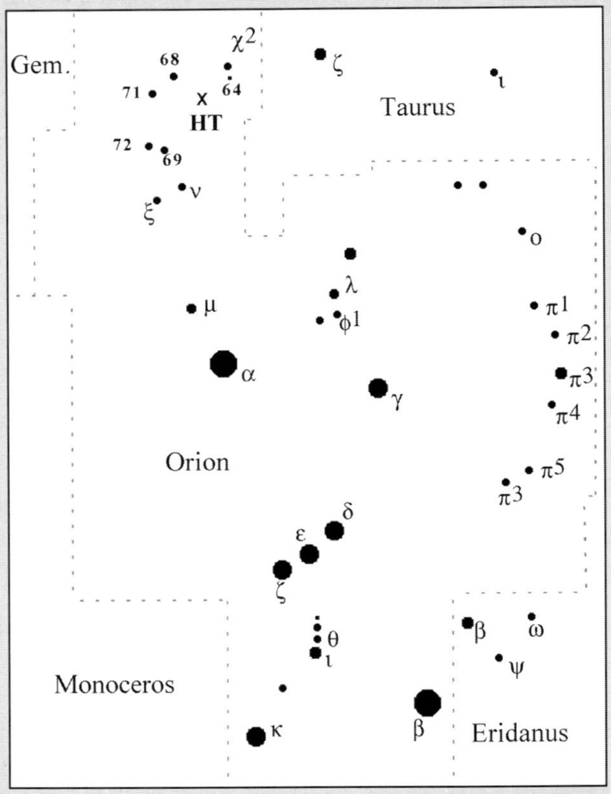

Tirion: Chart 11

Uranometria: Chart 137

35

NGC 2163 = Cederblad 62
Type: Reflection Nebula
Con: Orion

RA: $06^h\ 07.8^m$
Dec: $+18°\ 39'$
Mag: ~11
Dim: $3' \times 2'$
Dist: –
Disc: Edouard Jean-Marie Stephan, during his 1870–1884 observations with the 80-cm reflector at the Marseilles Observatory; Sven Cederblad included it in his 1946 *Catalog of Bright Diffuse Galactic Nebulae*

HERSCHEL: None

NGC: Extremely faint, extended, diffused, magnitude 11 star attached south.

ORION HARBORS A WEALTH OF celestial treasures. It is especially rich in bright nebulae, including the curious reflection nebula NGC 2163. Not many observers have seen it – not because they can't, but because they do not know it exists. Until recently, NGC 2163 was one of the hundreds of objects classified as "non-existent" in the 1973 *Revised New General Catalogue.* But NGC 2163 does exist, and it's a fine object for small telescopes. In fact, it is one of the sky's finest and brightest examples of a bipolar reflection nebula.

Edouard Jean-Marie Stephan (1837–1923) discovered this tiny ($3' \times 2'$) wonder with the 31-inch reflector at Marseilles Observatory in France. It was one of several discoveries he had made with that telescope between 1870 and 1884. In 1887 Stephan sent this list to John Louis Emile Dreyer, who was collecting data for the *New General Catalogue* (*NGC*), which was to be published the following year. Perhaps in haste, Dreyer mistakenly attributed the declination of NGC 1741, a galaxy in Eridanus, to NGC 2163. Dreyer did amend the mistake in his second *Index Catalogue* (in the Notes and Corrections section), but the error carried on into the future for one simple reason: the *Revised New General Catalogue* (*RNGC*) – which became a main source of data in modern catalogs; unfortunately it excluded the two *Index Catalogues* and so, the error of NGC 2163's declination remained unaltered. Actually, the *RNGC*'s authors introduced yet another error to NGC 2163 by

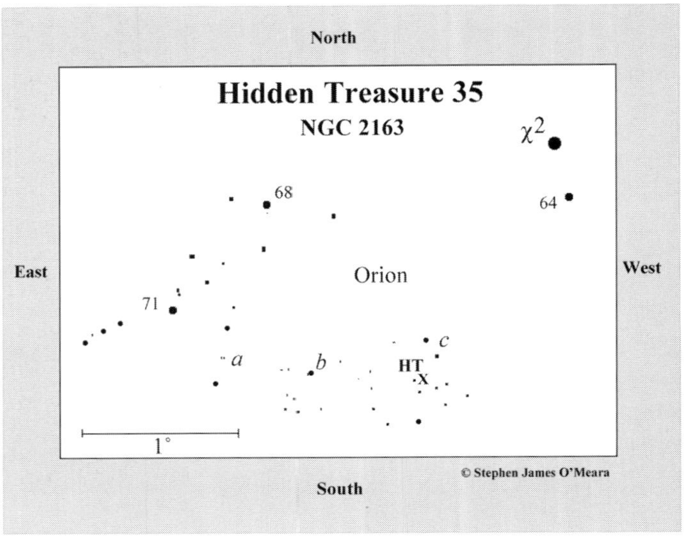

accidentally reversing the already erroneous sign of the object's declination. What's more, since the authors could not locate any deep-sky object at Dreyer's original NGC position on the Digitized Sky Survey (DSS), they listed NGC 2163 as non-existent in the RNGC.

The problem was solved when Brian Skiff recomputed Stephan's original position using precise coordinates for his offset star. What he discovered was a reflection nebula in the location of Cederblad 62 – just 0.45^s in right ascension and 13.7″ in declination from Stephan's original coordinates. Cederblad 62 is the 62nd object in Sven Cederblad's *Catalog of Bright Diffuse Galactic Nebulae*, which is a fairly complete compilation of the knowledge about bright galactic nebulae around 1945. In that work, Cederblad uses Dreyer's corrected position of NGC 2163. "Besides the excellent positional match," Skiff notes, "Stephan described NGC 2163 as 'elongated with [a magnitude 11 star] attached south.' And visually this nebula appears to extend more prominently north of the magnitude 11.5 (central) star." So there is no doubt that Cederblad 62, which appears on several modern star atlases, is, in fact, Stephan's long, lost NGC 2163.

In high-resolution images, NGC 2163 is an interesting bipolar nebula with two symmetrical funnel-shaped jets extending north–south from the central star. The star near the nebular apex is the 13th-magnitude irregularly variable emission-line star, LkHa208, which happens to be a fairly erratic young star of the UX Orionis type. NGC 2163 is a young star with strong bipolar winds that are interacting with the material surrounding the star. Light from the star is being reflected from this material producing two symmetrical funnel-shaped optical lobes. The entire object is still partially embedded within it's prenatal molecular cloud, part of which we happen to see as the dark nebula LDN 1574, explaining why the nebula's southern lobe appears more obscured than it's northern counterpart.

NGC 2163 lies in the far northern reaches of Orion, in the tip of the Hunter's Club, just $1\frac{3}{4}°$ southeast of 4.6-magnitude Chi² (χ^2) Orionis or $1\frac{3}{4}°$ west-southwest of 5.2-magnitude 71 Orionis. It's best to star hop to it from 71 Ori. Look 30′ southwest of 71 Ori for two roughly 7th-magnitude stars (*a*), oriented north-northwest–south-southeast with a close pair of 11th-magnitude stars between them. A 40′ hop west will bring you to a magnitude 8.5 star (*b*) with a very close 11.5-magnitude companion to the east-southeast. Now swing the telescope 45′ to the west-northwest, to a pair

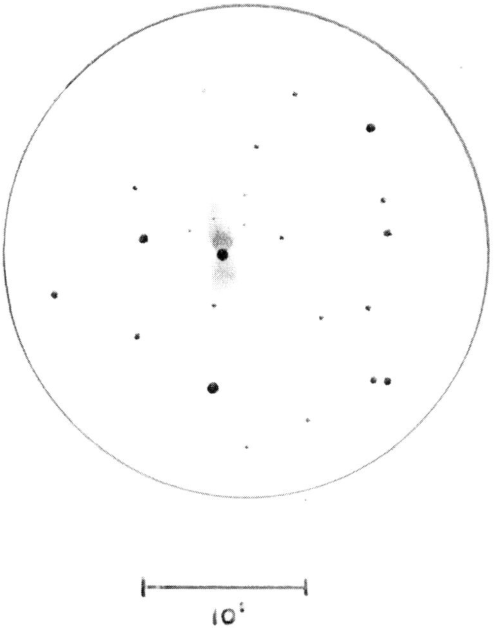

of roughly 8th-magnitude suns, separated by nearly 10′ and oriented east-northeast–south-southwest (*c*). A little more than 10′ to the southeast is a 9th-magnitude star with an 11.5-magnitude companion about 3′ to the west-southwest. That 11.5-magnitude "star" is NGC 2163.

In the 4-inch at 23×, the object is essentially stellar. Increasing the magnification to 72× will turn the "star" into a "tiny comet" with a dim, northward pointing tail some 1′ in extent. It is a high-surface-brightness object, so the nebula takes magnification well. At 216×, the nebula's dual nature is revealed, appearing like a little butterfly with asymmetrical wings. The northern component is clearly the brighter of the two and it appears separated from the star by a tiny spit of dust. The northern butterfly wing is longer on the eastern side and more feathery on the western edge. The southern wing is short and stubby, though it too appears slightly brighter on the eastern side; I cannot separate it from the illuminating star.

Jay McNeil (West Paducah, Kentucky), the discoverer of McNeil's Nebula in Orion,[1] is well aware of NGC 2163. "It has always been one of my personal 'off-the-beaten-path' favorites." He first spotted "this guy" in 1995, with a 10-inch f/5 Cave reflector and a 5.5-inch f/3.6 Comet-catcher reflector on the same night with powers ranging from 100× to 264×. "I noted it in the 10-incher," McNeil notes, "as a soft yet very distinct 1.5′ long fan-shaped glow having a 'fuzzy' 12th-magnitude star near it's southern apex – just as to be expected of a true reflection nebula, the object's surface brightness seems obviously higher towards the star. The faint southern extension was not as apparent as the northern extension. The nebula was fairly easily 'spotted' with the smaller scope."

By the way, McNeil alerted me to "another really cool obscure object that is almost identical in nature to NGC 2163, but easily spotted at powers over 100× in a 5.5-inch Comet-catcher." Its name? Gyulbudaghian 98 – 171 in Cygnus. McNeil says not to let the long name scare you. "This little guy is one of the sky's brightest examples of a cometary reflection nebula – after NGC 2261 Hubble's Variable Nebula [Caldwell 46], of course."

[1] On the night of January 23, 2004, McNeil imaged a new nebula in Orion. It was the first nebula discovery credited to an amateur astronomer since 1939. McNeil found the tiny (1′-wide) 15th-magnitude object on a charge coupled device (CCD) image of M78 taken from his backyard with a 3-inch refractor. Apparently, the nebula flares into view only when its illuminating star (possibly an FU Orionis variable) experiences an outburst. The period of the outbursts is unknown, but it is suspected to be very long.

The illuminating star is LkHa324 (V1982 Cygni), and it also lies at the very edge of it's prenatal cloud – which we partly see as the dark nebula LDN 988. Unlike NGC 2163, however, this curious object's opposing lobe of nebulosity is completely obscured at the visual wavelengths by the remaining cloud of dust. Its epoch 2000.0 coordinates are right ascension $21^h\ 03^m\ 54^s$, declination $+50^\circ\ 15'\ 10''$. Good luck!

Hidden Treasure 36

NGC 2169

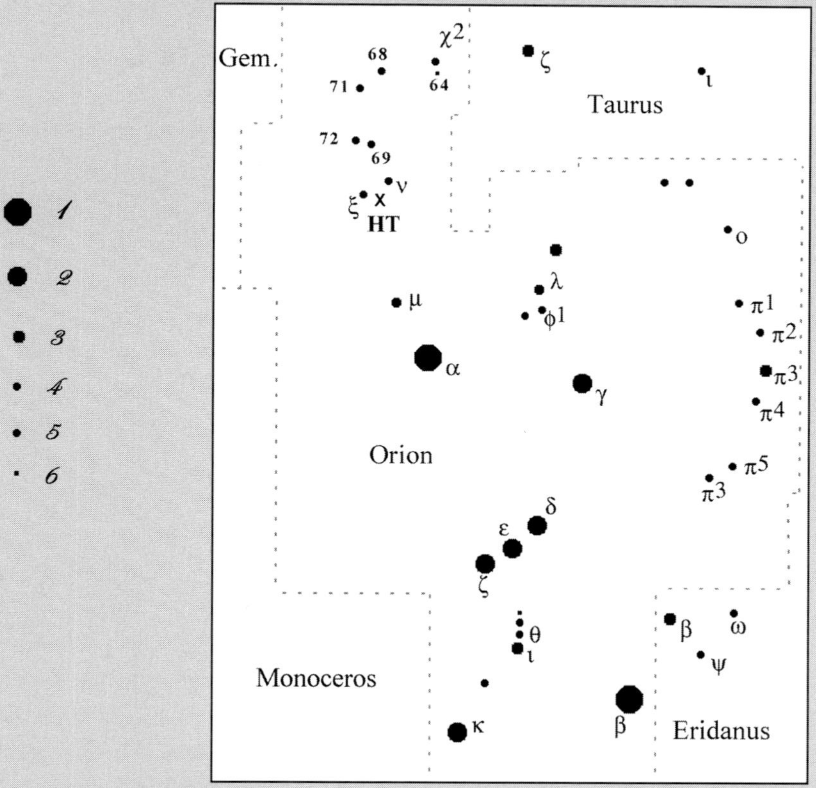

Tirion: Chart 11

Uranometria: Charts 181 & 182

36

Shopping Cart Cluster, Little Pleiades, 37 Cluster
NGC 2169
Type: Open Cluster
Con: Orion

RA: 06^h 08.4^m
Dec: $+13^\circ$ $58'$
Mag: 5.9 (O'Meara); 5.9
Diam: 6.0′
Dist: 3,000 light-years
Disc: Possibly Giovanni Batista Hodierna before 1654; William Herschel independently discovered it in 1784

W. HERSCHEL: [Observed October 15, 1784] A small cluster of pretty [bright] white stars. (H VIII-24)

NGC: Cluster, small, little rich in stars, pretty much compressed, double star (Struve 848).

NGC 2169 IS A BEAUTIFUL AND bright open star cluster in the northern reaches of Orion. The once overlooked Italian astronomer Giovanni Batista Hodierna arguably sighted it before William Herschel. In his 1654 *De Admirandi Coeli Caracteribus*, Hodierna lists an object in Orion's Right Arm that appears nebulous to the naked eye but resolves into stars in a telescope. And while the description seems vague, there is only one object in that location that would resolve into an obvious cluster in a simple Galilean refractor at 20× – the 6th-magnitude open star cluster NGC 2169. And while the nearby star cluster NGC 2194 is a much more popular object for observers with 8-inch and larger telescopes, it shines at magnitude 8.5, which is clearly too faint to see with the naked eye, nor would it be a resolved cluster in an antique scope at low power. But NGC 2169 is visible to the naked eye and its brightest stars appear as a trapezoid of tightly knit suns in my antique refractor; it looks like a magnified view of M42's Trapezium cluster without the nebulosity. So I have no doubt that Hodierna discovered this fine object.

While some might argue that Hodierna could not possibly have noticed the 6′-wide cluster as a hazy spot with the naked eye, he didn't have to. More likely, he saw the cluster's dim light combined with the light of the nearby 4th-magnitude stars Nu (η) and Xi (ξ)

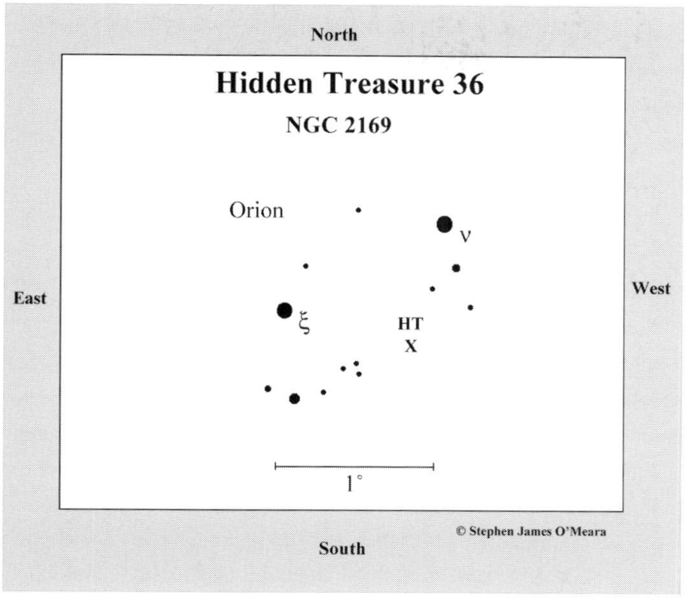

This 6th-magnitude gem is easy to find. Just raise your binoculars about 35′ southwest of the midpoint between Xi and Nu Ori (which comprise Orion's right hand). You should immediately see a tight double star centered in a fine mist. By the way, NGC 2169 lies very close to the middle radiant of the Orionid meteor shower, which peaks around October 18th; these meteors are debris left behind by periodic Comet Halley.

At 23× in the 4-inch the bright stars in NGC 2169 shine like glitter on a gift box. The cluster's 30-odd members are separated into two distinct groups of bright stars – one to the northwest, the other to the southeast. A bright pair of stars can be distinguished in each group and they are oriented nearly perpendicular to one another – sort of like the Double Double in Lyra. With magnification and south up, each pair can be seen in a group of stars shaped like the lower case Greek letter lambda (λ). The northwestern lambda is short and squat; the southeastern lambda is long and lean. These groupings are so bright, they should be a cinch to see from city skies.

With averted vision the lambda has a diffuseness about it suggesting more stars. And indeed it does. At 101×, NGC 2169 is a beautiful spectacle. A quick count shows 14 obvious suns with two dozen visible with concentration. The brightest member shines at 7th magnitude and is the famous double known as Struve 848. Although William Herschel first saw them, these stars were

Orionis, which, when seen together, have a misty, albeit illusory, quality to them. Had Hodierna turned his telescope toward that haze, he would have certainly noticed the beautiful cluster within.

In a 2001 American Astronomical Society paper, C. Yost and his colleagues from Appalachian State University determined that NGC 2169 is 15 million years young. The cluster is similar in age to NGC 1502 (Hidden Treasure 23) and NGC 663 (Caldwell 10), which are 5 and 9 million years young, respectively. Most of NGC 2169's stars are of spectral type *A*3 or younger, though a possible population of pre-main-sequence members could be present. Yost also notes a lack of nebulosity, either bright or dark, associated with it – a curiosity first reported in 1955 by J. Cuffey (Goethe Link Observatory) and S. McCuskey (Warner and Swasey Observatory). Apparently the dust and gas were used up most efficiently during the process of star formation, leaving very little for the interstellar medium.

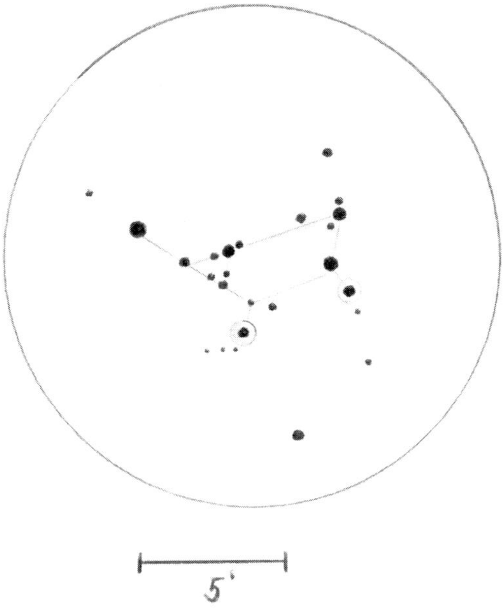

first measured by Wilhelm Struve in 1825. The magnitude 7.3 primary shines with a comely blue-white luster, while its companion – a magnitude 8.1 star 2.3″ to the east-southeast – is buff white.

On one very clear night in November 2003, the companion, oddly enough, appeared to have a dusty rose color. But this color shift may be more indicative of my local atmospheric conditions than anything else; that night, a thin volcanic smog emanating from the summit of Kilauea volcano must have tinted the otherwise clear air over my observing location; that night, to find clear skies, I had to drive down the volcano's southeastern flank and observe downwind of a sulfur-laden plume.

You need only a little imagination to see NGC 2169's box-like arrangement as the Trapezium (and its attendant stars) in M42, but greatly magnified. The cluster can also be viewed as a Little Pleiades. But NGC 2169 already has a great nickname – the Shopping Cart; and once you see the cluster in that way, there's no getting the vision out of your head. I suppose it could also be called the Baby Carriage, but that's up to you and what you want to see. For instance, the cluster is also known as the 37 Cluster, because the stars seem to spell out "37." The name originates from Alan Goldstein, in the 1981 March–April issue of *Betelgeuse*, the official newsletter of the National Deep-Sky Observers Society.

Larger telescopes will certainly show a greater background field. But I do not applaud the number of stars in NGC 2169 only its simplicity. Admiral William H. Smyth was particularly touched by the cluster, whose beauty and wonder caused him to ponder the very creation of the universe:

> These gatherings occurring indifferently upon the Via Lactea [Milky Way] and off it, awaken still more our admiration of the stupendous richness of the Universe, in every department of which there appears such a profusion of creation, if we may so express ourselves of the works of the ALMIGHTY, in which our utmost ken has yet never detected any redundancy, much less anything made in vain.

Then again, I was surprised to read these words by Walter Scott Houston: "Were [NGC 2169] not already in the NGC, I would never have recognized it as a cluster." Such is the way of human opinion, no matter what one feels, it is justified.

Hidden Treasure 37

NGC 2175

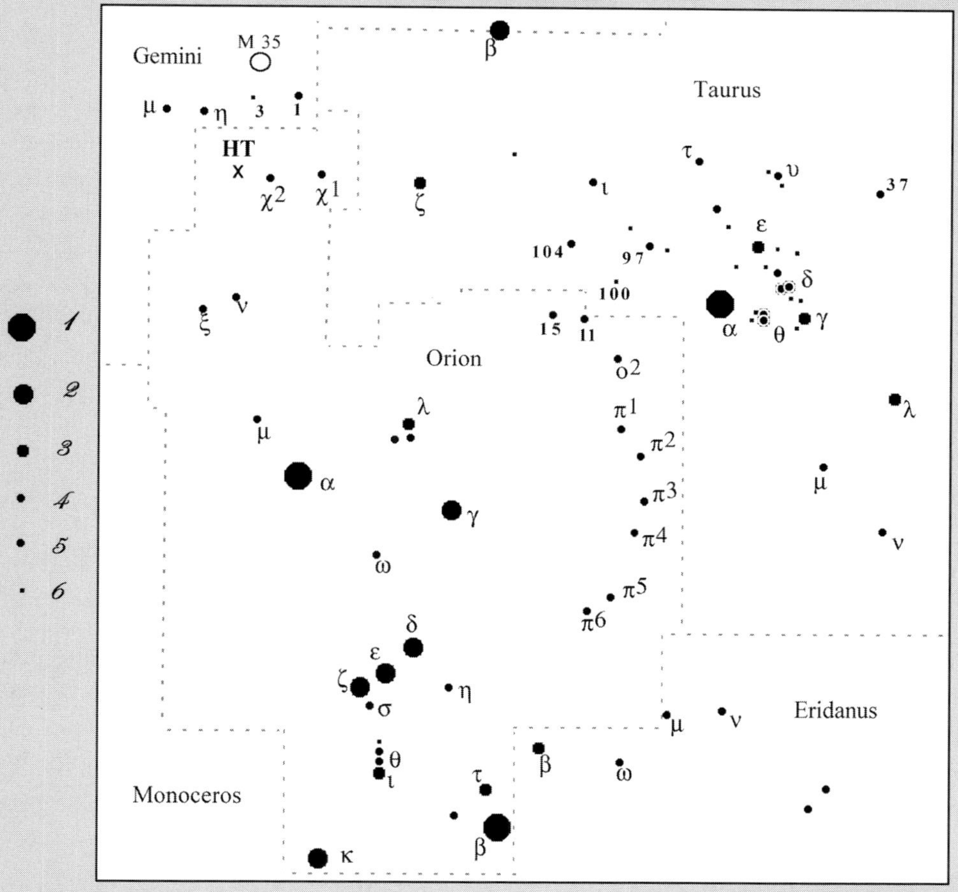

Tirion: Chart 5

Uranometria: Chart 137

37

NGC 2175
Type: Emission Nebula
Con: Orion

RA: 06^h 09.7^m
Dec: +20° 30′
Mag: 6.9 (O'Meara)
Diam: 40′ × 30′
Dist: 7,200 light-years
Disc: Carl Christian Bruhns ~1857; Giovanni Batista Hodierna possibly discovered the nebula before 1654, though his description is insufficient

HERSCHEL: None.

NGC: Magnitude 8 star with nebula (Auwers No. 21).

NGC 2175 IS A DELIGHTFUL NEBULA IN the northernmost recesses of Orion, near the border with Gemini. Under a dark sky, it is visible in 7 × 50 binoculars, but I wonder how many have seen it? I certainly was not aware of it until I swept it up in a comet search. Actually, the reason the nebula is easily overlooked is because it has also been a source of confusion. In most modern star charts and atlases, NGC 2175 is plotted as an open star cluster associated with some diffuse nebulosity (NGC 2174) to the north. But this is not the case.

NGC 2175 is a diffuse nebula discovered by Carl Christian Bruhns (1830–1881) around 1857, apparently while searching for comets with a small comet sweeper at Berlin Observatory. It's possible that the Italian astronomer Giovanni Batista Hodierna discovered it with a simple Galilean refractor and 20× sometime before 1654, though his description appears to be insufficient. Arthur Auwers first reported it as the 21st object in his list of new nebulae in William Herschel's *Verzeichnisse von Nebelflecken und Sternhaufen* (Königsberg, 1862). John Herschel included it in his *General Catalogue* as GC 1366, which later became NGC 2175. "So clearly," write Brent Archinal and Steven Hynes in their book *Star Clusters*, "the name NGC 2175 should be assigned to the nebula only."

Archinal and Hynes do note, however, that a possible cluster was discovered here by Per Collinder, who included it in his 1931 paper "On structural properties of open galactic clusters and their spatial distribution." If the cluster exists, its designation is Collinder 84. "The reality of the cluster itself is in some doubt," say Archinal and Hynes. "An examination of it visually or on the [Digital Sky Survey] shows a fairly low star density and no obvious cluster involved with the NGC nebula." If NGC 2175 is the nebula and

Collinder 84 is a possible cluster, what then is NGC 2174? It is a small wisp of nebulosity in the northern part of the larger complex that Jean Marie Edouard Stephan IX discovered. He described it as being "Extremely faint, between 3 very faint stars." (It lies at right ascension $06^h\ 06^m\ 24.7^s$; declination + 20° 39′ 44.1″; epoch 2000.0.)

Today we recognize NGC 2175 to be a part of Sharpless 2–252, an extended H-II region in the direction of the galactic anticenter. If we accept NGC 2175's distance to be 7,200 light-years, it extends some 84 × 64 light-years in true physical extent, which makes it about twice as extensive as the Great Orion Nebula well to its south. Had NGC 2175 been at the distance of M42, it would cover more than 3° of sky and shine at 5th magnitude. The H-II region is excited mainly by the magnitude 7.6 star HD 420888 (spectral type *O*6.5V) at the nebula's heart.

Several papers in professional journals discuss the stars apparently associated with the H-II region. As early as 1970, P. Pismis divided the stellar region into two groups, which are now commonly called NGC 2175main and NGC 2175small – but, as Archinal and Hynes point out, these should be called Collinder 84main and Collinder 84small. Pismis argued that Collinder 84small, "which lies in the northeast corner of the ionized region, has a distance of 10,000 light-years, and, therefore is more distant than Collinder 84main." But a more recent survey of the region by German astronomer L. K. Haikala (Max-Plank-Institute for Astronomy), which is published in a 1995 *Astronomy and Astrophysics* paper, shows that Collinder 84main and Collinder 84small are indeed at the same distance and are therefore probably associated with the same molecular cloud, though they formed at different epochs. Collinder 84main is about 1 million years young – about as young as the Trapezium cluster in the Orion Nebula – while Collinder 84small is on the order of 10 million years young. Star formation in the region, then, has been going on until quite recently.

As reported in a 2005 volume of *Astrophysics*, E. Ye. Khachikian (V. A. Ambartsumyan Byurakan Astrophysical Observatory) and colleagues studied the distribution of 120 *O*-*B*9-*A*2 stars and of the interstellar dust in the direction of Collinder 84main and Collinder 84small. They found 10 star groups (associations) at distances between about 1,300 light-years and 26,000 light-years. Three of these (at distances between about 1,300 and 3,300 light-years) are type-*B* associations. The remaining seven are *OB* associations. Most of the stars in each association are highly reddened – by 0.5 magnitude to 2.5 magnitudes. There is no dust in the space between the associations. Essentially, the researchers conclude, there is no dust inside those groups (associations) which lie at distances greater than 3,300 light-years.

To find the nebula NGC 2175, which is our hidden treasure, not the chaotic star fields or clusters within, I suggest starting at the 3rd-magnitude stars Mu (μ) and Eta (η) Geminorum, which are a little more than 2° southeast of the magnificent open cluster M35 in Gemini. In fact, if you have a polar-aligned equatorial mount, all you have to do is use low power and center M35, which has the same right ascension as NGC 2175. Now swing the telescope 3.8° to the south. If you are under a dark sky, try doing the same with your binoculars and look for a very pale diffuse glow 30′ in diameter, in a region rich with stars. If you prefer to star hop

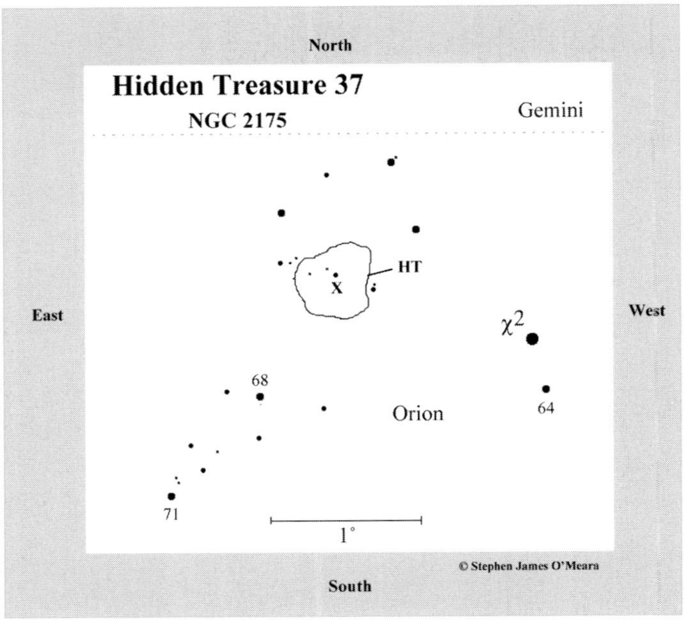

further, find Eta Geminorum, then look 5° to the southwest for 4.6-magnitude Chi^2 (χ^2) Orionis, and the 5th-magnitude star 64 Orionis immediately to its south–southwest. NGC 2175 is nearly $1\frac{1}{2}$° northeast of Chi^2 Ori. The nebula is large and diffuse, so it may be difficult to see from a suburban location. But its intensity under a dark sky is quite apparent, especially with averted vision. Again, consider that it might have been discovered with a simple Galilean refractor in the 1600s.

At 23× in the 4-inch, which gives a near 3° field of view, the nebula appears a vast irregularly round glow of pale uniform light, except for wherever a bright star (like the illuminating magnitude 7.6 star at the center) or group of stars appears. At these positions, with averted vision, the nebula seems to shine more brilliantly. I do not know if this is an illusion. Certainly, higher powers do little to enhance the view, except if you intend to seek out NGC 2174, which is about 10 north-northwest of the magnitude 7.6 central star, near an obvious arc of three roughly 11th- and 12th-magnitude stars oriented north–south. Also Hal Corwin notes that IC 2159 is also a part of the NGC 2175 nebula. "Bigourdan's long focus refractor," he says, "could not show him the entire nebulosity, so he picked up only the two bright knots in it, plus a fairly star-free section of it southeast of the center. It is this southeastern portion that carries the IC number."

Otherwise, the entire field is extremely rich in stars and ghostly nebulosity. It is one object that William Herschel never would have found, given its great size and low surface brightness. But NGC 2175 is one object that binocular observers can see perhaps better than most telescopic observers. It demonstrates one of the greatest ironies in astronomy – that bigger isn't always better.

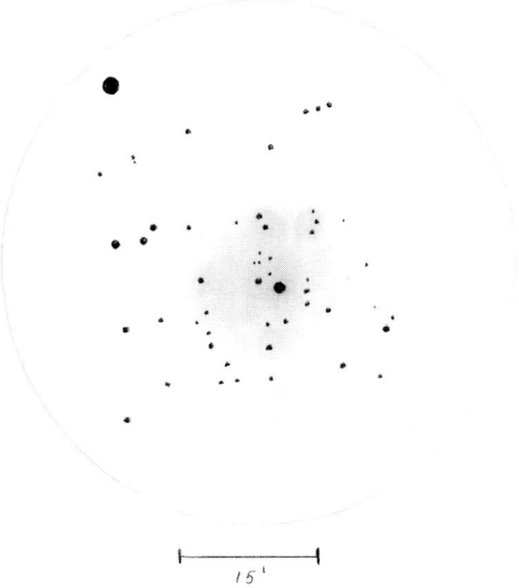

Hidden Treasure 38

NGC 2264

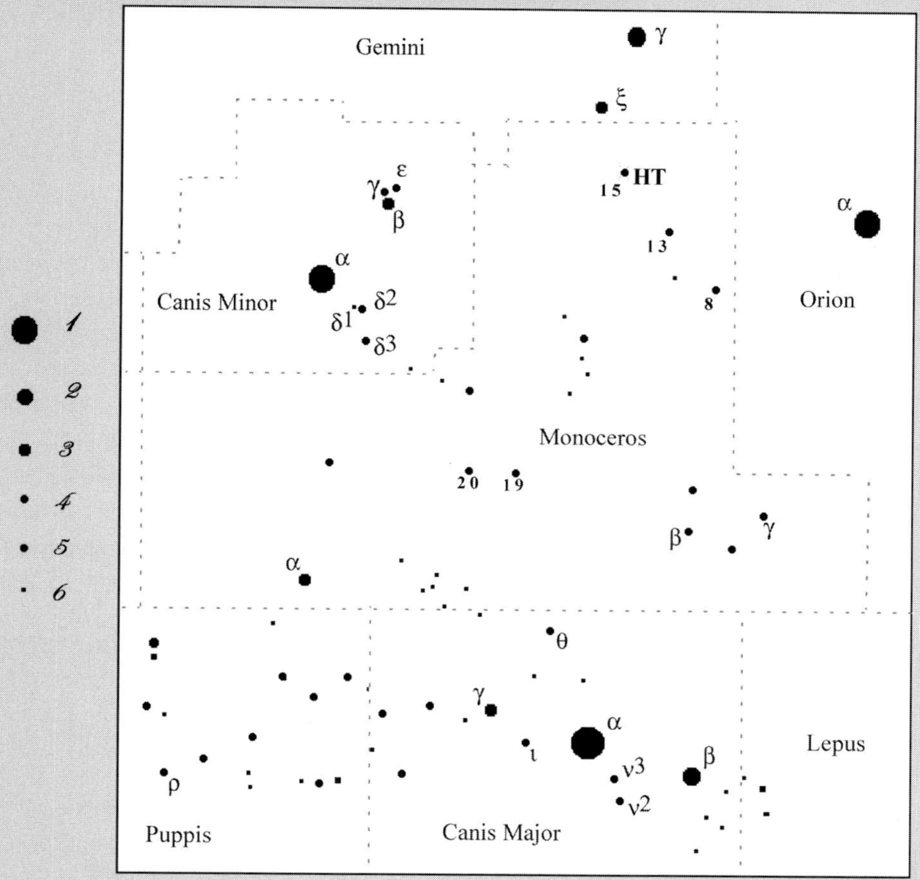

© Stephen James O'Meara

Tirion: Charts 11 & 12

Uranometria: Charts 182 & 183

38

Christmas Tree Cluster
NGC 2264
Type: Open Cluster and Nebula
Con: Monoceros

RA: 06^h 41.0^m
Dec: $+09°$ $54'$
Mag: 4.4 (O'Meara); 4.1
Diam: 40.0′
Dist: 2,500 light-years
Disc: William Herschel, 1784 (cluster); 1785 (nebula)

W. HERSCHEL: [Observed January 18, 1784] Double and attended by more than 30 considerably [bright] stars. (H VIII-5)

NGC: 15 Monocerotis, cluster, double star, nebula?

NGC 2264 IS AN EXTREMELY BRIGHT and extremely obvious open star cluster in Monoceros – one visible to the naked eye and well resolved in binoculars. Miraculously, it escaped the gaze of every observational astronomer until 1784, when William Herschel chanced upon it during one of his sweeps for new celestial objects. To this day, NGC 2264 is greatly overshadowed by its equally stunning neighbor NGC 2237–8,46, the renowned Rosette Nebula (Caldwell 49), and its associated open cluster NGC 2244 (Caldwell 50). Even the diminutive, but exceedingly interesting, NGC 2261, Hubble's Variable Nebula (Caldwell 46), made it into Patrick Moore's Caldwell list. But NGC 2264 did not. The irony: to find Hubble's Variable Nebula, most observers star hop to it from 5th-magnitude 15 Monocerotis, the head of Monoceros, the mythical Unicorn, and brightest star in NGC 2264. For these reasons, NGC 2264 is one of the classic hidden treasures.

When William Herschel discovered the cluster on January 18, 1784, he did not notice any nebulosity. But when he encountered this coarse scattering of stars again on December 26, 1785, he noticed that they "are involved with extremely faint milky nebulosity which loses itself imperceptibly." He cataloged the nebula (H V-27) separately from the cluster (H VIII-5). The history behind these objects got lost when William's son John decided to list the cluster and the nebula under a single listing (h 401) in his original observations, and as GC 1140 in his 1864 *General Catalogue of Nebulae*. When Johann Louis Emil Dreyer published a modified version of the *General Catalogue*, which he called the *New General Catalogue*, he followed John Herschel and listed both the nebula and the cluster as a single object – NGC 2264.

Today we know the Christmas Tree Cluster and the "faint milky nebulosity" is only part of a much larger star-forming region some 3 to 30 million years young. The Christmas Tree Cluster alone extends across more than 30 light-years of space. If this region dates to 3 million years, it was born when tectonic forces formed the Dead Sea (the lowest point on Earth), and when a female hominid we've named "Lucy" hit the ground walking (on two legs instead of four) in the Laetoli plains of northern Tanzania. If the star-forming region dates to 30 million years, then it emerged just as the first hummingbirds began savoring nectar in the flowers of Europe.

In long-exposure color photographs and CCD images, the cluster is surrounded by a symphony of nebulosity whose bright and dark folds seem to ripple across space like visible sound waves. The brightest clouds, which hug S Mon are a fascinating mixture of red fluorescent hydrogen and dark, obscuring dust lanes, some of which are close enough to bright stars to scatter blue light. Some of the wispy tendrils of nebulosity are Herbig–Haro objects, jets of matter ejected from newly formed stars still hidden within the nebula. Indeed, while about 250, mainly Type *O* and *B*, stars can be seen populating this bubbling cauldron of vapors, around 360 near-infrared sources are believed to be forming an embedded cluster in the giant molecular cloud complex behind the obscuring veils of dust.

NGC 2264 is at the core of the Monoceros *OB*1 Association (Mon R1), where a well-defined clustering of reflection nebulae lies at the same distance as the association. The region also includes the majestic, though dark, Cone Nebula, a dense 7-light-year-long stalagmite of obscuring matter. Had the ancients known of such a wonder, wrote Robert Burnham, Jr., of the Cone, they would not have called it by "any lesser name than 'The Throne of God.' Here, as in the Great Orion Nebula, even the modern observer is touched by a strange sensation of having been present at the drama of creation." No words, however, capture the spectacle of this dark cloud better than the image taken by the Hubble Space Telescope. Although it shows only the tip (2.5 light-years) of the Cone – a height that equals 23 million round trips to the Moon – the scene fires the imagination. Some have seen the dark cloud as some nightmarish creature thrusting its head through a sea of blood. To others the scene has religious overtones – the silhouette of Christ holding a child. But these dark shadows are actually pillars of dust and gas seen silhouetted against the raucous red sheen of emission nebulae – hydrogen gas set aglow by ultraviolet radiation gushing from the region's young hot stars. The same radiation is slowly eating away at the clouds, shredding the once dense interstellar material into ethereal tendrils of gas.

Although NGC 2264 is neglected, observers are not totally unaware of it. In fact, for those "in the know," this cluster is a popular winter target, especially in late December, because its shape resembles that of a well-trimmed Christmas tree. Indeed, Leland Copeland called NGC 2264 the Christmas Tree Cluster. To find it, first locate 2nd-magnitude Gamma (γ) Geminorum and 3rd-magnitude Xi (ξ) Gem in the feet of The Twins. Xi Gem also marks the northeastern end of a fainter 10°-long line of four stars

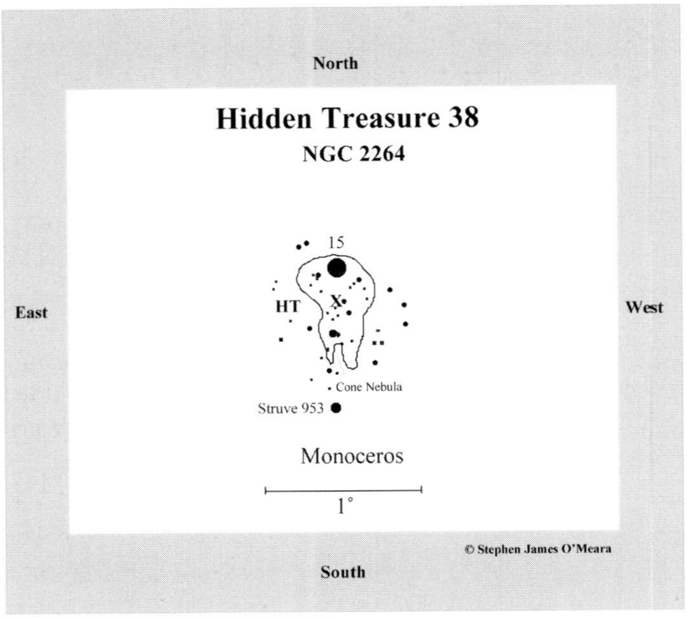

than this pale shade no matter how high the temperature goes." Of course, the color one sees depends on the sensitivity of one's eyes. For instance, Adm. William Henry Smyth saw the star as "greenish." (Bear in mind, though, that this man's vision would often ignite behind the eyepiece, allowing him to see the most vivid colors in the most mundane stars.) The cluster is a remarkable assemblage of some 50 diamonds set against a backdrop of unresolved Milky Way and patches of bright nebulosity. I could also detect a thin stream of nebulosity flowing a few arcminutes southwest from S Mon toward a small grouping of three 8th- to 9th-magnitude suns. This glow is most apparent around S Mon.

that flows to the southwest; the other three stars are 5th-magnitude 15 Monocerotis, 4.5-magnitude 13 Mon, and 4th-magnitude 8 Mon. You want to raise your binoculars to 15 Mon. The cluster will be immediately apparent.

In 7 × 50 binoculars from Hawaii, about a dozen stars are visible. In fact, the Christmas Tree Cluster and Rosette Cluster fit in the same field of view. The Christmas Tree Cluster, in fact, looks like the Rosette Cluster (without the nebulosity), only slightly warped, or dented. In my antique telescope, the Christmas Tree Cluster is bordered on either side by a loop of stars resembling butterfly wings.

The tree's base, 15 Mon, is a famous semi-regular variable star that also goes by the name S Monocerotis. In his book *Star-Hopping for Backyard Astronomers*, Alan M. MacRobert says this luminous young star, whose spectral type is *O*7, is "about as blue a star as you'll ever see; they don't get bluer

Finer and dimmer expanses of glowing gas can be seen to the north and northeast. The rest is truly hyper fine and would be best seen in a nebula filter. A pale reminder of the photographic splendor, a symphony of light and rippling texture, like the finest red silk blowing in an unsteady wind. I did, however, trace other sections of even fainter nebulosity throughout the entire region, especially around the 7th-magnitude star at the tip of the Christmas Tree asterism. Here is where the dark Cone Nebula resides, which, I suppose, is conspicuous by its absence. Certainly no dark drama unfolds before my eyes in the 4-inch. New York observer Sue French, however, notes that the dark finger of the Cone Nebula can just be detected in her 10-inch Newtonian reflector at 44×.

She also notes that an oxygen-III filter gives the nebula a minor contrast boost by helping isolate its greenish emission; other observers have also had success with other light-pollution-rejection filters as well.

Another equally bright star to the south is a fine double (Struve 953), which splits cleanly at 72×. It is comprised of a magnitude 7.1 primary and a magnitude 7.7 secondary 7″ to the north-northwest. I did not notice the blood red star at the cluster's "center" that many people refer to. At 23× it is nice to take my gaze away from the Christmas Tree and look beyond, where I see long bright arms extending to the northwest, northeast, southwest, and southeast, giving the cluster a little more definition. (These are the edges of the butterfly wings.)

A 16th-magnitude T-Tauri star near the Cone Nebula began experiencing "impossibly" long eclipses. The star, called KH 15D, disappears for more than 26 days out of every 48.36 days. It is near the 7th-magnitude star at the tip of the Christmas Tree asterism. The star is supposed to be magnitude 16 at maximum light, which is a cinch for CCD imagers. If you can see this star visually, you'll have witnessed a star possibly eclipsed by a planetary ripple.

By the way, did you happen to notice that William Herschel discovered the nebula on the night after Christmas? That the cluster is so nicely placed during that holiday season is a tacit reminder, of sorts, of how holy our nights can be when we look beyond the confines of the Earth and into the splendor of the universe.

Hidden Treasure 39

NGC 2301

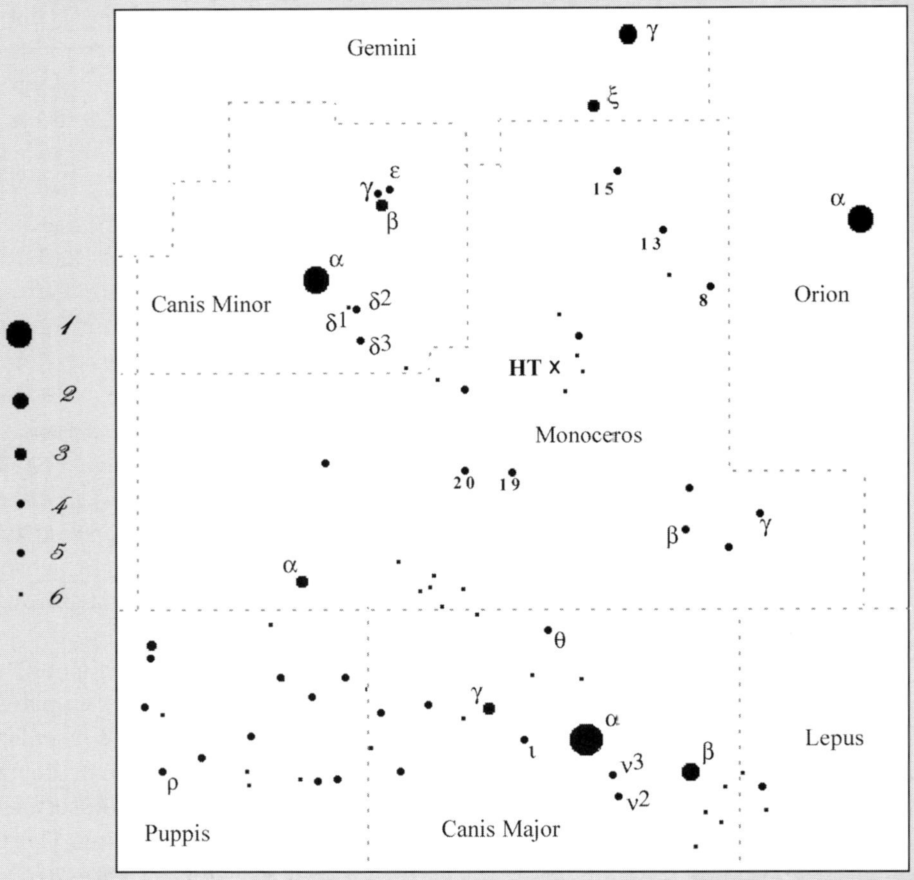

© Stephen James O'Meara

Tirion: Charts 11 & 12

Uranometria: Chart 228

39

Hagrid's Dragon
NGC 2301
Type: Open Cluster
Con: Monoceros

RA: $06^h\ 51.8^m$
Dec: $+00°\ 28'$
Mag: 6.0 (O'Meara); 6.0
Diam: 15.0′
Dist: 2,500 light-years
Disc: William Herschel, 1786

W. HERSCHEL: [Observed December 27, 1786] A very beautiful cluster of much compressed [faint] and [bright] stars above 20′ in diameter. (H VI-27)

NGC: Cluster, rich in stars, large, irregular figure, stars [bright] and [faint].

MONOCEROS, THE MYTHICAL UNICORN, is a relatively large but faint constellation that fills the void between Orion, his two dogs, and Hydra. It is also relatively new, though no one is certain as to when the constellation first appeared in astronomical texts. The German astronomer Jakob Bartsch charted it as "Unicornu" in a catalog that dates to around 1624, but a 1564 work alludes to it as "the other Horse south of the Twins and the Crab." Early Chinese skywatchers knew the four stars that make up the "body" of the Unicorn as the Four Great Canals. And the French scholar Joseph Scalinger found the Unicorn on an old Persian sphere. Johannes Hevelius was the first to detail the constellation's stars in his 1690 *Prodromus Astronomiae.*

And while Monoceros contains no stars brighter than 4th magnitude, the plane of our galaxy slices through the constellation from the northwest, so it is rich in open star clusters – including NGC 2244 (Rosette Cluster), NGC 2264 (Christmas Tree Cluster), and, of course, M50. Tirion's *Sky Atlas 2000.0* shows 21 open clusters in all within the borders of Monoceros. Arguably the most beautiful and illustrious of them is NGC 2301. This 110-million-year-young cluster is a surprisingly bright and obvious object, yet it has been in the periphery of observers' attention. Shining at 6th magnitude, it is a stunning gem, a remarkable treasure, perhaps the finest in Monoceros for small telescope users. The cluster is immediately striking in form and bountiful in stars, which shine at

8th magnitude and fainter. The cluster lies almost exactly on the galaxy's mid plane, and, as a consequence, is projected against a rich stellar background. Indeed in a photometric study of 900 stars in the cluster to magnitude 17, only about 80 members were found. So the visual appearance of the cluster is severely contaminated with field stars. If we accept the cluster's distance of 2,500 light-years, NGC 2301 spans 11 light-years of space.

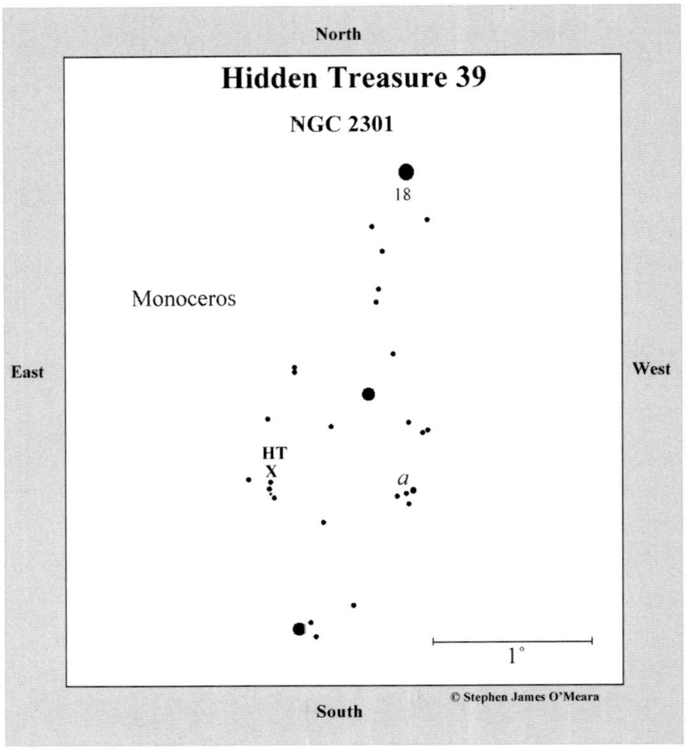

I came across this lovely object purely by chance. One November night in 2003, while waiting for clouds to pass, I noticed a hole opening in the southern sky toward Monoceros. After a few minutes of idle staring, I decided to grab my *Sky Atlas 2000.0* and see what deep-sky objects were accessible in that area of sky. That's when I noticed the tiny, almost invisible, yellow circle on the atlas marking the position of NGC 2301. To pass the time, I turned my telescope to the cluster.

To find it, I followed the same path I took to NGC 2264 (Hidden Treasure 38): first I located 2nd-magnitude Gamma (γ) Geminorum and 3rd-magnitude Xi (ξ) Gem in the feet of The Twins. Next I followed the 10°-long line of four stars comprised of Xi Gem, 5th-magnitude 15 Monocerotis, 4.5-magnitude 13 Mon, and 4th-magnitude 8 Mon. Just 6° to the southeast, the magnitude 4.5 star 18 Mon forms the southeast apex of a triangle with 13 and 8 Mon – note too that the Rosette Cluster (NGC 2244 [Caldwell 50]) will be inside that triangle to the northwest. Now, with your binoculars, center 18 Mon and then slowly sweep the field of view 2° to the south-southeast where you will find an obvious $1\frac{1}{4}$°-long diamond of four 6th- and 7th-magnitude stars. They are the brightest stars in the area. Note that the diamond's western star is actually three 7th- to 8th-magnitude stars in a row (*a*). The diamond's eastern "star" is NGC 2301.

In Chinese mythology, the appearance of a Unicorn is a good omen. The creature only comes to us when we are on an important mission. I guess the appearance of Monoceros in my case was a portent of good fortune, because I certainly had no regrets on making that celestial journey. NGC 2301 turned out to be a brilliant cluster, a dazzling

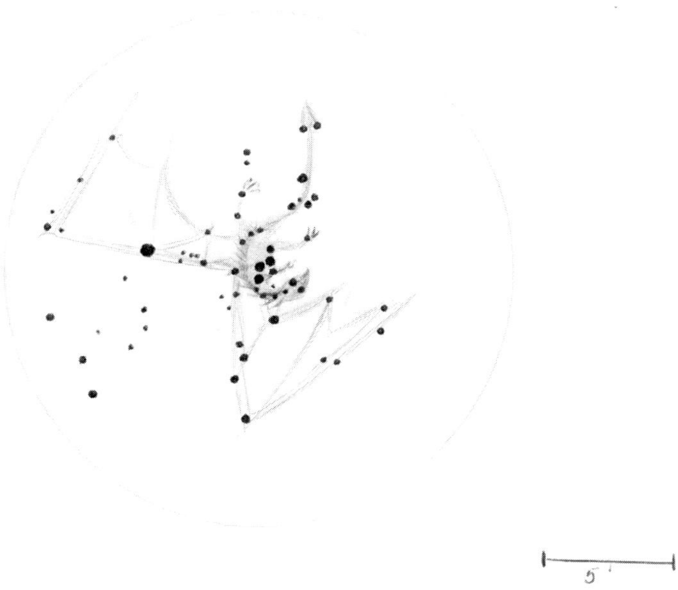

assortment of sparkling gems strung out in a wavy north–south line centered on a stunning double star. This form is easily detected in 7 × 50 binoculars, and the "three arms of a cross," first noted by the Rev. Thomas W. Webb, can be seen in my antique telescope.

In the 4-inch at 23×, the southern extension of the north–south oriented line is comprised of four roughly 9th-magnitude stars that form a brilliant arc leading from the south to the center. At 72×, the cluster's core is a tight bundle of five suns with a bright double star at the eastern end. The northern extension is sharply linear with a double star at the northern tip. The more southerly one has a warm hue. Smyth saw it as a "pale straw" color, Luginbuhl and Skiff call it "reddish," Hartung noted it as "orange."

A weaker arm extends to the east of the central grouping and an even dimmer extension loops to the west. Seen in unison, the cluster takes the form of a serpent body with wings. Indeed, Leland Copeland called it a "curving group with a flying wedge of stars." Likewise, with just a touch of imagination, I see a mythical Dragon in flight. The central gathering of five suns marks the beast's belly. The east and southern arms are the Dragon's raised wings. The straight northern extension is the creature's tail, and the looping northern extension is the Dragon's sinuous head looking Earthward for prey. I have since dubbed this asterism Hagrid's Dragon, after Norbet, Hagrid's baby Norwegian Ridgeback dragon in J. K. Rowlings' fictional *Harry Potter* series. Others have seen similar creatures or things of fancy in this pleasing cluster, which lends itself to all manner of flights of fancy. Members of the Albuquerque Astronomical Society have seen the cluster as a Romulan War Bird. Members of the Saguaro Astronomical Society have seen a Klingon Battle Cruiser. What do you see?

Hidden Treasure 40
NGC 2353

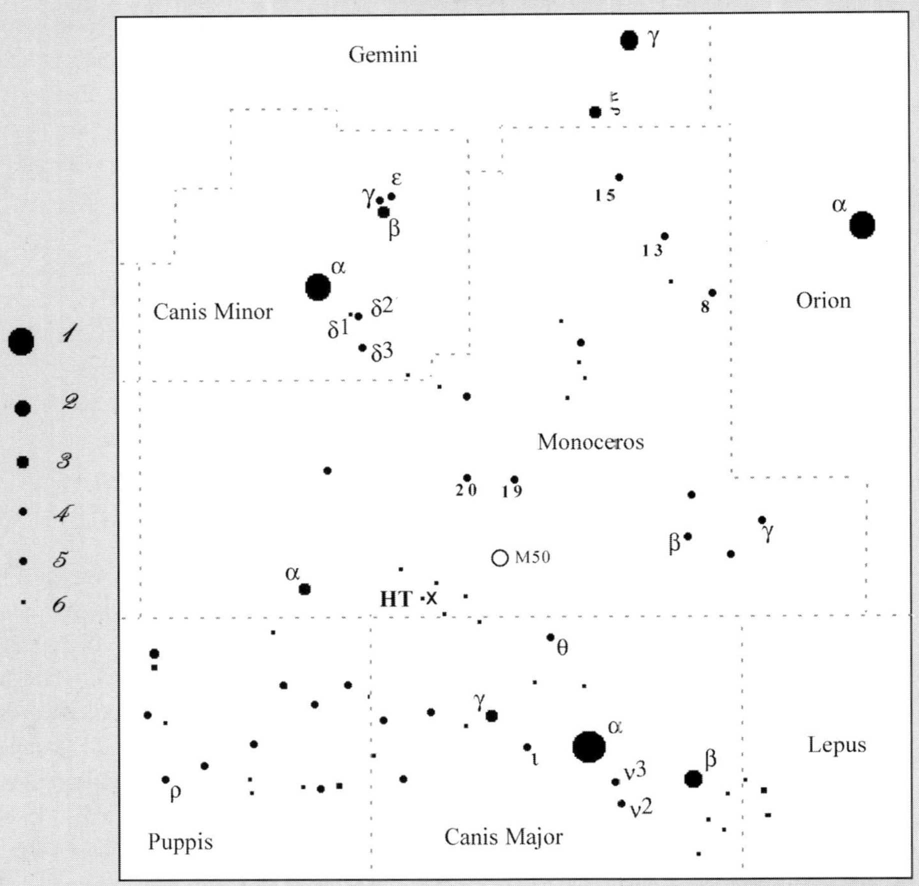

© Stephen James O'Meara

Tirion: Chart 12

Uranometria: Charts 273 & 274

40

Avery's Island
NGC 2353
Type: Open Cluster
Con: Monoceros

RA: 07^h 14.5^m
Dec: $-10°$ $16'$
Mag: 6.0 (O'Meara); 7.1
Diam: 18.0′
Dist: 3,400 light-years
Disc: William Herschel, 1785

W. HERSCHEL: [Observed January 10, 1785] An extensive cluster of scattered stars. (H VIII-34)

NGC: Cluster, large, very little compressed, one very bright star.

In November 2002, I sailed the border of Canis Majoris and Monoceros with my 4-inch refractor, trying to find NGC 2349, one of Caroline Herschel's lost treasures. Although I had no success with NGC 2349, I did discover a wonderland of riches – a field strewn with clusters and dim nebulosity that's scattered across more than 3° of sky. Chief among these new treasures was open cluster NGC 2353, a stunning island of stars, rich stellar sapphires and rubies. So many riches occupied this region that I decided to call it the Pirate's Paradise. I could have made any one of the objects a hidden treasure, but I limited the catch to the brightest and most alluring of them all – NGC 2353, my Avery's Island.

Captain Avery was a seventeenth-century pirate whose capture in 1695 of the *Ganj-i-Swai* – the largest ship belonging to the Great Mogul (emperor of the Mogul Empire in India) – made him a legend. Avery, also known as the Mock King of Madagascar, looted huge quantities of gold and silver from that ship, then wisely retired from his brief career as a pirate. It was rumored that Avery lived out his life on a tropical island (Avery's Island) in the possession of millions. "[B]ut it seems that he was swindled of most of his riches by merchants in the West Country," writes David Cordingly in *Under the Black Flag*, "and that he ended his days in poverty in the village of Bideford in Devon, 'not being worth as much as would buy him a coffin'." No matter, in deference to the legend, I decided to name an island of stellar riches after Avery and place it in a region that can only be seen in the imagination as a perfect retreat for retired pirates – those who want to live out the rest of their days looking out upon the boundless ocean, surrounded by their booty.

This section of sky really is a wonderment. Color images show numerous clusters and stellar groupings tangled in a vast network of reflection, dark, and emission nebulosity in an irregularly round area spanning more than six Moon diameters. The eye can only look upon this beautiful chaos with fuzzy thinking. Here, graceful folds of glowing gas are shaped into long wings by the dark nebulosity surrounding them. The wings are surrounded by van Gogh-like swirls of pastel patches and clumps of starlight. Throughout the region, blue reflection nebulosity shimmers like an afterthought – its hue as light as a warm breeze; a sensual highlight too subtle to ignore.

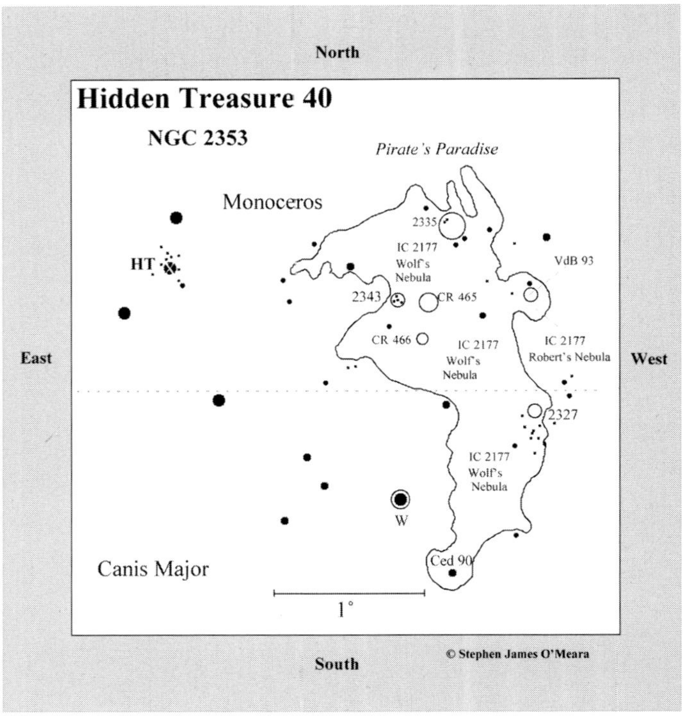

In 1949, a Georgian-born Armenian astrophysicist Viktor A. Ambartsumian (1908–1996) proposed that NGC 2353 belongs to the Canis Major *OB*1 Association – a region in the plane of the Milky Way where dust and gas are being compressed into new and massive stars. NGC 2353 had, in fact, once been called the nucleus of the CMa *OB*1 Association. Controversy ensued. Today we know that NGC 2353 is a moderately rich open cluster located *near the edge* of the CMa *OB*1 Association. We also know that while NGC 2353 is virtually at the same distance and direction as the CMa *OB*1 Association, the two systems are unrelated.

Age tells all. Fernando Comerón (European Southern Observatory) and his colleagues recently announced that star formation in the CMa *OB*1 Association was triggered by a single supernova that detonated about 1 million years ago. This event occurred at the edge of a dense cloud of gas and dust that measured about 3 light-years wide and had a mass of about 1,000 Suns. Then about 100,000 years ago, the propagating blast wave slammed into another cloud of dust and gas, triggering star formation that continues to this day in a region known as Canis Major R1. With an age of about 76 million years, NGC 2353 severely predates these events, but is still a young cluster. Its age is about midway between that of open cluster NGC 2264 (HT 38) and NGC 2301 (HT 39), both in Monoceros. NGC 2353, then, was born at a most interesting time in Earth history – in the late Cretaceous period, when dinosaurs (including Sauropods) were becoming increasingly varied – until their sudden extinction 66 million years ago.

CMa R1 is a lean star-forming region 100 light-years in extent. It is embedded in a curved region of dark, reflection nebulosity, which is part of a larger ring of emission nebulosity that includes IC 2177 and LBN 1036. The reflection nebulosity, though, may be nearer than the Association. And though it's uncertain, the larger ring of emission nebulosity may be a relatively old supernova remnant. Open cluster NGC 2353 lies on the northeast flank of that ring near LBN 1036. An orange 6th-magnitude star (HD 55879) dominates the cluster field. But again, this too is an illusion; the star is not related to the cluster, whose brightest member shines at 9th magnitude. Archinal and Hynes list the cluster as having 106 members, but twice that many may populate the cluster to magnitude 21.0, most of which populate the immediate 5′ of the core.

Herschel classified this bright group as a coarse scattering of stars. Admiral William Henry Smyth later called it, in his inimitable way, "A neat double star, on the [eastern] boundary of a loose cluster in the Galaxy . . . both silvery white. This is a very rich field of stars, in the which is a brilliant oval mass, bounded by a sapphire-tinted 6th-magnitude star, in the [southeast] quadrant, and the pair here measured a little north of it." Smyth also points out that Herschel failed to call attention to these and other doubles in the cluster. But perhaps Herschel did note these stars – and many other pairs. More than one-third of the cluster's stars are binary. Hartung noted this as well, saying it is a "fine open group of stars with a definite concentric or roughly spiral pattern . . . It contains many pairs and small star groups."

If you accept that the 6th-magnitude star superimposed on the cluster marks the position of NGC 2353, then the cluster, arguably can be seen with the naked eye. To find it, first locate Alpha (α) Canis Majoris

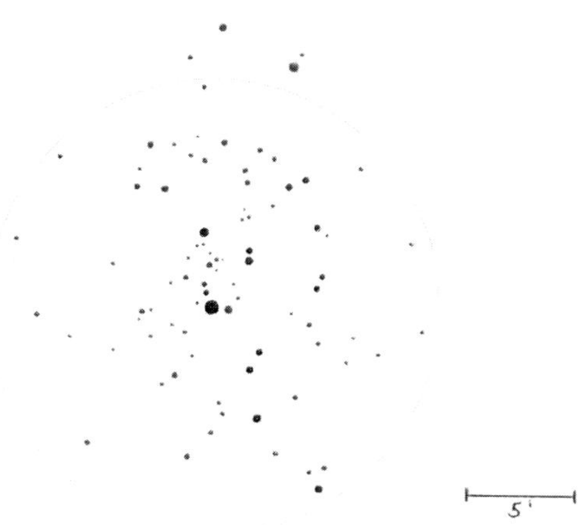

(Sirius) and the two 4th-magnitude stars marking the top of the Dog's head: Gamma (γ) and Theta (θ) CMa. Fifth-magnitude Mu (μ) CMa lies midway between and a little southwest of those two stars. Using your binoculars, draw an imaginary line from Sirius though Mu, then extend that line about 6° to the northeast; you should encounter three 6th-magnitude stars in an arc only $\frac{3}{4}^\circ$ long and oriented north-northwest–south-southeast. The middle of these three suns is NGC 2353. Look for a 6th-magnitude star caught in a web of nebulosity – in this case, the "nebulosity" is that of unresolved starlight.

In the 4-inch at 23×, the cluster is just as Herschel described it – coarse and scattered. The 6th-magnitude star superimposed on the cluster has an orangish cast and looks much the stranger. It lies on the southern outskirts of the cluster, whose dozen suns or so (at this power), seem to flow radially away from that orange sun. Two long arms extend to the northwest and southeast of a weak core of suns huddled around the orange 6th-magnitude star and three weak arms fan out to the northwest. With imagination, the cluster at low power looks like a Sauropod, like a Brontosaurus (an Apatosaurus, if you are a purist), galloping, with head down and neck and tail outstretched, on tiptoes.

At 72× the orange, 6th-magnitude star is centered on a pretty arc of four 9th- to 10th-magnitude suns, whose two northeastern members are the beautiful silvery double Struve 1052, which Smyth first noted. The primary shines at magnitude 8.8 and its magnitude 9.2 companion lies 20″ to the north-northeast. This arc is itself part of a larger assemblage of about two dozen suns in an area spanning about 5′, which at a distance of 3,400 light-years equals about 5 light-years of space, or a little more than the distance between our Sun and the Alpha Centauri system. The entire region out to 20′ (20 light-years) is awash with 11th-magnitude and fainter suns. I counted about 80 at 72×. If I step back with magnification, the cluster has a weak spiral structure whose S-shaped arms are oriented north-northeast–south-southwest. Note that the 6th-magnitude star to the north-northwest of the cluster also shines with a pale orange hue.

PIRATE'S PARADISE

If NGC 2353 is placed on the eastern edge of a 3° field of view, it is but one of five clusters visible in this exceptional Milky Way region. The most conspicuous of the other four clusters is 6.7-magnitude NGC 2343, a rich cluster about $1\frac{1}{2}°$ to the west-southwest; it contains 55 stars in an area of 6′. Just 15′ to its west is Collinder 465 – a roughly 10th-magnitude asterism of dim suns. And a little more than 10′ south-southeast of that cluster is Collinder 466 – an 11th-magnitude cluster of 25 stars packed in an area of 4′. Finally, nearly $1\frac{3}{4}°$ north-northwest of NGC 2343 is NGC 2335, another rich, 7th-magnitude cluster with 57 suns in an area of 7′.

All these clusters are aligned with the long curving nebula IC 2177, most commonly known as the Seagull Nebula, for its appearance in photographs. There is some confusion over the name of the irregularly round nebula forming the Seagull's Head. In Tirion's *Sky Atlas 2000.0*, it is called Gum1. In the *Millennium Star Atlas*, it is part of Gum 1, which equals IC 2177, which is the entire western side of the nebula complex (save for its southern tip, which is the emission nebula Cederblad 90). In the *Uranometria 2000.0*, it is listed as van den Bergh 93. Other references confuse it with NGC 2327. So the Pirate's Paradise is not without its mysteries.

The chart shows the region labeled correctly. Let's start with IC 2177, since its history is itself a bit mysterious. In the 1908 *Second Index Catalogue* Dreyer notes that IC 2177 was discovered by Isaac Roberts, and he references the *Astronomiche Nachriten* (3905). In that work, Roberts announces the discovery of a bright nebula surrounding the 7th-magnitude star BD–10 1848 (HD 53367). This nebula – which is, in fact, the brightest in the region – is the irregularly round patch that marks the Seagull's Head. The IC description confirms that fact: "Pretty bright, extremely large, irregularly round, very diffuse." Roberts makes no mention of the nearly 3°-long arc of nebulosity immediately to the south and east (the Seagull's Wings), so his exposures obviously did not pick it up. Considering these facts, IC 2177 should refer, then, only to the Seagull's Head.

If life were only that simple. When I precessed and plotted Dreyer's 1860 position for IC 2177, I discovered that it did not fall on

HD 53367 but rather in a dark spit 15′ to the southeast (right ascension 07^h 05^m; declination $-10°$ (42′) – between Robert's Nebula and the western flank of the Seagull's Wings. That position is the key to understanding the IC 2177 mystery. In a later *Astronomiche Nachritan* (4082), under the title "Extended nebulosities," Max Wolf, provides a compelling description of the 2.5-long nebula he imaged in the area southeast of Robert's Nebula – the Seagull's Wings. Clearly, while Dreyer uses only Robert's data as the principal source for the nebula's discovery (probably for the economy of space too), he acknowledged Wolf's contribution by splitting the difference in the two object's positions. If we accept that opinion – which by the way, is independently shared by Hal Corwin – IC 2177 represents both Robert's and Wolf's Nebulae. In other words, both the Seagull's Head and Wings are IC 2177. For this work, I have settled the matter by labeling the Head as "IC 2177 (Robert's Nebula)" and the Wings as "IC 2177 (Wolf's Nebula)."

Van den Bergh 93 (vdB 93) is yet another correct label for the Seagull's Head – in part. In a 1966 *Astrophysical Journal* article entitled "A study of reflection nebulae," Sidney van den Bergh (David Dunlop Observatory) published a catalog of BD and CD stars north of declination $-33°$, which are surrounded by reflection nebulosity. While Robert's Nebula is comprised of emission, dark, and reflection nebulosities, vdB 93 refers *only* to the reflection nebulosity surrounding BD -10 1848 (HD 53367). Van den Bergh classifies it as a very bright nebula with an apparent diameter of 20′ on the blue prints of the Palomar Sky Survey, and 17′ on the red prints.

Any source that lists NGC 2327 as the Seagull's Head is wrong. William Herschel discovered this nebula, observed January 31, 1785, and for some reason it has disappeared on many star atlases. Dreyer, in the 1888 *NGC* describes it as "Pretty bright double star involved in small, very faint, nebula."

I recommend anyone under a dark sky to try to detect these various nebulae. They are all truly neglected, obviously because of the distorted histories of the celestial players. When I first chanced upon this field without knowing its history I immediately spied the IC 2177 nebulosity. I called it "obvious and easy at 23×." Once I confirmed its existence on a star chart, I saw it clearly with 7 × 50 binoculars and averted vision.

Through a 10-inch f/5.6 Newtonian and powers varying from 59× to 141×, David Knisely (Prairie Astronomy Club) saw IC 2177 as a "long faint irregular diffuse band of haze." The nebula, he notes, is easier to see with an ultra-high-contrast (UHC) filter or an H-Beta filter, both of which boost the contrast of the nebula's core filaments. Similarly, using a 13-inch telescope at 62× and a UHC filter, Steve Gottlieb viewed the nebula as a "very large, very elongated strip of nebulosity." He found the southern portion difficult to trace, except for the "easily visible" nebula Cederblad 90, which surrounds a magnitude 8.5 star. Cederblad 90 is also known as vdB 94 – a "very bright" reflection nebula that appears 11.2′ in diameter in both red and blue prints of the Palomar Sky Survey.

You can spend hours using your scope to explore this vast region, which, aside from the clusters and nebulosity, is rich in topaz, sapphire, and ruby-colored stars. The most exotic of them all, however, is the fiery red carbon star W Canis Majoris. As with many carbon stars W CMa is also a variable star. Its brightness changes irregularly from magnitude 6.6 to 8.4. The progenitor of W CMa was most likely an *O*-type member of CMa *OB* 1 Association with a mass greater than 20 million Suns and a main-sequence lifetime less than 3 million years. Beholding this star is like seeing a drop of blood – a symbol of a pirate's red rage.

Hidden Treasure 41

NGC 2440

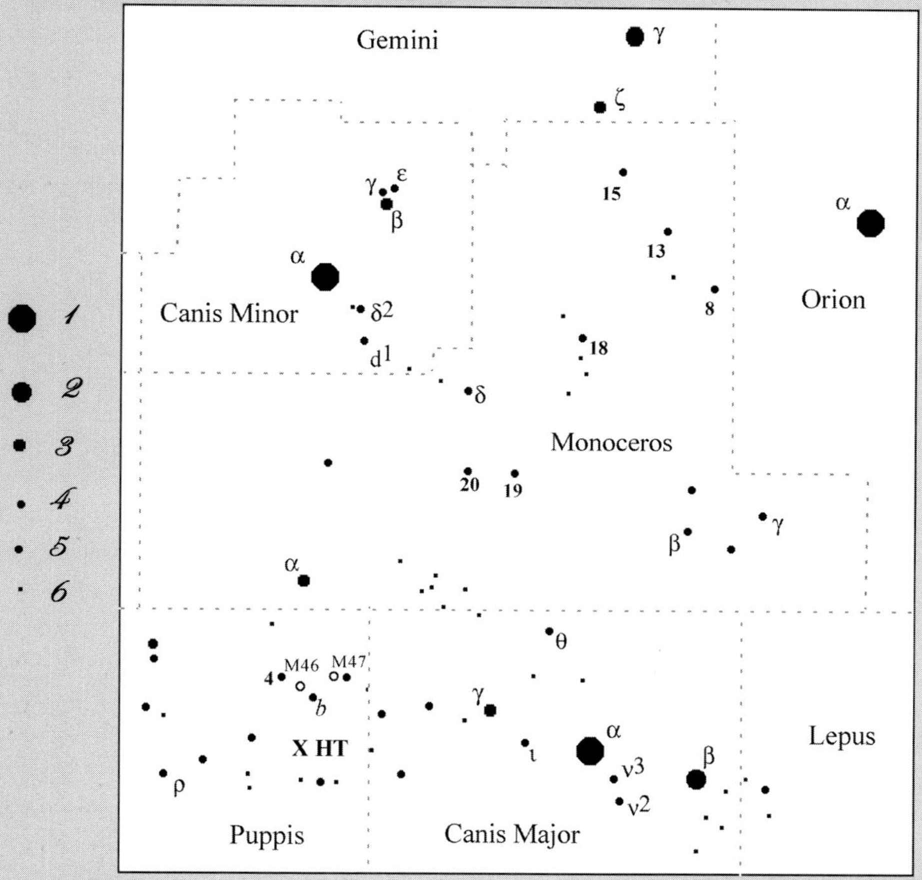

Tirion: Charts 12 & 19

Uranometria: Charts 319 & 320

41

Albino Butterfly, Kiss Nebula, Little Lips Nebula
NGC 2440
Type: Planetary Nebula
Con: Puppis

RA: $07^h\ 41^m\ 55.6^s$
Dec: $-18^\circ\ 12'\ 31''$
Mag: 9.1 (O'Meara); 9.4
Dim: $74'' \times 42''$
Dist: ~1,600 light-years
Disc: William Herschel, 1790

W. HERSCHEL: [Observed March 4, 1790] A beautiful planetary nebula of a considerable degree of brightness; not very well defined, about 12″ or 15″ in diameter. (H IV-64)

NGC: Planetary nebula, extremely bright, not very well defined.

NGC 2440 IS ONE OF THE MOST beautiful and bizarre butterfly-shaped planetary nebulae in the sky. When Rudolf Minkowski (1895–1976) first saw its form on an image taken with the 200-inch reflector atop Mount Palomar in 1964, he called it "an example of an object so complicated it defies description." The image showed a bright, somewhat elongated nebula with two distinct central condensations and fainter outer filaments, all of which was surrounded by a faint halo. NGC 2440's puzzling nature magnified in 1998, when J. Alberto Lopez (Universidad Nacional Autonoma de Mexico) and his colleagues discovered that three pairs of lobes were emerging at different position angles from the nebula's core. Sun Kwok (University of Calgary) notes that NGC 2440 has one of the most prominent multipolar structures of all planetary nebulae. "Many planetary nebulae," he says, "probably also have multipolar structures if one looks hard enough, but none is as easy to see as NGC 2440."

Now Hubble Space Telescope wide-field-camera images complicate matters further. They show the nebula's central star sitting in an irregular hollow surrounded by a shredded chalk-white ring. The shredded area marks the point where fast winds and intense ultraviolet radiation from the central star are eating away at the dense central ring, which is expanding away from the central star at 24 kilometers per second. Several bulbous white polyps (the butterfly's wings) swell away from the ring's hollow; the most prominent of these wings defines

the object's major axis, which lies perpendicular to our line of sight. The images also show "point-symmetric structure," which, as Kwok explains in his 2001 book *Cosmic Butterflies*, is a "technical term referring to blobs and filaments on one side of the nebula being reproduced exactly upon reflection onto the other side," like the letter *S*. These features, for the moment, defy explanation. When all these features are seen together, NGC 2440 looks like a sidewalk artist's rendering of a white rose after a rainstorm.

It is not certain whether the NGC 2440's lobes were ejected simultaneously or at different times. One likely scenario is that they formed when the central star underwent three episodes of mass loss. During each episode, a pair of lobes aligned along a different position angle owing to precession in the central disk. Each of these ejections might have been separated in time by about 10,000 years.

Since the HST images clearly separate the central star from the surrounding inner ring, astronomers have been able to make the most accurate estimate yet for the star's temperature: about 200,000 °C – more than 30 times hotter than our Sun – making it one of the hottest stars known in our galaxy. Yervant Terzian (Cornell University) calls NGC 2440 a proto-planetary, or a very young planetary nebula, which makes sense, given the planetary's diminutive size and the presence of infrared dust in its envelope.

But Kwok has a different opinion: "I won't call NGC 2440 a proto-planetary nebula," he says, "because it has ionized gas. Unfortunately the term proto-planetary nebula has been abused to refer to anything unusual. It is probably a planetary nebula with a high central star mass, as implied by its high central star temperature. Since a high mass star evolves very fast, it is therefore dynamically young. However, in terms of evolution, NGC 2440 (like NGC 7027 [Hidden Treasure 104]) is 'old' (i.e. evolved) for its mass. This is a technical distinction. Most people call NGC 7027 'young' and it is fine (I do that too). The presence of dust is common in almost all planetary nebulae, but for younger ones, the dust has not yet had a chance to disperse, so it is more prominent."

NGC 2440 lies 4.4° almost exactly due south of another planetary nebula – NGC 2438 – in the bright open star cluster M 46, also in Puppis. If you have an equatorial mount, you could try centering NGC 2438 in your scope at low power, then swinging the scope 4.4° to the south. But I would suggest using a moderate magnification (like 72×), because while NGC 2440 is bright, it looks stellar at low powers. If you want to star hop to it, start at M46, where, just $\frac{1}{2}$° to its southwest, you'll find a 5th-magnitude

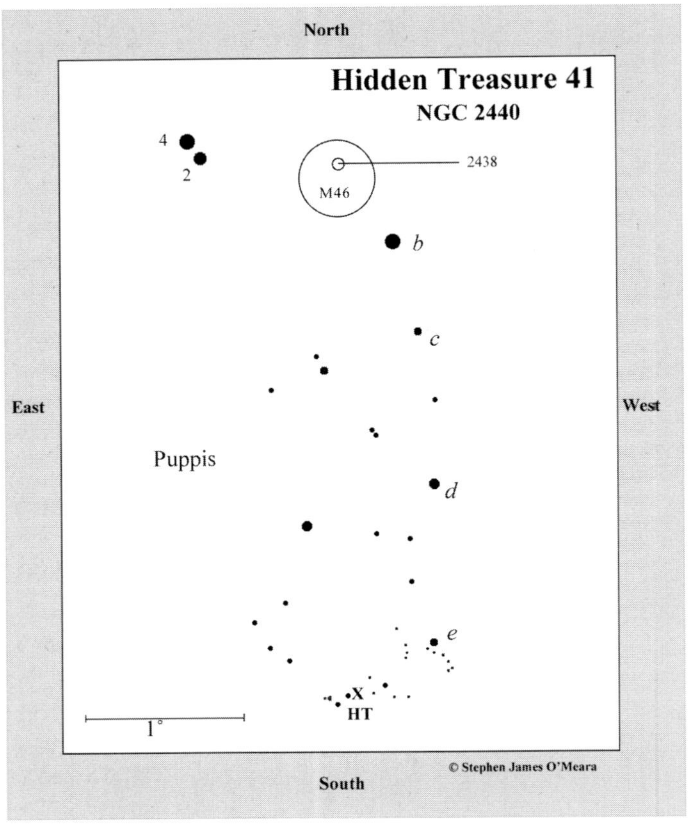

star (*b*). From Star *b*, hop 35′ to the south-southwest to a solitary 7th-magnitude sun (*c*); it marks the northwest corner of a 1°-wide trapezoid of similarly bright suns; use binoculars to confirm the trapezoid. Now, center the trapezoid's southwest corner star (*d*) and move 1° south to another solitary 7th-magnitude star (*e*). NGC 2440 is about 40′ to the southeast. The planetary is the westernmost "star" in a roughly 10′-minute-wide arc of three 9th-magnitude stars, oriented northwest–southeast. The surrounding star field is rich, but just concentrate on looking for that asterism of three close "stars."

Again, at 23×, the nebula is virtually stellar. Reverend Webb agreed, saying it looks like a "dull" 8th-magnitude star at 64×: with more power, "small, brilliant, undefined, surrounded with a little very faint haziness." At 72× in the 4-inch, NGC 2440 simply displays a bright core surrounded by a dim crown of light. It is small (1.2′) but apparent, especially with averted vision. Use averted vision and see if you cannot discern a slight bluish cast to the nebula. Admiral Smyth noted this color too, calling it a "pale bluish-white." Smyth also notes that Herschel registered the nebula as being only about 12″ or 15″ in diameter. "But the inference from such a supposition is vast" he writes, "'Granting,' says [Herschel], 'these objects to be equally distant from us with the stars, their real dimensions must be such as would fill, on the lowest computation, the whole orbit of Uranus'." But as we know today, Herschel was a factor of a thousand off; in true physical extent, NGC 2440 measures 0.6 × 0.3 light-years.

The nebula's character changes at 101×, which shows the nebula's crown ever so slightly out of round, being oriented north-northeast–south-southwest. The inner disk has irregularities with averted vision, but the disk is still too small to differentiate the suggestive details that wag in and out of view. At 168×, the crown is clearly elongated. The core is also more defined, with its western edge appearing as a bead of light. With time and concentration, a tiny dark hole

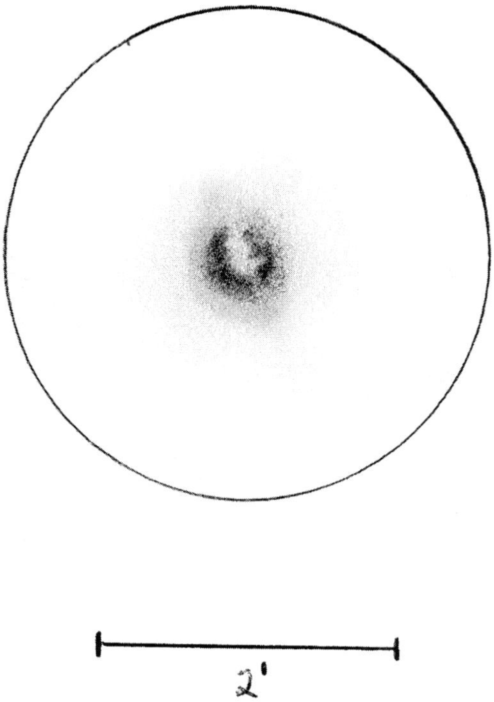

emerges at the core, creating an annulus, whose northeast edge now appears beaded. The more power you can use on this nebula, the better. It takes magnification well. You will only be limited by atmospheric turbulence. I find the view best in the 4-inch at 301×, which shows the inner "ring" or annulus very well. It's not symmetrical; it looks as if the ring has split and slipped in opposite directions along the nebula's major axis, like land along a slip fault after an earthquake.

There also appear to be "arms" inside the outer halo, so the planetary really looks more like a galaxy than a planetary. Others have noted this appearance as well. Of course, the planetary's visual appearance pales in comparison to that imaged by the HST, which shows the nebula's wicked turbulence. Even at high power in the 4-inch, the nebula is quite sublime. I call it the Kiss Nebula because the view mimics that of a lipstick stain from a woman who placed a kiss on a glass window looking out to the stars. Luginbuhl and Skiff claim that the magnitude 14.3 central star is visible at 200× in a 10-inch telescope. To see it, you will need enough magnification to separate the star from the inner edges of the central annulus. Good luck!

By the way, beware of a false comet nearby: the 10th-magnitude open star cluster NGC 2432 lies about 1° to the south-southwest of NGC 2440. It is plotted in the first edition of Tirion's *Sky Catalogue 2000.0* but not in the second edition.

Hidden Treasure 42

NGC 2451

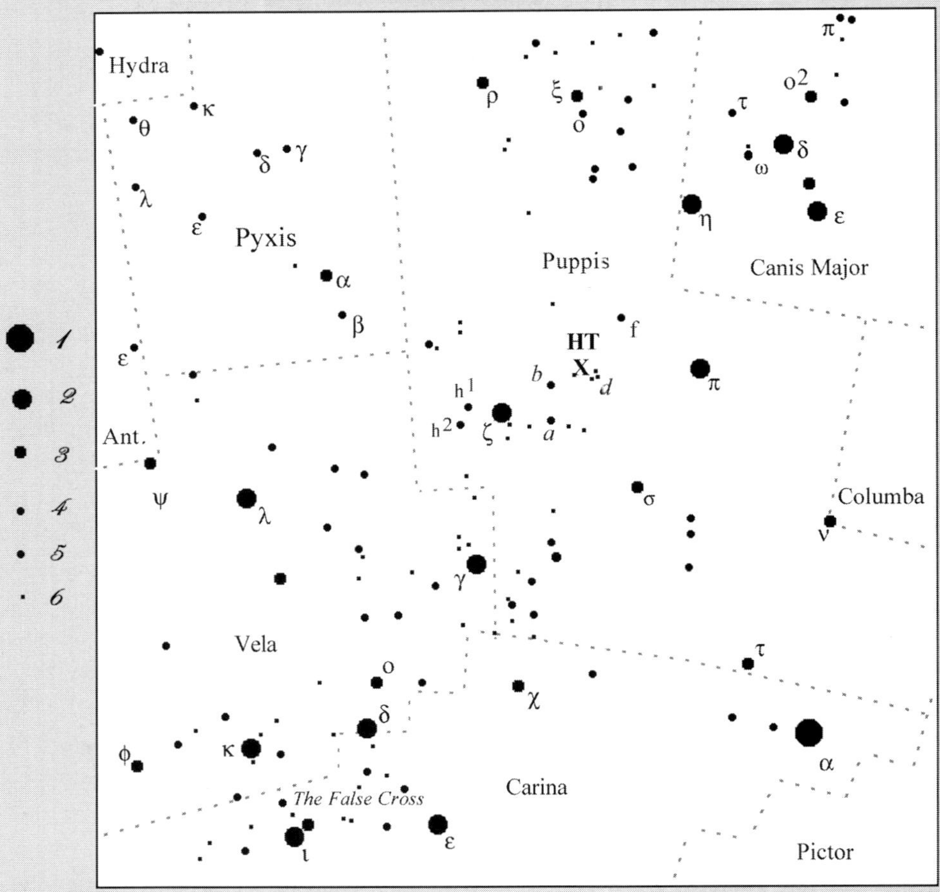

© Stephen James O'Meara

Tirion: Chart 19

Uranometria: Chart 362

42

Stinging Scorpion
NGC 2451
Type: Asterism superimposed on the Puppis Moving Group and possibly a more distant cluster
Con: Puppis

RA: $07^h\ 45.4^m$
Dec: $-37^\circ\ 57'$
Mag: 2.7 (O'Meara); 2.8
Diam: 50.0′
Dist: ~600 light-years (Puppis Moving Group or NGC 2451-A)
Disc: Giovanni Batista Hodierna, before 1654

J. HERSCHEL: The chief star (4th magnitude) of an orange colour, of a very large and very diffused cluster of large stars, too loose to be a fit object for the ordinary magnifying power. (h 3099)

NGC: Cluster, *very* large, very little compressed, includes a magnitude 4.5 star.

SOMETIME PRIOR TO 1654, WHEN Italian astronomer Giovanni Batista Hodierna published a catalog of some 40 deep-sky objects that he found with a 20× Galilean refractor, the astronomer placed on one of his star charts an object in the position of NGC 2451. John Herschel later cataloged it as a diffuse cluster of bright stars. Photometry on the stars in the region, performed in the mid 1960s, supported the notion that NGC 2451 was indeed a cluster, perhaps a part of an old *OB* Association that includes open clusters Collinder 135, 140, and 147. And that's how matters stood until the mid 1980s, when astronomers began debating whether NGC 2451 is a real star cluster.

The issue was convincingly resolved in a 1994 *Astronomy and Astrophysics* paper entitled "NGC 2451: what is it?" In that work, German astronomers Siegfried Roser and Ulrich Bastian determined via proper-motion studies that the stars classically known and observed through amateur instruments are not physically related, so they do not form a true cluster. However, in a 4°-square area around NGC 2451, they did find a group of 24 stars having common proper motion. Furthermore, when plotted on a color-magnitude diagram, the stars

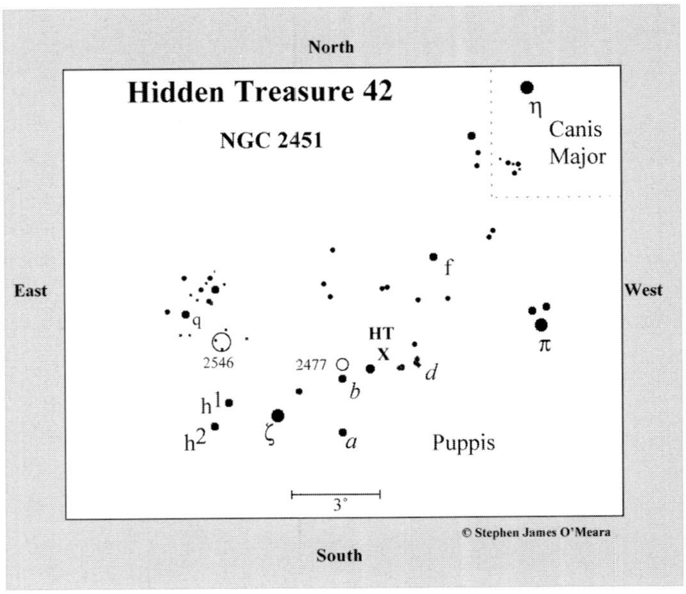

display a remarkably well-defined main sequence. The authors conclude that these stars, which they refer to as the Puppis Moving Group, form a star cluster at a distance of about 700 light-years. The authors also stress that this cluster is *not* the non-existent NGC 2451: "We emphasize that the [Puppis Moving Group] is not NGC 2451. The cluster originally suspected by Herschel and numbered NGC 2451 by Dreyer is a group of 20 or so stars of magnitudes 5 to 8 surrounding the 3rd-magnitude giant [c Puppis] in a 40 arcminute field. Only 4 of these belong to the [Puppis Moving Group]."

Roser and Bastian also suspect a second cluster in the area at almost twice the distance. Imants Platais (Yale University) and his colleagues, who presented new proper motion studies of the stars in the region of NGC 2451, came to the same conclusion. They "definitely found" the Puppis Moving Group, which they refer to as NGC 2451-A, at 600 light-years. The group also detected a second, more distant, grouping (NGC 2451-B) at about 1,300 light-years. Both groups are sparsely populated and appear to be similar in age to the Pleiades, being perhaps about 100 million years young. (It's important to note that, as Archinal and Hynes mention in *Star Clusters*, calling these groups NGC 2451-A and NGC 2451-B is unfortunate, since they are not listed that way in the *NGC*.) Hipparcos data show that the Puppis Moving Group definitively does exist, while the more distant cluster *may* exist.

While studying NGC 2451 can be an Escheresque ball of confusion on a professional level, observing NGC 2451 visually, especially in small apertures with wide fields of view, is nothing but a delight. It is also easy to find. Look 10° (a fist width held at arm's length) southeast of the Greater Dog's tail star, 2nd-magnitude Eta (η) Canis Majoris, or 4° northwest of 2nd-magnitude Zeta (ζ) Puppis. It lies about 1° east of a tight grouping of three 5th- and 6th-magnitude suns – d^1, d^2, and d^3 Puppis – that appear as one star (d Puppis), which, confusingly, is the westernmost of three similarly bright suns forming a slightly larger triangle in the same area.

Of the seemingly endless open clusters distributed throughout the Puppis region of the Milky Way band, NGC 2451 has found a special place in my heart. It is visible to the naked eye as a 3rd-magnitude object – a star surrounded by a dim haze – a little more than $1\frac{1}{2}$ Moon diameters across. In 7 × 50 binoculars, about 15 or more stars are visible

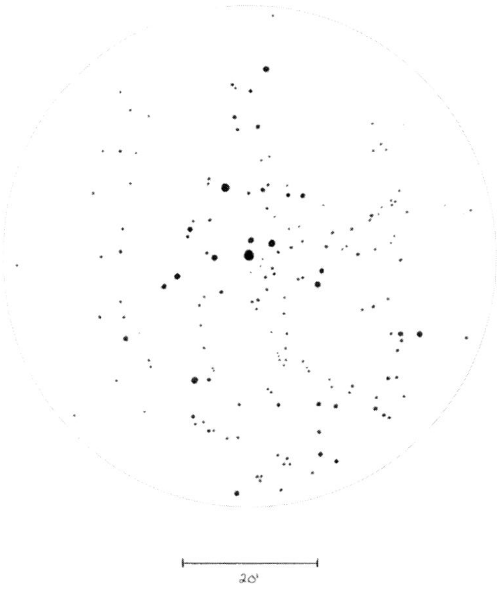

immediately at a glance. The antique telescope reveals more than two dozen. These stars seem to burn in adoration of the brightest luster member, c Puppis, a yellow giant of magnitude 3.6. Its age has been estimated at 36 million years, and it is receding from us at 26 kilometers per second.

I have long seen NGC 2451 as a scorpion approaching me nearly head-on, with two outstretched claws and an upright tail ready to sting. It's a fitting description, because this stinging scorpion, which travels the sky with NGC 2477 (Caldwell 71) about $1\frac{1}{2}°$ to the southeast rises just as M71 in the celestial Arrow sets. So we can create our own myth. Ages ago Sagittarius shot Sagitta, the Arrow, toward Puppis, killing our tiny arachnid. Feeling sorry for the innocent creature, Zeus placed the cluster in the company of NGC 2477, as far as possible from the Arrow in the night sky. Thus the Arrow will never again slay the stinging scorpion cluster – just as big brother Scorpius will never again sting the mighty Hunter, Orion.

I have also seen the cluster looking like a kitten playing with a ball of string. The kitten's head is comprised of the group's prominent central triangle of stars: the northern and western stars in that triangle mark the kitten's ears, while brilliant c Puppis marks the kitten's nose. Take the time to study the color of c Puppis. Luginbuhl and Skiff say it is "very red." Hynes and Archinal, as does John Herschel, claim it is "orange," and I detected a golden yellow hue, like that of wheat being illuminated by the setting Sun. The "ball of string" is an attractive collection of 10 stars (ranging from 8th- to 11th-magnitude) within 20′ of c Puppis, and oriented northeast–southwest. The brightest of these stars form a diamond, which is comprised of three obvious groups of stars: three 11th-magnitude suns in an arc immediately southwest of c Puppis and concentric with it, followed by a prominent diamond-shaped clustering of four roughly 8th-magnitude stars, and finally a near-equilateral triangle 15′ to the southwest; the southern apex of that triangle is a bright blue 7th-magnitude gem.

As is common in the Puppis region, many of the stars are clustered into pairs or groupings, which add to the complexity of the scene, while the region's colorful stars add to the beauty of the scene. NGC 2451 is best seen at low powers, and it will satisfy any binocular user. My wide-field drawing of NGC 2451 illustrates what is most likely an illusion, but one that is titillating nonetheless – namely, the brightest stars in the region are arranged into weak spiral arms centered on the core component of NGC 2451. The most prominent arm extends to the west before curving north. Two weaker arms extend to the south before curving north, and a weak linear arm projects to the north. Only the eastern arm is a bit disjointed, arcing to the southeast before sharply curving north.

While NGC 2451 is best seen from southerly locals, I used to enjoy sweeping it up in binoculars from my apartment in West Roxbury, Massachusetts, which was high on a hill overlooking light-polluted Boston. From that location, NGC 2451 was only 10° above the horizon. And while I did not see it as well as I had from more southerly locals, its beauty was not diminished, at least not in my eyes and certainly not in my heart. With its stars scintillating wildly at that low altitude, the asterism looked, well, as playful as a kitten.

Hidden Treasure 43

NGC 2467

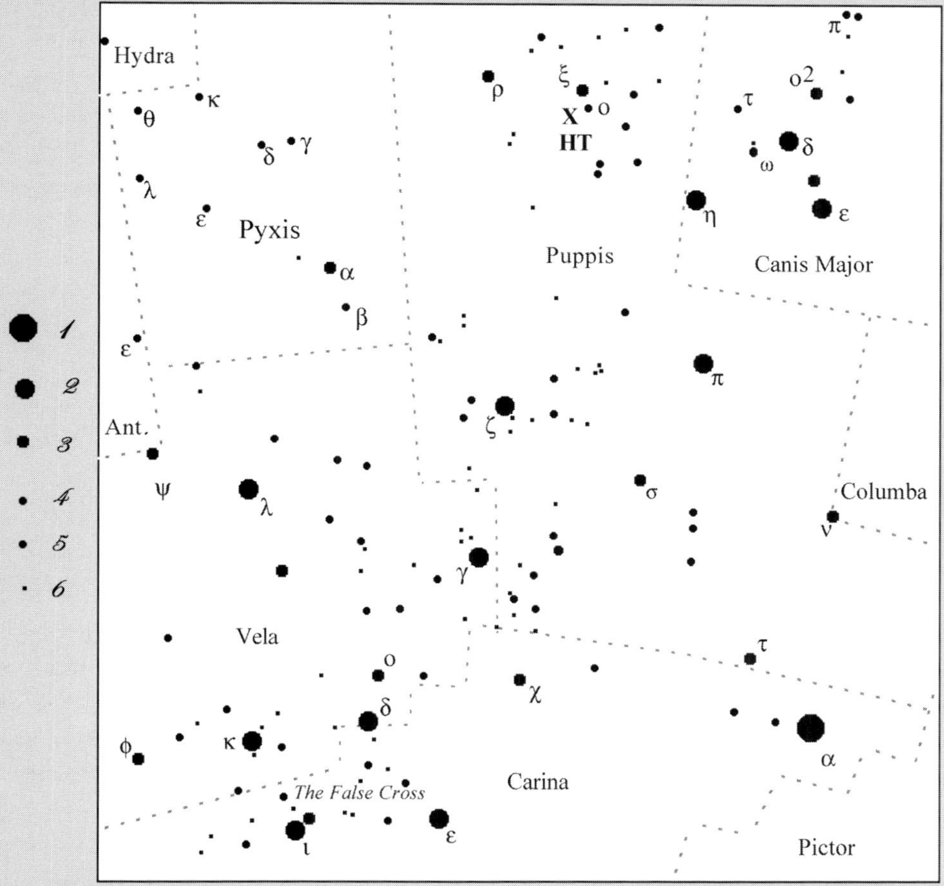

Tirion: Charts 19 & 20

Uranometria: Chart 320

43

NGC 2467 = Sharpless 2–331
Type: Emission Nebula
Con: Puppis

RA: 07^h 52.5^m
Dec: $-26°$ $24'$
Mag: 8.0 (O'Meara); Nebula
Dim: $16' \times 12'$
Dist: 13,700 light-years
Disc: William Herschel, 1784

W. HERSCHEL: [Observed December 9, 1784] Large pretty bright, round and easily resolvable, 6′ or 7′ diameter, a faint red color is visible. A star of 8th magnitude is not far from the center but not connected. (H IV-22)

NGC: Pretty bright, very large, round, easily resolvable, magnitude 8 star in the middle.

NGC 2467 IS AN HISTORICAL headache, in that, depending on the source, the object is listed as either an open star cluster, a nebula, or a star cluster with nebulosity. But we need no medication to alleviate the suffering other than the truth. The solution is simple. William Herschel, the object's discoverer, classified it as a nebula. In fact, he classified it as a planetary nebula, because it appeared round. Today we use Herschel's term, "planetary nebula," to describe a specific type of nebula – a shell of gas blown off a dying star. But that is not what Herschel had in mind back in the eighteenth century; the man had no means at his disposal to fathom the true nature of this gaseous object. His description applies to the object's physical appearance not its astrophysical nature. In other words, the object was round and looked like Uranus, the gas-giant planet he discovered.

Note that Herschel also makes no specific mention of a star cluster, though he does state that the nebula is "resolvable." Herschel believed all nebulae would ultimately resolve into stars, given sufficient aperture and magnification. Therefore, his comment should not be taken out of context. Besides, Herschel had three classes reserved for star clusters, and he did not place our target in any of them. That said, if we are to be true to history, NGC 2467 is the emission nebula commonly known today as Sharpless 2–311 – end of story! Larry Mitchell, a Herschel historian, also wants us to note

Herschel's profound observation of the nebula's red color.

The "cluster" commonly mislabeled NGC 2647 today was first identified by Per Collinder in 1931; it carries the name Collinder 164 (Cr 164). According to Archinal and Hynes, it is 13,000 light-years distant, shines at magnitude 7.1, and has some 50 stars. But in an ironic twist, recent studies now argue that the cluster does not exist, at least not as we imagine it.

In 1989, Argentinean astronomer A. Feinstein (National University of La Plata) and his colleagues announced that new photometry of the stars in the field of NGC 2467, which was assumed to be the nucleus of the Puppis I Association, identifies no unique star group. The stars that Herschel "resolved" in the nebula are but a chance alignment of superimposed star groups in the rich Puppis Milky Way. "From our observations of NGC 2467," the astronomers explain, "it is concluded that it is not an open cluster."

There are two principal star groups here: (1) a young and very distant swarm of *O*-type stars *beyond* the Puppis *OB*2 Association, which delineates a spiral feature in the Milky Way about 13,700 light-years distant, and (2) a closer group (with later *B*-type stars) about 8,150 light-years distant – the same distance as another spiral feature in the Milky Way, the Puppis *OB*1 Association. The view is further complicated by the superimposition of two 9th-magnitude open star clusters, Haffner 18 and 19 – which lie about 10′ and 2′ north-northeast of the center of the NGC 2467 nebula complex, respectively. These clusters are, in fact, embedded in the NGC 2467 nebula complex, which itself is part of the Puppis *OB*2 Association. The densest part of the nebula is being excited by ultraviolet radiation streaming from the 9th-magnitude, *O*-type star HD 64315, which is located only 1′ from the nebula's center.

The whole complex is a young object with an estimated age of about 2 to 3 million years – around the time when many marine creatures, such as bivalve mollusks, suddenly died out all over the planet. Narciso Benítez (Johns Hopkins University) and his colleagues believe that a supernova explosion 130 light-years distant (perhaps from the Scorpius–Centaurus Association) could have been the culprit. Such a nearby explosion would have showered the Earth with enough cosmic rays to have thinned its ozone layer by 60 percent. An increase of intense ultraviolet radiation from the Sun could have killed off plankton, which, in turn, would have led to widespread death to the bivalves that live off them. When we look at NGC 2467, and its neighboring stars, we are seeing a moment of creation at a time when widespread death was occurring here on Earth. Early man may have been a witness to this event.

To summarize, emission nebula NGC 2467, open clusters Haffner 18 and 19, and the Puppis *OB*2 Association, all belong to one and the same spiral feature, which lies at a distance of about 13,700 light-years from the Sun. In the same line of sight is a group of hot young stars residing in a Milky Way field beyond that spiral feature, as well as another group of suns, in another spiral feature, which includes the Puppis *OB*1 Association, some 5,500 light-years closer. So much for the simplicity of the two-dimensional sky we adore from our simple observing platform on Earth.

To find this stellar menagerie, first locate 2nd-magnitude Eta (η) Canis Majoris (Aludra) – the tail of Canis Major, the

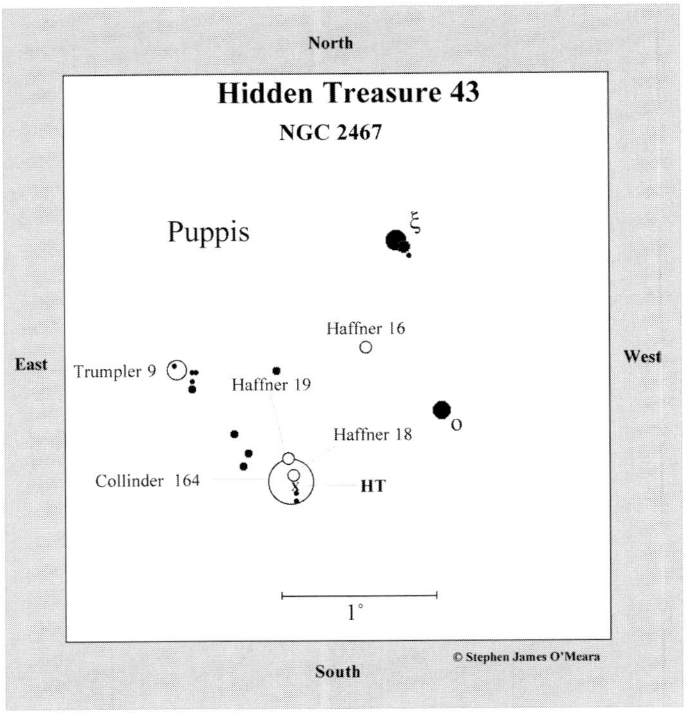

star. If you are under a dark sky, use averted vision and see if you do not see a halo of nebulosity surrounding that 9th-magnitude sun. I can see its glow in 7 × 50 binoculars from Hawaii.

At 23× in the 4-inch, the nearly 3° field is a fascinating assortment of geometrical patterns of stars (straight lines, star pairs, triangles, and arcs), clusters, and nebulosity. It's a hypnotic field. The more you look at it, the more you see. Averted vision brings out the ghostly glow of NGC 2467, which forms a 5′-wide circular halo around HD 64315.

Great Dog. About 10° (one fist) to the east-northeast you'll see 3rd-magnitude Rho (ρ) Puppis. Note that Rho Pup also marks the end of a jagged stream of 3rd- and 4th-magnitude suns between it and Eta CMa. Now look for Xi (ξ) Puppis, another 3rd-magnitude star 4° to the west; it has a close 5th-magnitude companion to the south-west. If you center that pair of stars in your binoculars, then move a little more than 1° to the south-southwest, you'll hit 4th-magnitude Omicron (o) Puppis. Just $1\frac{1}{4}$° to the southeast, look for an obvious 15′-wide acute triangle of 7th- to 8th-magnitude suns, whose apex points to the south-west. An additional 7th-magnitude sun lies about 10′ to the northeast of the northernmost star in the triangle, and a 9th-magnitude star lies about 10′ due west of the triangle's apex. The latter star is (HD 64315) – NGC 2467's main illuminating

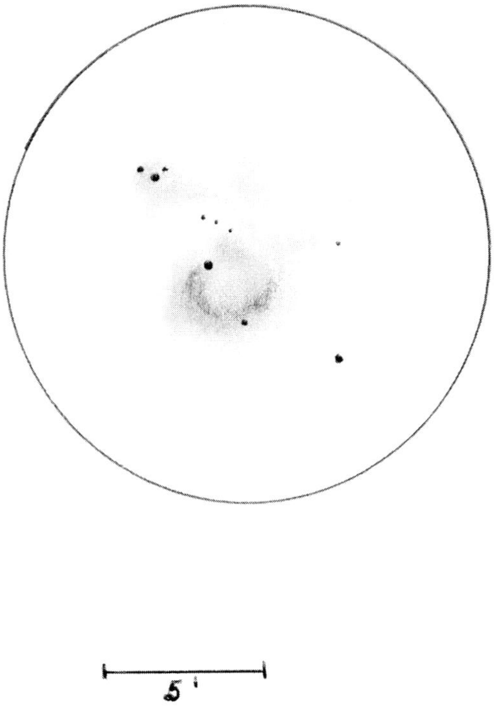

This is the nebula's brightest section and it takes magnification well. I found that 182× gave a very comfortable view. At that power, NGC 2467 looks like a miniature version of the Helix Nebula (Caldwell 63) in Aquarius (a true planetary nebula) – a wide ring of light with multiple edges and a prominent hollow. NGC 2467's "ring" is highly asymmetrical in brightness. The most intense section is a well-defined arc of nebulosity about 3′ or 4′ southwest of HD 64315 – which, by the way, is not centered in the hollow, but lies on the northern fringe of it.

The nebula is more extensive and obvious in photographs than it is in a small telescope. But if you take the time, and breathe regularly as you peer into the eyepiece, dim tufts of light can be seen flaring off the ring like prominences. For instance, about 3′ from HD 64315 to the north is a little loop of nebulosity whose northern flank brushes against a warped Y-shaped asterism of stars. Little ruffles of nebulosity can also be seen extending from the ring to the west, northwest, and southeast. Overall, the brightest part of the nebula looks like a disintegrating smoke ring. All these features should stand out better if you use an ultra-high-contrast (UHC) filter.

Hidden Treasure 44

NGC 2547

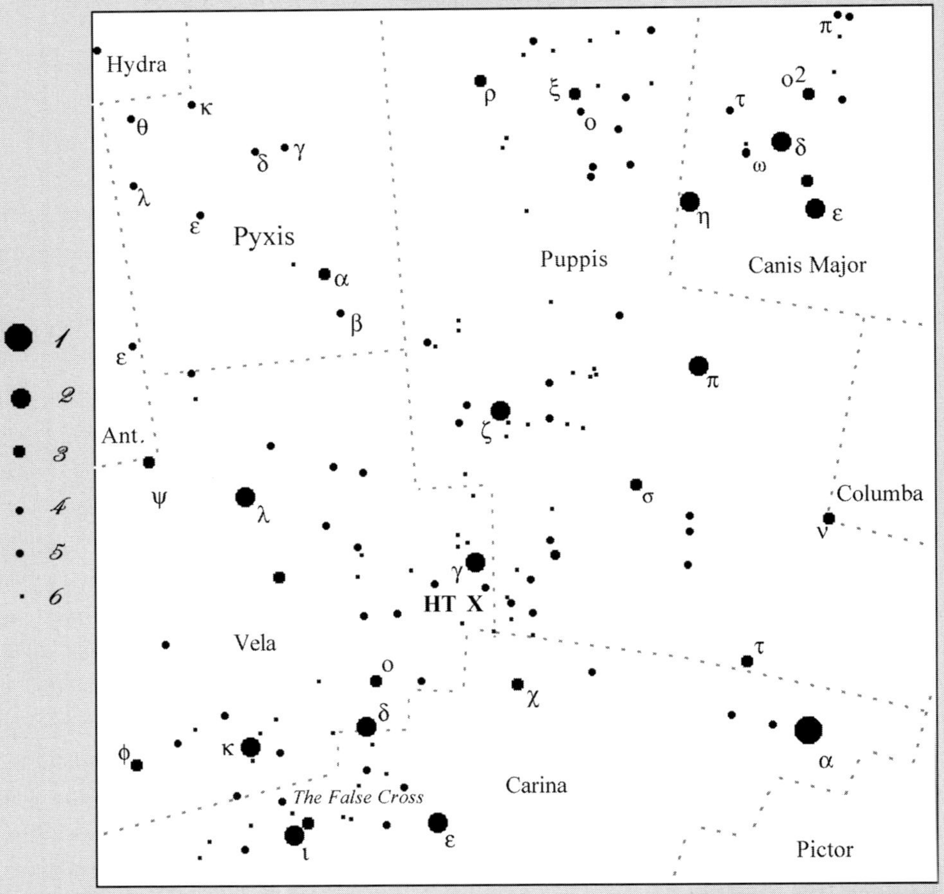

© Stephen James O'Meara

Tirion: Charts 19 & 20

Uranometria: Chart 396

44

Golden Earring, St. Peter's Cross, Malus
NGC 2547
Type: Open Cluster
Con: Vela

RA: 08^h 10.2^m
Dec: $-49°$ 13′
Mag: 5.0 (O'Meara); 4.7
Diam: 25.0′
Dist: 1,400 light-years
Disc: Abbe Nicolas Louis de Lacaille, listed in his 1755 catalog

J. HERSCHEL: Chief star about 7th magnitude, of a very large, loose, brilliant cluster of very scattered stars, 1 of 7th magnitude, 2 of 8th magnitude, rest 9th to 6th magnitude. Fills more than field; 100 to 150 stars. (h 3117)

NGC: Cluster, bright, large, little compressed, stars from magnitude 7 to 16.

VELA IS THE BROAD SAIL OF ARGO, the mythological ship that carried Jason and his Argonauts on a quest to find the magical Golden Fleece. The tale is one of the most familiar and enduring legends from Greece. Its roots date to Pindar's fourth Pythian ode, which the medieval Alexandrian poet Apollonius Rhodius (Apollonius the Rhodian) expanded into an epic sometime in the third century BC. In Appollonius's *Argonautica*, we find Jason and his crew taking shelter from a storm "on the calm shore of Magnesia on the mainland and the tomb of Dolops; on the third day they put forth the ship, spreading on high the broad sail [Vela]."

Although the tale portrays Jason and his followers as heroes, one can also jump the fence and argue that "Jason and the Argonauts" is the greatest pirate story ever told. That assertion did not go unnoticed by Hollywood: "I know you've come for the Fleece," King Aeetes says to Jason, in the 1963 cinematic interpretation of the myth, "and I know too that if you cannot win it by bargaining, then you plan to steal it . . . We can no longer welcome or entertain you but treat you as you deserve. As thieves, as pirates, as murderers."

One country's hero is another country's pirate. The parallels between Jason as hero

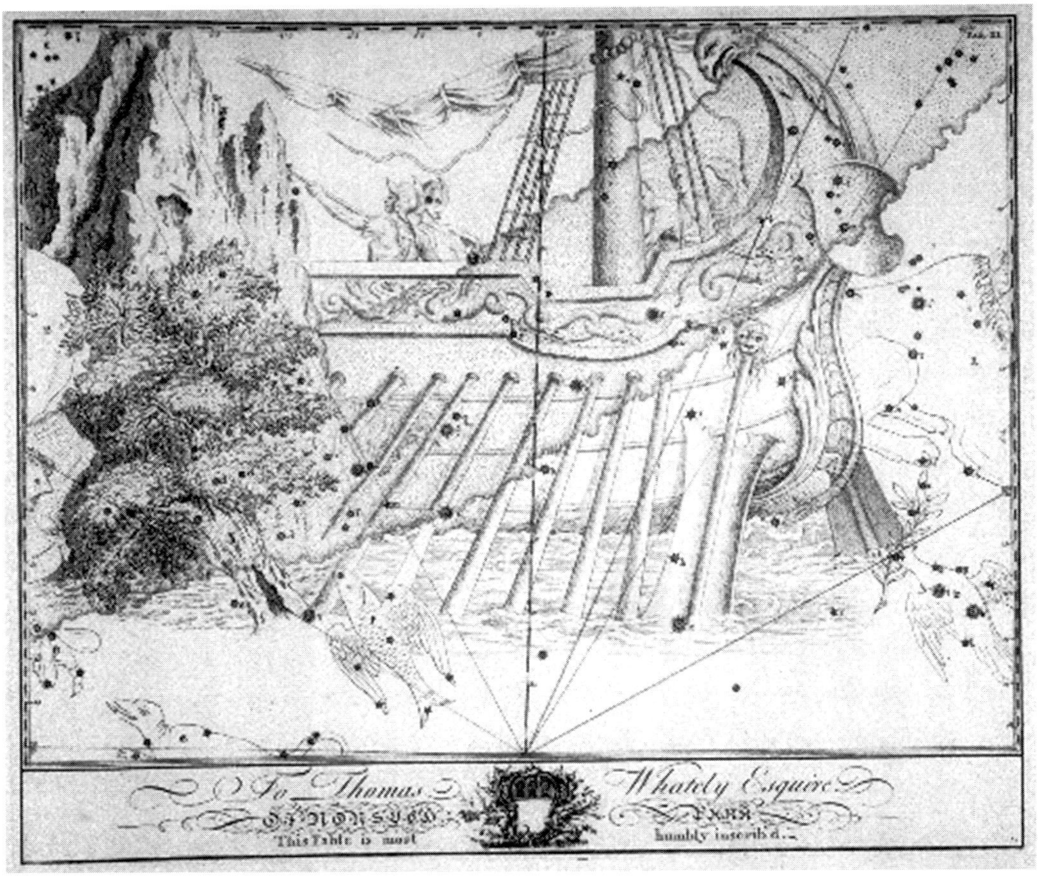

versus Jason as pirate are inescapable. What are pirates but legendary seafarers who make long voyages in search of hidden treasures and have adventures along the way. Some pirates, like Sir Francis Drake (the greatest British mariner of the Elizabethan age) were *privateers* – seafarers licensed by the sovereign to attack and plunder the riches of the enemy. Jason and the Argonauts, then, may be the first tale ever told about a privateer.

Certainly there was no shortage of danger or killing in the story, which parallel's Hercules' 12 labors. Also consider Jason's mysterious liege, Medea – a woman betrayed in the end by her prince. But Medea is a heartless sorceress, a witch whose thirst for blood surpassed that of any pirate known today. Among her many evil deeds, Medea murdered her own two children, as well as her little brother – who she dismembered, limb by limb, before tossing each body part, one by one, into the sea; she did this to stop her father who was in hot pursuit of Jason and the stolen Fleece. Is it surprising, then, that Jason tired of Medea's methods of dealing with issues and married another?

In *The New Patterns in the Sky,* Julius Staal provides a detailed account of how the story of Jason and the Argonauts is connected with the the Sun's trek through the zodiac. "With a star chart or planisphere,"

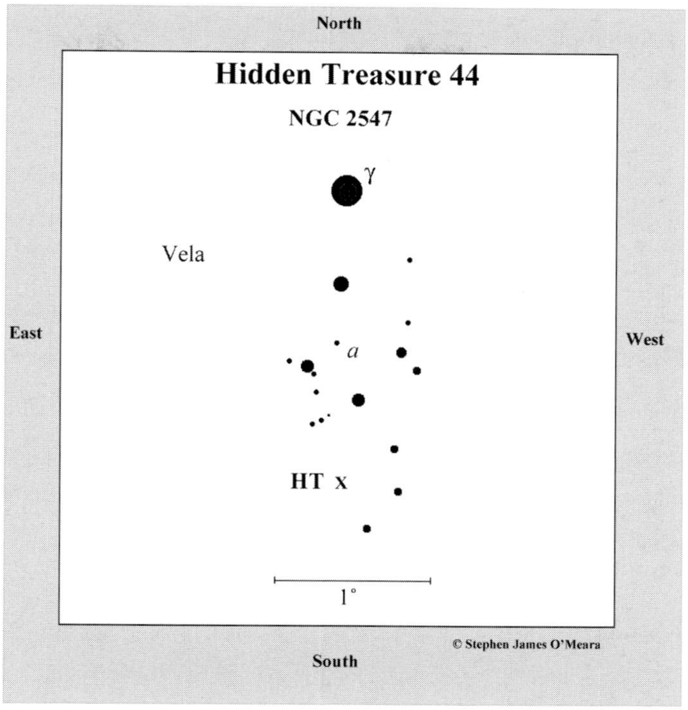

Staal says, "the whole story can be followed quite clearly." Among the vast cast of players, we see Jason as Ophiuchus, the Curer of Injustice. Perseus is King Aeetes. Algol, the Demon Star, represents Medea. And the severed pieces of Medea's brother are the stars of Auriga floating in the Milky Way. Argo Navis is the Ship Argo – which, by the way, was also known to the ancient Romans as *Navigium Praedatorium* . . . the Pirate Ship.

During an astronomical expedition to the Cape of Good Hope in southern Africa, the Abbe Nicolas Louis de Lacaille (1713–1762) did some dismembering himself. Sometime during his stay from April 19, 1751, to March 8, 1753, the astronomer broke Argo Navis into three principal pieces: Carina (the Keel), Puppis (the Stern), and Vela (the Sail). He also created the now defunct Malus (the Mast), and Pyxis Nautica (the Mariner's Compass), which still exists, but simply as Pyxis.

Lacaille also cataloged, among other things, 42 "nebulae," which he placed into three classes: "nebulae without stars" (Class I); "nebulous clusters" (Class II); and "stars accompanied by nebulosity" (Class III). NGC 2547 is the second object in Lacaille's Class III and hence is denoted Lac III-2. This means that while Lacaille could see stars within the "nebula," he could not fully resolve it with his $\frac{1}{2}$-inch 8× telescope. He describes it as "five faint stars like the letter T in nebulosity." James Dunlop became the first person fully to resolve the cluster, which he included as No. 410[1] in his 1827 catalog. Through his 9-inch f/12 reflector, he saw "[a] curiously arranged group of pretty bright small stars of [mixed] magnitudes . . . There is no nebulosity in this place."

Dunlop also drew the cluster showing an oval of stars dangling from two rows of parallel suns, like a golden ring dangling from a pirate's ear. Ironically, to find this hidden treasure, all you need to do is look 1° south of the 2nd-magnitude gem Gamma (γ) Velorum to a diamond-shaped asterism of four 5th-magnitude suns (*a*). Seen together in 7 × 50 binoculars, Gamma Vel and the four diamonds look like a chandelier earring with a glittering topaz stud

[1] In his 1847 "Results of Astronomical Observations," John Herschel misidentifies the cluster as Dunlop 411.

and four gemstones set in a diamond pattern. NGC 2547 is another tier to the chandelier, lying a little more than 30′ south-southeast of the southernmost point in the diamond.

Return, however, to Gamma Vel, because it is a stunning binocular double embedded in the Gum Nebula – a 40°-wide remnant of a supernova that may have erupted about a million years ago. The 2nd-magnitude primary (Gamma^2 Vel) has a 4th-magnitude partner (Gamma^1 Vel) 41″ to the southwest. The reason the fainter component is Gamma^1 is because it precedes Gamma^2 in right ascension. The Hipparcos satellite has revealed that the two are not gravitationally bound. Also, with a distance of 840 light-years, Gamma^2 is nearly half the distance of Gamma^1, which belongs to a relatively narrow "sheet" of associated stars in front of a dense dark cloud; this sheet of stars includes the hot (25 million years young), stars in the Vela *OB*2 Association and NGC 2547. In 2000, M. Pozzo (Keele University) and his colleagues also announced the discovery of a population of low-mass, pre-main-sequence stars in this region.

To the naked eye, Gamma Velorum is a brilliant gem, a white sapphire whose radiance has a calming effect on the soul. But that's an illusion. Gamma^2 Vel is the brightest and perhaps the most massive Wolf-Rayet star known. Wolf-Rayets are extremely luminous and hot stars in an advanced stage of evolution. They shed mass in the form of a powerful stellar wind, which rushes out at speeds topping 3,000 kilometers per second. Gamma^2 Vel's shedded matter is colliding with the stellar wind from a spectroscopic *O*-type giant companion, which has an orbital period of 78.5 days. Observations with the *Copernicus* and *IUE* satellites have identified an accruing disk of dense matter between the two stars.

The *O*-type giant has a surface temperature of 35,000 K, a total luminosity of 200,000 Suns, and a mass of 30 Suns. By comparison, the Wolf-Rayet star is about half as luminous and contains less than 10 solar masses, but is perhaps as hot as 60,000 K. That's the way we see things today. A few million years ago, the story would have been different. The Wolf-Rayet star would have been the more massive of the two, having a mass of about 40 Suns, but as it is more highly evolved, the star has stripped itself to a quarter of its size. Recent studies reveal that the Wolf-Rayet companion is losing mass at a rate of 8 million solar masses per year. Since Wolf-Rayet stars are doomed to perish within a few million years, astronomers have taken a keen interest in Gamma^2 Vel, which may be about 4 million years old and about to end its life in a supernova explosion.

Amateurs using a moderately large telescope and an eyepiece spectroscope may want to check out Gamma^2 Vel, because its spectrum is "incomparably the most brilliant and striking in the heavens." Gamma^1 and Gamma^2 Vel are also one of the few fine double stars that can be spied in broad daylight through a telescope. At night, this pair lies in a stunning patch of Milky Way. Two other nearby stars add to the beauty of the scene: one shines at 8th-magnitude and lies 62″ to the south-southeast of Gamma^2 Vel; the other is 9th-magnitude and lies 93″ to the southeast of Gamma^2 Vel.

When you've finished admiring the Gamma Velorum field, drop your sights southward to NGC 2457. I was really surprised by its beauty. Even 7 × 50 binoculars will show Lacaille's T-shape asterism within its form. And my antique telescope resolves

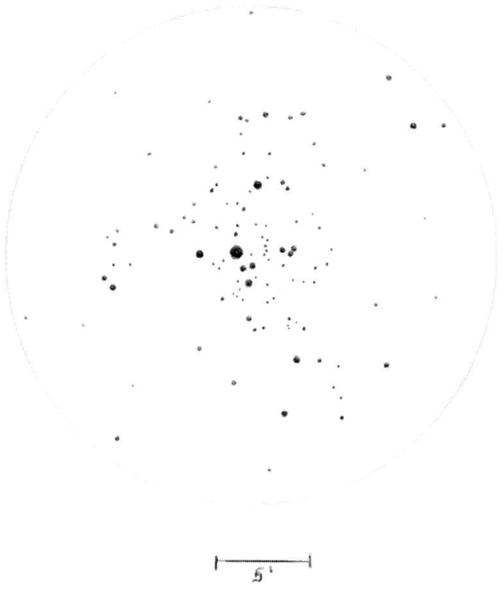

it into about a dozen suns. At 23× in the 4-inch, the cluster is very bright and condensed and, when viewed with north up, it looks like an upside down crucifix, which is why I also call it St. Peter's Cross. Many historians believe Nero ordered St. Peter's crucifixion in about AD 64. While there are no contemporary accounts of his life or death, it is said that Peter asked to be laid on the cross head down because he was not worthy to die in the same manner as Christ. The basis for this story can be traced, in part, to Seneca, who records in a letter that he had seen criminals being crucified upside down at around the time of Peter's execution.

Archinal and Hynes list 112 members in an area nearly the size of the full Moon. Many of them are within the grasp of small telescopes; my drawing, for instance, shows about 100 stars ranging from 7th to 13th magnitude. If the cluster is 1,400 light-years distant, these stars are packed into an area measuring only 10 light-years across. The night sky from an imaginary planet around one of these suns would be spectacular indeed. At 72×, the cluster reveals many star pairs and asterisms – including several streams of stars that loop and curl around the cross like hair blowing in the wind. Several of these form loose loops or broken spiral structures, especially at the four major cardinal directions.

Returning to our story of Jason and the Argonauts, the crucifix can instead be seen as a miniature version of Malus, the Mast, which is surrounded by a canvas of misty starlight – a miniature Vela, the Sail. Drama is added to the scene if you focus on the fact that the mast is bent – as if it is holding fast against Aeolian winds of hellish force. Those loose loops of starlight, then, are the Sail's rigging, which has snapped free and is being tossed asunder by the devilish winds.

By the way, to the early Arabian skywatchers, Gamma Vel was *Al Suhail*, which, unfortunately, is also the name for Lambda (λ) Velorum. The consequent arguments about the identity of these stars "were the occasion of much profanity among the disputatious Arabs," Allen notes in his *Star Names*. Today, the star is popularly known as *Regor*, which is not Arabic; nor is it Roman nor Greek. It's "Roger" spelled backwards. It memorializes astronaut (Argonaut) Roger Chaffee who died in the tragic *Apollo 1* fire. Gamma Velorum was one of the stars that the Apollo astronauts would ultimately use to navigate across the sea of space to the Moon – America's greatest adventure.

Hidden Treasure 45

NGC 2539

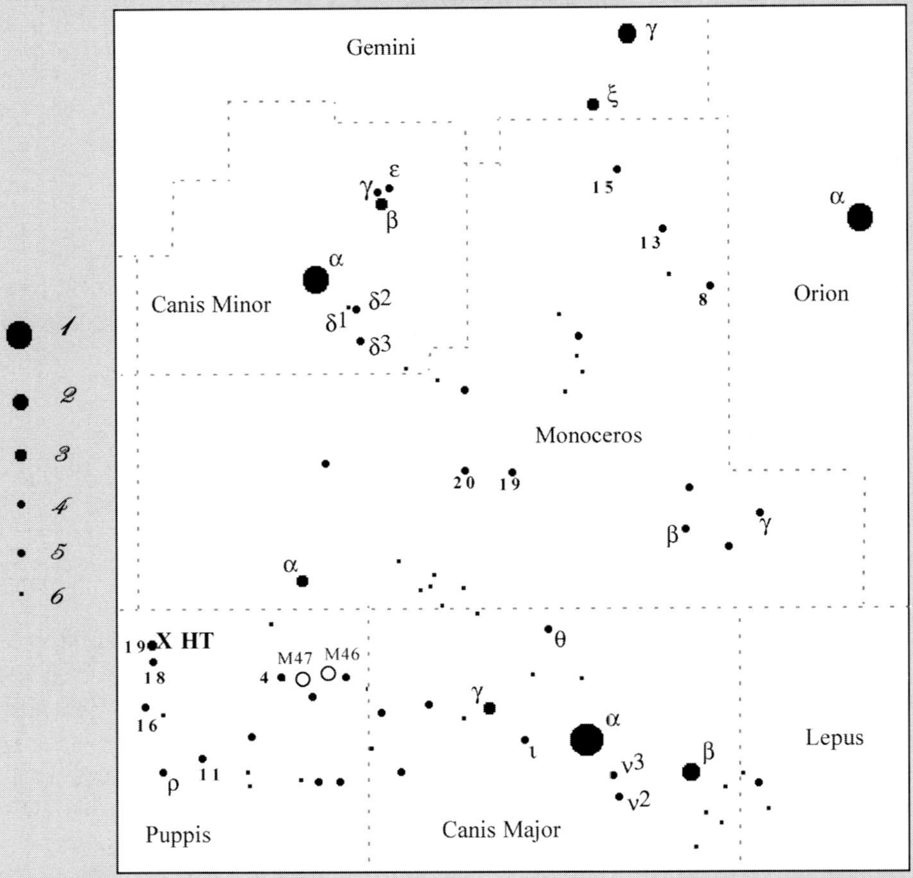

© Stephen James O'Meara

Tirion: Chart 12

Uranometria: Charts 275 & 276

45

"The Dish" Cluster
NGC 2539
Type: Open Cluster
Con: Puppis

RA: $08^h\ 10.6^m$
Dec: $-12^\circ\ 49'$
Mag: 6.5 (O'Meara); 6.5
Diam: 15.0′
Dist: ~4,000 light-years
Disc: William Herschel, 1785

W. HERSCHEL: [Observed January 31, 1785] A considerably rich cluster of coarse scattered stars above 20′ in diameter. (H VII-11)

NGC: Cluster, very large, rich in stars, little compressed, stars from magnitude 11 to 13.

NGC 2539 IS A MODERATELY condensed open cluster tucked away in the obscure northeastern quadrant of Puppis, just south of the point where the southeastern corner of Monoceros meets the Serpent, Hydra. It is a lonely recess of sky, whose dim stars lie in the periphery of our vision. Although the cluster teeters on the verge of naked-eye visibility, NGC 2539 is often overlooked because there are popular Messier objects in the area: 7° to the north-northeast is open cluster M48, in Hydra; and more than 7° to the west-southwest is M46 in Puppis and its equally stunning companion M47. What's more, NGC 2539 hides in the shadow of 19 Puppis, an equally lonely 5th-magnitude sun that lives on the outskirts of the Milky Way.

NGC 2539 lies about 4,000 light-years from our Sun and 760 light-years from the galactic plane, so it is affected little by interstellar dust, making it a preferred object for professional study. Astronomers recently examined NGC 2539 as part of a long-term project called the WIYN Open Cluster Study – a coordinated effort by astronomers (who were at one time, or are currently, affiliated with one of the institutions responsible for the scientific mission of the 0.9-meter WIYN telescope at Kitt Peak National Observatory) to establish a standard database of open cluster photometry, astrometry, and spectroscopy. The WIYN data suggest that NGC 2539 is about 650 million years old, making it similar in age to the Hyades (Caldwell 41), the Praesepe (M44), and NGC 6633 (Hidden Treasure 92).

The WIYN finding, which was reported at an American Astronomical meeting in 2004, agrees well with the 630 million year age

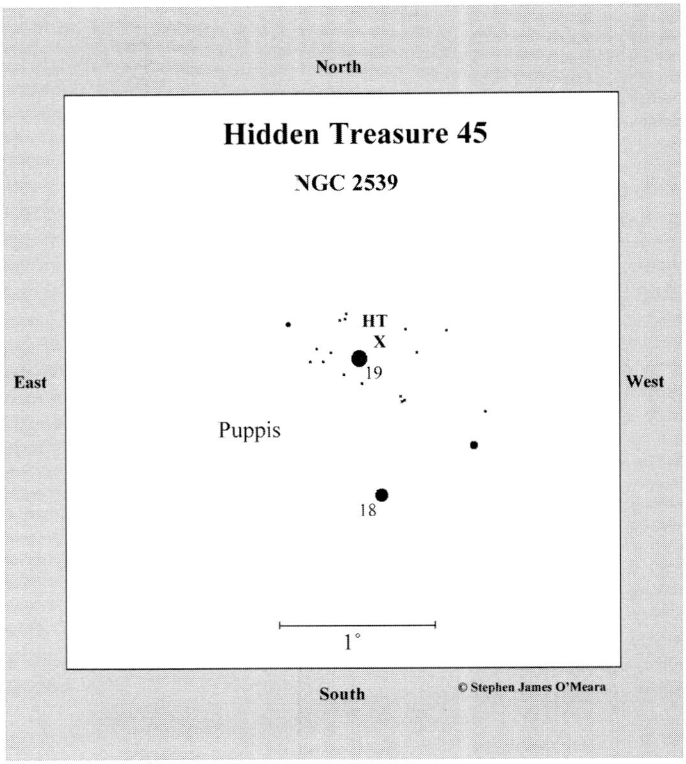

estimate independently determined by Argentinean astronomer E. Lapasset (National University of Cordoba) and his colleagues in a 2000 issue of *Astronomy and Astrophysics*. The cluster's color-magnitude diagram displays a well-defined main sequence consisting of Type *A* and *F* stars. It also reveals 11 evolved stars, eight of which are typical red giants. Three possible blue stragglers have also been identified in the field, but these, Lapasset says, are most likely foreground objects.

In their book *Star Clusters*, Archinal and Hynes list NGC 2539 as having about 59 stars in an area 15′ wide, which is about what most visual observers see. Lapasset's data, however, reveals 169 members brighter than 15th magnitude out to 21′. Given the latest data, NGC 2539, then, measures 24 light-years across, making it slightly larger than the Hyades but 760 times farther away. Had NGC 2539 been at the same distance as the Hyades, it would have rivaled that great cluster in grandeur and brightness. Its beauty would have burned into the souls of our ancestors, who would have incorporated it into their myths and legends of the sky.

To find this hidden treasure, all you have to do is locate 19 Puppis, which marks the eastern end of a right triangle with 4th-magnitude Alpha Monoceros and the naked-eye double star 2 and 4 Puppis (the stars immediately east-northeast of M46). The southeastern lip of NGC 2539 kisses 19 Puppis; in 7 × 50 binoculars, the scene looks like a veil of morning mist rising before a burning yellow Sun. With just a little concentration and averted vision, the 6th-magnitude glow appears mottled. My antique telescope resolves many members, and the scene is like that of agitated honeybees swarming about a hive. "[T]he swarming bees have begun to lose patience," writes Maurice Maeterlinck in his 1954 book, *The Life of the Bee*, "the hive whose black and vibrating waves are bubbling and overflowing, like a brazen cup beneath an ardent sun."

I find the binocular view of these glowing clouds of dim starlight exhilarating, especially when they hug bright stars; when I see NGC 2539 and 19 Puppis together,

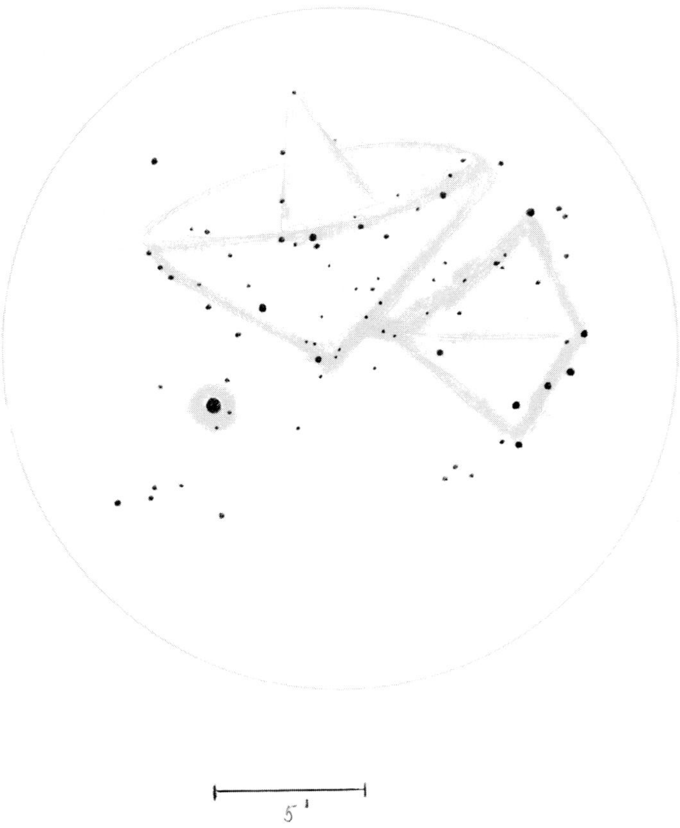

like standing high on a hill overlooking suburbia at night. This view differs somewhat from Adm. Smyth's who says, "It consists of a large and loose, but rich, group of small stars pretty equally strewed over the field, with a close double star in the middle, and a bright yellow one of the 6th magnitude to the [southeast]." The Reverend Webb also commented on the yellow sheen of 19 Puppis.

At 72×, many pairs of stars and other geometrical patterns appear strewn helter-skelter across the field like jacks. Take your time to scan the cluster, piece by piece, because it's filled with asterisms. It's a virtual playground for the eye, so let your imagination fly. When I did, I saw all those groupings mesh into one arrangement – that of a giant radio telescope, like the one at Harvard now used in the Search for Extraterrestrial Intelligence. Add the lemony light of 19 Puppis, and the scene looks frighteningly familiar. Here is a *G*-type star, like our own Sun, only 185 light-years distant. Its position next to our radio telescope cluster, 4,000 light-years beyond, illustrates what Harvard astronomer Harlow Shapley called "internationalism" in his 1963 book, *A View from a Distant Star*:

> The bright stars of the Southern Cross, shine equally on . . . all the countries of the Southern Hemisphere. There is no discrimination. Among those stars are some which are continuously radiating messages of special importance to civilized and inquiring mankind. They are,

something wells up inside me. It's a moment of awakening, like seeing a dark and misty forest lanced by sunbeams in the early morning light. It's as if these inanimate objects are breathing in cold air, allowing us to see their breath.

In the 4-inch at 23× with direct vision, the cluster looks like a tapered candle flame. With averted vision, its oval shape and prickly body remind me of NGC 3532 (Caldwell 91), the Pincushion Cluster. The brightest stars shine at 9th magnitude, and a beautiful 11th-magnitude double star lies near the cluster's center 10′ northwest of 19 Puppis, which itself has a powder-blue companion to the west-southwest. The entire cluster is broken into patches of starlight. It's for instance, sending this sort of message: 'Our ages are great [I am quoting the stars], our distances stupendous; the total of time and space, in which you humans on the earth are momentarily involved, is of a magnificence that dwarfs your brief life-span and your space range to insignificance. But look at us stars again, you humans, and you will see that you are part of a grander scheme than your prophets of old imagined, for you share with us the laws that rule the galaxies, and share the materials of which they are made. You are made of star stuff. We say to you : "Man may transcend the inanimate world through his desire and power to comprehend. He can match his mind and spirit against the eternals of space, of time, of cosmic energy; and, with an appropriate balance of pride and humility, he can face the cosmic facts, as becomes the high-born animal that he is'."

So go out and look at 19 Puppis and "The Dish" Cluster and take the time to dream of the possibility that we are not a lone and dusty mote in the void. We are part of an international community whose players exist in the twilight of the imagination, which we call the universe.

Hidden Treasure 46

NGC 2546

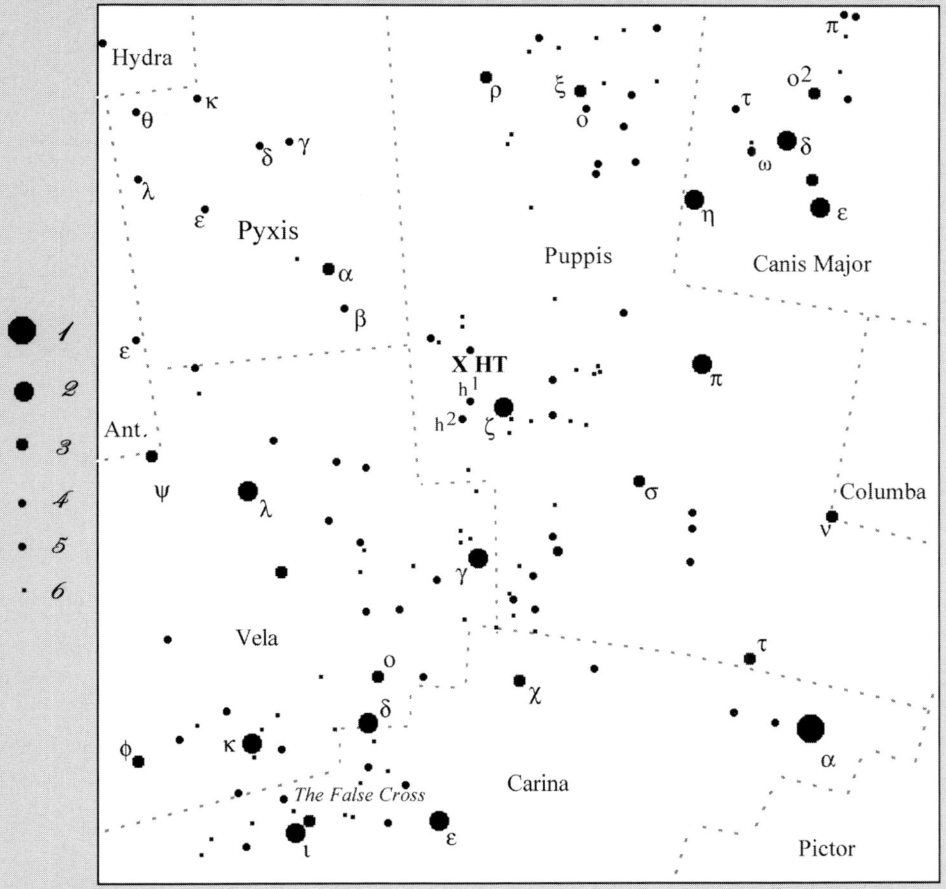

© Stephen James O'Meara

Tirion: Charts 19 & 20

Uranometria: Chart 362

46

Heart & Dagger, "Wounded Heart" Cluster
NGC 2546
Type: Open Cluster
Con: Puppis

RA: 08^h 12.4^m
Dec: −37° 37′
Mag: 5.3 (O'Meara); 6.3
Diam: 70.0′
Dist: 3,200 light-years
Disc: Abbe Nicolas Louis de Lacaille, listed in his 1755 catalog

J. HERSCHEL: A cluster VIII class of about 20 bright stars in an oblong, 8′ long, 3′ broad. (h 3116)

NGC: Cluster, bright, large, little compressed, irregularly extended, stars from magnitude 9 to 12.

NGC 2546 IS YET ANOTHER BEAUTIFUL open cluster in Puppis, and yet another inspired discovery by Lacaille. He lists it as the 4th object in Class II ("nebulous clusters"), which represents, in this case, the view as seen with the naked eye and with his $1\frac{1}{2}$-inch 8× scope. "One sees with the naked eye two neighboring confused groups of stars; but with the telescope they are small distinct stars, in very great number and very close together." As Lacaille explains, "The second class of nebulae comprises [s]tars which are only nebulous in appearance to the naked eye, but when seen in the telescope, show up as a cluster of distinct [s]tars, although very close to each other."

As with many clusters in the region, NGC 2546 is young; at 30 million years, it is the same age as NGC 2264 (Hidden Treasure 38), the Christmas Tree Cluster, in Monoceros. The group is very coarse and has 40 members of 7th magnitude and fainter, which extend across more than two Moon diameters of sky. If we accept NGC 2546's distance as 3,200 light-years, its true physical diameter is nearly $64\frac{1}{2}$ light-years across, which is more than twice as large in true physical extent as NGC 2264, which is 700 light-years closer.

To find the cluster, just look 5° due east of NGC 2451 (Hidden Treasure 42), or 2° northeast of 2nd-magnitude Zeta (ζ) Puppis. And it can be seen as far north as Toronto in Canada, where it is 9° above the horizon at southern culmination. While most catalogs list the cluster's magnitude as 6.3, it is a magnitude brighter. But don't be fooled into thinking you'll be looking for

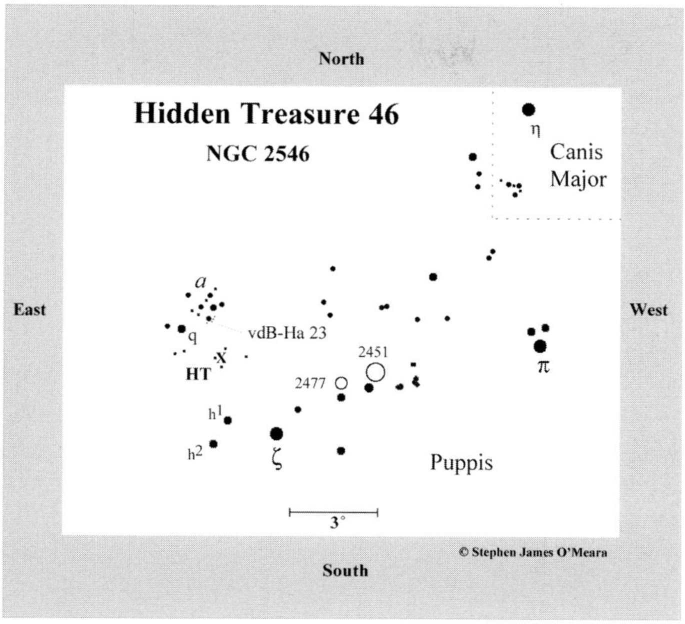

something very obvious. You must imagine the light of a 5th-magnitude star spread over an area larger than two Moon diameters across. While the cluster can be seen as a nebulous glow with the naked eye from a very dark site with a good southern horizon, it can easily disappear with any sky glow.

The view is further complicated by the fact that it lies in one of the most glorious sections of the Puppis Milky Way; it is certainly one of my favorites, being so lush with dense patches of starlight. Note, however, that Lacaille saw the cluster with the naked eye as two confused groupings. The other grouping must have been the prominent Lambda (λ)-shaped asterism of suns (*a*) just 2° to the northeast of NGC 2546. This very impressive asterism packs a dozen suns, ranging from 5th- to 7th-magnitude, in an area equal to that of NGC 2546. In essence, when you look up at that part of the sky with the naked eye, you see two nebulous patches – one bright (an asterism) and one dim (the cluster) – set against a glittering backdrop of Milky Way. What's more, the 40′-wide cluster van den Bergh-Haffner 23 (vdB-Ha 23) lies about $\frac{1}{2}°$ due south of the asterism's southwestern tip, and its light mingles with the asterism's. So Lacaille's description of a confused star field is right on target.

That Lacaille detected NGC 2546 with the naked eye is testament to the observer's prowess. I have found that even with 7 × 50 binoculars, if you sweep past the cluster too quickly, it escapes your view. I suspect that Lacaille, during his telescopic sweeps of the heavens, came across the brighter Lambda asterism first, then, while examining it visually with the naked eye, caught sight of NGC 2546 with his averted vision. When he trained his telescope on those neighboring patches of light, he saw them melt into a tangle of clusters comprised of distinct starlight. While we credit Lacaille for the discovery of only NGC 2546, we should not discount the fact that he may have also been the original discoverer of vdB-Ha 23 or the equally fascinating Lambda-shaped asterism, or both.

James Dunlop carried Lacaille's torch of confusion into the nineteenth century. The 563rd entry in his 1827 catalog has been identified as NGC 2546. But is it? When Dunlop's coordinates for that entry are precessed to epoch 2000.0, we find ourselves looking about halfway between NGC 2546 and vdB-Ha 23. Dunlop's description of the object – "A large cluster of stars of [mixed] magnitude, rather extended figure, not rich

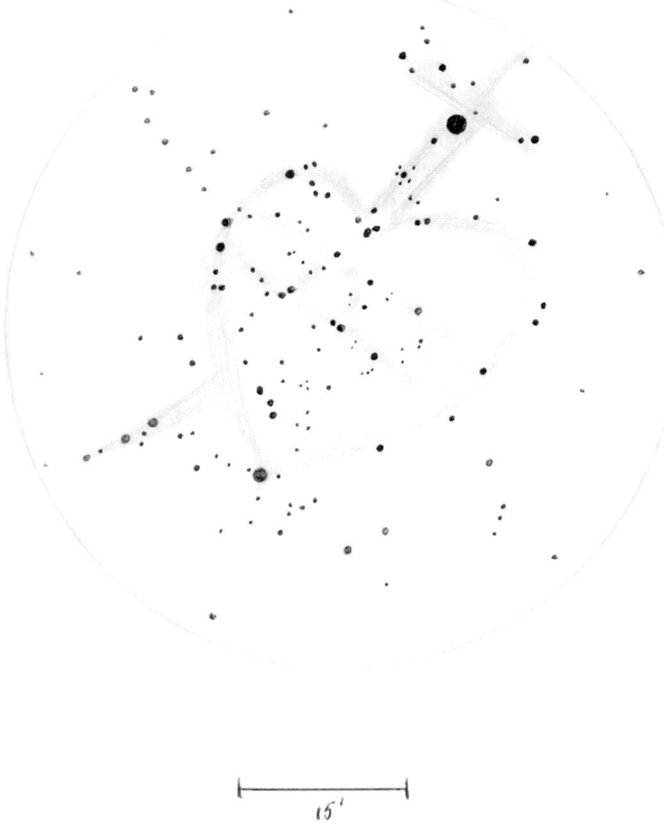

in very [faint] stars" – could match either. But the last phrase, "not rich in very [faint] stars," causes my opinion to lean closer to vdB-Ha 23; while both NGC 2546 and vdB-Ha 23 are large and extended, NGC 2546 is rich in faint stars while vdB-Ha 23 is not.

The Lambda-shaped asterism (and perhaps vdB-Ha 23) appears to be in the Local Interstellar Medium, which flanks part of the Local Bubble – a great cavity some 300 light-years across, within which our Sun and many nearby stars reside. The "Bubble" is actually hourglass-shaped, being narrowest toward the galactic plane and open-ended above and below it. The Puppis–Vela component is sheet-like and extends in galactic longitude from 245° to 275°, in galactic latitude from −15° to +5°, and in distance out to about 3,260 light-years. Although the stars in the asterism are not physically related as a moving group, some (or all) of its stars may belong to the same Milky Way structure as NGC 2546 but about 700 light-years closer.

In 7 × 50 binoculars, the bulk of the cluster covers nearly a degree in size and appears elongated, oriented south-southeast–north-northwest. It is nestled inside an acute triangle of 6th-magnitude suns. In the antique telescope the stars are quite prominent. The surrounding triangle becomes a trapezoid if you add a pair of 8th-magnitude suns to the east-southeast.

At 23× in the 4-inch, it's a magnificently rich object with large bubbles of irregularly bright suns at 90° angles. It's as if some invisible force had blown air into the center of this cluster causing it to froth up like foam. The southeastern and northwestern flanks of the cluster contain the brightest stars. Those at the southeastern end form a disorderly, zigzagging line, while those at the northwestern end are arranged in a warped "J."

Defocus the cluster ever so slightly and relax your gaze. Do you see the stars inside the trapezoid as a heart? It contains several beautiful pairs of stars; four of which are part of a sinuous vine of 10th- to 12th-magnitude stars, oriented east–northwest, which wrap the heart like a crown of thorns (the Christian symbol for the "Wounded Heart" of Christ). Now, if you include the bright "J" of stars to the northwest, and follow its shaft, you will see another row of suns slicing into the heart from northwest to southeast; it exits the heart at a pair of 9th- and 10th-magnitude stars. Here we have a pirate's tattoo – a heart and dagger. The crown of thorns now becomes that playful ribbon reading "Mom."

The cluster loses its luster at high powers. But do look at the western-most 6th-magnitude sun in the acute triangle, because it is a beautiful golden nugget at the end of the dagger's hilt. Also, immediately southeast of that gem is a roughly 7′-long string of suns (the hilt's hand guard) that includes a star-shaped clustering of 11th-magnitude suns on the northeast end and a 10th-magnitude double star at the southwest end. The asterism looks like a tiny skeleton key . . . the key to the heart.

Hidden Treasure 47

NGC 2683

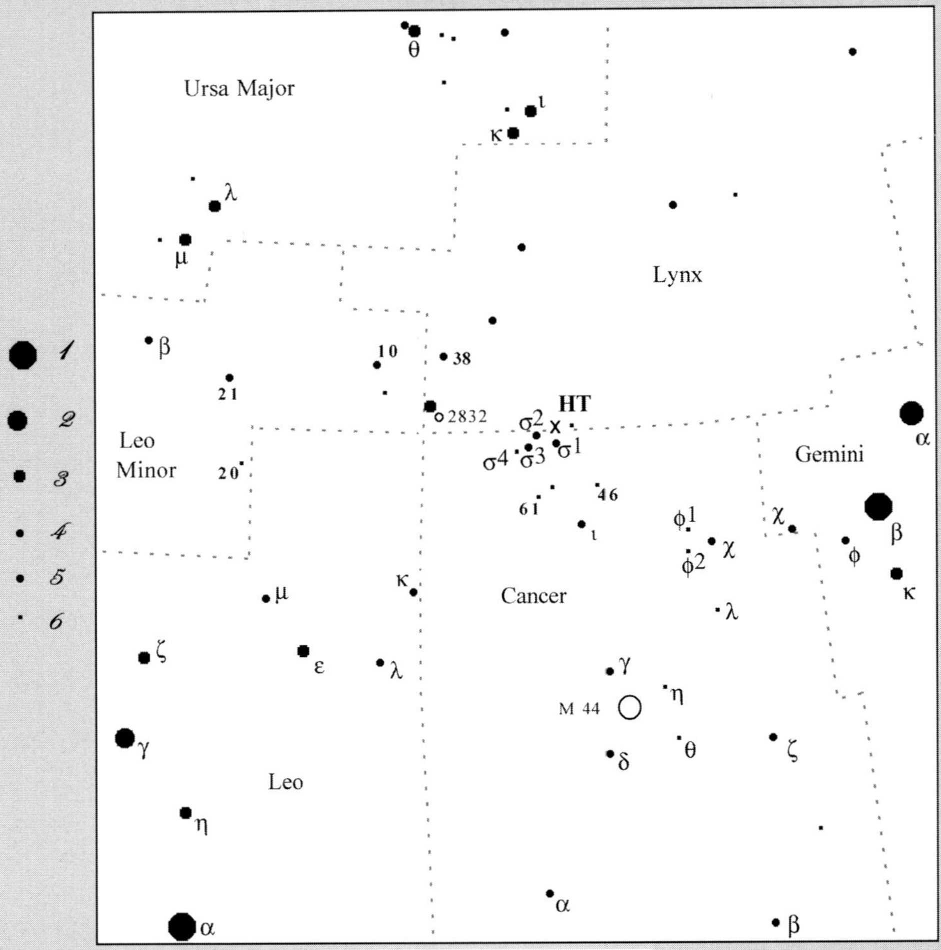

Tirion: Chart 6

Uranometria: Chart 102

47

UFO Galaxy
NGC 2683
Type: Spiral Galaxy (Sb)
Con: Lynx

RA: 08^h 52.7^m
Dec: +33° 25′
Mag: 9.7 (O'Meara); 9.8
Dim: 9.1′ × 2.7′
SB: 12.9
Dist: 19 million light-years
Disc: William Herschel, 1788

W. HERSCHEL: [Observed February 5, 1788] Very brilliant, much extended in a direction south preceding to north following, 8′ length, 3′ breadth, beautiful. (H I-200)

NGC: Very bright, very large, very much extended toward position angle 39°, gradually much brighter in the middle.

TO LELAND S. COPELAND, THE amateur astronomer who nicknamed NGC 2264 (HT 38) the Christmas Tree Cluster, NGC 2683 was the "forerunner of the galactic host of the spring and early summer." This near edge-on spiral system is the brightest of some two dozen galaxies populating the obscure constellation of Lynx, which, by the way, is not named after a creature from ancient Greek mythology. Johannes Hevelius (1611–1687) created the grouping to fill a narrow corridor of seemingly empty space; he explained the title by saying that anyone who wanted to see the constellation's inconspicuous members needed the keen eyes of a lynx.

The stars of Lynx extend from Camelopardalis (another dim and inconspicuous constellation) in the north to Leo in the southeast; it also slinks past the toes of Ursa Major and the extreme northern flanks of Auriga, Gemini, and Cancer. Little wonder, then, that few observers venture into this inner sanctum of the naked-eye sky, especially since many reference books say the constellation has little to offer, except, say, for a few double stars and the dim and distant globular cluster NGC 2419 (Caldwell 25), the Intergalactic Tramp.

But what a wonder observers are missing by not turning their telescopes to NGC 2683, especially those with moderate-to large-sized telescopes. The staff at the Astronaut Memorial Planetarium and Observatory, nicknamed NGC 2683 the UFO Galaxy because its visual appearance is reminiscent of the silver saucer so commonly reported by UFO enthusiasts. But in color images taken with large telescopes, this galaxy looks more like the basal surge of an atomic explosion rippling out from ground zero one second after impact. Cresting waves of dusty debris, dappled with blue and red incendiary embers, ripple out from a dense cloud of yellow light igniting from a spark at the galaxy's core. That core looks exact and precise – a punctuation mark set down with clarity of purpose. Surrounding this rippling madness is a foggy lens of light seen obliquely 11° from edge-on.

The large angular diameter of the main disk is surrounded by a host of faint anonymous galaxies. The stellar content is rather poorly resolved, even in the outer regions, for a galaxy with such a low redshift. Star counts around the halo show it to have a globular cluster system at least as populous as that of the Milky Way with about 100 (±31) globular clusters detected so far.

On early photographic plates, astronomers noticed that the galaxy's tiny nucleus is centered in a peanut-shaped bulge. In fact, only part of the nucleus peeks above the surrounding core. Its lower part is being blocked by dense clouds of dust in the near side of the disk – which is why, in color images, the galaxy's core appears yellow; just as dust in our atmosphere can warm the setting Sun's light, the intervening dust and gas in NGC 2683 warms our view of its core. The galaxy contains many filamentary arms, whose spiral structure is defined by those lanes of dust. Faint outer arms exist, as do pink H-II regions (the red incendiary embers in our hypothetical atomic blast).

Despite its photographic might, NGC 2683 is a modest-sized galaxy, having a linear diameter of 38,500 light-years. It shines with a luminosity of 6 billion Suns and has a total mass of 60 billion Suns. It belongs to the Leo Spur of Galaxies and is receding from us at a speed of 415 kilometers per second. Other members of this clan include the hidden treasure galaxies NGC 2841, 2903, 3184, 3344, 3521, 3621, and 3628. In 1999, E. Pompei (University of Trieste) and his colleagues announced the discovery of a stellar system within NGC 2683 that rotates against the spiral pattern. They also found opposing gas flows in the disk. These data imply that either a companion has merged with the galaxy, or perhaps a fast-moving bar near the galaxy's minor axis is causing gas to fall into the bar. In the latter case, such infalling gas flow can cause episodes of starburst. NGC 2683 has an unresolved radio core and X-ray emission has also been observed from its central region, so the galaxy could be a candidate for a previously unknown active galactic nucleus. Then again, a single supernova remnant could also account for the latter two observations.

To find this extragalactic jewel, draw an imaginary line from Beta (β) Geminorum (Pollux) and Epsilon (ε) Leonis in the Lion's Sickle. Now look about halfway along that line and a little north, where you will find the 4th-magnitude double star Iota (ι) Cancri. Almost 10° to the north is the 3rd-magnitude star Alpha (α) Lyncis, which marks the northern apex of an isosceles triangle with Iota Cancri and Epsilon Leonis. Now set your gaze halfway between Alpha Lyncis and Iota Cancri, and a tad northwest. There you'll find a roughly 1°-long chain of

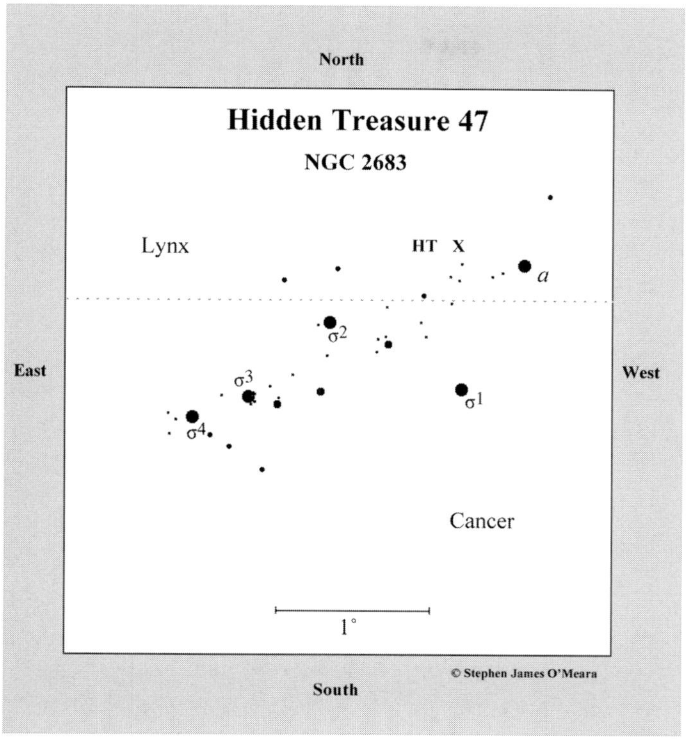

94 light-years, which is more than 200,000 times closer than our target; now you use your mind to abolish the two-dimensional aspect of the sky.

The galaxy is definitely visible in my antique telescope as an elliptical haze, just 10′ north of a little triangle of 10th- and 11th-magnitude stars and about 3′ west of a 12th-magnitude sun. At 23× in the 4-inch, it is a beautiful silver needle, elongated northeast to southwest, with distinct sheen, like moonlight glinting off a sword. The galaxy is easier to see than the edge-on systems NGC 891 (Caldwell 23) in Andromeda and NGC 4565 (Caldwell 38) in Coma Berenices; although NGC 4565 and NGC 2683 are equals in magnitude and surface brightness, NGC 2683 is slightly more open, so we see more light from its disk.

three 5th- and 6th-magnitude suns: Sigma^2 (σ^2), Sigma^3 (σ^3), and Sigma^4 (σ^4) Cancri; from a dark-sky site, these stars are visible to the unaided eye as a mottled stellar haze. Sigma^2 Canci, marks the eastern end of a $1\frac{1}{2}°$-wide triangle with two other 6th-magnitude stars – Sigma^1 (σ^1) Cancri, which lies to the southwest, and Star *a*, to the west-northwest, just across the border in Lynx. Seen together, all these stars form a $2\frac{1}{2}°$-long asterism that, with imagination, looks like a pirate tripping, or falling down drunk; NGC 2683 is the pirate's bottle of rum flying out of one of his outstretched hands (Star *a*). The galaxy is just visible in 7 × 50 binoculars under a dark sky; it should be more readily visible in 10 × 50s. Don't be fooled by Sigma^2 Cancri; it is the brightest star in a tight triangle of suns, which, like M40 in Ursa Majoris can look fuzzy. Star *a* lies at a distance of

Details start to appear at 72×, which, in small apertures, is about the maximum power you'll want for comfortable study. The nucleus gleams like a brilliant diamond flanked on either side by cultured pearls (a tempting treat for the eye of any pirate). The two "pearls" mark the locations where dust piercing starlight is packed tightly enough together to overpower the region's dense shrouds of dust. But when I reverse my thinking, and look for lanes of darkness rather than light, I see a distinct ripple of darkness on either side of the core. Spiral structure in the inner lens may be inferred in larger telescopes, but in the 4-inch all I see can see are irregular patches of light

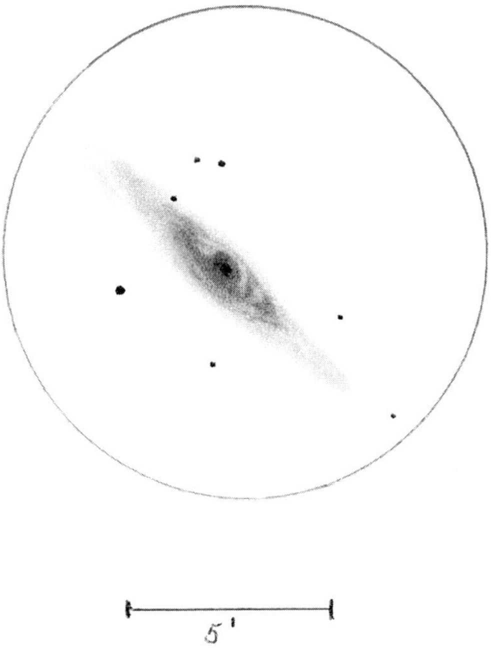

and dark. But with time and concentration, see if you cannot see the galaxy's northeast end curl gracefully to the east, until it gradually fades into the sky background. Here is one of the galaxy's outer arms. I do not see a similar hook on the southwest end, though, even though that region is brighter.

Although the galaxy is very stunning photographically, it is a delicate "flower" visually. It takes much time to see the details I show in my drawing. Larger apertures will have an easier time with it. Supernova hunters should not be fooled by a 13th-magnitude star 2′ northeast of the nucleus; the most prominent dark ripples I mention above lie between this star and the core.

Before you leave the area, return to Alpha Lyncis. Just 40′ to the southwest is NGC 2832, a 12th-magnitude galaxy (3.0′ × 2.1′ across). It is the brightest member of the galaxy cluster Abell 779. Steve Coe, observing in the central mountains of Arizona, reports that through a 13-inch reflector, he could see NGC 2832 as a "pretty bright, pretty large and round galaxy which is much brighter in the middle at 100×. On an excellent night using 165× and 220×, I could also see nine companions around NGC 2832. In moments of good seeing many extremely faint members make the field mottled or 'lumpy'." Give it a try.

Hidden Treasure 48

NGC 2655

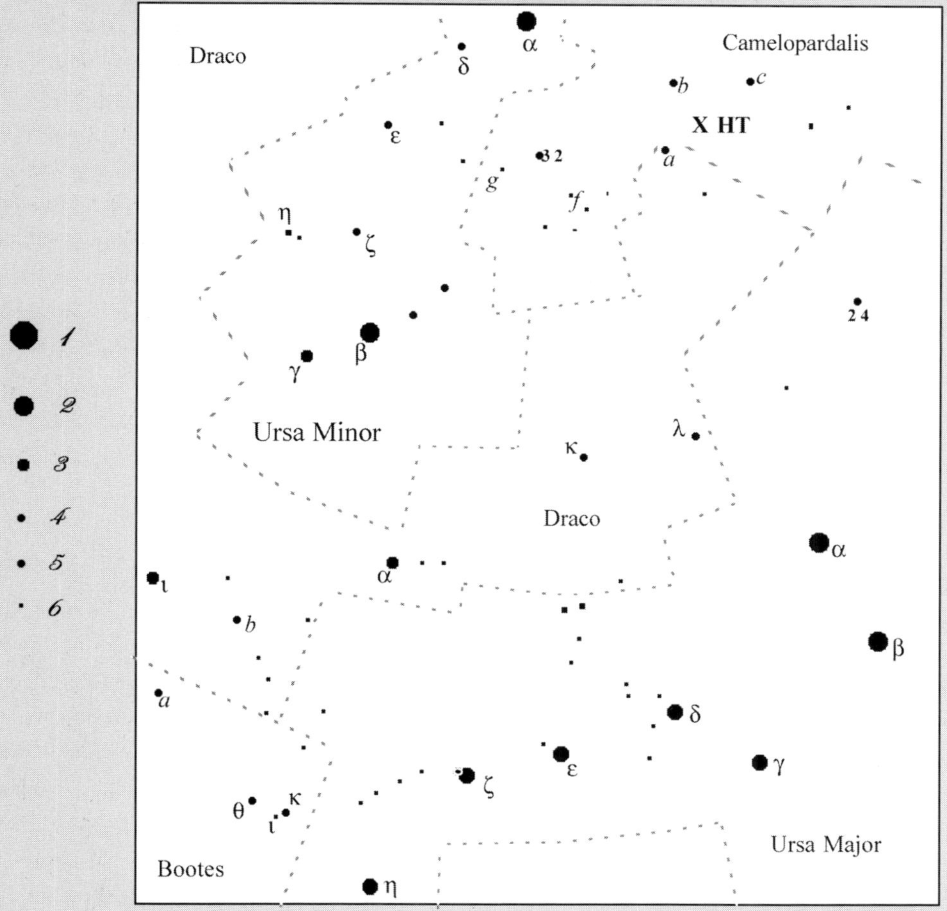

© Stephen James O'Meara

Tirion: Charts 1 & 2

Uranometria: Chart 7

48

NGC 2655
Type: Mixed Lenticular Galaxy (SAB0a)
Con: Camelopardalis

RA: 08^h 55.6^m
Dec: $+78°$ $13'$
Mag: 10.2 (O'Meara); 10.1
Dim: $5.9' \times 5.3'$
SB: 13.7
Dist: 80 million light-years
Disc: William Herschel, 1802

W. HERSCHEL: [Observed September 26, 1802] Very bright, considerably large, a little extended, with a suddenly much brighter middle. (H I-288)

NGC: Very bright, considerably large, little extended toward position angle 90°, gradually, then suddenly very much brighter in the middle.

NGC 2655 IS A VERY NEGLECTED extragalactic wonder that circles the North Celestial Pole a little more than 10° from Polaris. It is circumpolar from most of the Northern Hemisphere and is visible on any clear night. How it must move through the night, as quiet as a librarian's whisper, for so few observers to take notice. But this beauty is as bright and obvious as NGC 5866 (Hidden Treasure 75) in Draco, arguably the most targeted non-Messier galaxy in the heavens. Luginbuhl and Skiff say that even a $2\frac{1}{2}$-inch refractor will show NGC 2655 as a "small round spot with a bright center."

This hidden treasure belongs to the Ursa Major Cloud of Galaxies and is traveling away from us at a dizzying speed of 1,445 kilometers per second. It is a sizable system, having a true linear diameter of 135,000 light-years and a total luminosity of 40 billion Suns. Brent Tully catalogs it as a mixed lenticular system, meaning that it may have the presence of a bar, but astronomers now tend to classify it as an Sa galaxy – a transitional lenticular system with spiral structure. Halton Arp includes NGC 2655 as the 225th object in his *Atlas of Peculiar Galaxies* – a selection of unusual galaxies, interacting pairs, or larger groups – which Arp compiled in 1966 using photographs from the Palomar 200-inch telescope. He classified it as a galaxy "with amorphous arms."

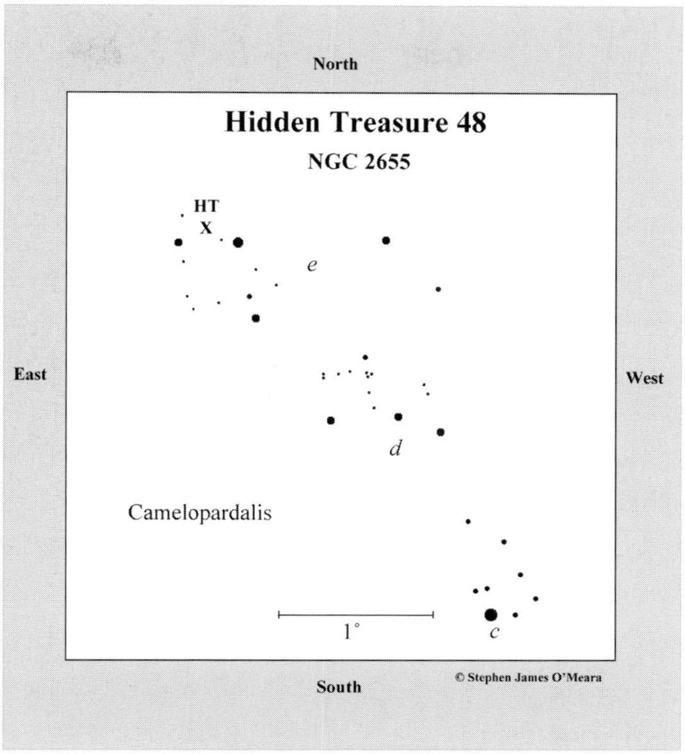

NGC 2655 is very peculiar. In photographs, the galaxy displays a smooth oval disk, seen 29° from face-on, with a bright inner lens and an intense starlike nucleus. Very faint and diffuse outer arms extend beyond the galaxy's shapeless outer halo in smooth whorls. After studying that intense nucleus, Russian astronomer O. K. Sil'chenko (Special Astrophysical Observatory) and his colleagues announced in 1990 that NGC 2655 may be a new nearby Seyfert galaxy – one with a hot, fast-moving disk of gas near the galactic center. That disk has a diameter of about 1,600 light-years and has a mass of about 10 million Suns. A probable black hole lurks at the heart of that swirling madness. If true, as the material falls into the black hole, frictional heating should cause it to turn into a concentrated plasma, which will be propelled out of the nucleus by intense and twisted magnetic fields at nearly the speed of light. Indeed, NGC 2655 contains strong jetlike features extending 6,000 light-years southeast of the nucleus, and a weaker jet extending about 15,000 light-years to the northwest. A distinct clump of matter has also been detected along the northwestern jet, which may have been ejected from the galaxy's nucleus.

Deep images, show that NGC 2655's amorphous outer envelope has faint spiral structures or ripples, with a number of dust patches, especially along the galaxy's minor axis. Such images, for instance, taken in 1996 with the 140-inch WIYN telescope atop Kitt Peak in Arizona have revealed strong dust lanes in parts of the galaxy slightly closer to us. The dust may follow gaseous material that orbits the galaxy "above" the plane, on the side facing us. If true, the dust is most likely obscuring our view of the true nucleus. It also indicates that NGC 2655 may have recently captured material from a companion galaxy. Then again, it may have already swallowed such a companion entirely.

How amazingly simple the view through a telescope is compared to the utter chaos that exists deep behind the veil. Looking at NGC 2655 through a small backyard telescope is like standing in front of a closed door while a silent murder is going on behind it. One reason this amazing galaxy is neglected is that it is a challenge to find. Plan to spend some time on the hunt.

Also, unless you are under a dark sky, be sure to have binoculars at your side. To start, imagine a line from Alpha (α) Ursae Majoris (Dubhe) to Alpha (α) Ursae Minoris (Polaris). About one-third of the way from Polaris to Dubhe, and a tad west, is a 4th-magnitude star (*a*). It is the brightest star in the area, so there is no mistaking it. Now use your binoculars to locate a 5th-magnitude star (*b*) 4° to the west-southwest. It is the brightest of three equally spaced stars oriented north-northwest–south-southeast. Next look about 4° south-southeast of Star *b* for another 5th-magnitude sun (*c*); it is the brightest star in a little hook of seven suns. (Use your binoculars to confirm that Stars *a, b,* and *c,* form a roughly $6\frac{1}{2}$°-wide isosceles triangle.)

Once you confirm Star *c* in binoculars, locate it in your telescope at low power. The entire hook asterism fits nicely in a 1°-wide field of view. If you move $1\frac{1}{2}$° north-east from the northeastern tip of the hook, you should see a 45′-long arc of three 7th- and 8th-magnitude suns (*d*), oriented east–west. A sweep 1° to the northeast will center a 1°-wide, isosceles triangle comprised of 7th- and 8th-magnitude suns (labeled *e*); center the triangle's northeastern star, which is the brightest. NGC 2655 is only 15′ east-northeast of that star; another 7th-magnitude star lies about 10′ to its south-east.

I could not see the galaxy in 7 × 50 binoculars, and my antique telescope shows it just barely. But it's a nice object in the 4-inch at 23×, appearing as a soft roundish glow with a delicate core, all of which is visible with direct vision. The galaxy swells into a slight egg-shaped wonder with averted vision, with the major axis oriented east–west. The galaxy's intensity is enhanced at

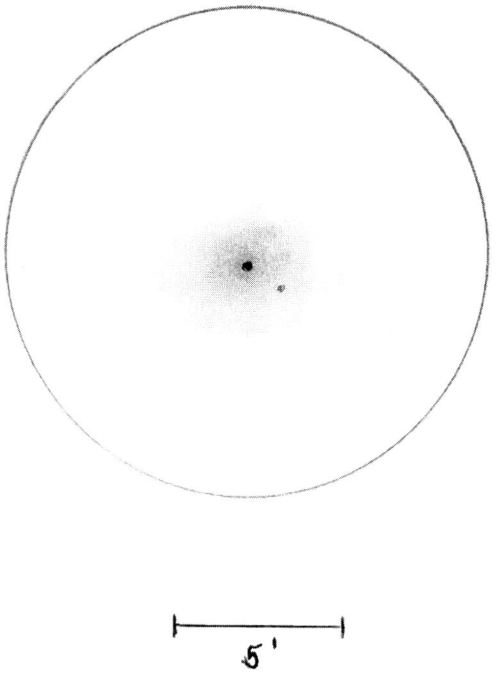

72×, but I was surprised not to see the galaxy's famed stellar-like Seyfert nucleus. Instead, I saw a fuzzy bead of light surrounded by an elliptical inner lens with a dim and diffuse outer halo. The galaxy takes magnification well, and, at 168×, I noticed that the fuzzy bead at the galaxy's core breaks down into two components – a starlike nucleus surrounded by a tight collar of light.

Overall, NGC 2655 looks more like a planetary nebula than a galaxy. Most amazingly, however, with time and effort, and after studying the galaxy with different magnifications and direct and indirect vision, I could see the western flank of the bright inner lens appearing dappled. Here is where the galaxy's intriguing dust loops and ripples are strongest. And though I could not see any definite dark features, the dust must be responsible for the mottling effect.

Observers using 10-inch telescopes see about as much detail. And Luginbuhl and Skiff have seen the inner core with "bright mottlings along the major axis at high power through a 12-inch reflector." The dark loop has been observed in 18-inch telescopes, which are common at star parties nowadays. Some observers, like Canadian amateur Roland Prevost (Ottawa Astronomy and Observers Group), note that while NGC 2655 would be "incredibly easy" to see in a dark sky, it may require "averted vision" in suburban skies. Good luck!

Hidden Treasure 49
NGC 2841

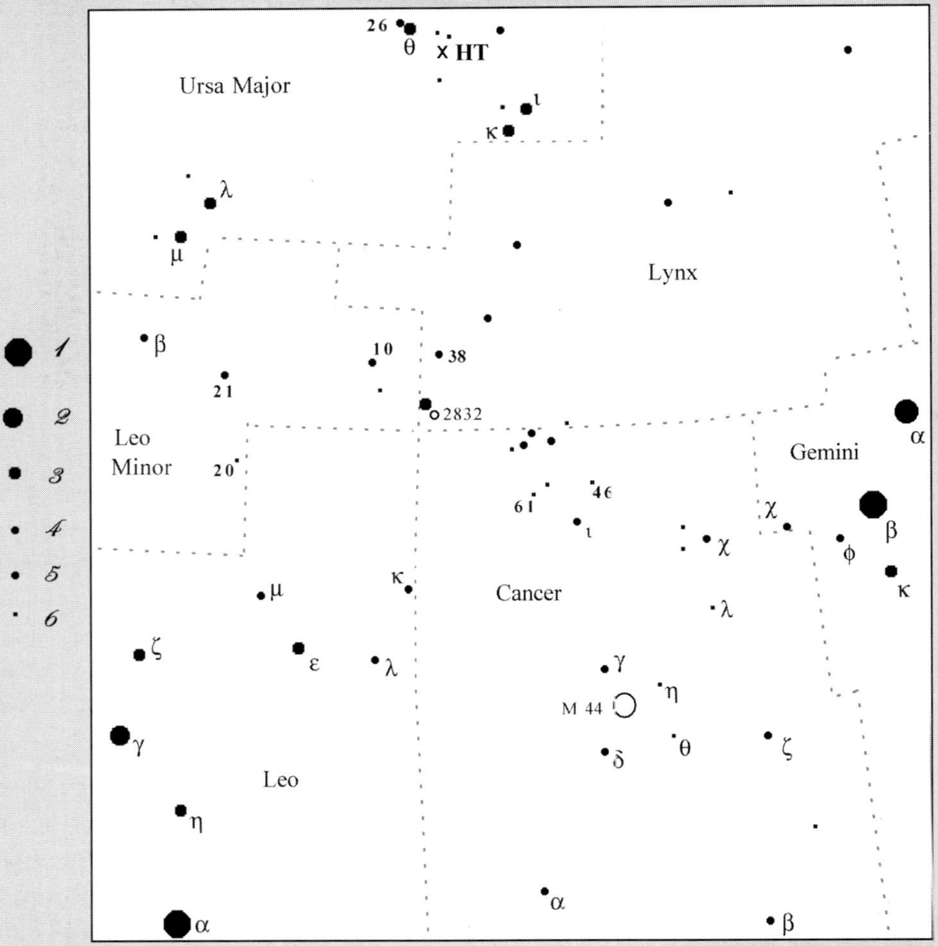

© Stephen James O'Meara

Tirion: Charts 2 & 6

Uranometria: Chart 44

49

Tiger's Eye Galaxy
NGC 2841
Type: Spiral Galaxy (Sb)
Con: Ursa Major

RA: 09^h 22.0^m
Dec: $+50°$ $59'$
Mag: 9.0 (O'Meara); 9.2
Dim: $6.6' \times 3.4'$
SB: 12.4
Dist: 46 million light-years
Disc: William Herschel, 1788

W. HERSCHEL: [Observed March 9, 1788] Very bright, a little brighter middle, chevelure, branches, milky nebula, 6′ length, 4′ breadth. (H I-205)

NGC: Very bright, large, very much extended toward position angle 151°, very suddenly much brighter in the middle equal to a magnitude 10 star.

NGC 2841 IS ONE OF THE FINEST examples of a multiple-arm, intermediate-type (Sb) spiral galaxy in the heavens. It belongs to the Leo Spur of galaxies and stands up well against many a Messier galaxy. In true physical stature, NGC 2841 is a modest one, having a linear diameter of only 38,500 light-years and a total mass of 60 billion Suns. A bright starlike Seyfert nucleus punctuates the galaxy's amorphous central lens, which otherwise resembles an elliptical or lenticular galaxy; in other words, the lens is devoid of dust or spiral structure. Optical and near-infrared observations of the nucleus, however, suggest that it is surrounded by a ring of enhanced star formation between 6,500 and 23,000 light-years out from the central region. This is also an area of enhanced H II activity which also contributes to the overall bluer color of this region. Otherwise, the galaxy's bulge is comprised of very cold dust and a population of old suns.

A series of thin dust lanes spirals outward from the periphery of the lens. The dust separates luminous and very intricate spiral filaments, which we see tilted 26° from edge-on. At first glance the filaments look like complete spiral arms, but high-resolution images show them to be thin broken segments – none of which can be traced as individual arcs for more than 30°. The Hubble Space Telescope has since resolved these fragmentary arcs into a multitude of stars

and tiny H-II regions, distributed regularly about the disk in a spiral pattern. The HST image shown here depicts the southeast portion of the galaxy's disk as a curdled mass of starlight with the spiral wonder of a fingerprint.

Twenty-nine Cepheid variables were also found in the HST images of NGC 2841, which Lucas M. Macri (Harvard-Smithsonian Center for Astrophysics) and colleagues used to secure a relatively accurate distance of 46 million light-years. That distance is also supported by HST observations of a fortuitous Type Ia supernova (1999by), which flared to prominence 96″ west and 91″ north of the galaxy's nucleus. British amateur Ron Arbour discovered the 15th-magnitude star on April 30, 1999 (Universal Time); Erno Berko in Hungary independently discovered it that same night. The supernova achieved maximum brightness (13.2) sometime around May 12, 1999. Since the peak light output from Type Ia supernovae is always approximately equivalent to an absolute blue sensitive magnitude of −19.6, astronomers can use the inverse square law to infer the supernova's distance and therefore the distance of its parent galaxy. Actually, the discovery of 1999 was not *shocking* news. NGC 2841 had long been a popular target of supernovae hunters, because the galaxy has hosted three supernovae in the past: 16th-magnitude 1972r, discovered 46″ west and 70″ south of the nucleus; 1957a, a 14th-magnitude object 106″ west and 73″ north of the nucleus; and 13th-magnitude 1912a, which appeared 50″ west and 20″ north of the nucleus. Both SN1957a and SN1999by were Type I events, and, interestingly, both were subluminous objects. In fact SN 1999by was one of the least luminous Type Ia events ever observed. Type I supernovae show an absence of hydrogen lines in their spectra. They are the product of an explosion generated by a white dwarf component in a binary star system which draws material from its companion. When enough foreign material builds up on the surface of the dwarf and exceeds the critical mass of the star it explodes.

Finding NGC 2841 is more like taking a leisurely stroll to a neighbor's house rather than going on a long treasure hunt; the galaxy lies only about $1\frac{3}{4}°$ west-southwest of 3rd-magnitude Theta (θ) Ursae Majoris, in the Great Bear's right foreleg. From Theta UMa, first sweep $1\frac{1}{4}°$ west to a 6th-magnitude star (*a*), then move 40′ to the southwest, where you'll find another 6th-magnitude sun (*b*) with an 8th-magnitude companion about 4′ to the northeast. These latter two stars are true companions, lying 95 and 97 light-years distant, respectively. NGC 2841 (46 million light-years distant) is only 20′ south-southeast of that pair. Use your imagination to fathom the great gulf

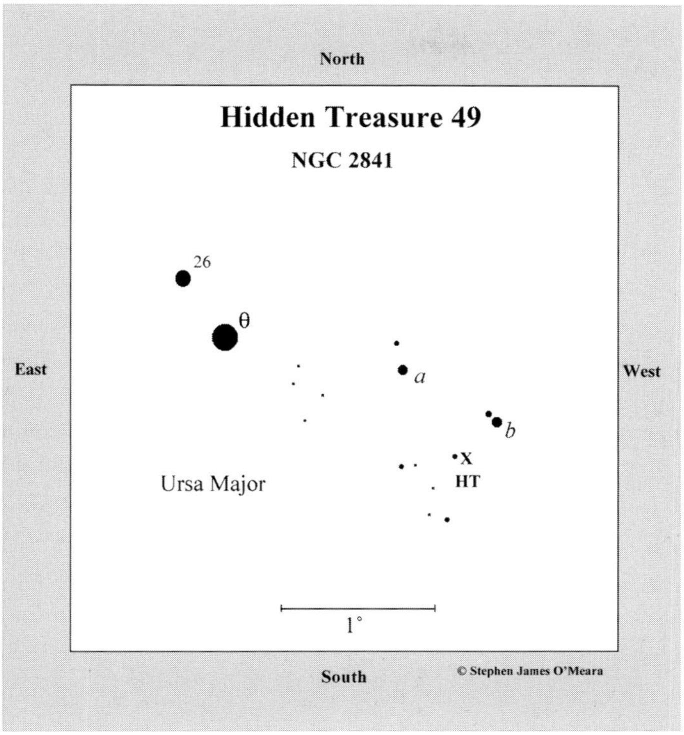

whose sudden appearance in the eyepiece would stir me spiritually.

Now, whenever darkness falls over Hawaii, I can see NGC 2841 with averted vision in 7 × 50 binoculars, looking like the dim companion to a 9th-magnitude star about 5′ to the northeast. In my antique scope the galaxy looks like a dim vapor trail, oriented northwest–southeast, drifting by that 9th-magnitude star. In the 4-inch, 23× will show NGC 2841 as an elongated glow with a sharp stellar nucleus, which looks like a moonlit stone centered in a pool of vapors. Be aware that a prominent 11th-magnitude star can be seen northwest of the nucleus in the disk; that star can cause you to think that a supernova has erupted.

The galaxy can be viewed as a false comet in two respects: (1) at low power, the galaxy and its close 9th-magnitude companion resemble the fuzzy double star M40, which John Hevelius mistook as cometary in nature in 1660. And (2) at higher powers, the galaxy disk displays a sharp nucleus surrounded by the soft cometary glow of unresolved starlight. At 72×, for instance, the galaxy's starlike nucleus lies within a slightly elliptical inner lens, that looks like the comet with a tail seen nearly head on. At that power, with concentration, the galaxy's northwestern side appears mottled. The rest of the disk is milky smooth – a view shared by the galaxy's discoverer,

of space that separates the galaxy from the stars in our own Milky Way.

For a 9th-magnitude galaxy, NGC 2841 is deceivingly bright. I could see its central lens in astronomical twilight through the 4-inch at 23×. I wasn't surprised. More than a quarter-century before, I had encountered this galaxy while hunting for comets with the 9-inch f/12 Clark refractor at Harvard College Observatory. Despite the light-polluted skies of Cambridge, Massachusetts, I could clearly see the galaxy's bright central lens and quickly became enamored by its cometary nature. Finding the galaxy gave me hope that I would some day discover a 9th-magnitude comet from that city location. And though I did not meet with success in that endeavor, I appreciated discovering new celebrities, like NGC 2841,

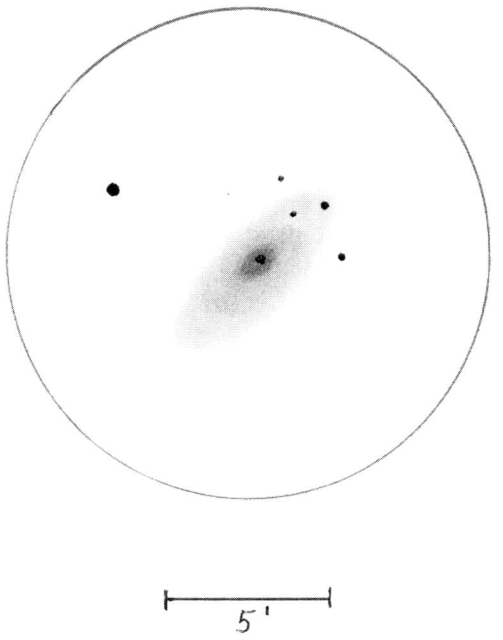

William Herschel, who called it a "milky nebula." Admiral Smyth went a step further saying that the object has a "pale creamy whiteness."

The mottled northwest section of the galaxy appears brighter than the southeast side, so your eyes may be drawn to that section first. The reason for these appearances is evident at 101×. Not one but four stars lie in the galaxy's northwest quadrant: the 11th-magnitude sun already mentioned, a 13th-magnitude star abuts the galaxy's west-northwest flank; a 13.5-magnitude star hugs the northeast flank; and a 14th-magnitude sun lies in the disk's northeast quadrant. In the southeastern side of the disk, however, some mottling appears to be true knots and lanes in the disk structure.

There also appears to be a bar running along the galaxy's major axis, but this is just an illusion. If you look at a line drawing of a circle with radial spokes, the radial lines are all equidistant when viewed face on. But if you view the circle obliquely, almost edge-on, you will see a "bar" of light appear along the axis perpendicular to your line of sight – like the band of light that characterizes Tiger Eye – a gemstone frequently found in pendants and once craved by Pirates, who believed the stone would offer them protection during their journeys and strengthen their convictions and confidence. Ironically, seen in photographs there is a bright linear strip on the galaxy's minor axis, that just happens to be in the same place as a dark dust lane. Spectra of this strip with the Keck II telescope show it to be identical to spectra of the bulge area. This indicates the emission from this optical strip is the product of scattered light from the bulge. It is thought this scattering cloud could extend to over 1,600 light-years above the plane of the galaxy. This is similar to the heights of other dust features found in similar galaxies.

My drawing of NGC 2841 shows the extremely delicate details that can be spied over time. It is like trying to render the fine filigree adorning a Victorian wedding gown. To see these details, you may want to do what I did, and that is to start early in astronomical twilight and draw the brightest features first. Fainter details will emerge gradually thereafter. Do you see the inner lens as uniform in brightness? Also look for a dense knot or arc of material immediately southwest of the lens; it appears to be separated from the lens by a wafer-thin lane of dust, which is just at the limit of vision in the 4-inch. Another arc, a bit more diffuse, lies at the galaxy's extreme southeastern edge. Observers using larger telescopes have also noted seeing a dark patch or lane preceding the nucleus. The Third Earl of Rosse was so duly impressed with the view of NGC 2841 in his great Leviathan telescope at Birr Castle in Ireland, that he likened it to the "Andromeda Nebula."

But trying to spot the finer details in NGC 2841, whose disk is fragmented into countless arcs of light and dark, is like trying to spot a tiger in the grass – the tiger's color and stripes mimic the way light and shadow play in the grass so that the tiger seems to be everywhere at once. Indeed, as soon I pick up some faint detail in NGC 2841 with averted vision, my eye tries to zero in on it with direct vision. But that only causes that detail to fade from view, until my averted vision picks up another faint detail somewhere else in the disk, and the process starts over again, and again, *ad infinitum.*

Hidden Treasure 50

IC 2488

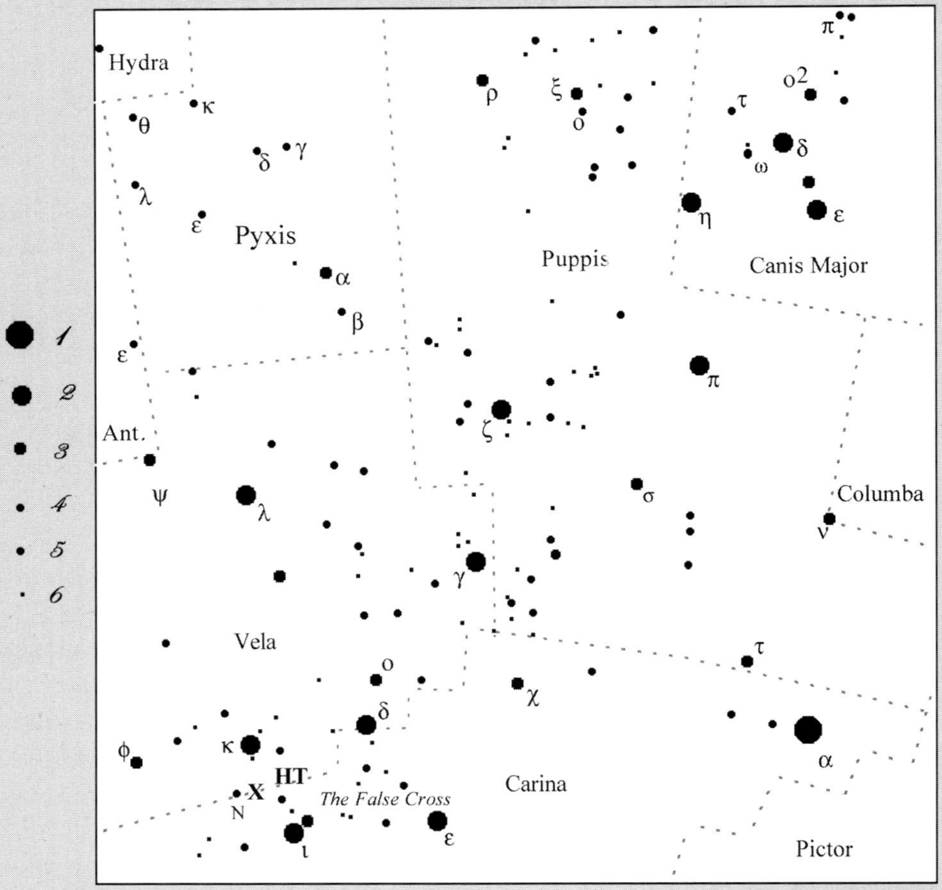

© Stephen James O'Meara

Tirion: Chart 25

Uranometria: Charts 425 & 426

50

Strings of Pearls Cluster, Hoopskirt Cluster
IC 2488
Type: Open Cluster
Con: Vela

RA: 09^h 27.4^m
Dec: −56° 57′
Mag: ~6.0 (O'Meara); 7.4
Diam: 18.0′
Dist: 4,700 light-years
Disc: Abbe Nicolas Louis de Lacaille, listed in his 1755 catalog

HERSCHEL: None.

IC: Cluster, coarse.

DESPITE BEING VERY BRIGHT, IC 2488, is a little-studied open cluster in the deep southern sky. Open clusters, especially those containing stars in advanced stages of evolution, are particularly valuable to astronomers testing models of stellar evolution. Stars in open clusters are all born at about the same time, so they are all about the same age. But just glance at a cluster and you'll see that the stars vary in brightness. If we could bring all these stars to the same distance as the Sun, then plot their brightnesses and temperatures, we'd discover that they would run in a continuous series from bright to faint, from hot to cool, from massive to light, and from larger to smaller. This series, called the *main sequence*, represents the hydrogen-burning phase of a star's evolution. The main sequence, is, as the late Harvard astronomer Cecilia Payne-Gaposchkin called it, the backbone of the famous Hertzsprung–Russell diagram of stellar evolution.

As the cluster ages, the first stars to leave the main sequence are the massive hot, blue stars, which burn their fuel more rapidly than less massive stars; they are also the first to become red giants – stars that have exhausted their core supply of hydrogen and are now fusing hydrogen to helium in a shell outside their cores, or doing both. Of course, as the cluster ages, more and more stars use up their fuel and become red giants. The Hertzsprung–Russell diagram then is like a snapshot of the evolutionary state of a cluster. It is a reflection of a cluster's age; the older an open star cluster gets, the redder its Hertzsprung–Russell diagram becomes.

Young IC 2488 should be of special interest to astronomers because it contains several red-giant members. In a 2003 *Astronomy and Astrophysics* abstract, Argentinean astronomer Juan José Clariá (National

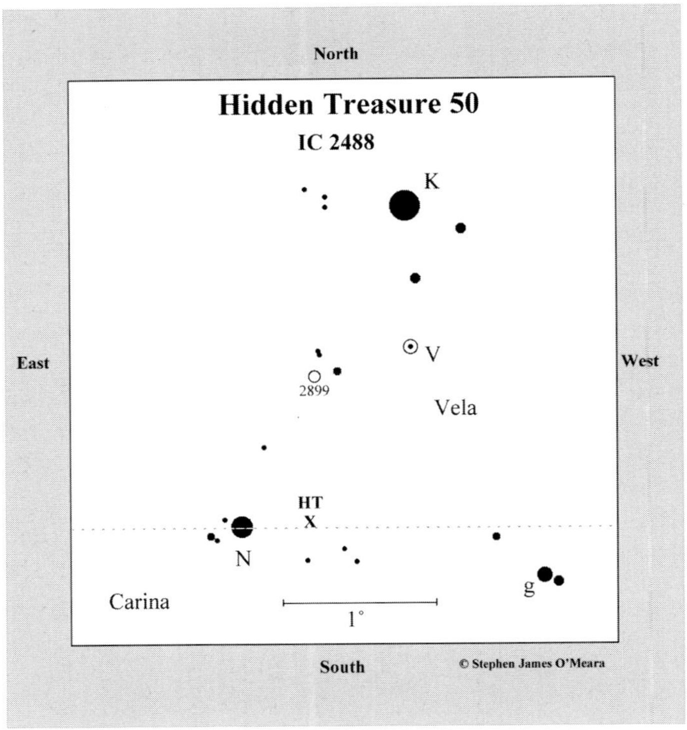

Epsilon (ε) and Iota (ι) Carinae. I also call this form the Juggler, because 2° northeast of Iota Carinae (the southeastern arm of the False Cross), you will find a roughly 3°-wide triangle of 3rd- and 4th-magnitude stars – g and h Carinae and N Velorum – which look like three balls being juggled; the 2nd-magnitude Omicron (o) Velorum Cluster (Caldwell 84), is a larger ball being juggled above the False Cross' northwestern arm.

Set your sights on 3rd-magnitude N Vel, the brightest and northernmost of the three balls being juggled; it's a wonderfully orange *K*5 star. Our target, IC 2488, is a mere 30′ (about one Moon diameter) due west of that tangerine star. James Dunlop, the "gentleman astronomer, from Parramatta, New South Wales, listed it as the 330th object in his 1827 *A Catalogue of Nebulae and Clusters of Stars in the Southern Hemisphere.* He describes it as a "faint cluster of [faint] stars of mixed magnitude, with two or three pretty bright stars in it." Dunlop also notes that the cluster is probably Bode's 485 Argus which, he says "is described as a [faint] star surrounded by nebula." But it was the Abbé Nicolas Louis de Lacaille (1713–1762) who first discovered this cluster during his famous expedition to the Cape of Good Hope in southern Africa. From April 19, 1751, to March 8, 1753, Lacaille cataloged, among other things, 42 "nebulae," which first appeared as a list in 1755, in the *Memoirs* of the French Royal Academy

University of Córdoba) and his colleagues estimated the cluster's age to be 180 million years – old enough for certain massive stars to have evolved into red giants; the most massive blue giants will have consumed all the hydrogen in their cores after only about 10 million years.

To find this interesting object, one needs only to know the False Cross, a celestial signpost as familiar to many Southern Hemisphere observers as the Southern Cross itself. The False Cross is an asterism that looks like Crux, is at nearly the same declination as Crux, has the same orientation in the sky as Crux, but is slightly larger than Crux. It is also about the same distance west of the great Eta (η) Carinae Nebula (Caldwell 92) as the Southern Cross is east of it. The False Cross is comprised of four 2nd-magnitude stars – Delta (δ) and Kappa (κ) Velorum, and of Science. In that work, entitled *On the Nebulous Stars of the Southern Sky*, Lacaille describes the nature of these objects:

> I first observe that three kinds of nebulae can be distinguished in the heavens; the first is no more than a whitish, ill-defined area, more or less luminous and of a very irregular shape: these patches are quite similar to the nuclei of faint, tail-less comets.
>
> The second class of nebulae comprises Stars which are only nebulous in appearance and to the naked eye, but when seen in the telescope, show up as a cluster of distinct Stars, although very close to each other.
>
> The third class of nebulae comprises Stars which are actually accompanied by or surrounded with white patches or by nebulae of the first class.

Lacaille listed our target as the 4th object in his third class. Through a $\frac{1}{2}$-inch telescope at 8×, he described it simply as a "faint star surrounded by nebulosity." Dreyer included the object in his 1908 *Second Index Catalogue* as IC 2488, where it is simply described as a "coarse" cluster. Interestingly, Dreyer did not credit Lacaille with the discovery. He instead refers the reader to Harvard astronomer Solon Bailey's *Catalogue of Bright Clusters and Nebulae*, which contains 13 clusters not in the *NGC*. Dreyer was not trying to slight Lacaille; he was prudently acknowledging the source of his position for the cluster, which Bailey obtained from photographic plates taken with the 24-inch Bruce telescope at Arequipa, Peru.

Lacaille's simple description of IC 2488 is apt for his aperture. In 7 × 50 binoculars from Hawaii, the cluster clings to N Vel like a piece of ill-defined lint. The antique telescope makes it look grainy. With time, its brightest members pop into view like popcorn. At 23× in the 4-inch, IC 2488 is a weak assemblage of some 30 suns (10th-magnitude and fainter) spread out in a northwest–southeast direction across about 15′ of sky. It's also divided into two disparate groups of mismatched suns. A bright 5′ "core" of about a dozen 10th- to 11th-magnitude suns lies a bit closer to N Vel than does a wider wash of stars spreading out to the northwest. At this power, this core appears to have several arms radiating out from the core, especially to the east-northeast, southeast, south, and southwest. At 72×, the coarse cluster spreads out and loses its appeal. Still, the "core" is most alluring, because it is comprised of two striking parallel rows of stars, many of which are in pairs. The rows are oriented north–south and look like two strings of pearls

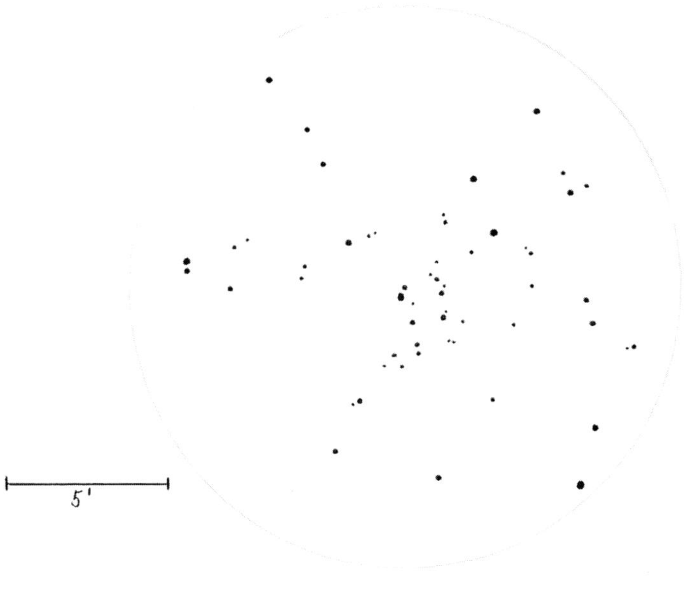

being lifted from a chest of scattered pearls. Those with larger apertures observing from more southerly locations will probably get a much richer view, since the cluster contains some 70 members out to 18′, which at a distance of 4,700 light-years, equates to a true physical extent of 25 light-years.

Celestial pirates with large apertures may want to dig deep into this little treasure chest, because hidden among the scattered jewels of IC 2488 is the planetary nebula RCW 44, also known as VBRC 2 and ESO 166-PN21. First discovered by James D. Wray in 1966, the planetary lies 51′ northeast of the cluster center – (epoch 2000.0: right ascension $08^h\ 23^m\ 43^s$; declination $-43°\ 12'\ 45''$) – and measures a diminutive 2′ across. If this nebula were at the same distance as IC 2488, it would have a true linear diameter of nearly 3 light-years with an upper age limit of nearly 19,000 years. Alas, Clariá and his colleagues note that the planetary is most likely not related to the open cluster. So when Dunlop wrote that "the cluster is about 5′ diameter, irregular figure, no nebula," he was, unknowingly and most probably, correct. By the way, Australian amateur Albert Brakel (Canberra Astronomical Society) calls it the Hoopskirt Cluster because it has a rough wedge shape, with longest dimension about 20′.

Before leaving the area, look a little less than 1° due south of Kappa (κ) Velorum, the northeastern tip of the False Cross. Here you'll find the bright binocular variable V Velorum, a classical Cepheid whose brightness fluctuates between magnitude 7.2 and 7.9 every 4.37 days.

Hidden Treasure 51
NGC 2903

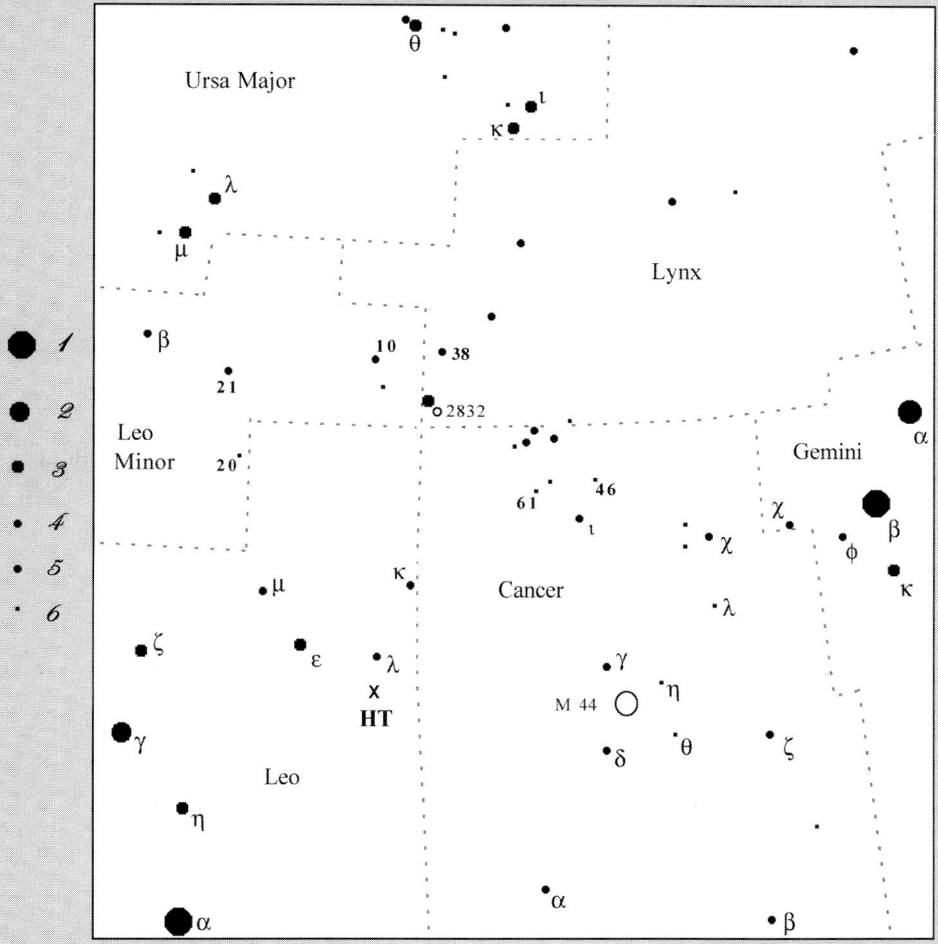

Tirion: Chart 6

Uranometria: Chart 113

51

NGC 2903,05
Type: Mixed Spiral Galaxy (SABbc)
Con: Leo

RA: $09^h\ 32.2^m$
Dec: $+21°\ 30'$
Mag: 9.0 (O'Meara); 9.0
Dim: $11.6' \times 5.7'$
SB: 13.4
Dist: 20 million light-years
Disc: William Herschel, 1784

W. HERSCHEL: [Observed November 16, 1784] Two at 1′ distance. Both considerably bright, considerably large and appear like one, much extended. (H I-56 & H I-57)

NGC: Considerably bright, very large, extended, gradually much brighter in the middle, [mottled], south preceding of 2.

I WAS BORN 172 YEARS TO THE DAY that Sir William Herschel discovered NGC 2903, a dramatic spiral galaxy in the Sickle of Leo. Actually, Herschel believed he discovered two objects here. As William Henry Smyth tells us in his *Cycle of Celestial Objects*, Herschel saw it as a "double nebula, each having a seeming nucleus, with their apparent nebulosities running into each other." Smyth's own observations revealed that the object's "upper or south part, is better defined than the lower; it requires, however, the closest attention and most patient watching, to make it a bicentral object." The Rev. Thomas W. Webb carried this visual impression into the nineteenth century: "Long, with two nuclei, rather faint." It took the Third Earl of Rosse's great 72-inch Leviathan at Birr Castle, however, to resolve it into an oblique spiral. Even then, no one knew of this object's grand nature until after 1923. In October of that year, Edwin Hubble used the 100-inch Hooker reflector atop Mount Wilson in California to study individual Cepheid variable stars in what was then known as the Great Andromeda Nebula. The results of his study established for the first time that "spiral nebulae" were actually extragalactic systems lying far beyond the confines of the Milky Way.

NGC 2903 lies 20 million light-years distant in the Leo Spur of Galaxies. It is a moderately large system measuring 60,000 light-years in diameter. It has a total mass of 80 billion Suns. A very bright nucleus punctuates the galaxy's short, bright bar, from which a complex pattern of partially resolved arms unwind. The inner lens contains two, thick spiral arms that have an exceptionally high surface brightness. Two thin, regular spiral arms are also present in the outer regions. In general all arms show color differences from that of the disk. The northwest arm divides into two branches and interestingly one branch seems to have only old population stars while the other shows evidence of young stars and gas.

We see the galaxy 24° from edge-on, offering us fine views of the galaxy's delicate dust filaments, which thread across its luminous regions in a most intricate manner. Seen collectively, they form weak spiral fragments throughout the entire disk. These fine details are reminiscent of those seen in NGC 253 (Caldwell 65), the Silver Coin Galaxy in Sculptor, but they are not as fine. But that's only because NGC 2903 lies at twice the distance of NGC 253.

Yet, the details in NGC 2903 are quiet impressive. Dust appears as spiral fragments and the dust lane along the bar on either side of the nucleus is located on the leading edge of the bar relative to the direction of rotation. This is a common characteristic of dusty barred spiral galaxies. They are thought to be shocks in the vicinity of the bar that are caused by the bar's rotation. The nucleus is composed of about eight giant H-II regions, which lie within 2,000 light-years of the core. These are part of a circumnuclear ring of active star formation. Indeed, the ultraviolet spectrum of the region has the deep, complex, absorption features of a starburst galaxy. Then again,

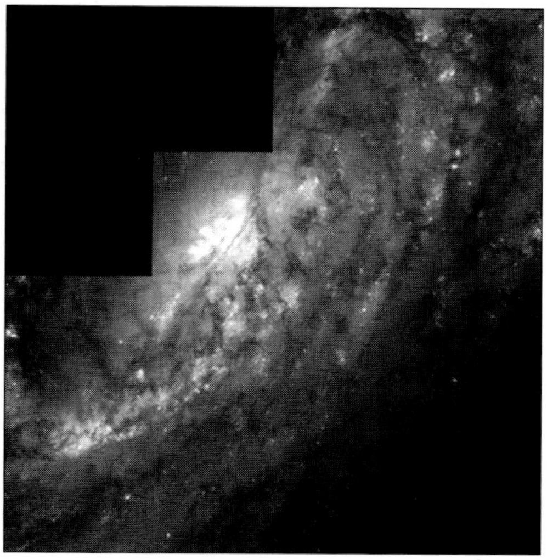

the region is also populated by a mixture of early- and late-type stars. When the Hubble Space Telescope inspected this region with a high-resolution near-infrared camera, it too found a series of hot spots forming a ring within 2,000 light-years of the core. These appear to be individual clusters or groups of bright young stars not coincident with the H-II regions. The star-formation rate in this "hot-spot" galaxy (named so for the intense star formation that is occurring within the complex hot-spot structures in the nuclear regions) has increased during the lifetime of the galaxy – only slightly in the arms but spectacularly in the nuclear region.

A color view of that HST image shows the hot young clusters as a fringe of blue filigree that appear to be stitched to the galaxy's old ochre-colored core. It's also possible to follow some fine threads of blue starlight from the circumstellar ring into the very core, where a maelstrom of stellar heat seems to be radiating in a tiny navel of yellow light. The bright blue clusters in the ring range between 6.5 and 9.5 million years in age and have luminosities similar to those found in the famous interacting galaxies

NGC 4038–NGC 4039 (Caldwell 60 and 61), more popularly known as the Antennae, in Corvus.

In a 2001 *Monthly Notices* of the Royal Astronomical Society, Almudena Alonso-Herrero (University of Hertfordshire) and his colleagues note that this population of young hot stars accounts for some 7–12 percent of the total stellar mass in the central 2,000 light-years of NGC 2903. They also found that the H-II regions in that ring of star formation have luminosities close to that of the supergiant H-II region 30 Doradus, in the Large Magellanic Cloud. The H-II regions are younger than the star clusters and will probably evolve into bright infrared stellar clusters. Star formation in the central kiloparsec of NGC 2903 is more elevated than in normal galaxies (those without bars). But this is not uncommon. Many barred spirals have similar star-forming rings within 1 kiloparsec of the galaxy's nucleus – a result of the central bar's gravity, which pulls material into the region, expediting the process of star formation.

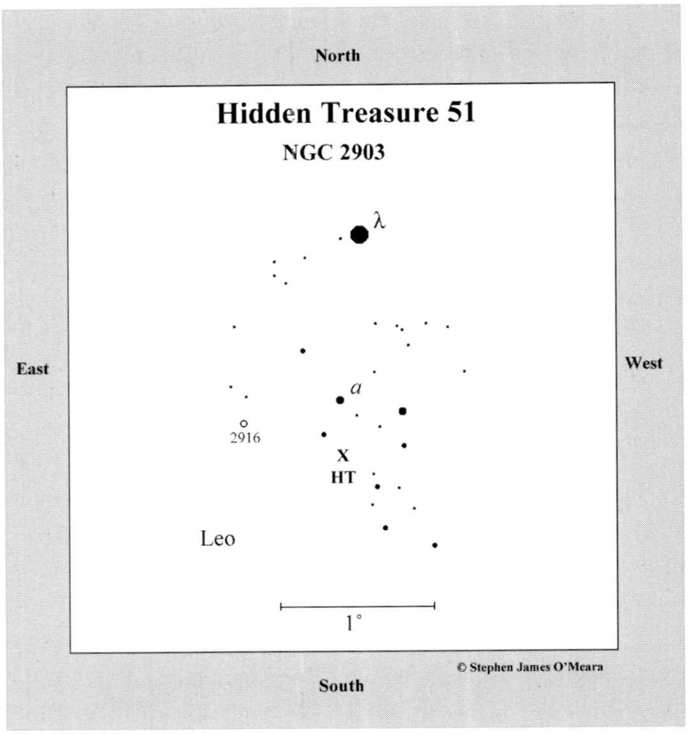

To find this semi-barred wonder, simply focus your gaze on the northwest tip of Leo's Sickle – at 3rd-magnitude Epsilon (ε) and 4th-magnitude Lambda (λ) Leonis (Alterf). The galaxy is only $1\frac{1}{2}°$ south of Lambda and 20′ due south of a magnitude 7.4 star (*a*). From a dark sky, the galaxy is visible in 7 × 50 binoculars. My antique telescope shows it as a slightly fuzzy star.

At 23×, the galaxy has a sharp nucleus inside a bright central lens, which is surrounded by a diffuse, elliptical glow. The view becomes most intriguing at 72×. It shows two knots, one north, and one south, of the starlike nucleus. This view got me thinking about Herschel's discovery description: "Two at 1′ distance. Both considerably bright, considerably large and appear like one, much extended." My observation is reminiscent of Herschel's, if the bright core is disregarded. Perhaps, Herschel saw the core as a star and the two enhancements in the bar as separate objects. Then again, if he did see it that way why didn't he say so? *Sky & Telescope*'s late Deep-Sky Wonders columnist Walter Scott Houston thought it would be an interesting experiment to view NGC 2903 with a long-focus instrument at medium to high

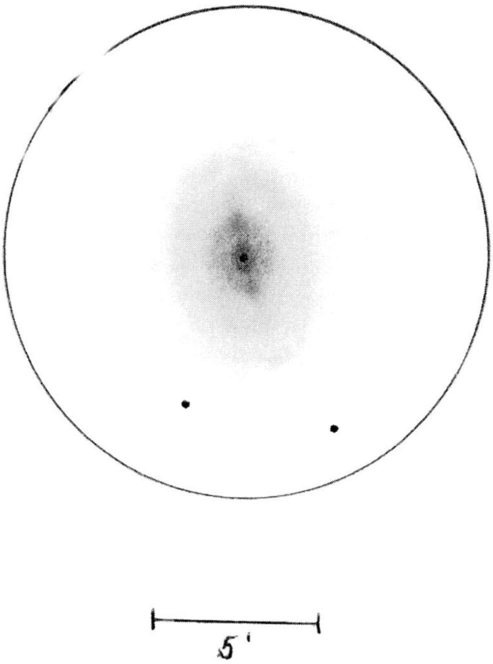

magnification, gradually reducing the aperture until the object becomes nearly invisible in the hope of inducing the "double" appearance. Houston, himself, could not detect such an appearance, despite experiments with various apertures.

At 72× and 101×, the core remains quite bright, like a sunlit jewel laying in an elliptical bed of cotton. This soft inner lens is quite defined, having a uniform texture until it reaches a sharp outer boundary. Use your imagination and try to see this sharp ring of light as that fantastic loop of star formation encircling the galaxy's core, which appears as a smooth oval of uniform light. With time and patience, I can also detect an even fainter shell beyond that inner iris. I can also detect two distinct arms in an S-shaped pattern.

If you're using a small telescope, try using these moderate powers with keen averted vision; in other words, direct your gaze at the very edge of the field of view but concentrate on the galaxy at its center. See if NGC 2903 does not swell in brightness. Here are the thick, main arms that fly off the bar like water from a reciprocating lawn sprinkler. Now lower the magnification once again and use averted vision. Can you see the S-shaped arms branded? The view at 130× reveals the dim outer halo of spiral arms with averted vision, which turns an S-shaped object into a complete oval with an S-shape branded on it.

Houston detected several of the galaxy's 70-odd H-II regions with a 10-inch telescope, and the galaxy's more extensive outer arms have been glimpsed as an oval haze in a 14-inch telescope. If you want a challenge, try for the 12th-magnitude NGC 2916 just northeast of NGC 2903. It's a pip of light that could easily be mistaken for a star.

Hidden Treasure 52

NGC 3184

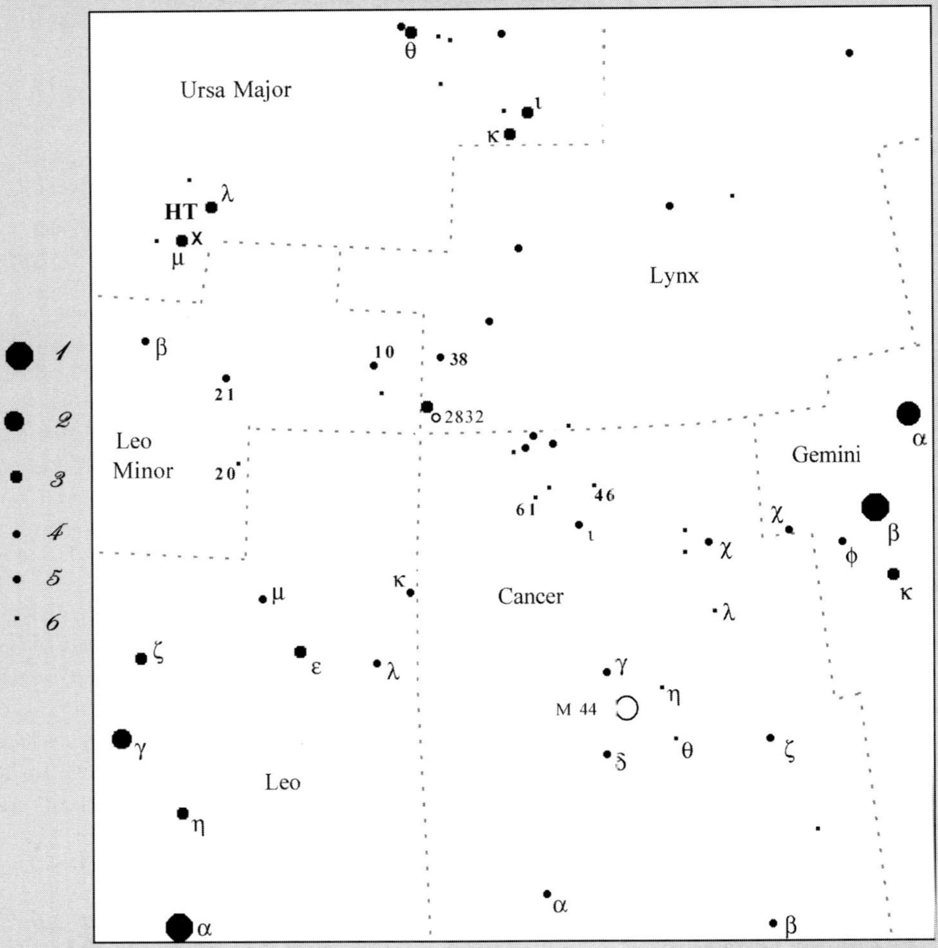

Tirion: Charts 2 & 6

Uranometria: Chart 72

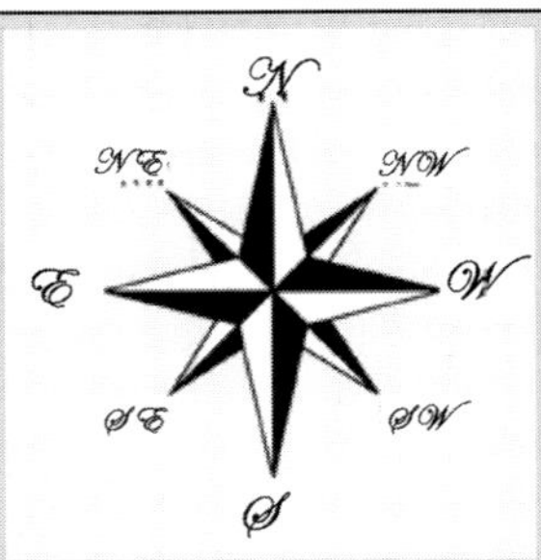

52

Little Pinwheel Galaxy
NGC 3184
Type: Mixed Spiral Galaxy (SABcd)
Con: Ursa Major

RA: 10^h 18.3^m
Dec: $+41^\circ$ 25′
Mag: 9.4 (O'Meara); 9.8
Dim: 7.5′ × 7.0′
SB: 14.0
Dist: 36 million light-years
Disc: William Herschel, 1787

W. HERSCHEL: [Observed March 18, 1787] Considerably bright, round, very gradually brighter in the middle, 8′ diameter, considerable stars in it, unconnected. (H I-168)

NGC: Pretty bright, very large, round, very gradually brighter in the middle.

NGC 3184 IS A STRIKING, NEARLY face-on galaxy tickling the left hind toes of Ursa Major, the Great Bear. Its long and swooping arms unwind in a grand spiral fashion from the rim of a small, smooth inner disk. In high-resolution images, this tiny disk contains two faint spiral dust lanes, which start at the galaxy's small, but very bright, nucleus. The leading edge of these dark spirals blends into the trailing edge of the galaxy's two main luminous arms. The luminous arms, in turn, branch into several thick fragments that spread across much of the galaxy's outer disk. Dust lanes can be traced throughout the spiral pattern, and the northern flank consists of a series of stellar ripples, that look as if someone has put a serrated carving knife to that end of the galaxy and skillfully filleted the matter.

NGC 3184 is near enough that individual stars begin to resolve out of the background of the arms. In color photographs, the core is distinctly yellow, while the stars in the arms have a striking blue patina – the telltale sheen of hot, young stars. The arms are further enhanced by the delicate pink flecks of H-II (star-forming) regions. It was from one of these pink star-forming regions that a massive star erupted in 1999 as a supernova (1999gi). Japanese supernova hunter Reiki Kushida (Yatsugatake South Base Observatory) discovered this new star on December 9, 1999, 4″ west and 61″ north of the nucleus at magnitude 14.5. At the time, it

52

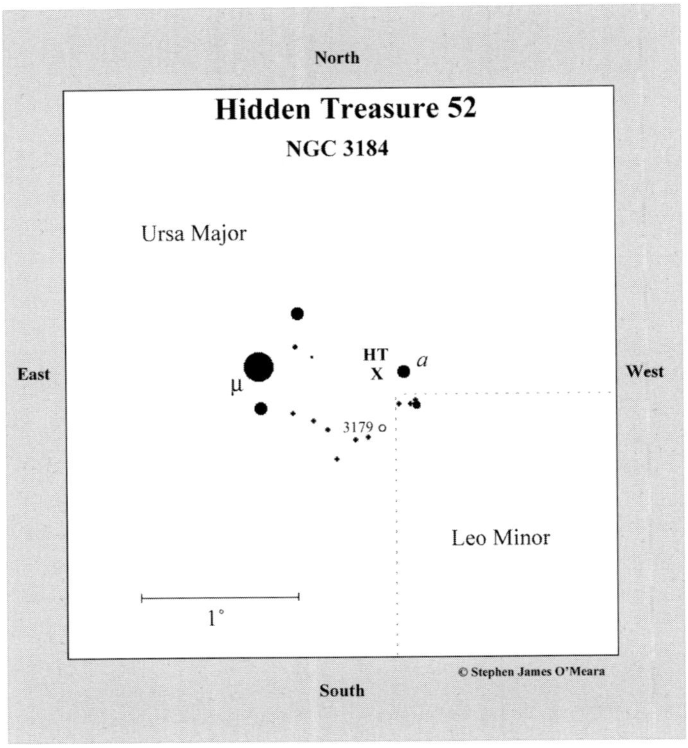

It is a very bizarre galaxy for small telescopes and I'll explain why. First, the galaxy is just visible in 7 × 50 binoculars as a threshold object – a nice round haze bordered by two stars. Then, when I sweep it up in the 4-inch at 23×, I get excited, knowing that this face-on galaxy is a good place to supernova hunt. But once I begin putting magnification on it, the galaxy loses its luster and appeal.

At 23× in the 4-inch, the galaxy is a typical face-on spiral – a round diffuse glow with a bright core. The size swells with averted vision. It does appear very similar to a comet just beginning to shine and would be easily swept up

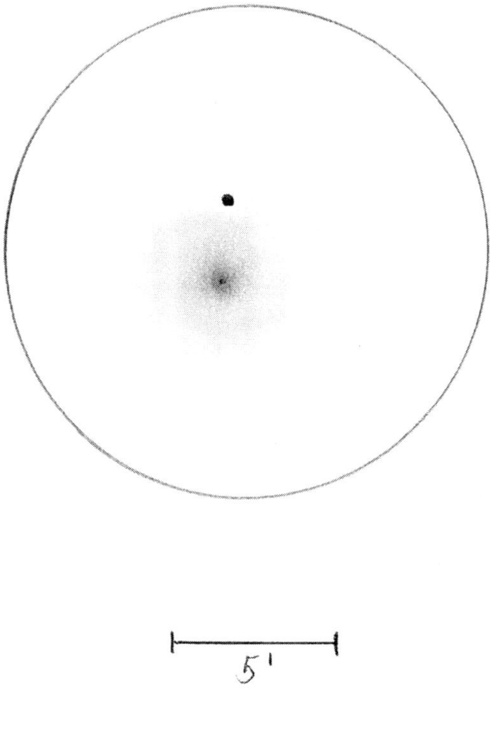

was her eighth supernova discovery. The galaxy should be on everyone's watch list. Exactly 62 years earlier, a magnitude 13.5 supernova erupted in the galaxy 5″ east and 149″ south of the nucleus. The galaxy experienced two other supernovae explosions – both in 1921: one, again, in December, the other, in April. In a 2002 *Astronomical Journal,* Douglas Leonard and his colleagues determined a new distance to NGC 3184 based on observations of Supernova 1999gi; their value, 36 million light-years, places the galaxy 8 light-years beyond that listed in Brent Tully's *Nearby Galaxies Catalog.*

To find NGC 3184, just center 3rd-magnitude Mu (μ) Ursae Majoris (Tania Australis) in your telescope at low power and look 45′ to the west. The galaxy is only 10′ east-southeast of a 6th-magnitude star (*a*).

in a comet hunt, as long as the sweep-rate is slow. With 72×, the galaxy appears to have a very bright and starlike nucleus. But with a little bit of study, it becomes apparent that this bright star is a roughly 11th-magnitude field star immediately north of the true nucleus, which is visible as a tiny, and much fainter, pip of light.

When William Herschel discovered NGC 3184 on March 18, 1787, he described it as being "bright, round, very gradually brighter in the middle, 8′ diameter, considerable stars in it, unconnected." William's son, John, also observed it, but he classified it as a double object (688 = 689), and so NGC 3184 is also equal to NGC 3180. Neither of these observers wrote anything about observing NGC 3180, however. But Lord Rosse noted that NGC 3180 is within 5″ of NGC 3184, adding that it is "connected with h 688." Lord Rosse also described NGC 3180 as being very faint and elongated. Hal Corwin explains that NGC 3180 is a star cloud or H-II region in NGC 3184's northwestern arm. Dreyer's position of NGC 3180, he says, "fits the star cloud better, but the H-II region is brighter, though smaller. The number may well apply to both objects or simply the general area of the arm where they are found." But there's more. NGC 3181 is also in NGC 3184. Corwin says it is the brightest H-II region in NGC 3184, located southwest of the nucleus.

While NGC 3184 can indeed be spied in small apertures, it is only by staring at this galaxy with fits and starts that I can discern any detail. I struggle with this object at high power: I truly struggle. Yet, this galaxy is an overwhelming favorite among observers (see Appendix D). Obviously it is an object best suited for detailed study by those owning larger telescopes than a 4-inch. Actually, most of the reports I've read of views through larger instruments do not add much to the scene. Houston, for instance, says that through an 8-inch, the galaxy looks like a "pale disk about 5′ across with a bright center. At 30× it looks much the way the well-known spiral M33 does in my 2-inch finderscope," which is why I call it the Little Pinwheel. Luginbuhl and Skiff say that the galaxy appears horseshoe-shaped at 200× in a 10-inch telescope. Most surprising is that they do not see a distinct core or nucleus in a 12-inch telescope: "the galaxy showing almost no general brightening to the center." Tokuo Nakamoto of Temple City, California, glimpsed a hint of spiral structure with a 13-inch.

The true beauty of NGC 3184 remains in the photographic image. In shape and size it is a near dead-ringer of the Great Pinwheel galaxy M101, in the same constellation. Both galaxies measure about 60,000 light-years in true physical extent. But, if we accept that NGC 3184 is 36 million light-years distant, it is twice the distance of M101, so it appears fainter in the sky. You can imagine then, how dim and inconspicuous M101 would be at the same distance. You can also imagine just how glorious NGC 3184 would be at M101's distance. NGC 3184 would, no doubt, be one of the most observed extragalactic glories in Ursa Major.

While we see M101 exactly face-on, we see NGC 3184 at 20° from face-on. The galaxies also belong to different groups. M101 is part of the Coma–Sculptor Cloud of galaxies, while NGC 3184 travels through space with NGC 2903 (Hidden Treasure 51) and others in the Leo Spur of Galaxies. It is receding from us at a speed of 598 kilometers per second and shines with a total luminosity of 8.5 billion Suns – about $\frac{1}{3}$ the velocity of M101.

Hidden Treasure 53

NGC 3228

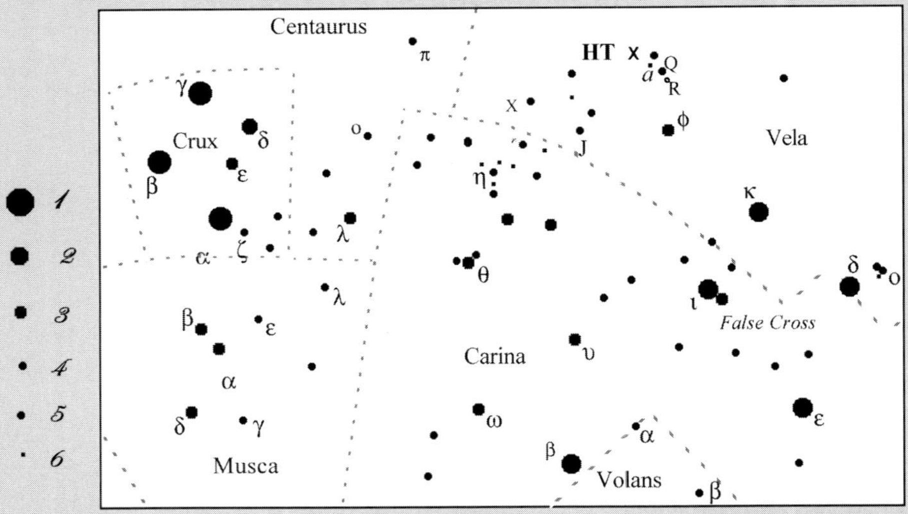

Tirion: Charts 20 & 25

Uranometria: Chart 426

53

Queen's Cache cluster
NGC 3228
Type: Open Cluster
Con: Vela

RA: 10^h 21.4^m
Dec: −51° 43′
Mag: 6.0 (O'Meara); 6.0
Diam: 5.0′
Dist: 1,600 light-years
Disc: Abbe Nicolas Louis de Lacaille, listed in his 1755 catalog

J. HERSCHEL: A group of 9 large, and a few scattered small stars. (h 3245)

NGC: Cluster, 9 [bright] and a few [faint] stars.

NGC 3228 IS A TINY GEM OF AN OPEN cluster adorning the inner relatively dust-free edge of the Carina–Cygnus arm of the Milky Way. At a distance of 1,600 light-years, the cluster's paltry 23 or so members span an almost insignificant 2 light-years of space. Yet the cluster is simply adorable in small apertures. Abbe Nicolas Louis de Lacaille discovered it during his investigation of the southern skies from the Cape of Good Hope in southern Africa and included it in his 1755 catalog of new "nebulae." It was the seventh "nebulous cluster" he discovered with his $\frac{1}{2}$-inch 8× telescope, calling it a "group of four or five stars, very [faint] and very compressed."

While the cluster's brightest members shine between 8th and 9th magnitude, the most prominent ones form a tight equilateral triangle only about 3′ across; it's impressive that Lacaille was able to resolve the stars that he did, especially during a sky survey with such a tiny instrument. When the nineteenth-century southern sky explorer James Dunlop observed this object from Paramatta, New South Wales, with a 9-inch f/12 reflector, he added another layer to the verbal description, calling it a "group of eight or ten pretty bright [faint] stars, with very [faint] stars, about 6′ diameter." Today, photographs of the region show a beautiful assortment of two dozen bright jewels lying on a carpet bristling with dim diamond dust; the dimmer stars, of course, belong to the intense backdrop of the rich Vela Milky Way.

Dunlop also records that NGC 3228 is "11 Roboris Caroli." Robur Carolinum, Charles' Oak, is a now-defunct constellation created by Edmund Halley in 1677, during his stay on the small South Atlantic island of St. Helena. It commemorates the Royal Oak that Charles II (Halley's royal patron) hid in with Colonel Carlis after Charles failed to reclaim his crown during the 1651 Battle of

53

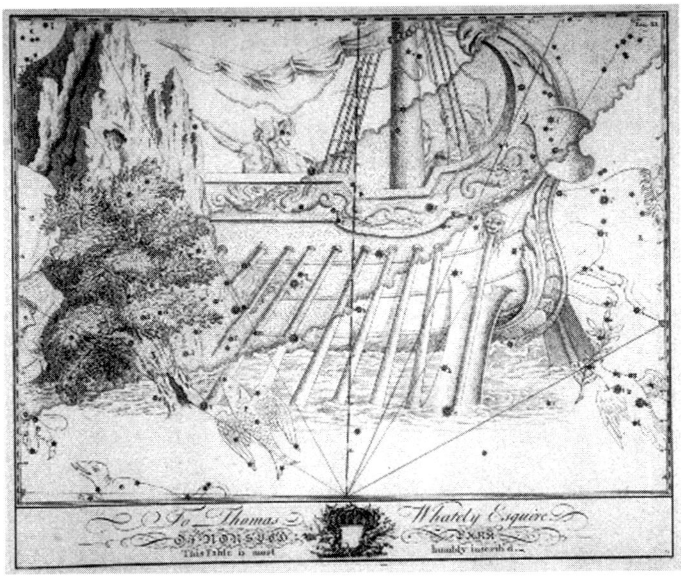

To find this young grouping, I prefer to start with the naked eye and look for magnitude 2.5 Kappa (κ) Velorum, the northeastern tip of the False Cross. Just 5° to the east-northeast is magnitude 3.5 Phi (ϕ) Velorum. A simple sweep 4° northeast will bring you to a near-perfect, 2°-wide binocular version of the W of Cassiopeia (*a*). NGC 3228 is visible in 7 × 50 binoculars as a 6th-magnitude fuzzy star 2° west and slightly north of 5th-magnitude Q Velorum – the brightest and westernmost star in the W. Here we have a mythical queen seated in her throne, a cache of jewels at her feet. Again, I marvel at Lacaille's ability to separate four or five

Worcester – the final battle of the English civil war. (It was a clever creation, because it secured for Halley, by the King's express command, his master's degree from Oxford upon his return in 1678.) Lacaille, however, found the new constellation appalling, since it ruined Argo Navis. Indeed, to form Robur Carolinum, Halley removed 12 of the brightest stars in the famed ship Argo Navis. Alas, Halley's celestial oak fell prey to the axe of time – apparently, because no one could truly point a finger at "the" oak in which Charles had hid. Argo Navis also got the axe, though the ship did not sink from view – it was merely chopped to pieces, with Vela, the Sail, being one piece. The Oak too is not lost, to this day, many English pubs are named the Royal Oak.

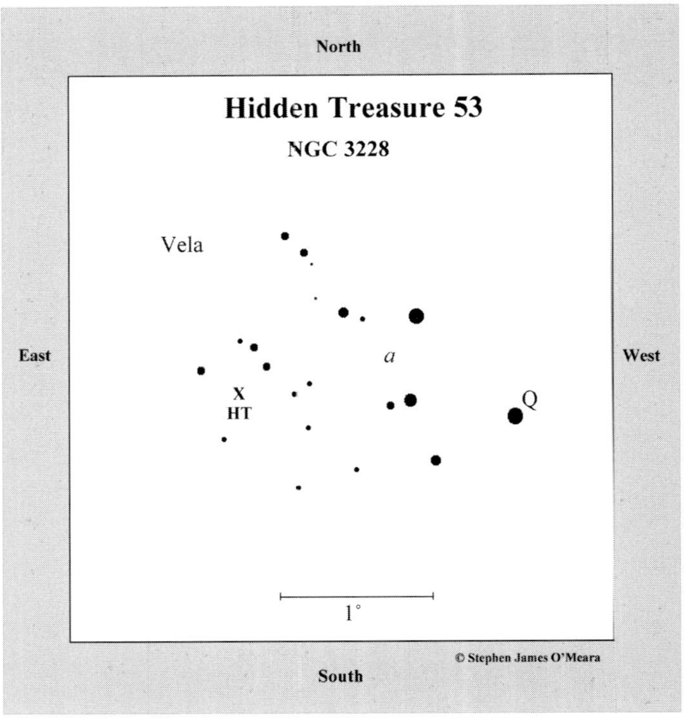

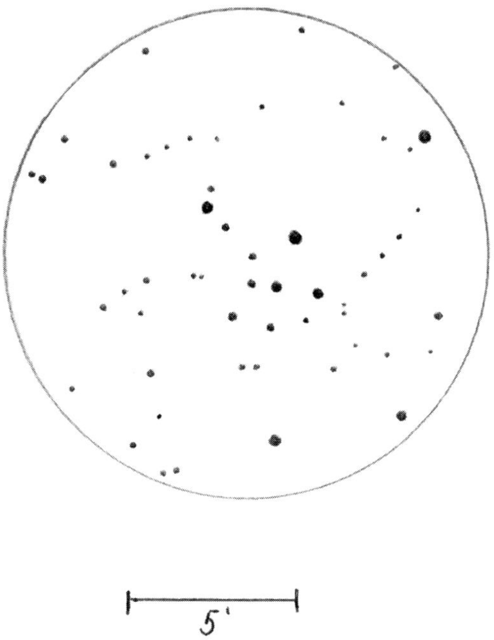

stars here, because at a glance in modern 7 × 50 binoculars, the cluster appears very tight and compressed and its components are seen on the verge of resolvability. With my antique telescope, the cluster's core breaks apart into a slightly speckled wonder with wide arms.

At 23× in the 4-inch, the equilateral triangle is part of a prominent central diamond. That diamond, in turn, forms the base of a capital J-shaped asterism, whose long arm ends at a line of three 8th-magnitude stars, oriented northeast–southwest, 30′ to the northwest. Some 70-odd stars can be readily seen, though most are dim field stars. The cluster does not gain much with increased power. It is definitely an object for small telescopes and low power.

Hidden Treasure 54

NGC 3293

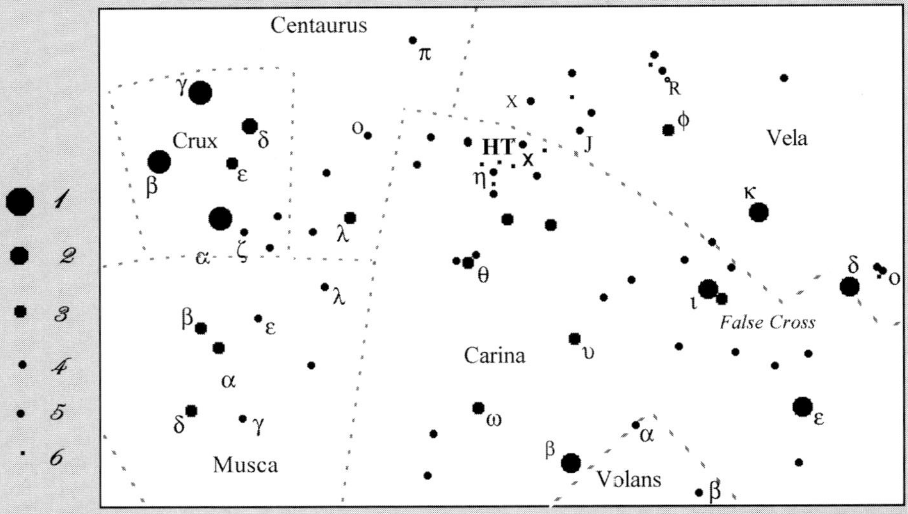

© Stephen James O'Meara

Tirion: Chart 25

Uranometria: Chart 427

54

Spider Spit Cluster, Little Jewel Box
NGC 3293
Type: Open Cluster
Con: Carina

RA: $10^h\ 35.8^m$
Dec: $-58°\ 13'$
Mag: 4.6 (O'Meara); 4.7
Diam: 5.0′
Dist: 8,000 light-years
Disc: Abbe Nicolas Louis de Lacaille, listed in his 1755 catalog

J. HERSCHEL: A fine bright rich not very large cluster. (h 3276)

NGC: Cluster, bright, rich in stars, pretty large.

IF THERE IS ONE OBJECT IN THE SKY more deserving of the Jewel Box moniker than Kappa (κ) Crucis (NGC 4755 [Caldwell 94]), it is open cluster NGC 3293 in Carina. This 4th-magnitude treasure chest packs 93 stars, 8th-magnitude and fainter, in an area only 5′-wide – almost half the apparent diameter of the Trapezium in the Great Orion Nebula (M42). And while the Jewel Box in Crux is twice as large as our hidden treasure and contains three times as many members, its stars do not punch the eye the way those bright and tightly packed ones in NGC 3293 do, especially in small telescopes.

Let's face it, part of the Jewel Box's attraction is its location. It shines like a diamond ring at the end of the eastern arm of the famous Southern Cross, and its beauty helps to illuminate the bleak shores of that Black Sea of nebulosity known as Coalsack Nebula to its south. NGC 3293, on the other hand, suffers the misfortune of lying only about 2° north-northwest of the core of the Eta (η) Carinae complex, a vast and alluring network of glowing gas, dark dust, and hot *O*- and *B*-type stars, many of which lie in clusters. Not only is NGC 3293 some 24 times smaller than the Eta Carinae Nebula and its associated starlight, but it is also surrounded by more than a dozen other open clusters in Carina, including 2nd-magnitude IC 2391 (the Southern Pleiades [Caldwell 85]), 3rd-magnitude NGC 3532 (the Pin Cushion Cluster [Caldwell 91]), and 4th-magnitude NGC 3114 – all dynamic objects, all vying for one's attention, and all deserving of one's time.

Yet NGC 3293 is arguably *the* most striking object in the entire region. And while it lies on the fringe of the Eta Carinae complex, it contributes some illumination to the clouds of gas in the region – as any photograph or wide-field telescope under dark skies will show. Some of the cluster's stars illuminate a river of reflection nebulosity in the northern part of NGC 3293, while immediately to the cluster's southeast lies a dark nebula, which appears to be part of the prominent dust lane that runs through the Eta Carinae Nebula. D. G. Turner (David Dunlop Observatory) and his colleagues have raised the possibility that NGC 3293 may be associated with other objects in the Carina Nebula complex, which is the heart of the Carina *OB*1 association.

NGC 3293 is very young ($\sim$10 million years) and contains a large number of *B*-type giants and supergiants, as well as an *M*-type supergiant. Its age is similar to that of the Double Cluster in Perseus (Caldwell 14). Turner and his colleagues suspect that NGC 3293 may be physically related to IC 2581, another 4th-magnitude cluster in Carina, about $1\frac{1}{4}^{\circ}$ to the northwest. Turner and his colleagues see them as a Southern Hemisphere analog to the Double Cluster, though more data need to be collected to confirm this theory (see also Hidden Treasures 25 and 26).

Based on their studies of the region, Turner and his colleagues suggest that the star-forming process in the Carina *OB*1 Association, which was initiated in the northwest section of the region, led to the creation of NGC 3293 and IC 2581. Later episodes of star formation followed in regions lying progressively eastward and southward of these two clusters. "This rather simple picture does not account for all the observations," the researchers warn, especially since there are some anomalous clusters in the region. No matter, it appears that NGC 3293 and IC 2581 represent one of the oldest areas of star formation in the association. But as G. Baume (University of Padova) and his colleagues report in a 2003 *Astronomy and Astrophysics* paper, 19 stars with signs of H-alpha emission in the region of NGC 3293, indicate that star formation is still active there.

Although Lacaille is credited with discovering the cluster NGC 3293 (it is the eighth "nebulous cluster" in his 1755 catalog), the object is part of the naked-eye backdrop of the Milky Way and no doubt was spied as an innocuous star by numerous wonder-struck aboriginal stargazers – history's first celestial treasure hunters. When James Dunlop observed it on April 29, 1826, he saw "a very small cluster of very small bright stars; round figure, about 4′ diameter; rich in extremely small stars resembling faint nebulosity."

Both Lacaille's and Dunlop's reference to nebulosity is curious, only because, as stated earlier, nebulosity is involved with the cluster at its northern edge. In their 1975 "Catalogue of Southern stars embedded in nebulosity," van den Bergh and Herbst list two reflection nebulae here: BHe 42A and BHe 42B. The first nebula surrounds the illuminating star and measures 4′ across on red plates. The second nebula lies just to the north and west and measures 1.6′ on the blue plates, where it is strongest. Colin S. Gum, also lists them as Gum 30 in his "A survey of Southern H II regions." He gives the nebula a maximum diameter of 40′ and rates it as being moderately bright against the background Milky Way. Interestingly, few observers have seen the nebulosity, and I have to wonder why.

To find this stunning cluster and mysterious nebulosity, look about 2° north-northwest of the Great Eta Carinae Nebula for two 4th-magnitude stars separated by

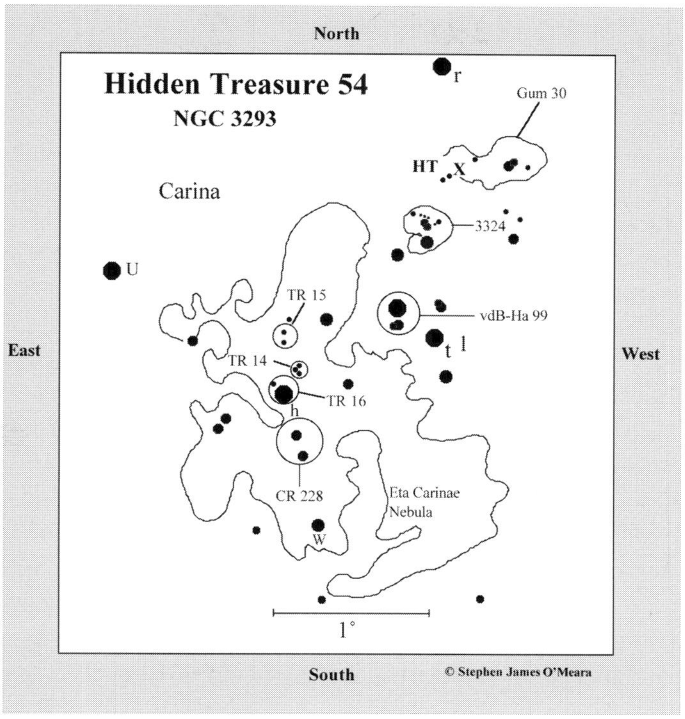

some 40′. The northern star is r Carinae; the southern one is NGC 3293. In 7 × 50 binoculars, the cluster is a fuzzy ellipse of light hanging on a ½°-long branch of four 8th-magnitude stars like "spider spit" – that foamy white embryonic goo from which all manner of insects evolve in the wild. The branch arcs gracefully from the east to the northwest and with some fainter stars forms a true branch or wishbone pattern. The cluster's core is very bright with a slightly swollen disk, one star is clearly separated to the north. With time the central disk breaks up into a swarm of extremely congested suns. With averted vision, the swarm seems to tremble and swell erratically, as if it were made of moldable clay, and something inside is trying to push its way out. In the antique telescope, the core is more defined and looks stretched, like taffy, from the northwest to the southeast.

The cluster looks similar at 23×, only now the cocoon breaks down into a sizzling swarm of stars. It's hard to judge which stars are members, but no matter, some 40 to 50 suns burn into view with any increase of magnification. They are a dazzling assortment of colorful orbs, especially one golden star – the one type-M member of the cluster which has aged the fastest. This star is the southernmost of three, which form a northeast–southwest trending line from the bright

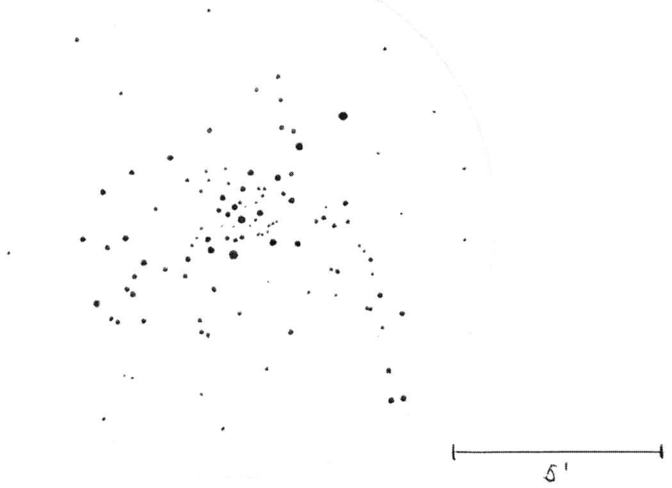

star at the cluster's core. It is around this bright star that the cluster's fainter members swarm like bugs to a light. A long arm extending to the east, and a stubby one, end at a distinct triangle to the west. If you've got a pirate's eye, might not the cluster (with its strong central axis, arms, and sharp surrounding starlight) be a sterling silver crucifix wrapped in a crystal bead rosary – a treasure well worth pocketing on the way to the gallows.

If you cannot see the nebula described above, don't fret, few have. But I find it is not apparent under magnification. It requires low power and averted vision, as well as the ability to differentiate nebulosity from Milky Way, which takes an experienced yet discerning eye. Try first in binoculars. If you fail to do so, move your scope a little more than 30′ to the southeast, where you will find a 6th-magnitude star immersed in a tiny but bright reflection nebulosity (NGC 3324). Dunlop discovered the nebulosity, which is described in the original *NGC* as "pretty bright, very large, irregularly faint, double star involved." Note that there has been no mention of a cluster. Yet, today, most sources and star charts label NGC 3324 as a cluster.

Finally, I think it's important to note that although NGC 3293 has a true physical diameter of 12 light-years (about the same as the Jewel Box's), it is nearly twice the distance of the Jewel Box. If we could haul in NGC 3293 and position it next to the Jewel Box, we'd find it having about the same apparent diameter but shining nearly a full magnitude brighter – almost as bright as Delta (δ) Crucis, the star that marks the western arm of the Cross.

Hidden Treasure 55

NGC 3344

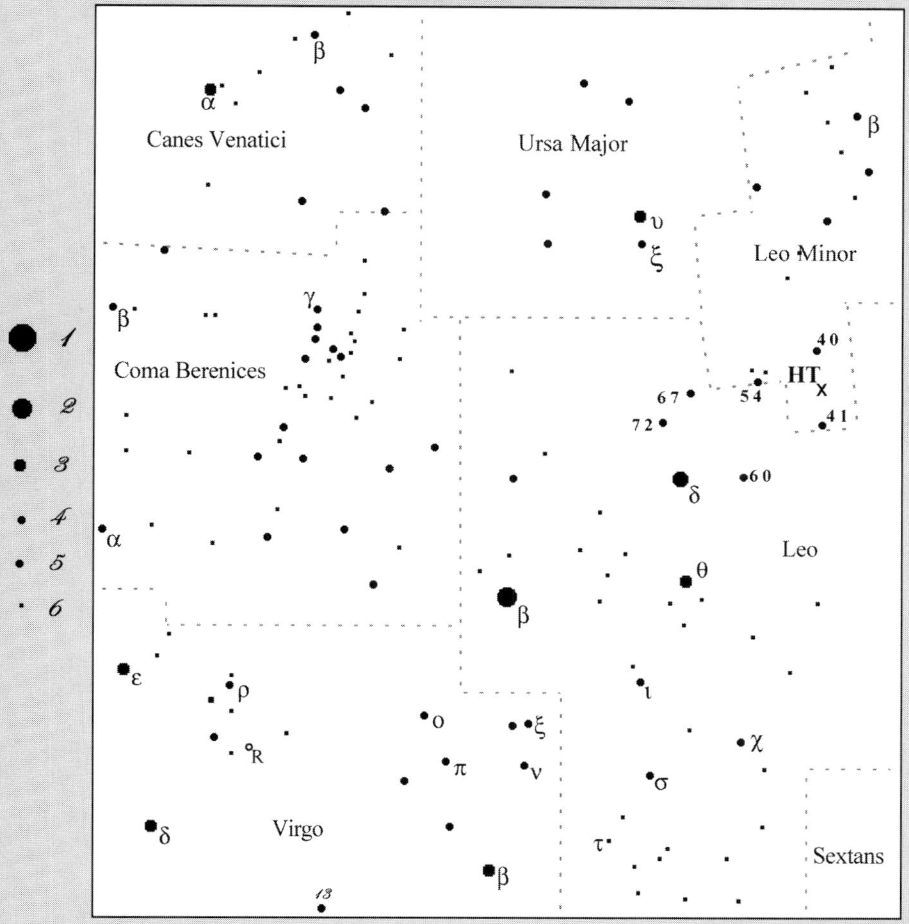

© Stephen James O'Meara

Tirion: Chart 6

Uranometria: Chart 145

55

Sliced Onion Galaxy
NGC 3344
Type: Mixed Spiral Galaxy (SABbc)
Con: Leo Minor

RA: 10^h 43.5^m
Dec: +24° 55′
Mag: 9.3 (O'Meara); 9.9
Dim: 6.7′ × 6.3′
SB: 13.8
Dist: 20 million light-years
Disc: William Herschel, 1785

W. HERSCHEL: [Observed April 6, 1785] Considerably bright, considerably large, milky, just preceding 2 stars. (H I-81)

NGC: Considerably bright, large, gradually brighter in the middle, star involved, 2 stars following.

NGC 3344 IS THE LARGEST AND brightest galaxy in the Little Lion, one of the most neglected proving grounds for extragalactic treasure. Johannes Hevelius created the constellation from 18 stars between Leo and Ursa Major. So, just as the Great Bear has its companion, so too then does the Lion. The galaxy is a relatively isolated system with no other relatively bright galaxies within 3° of it. It's a stunning open system seen only 23° from face-on. It's about as open as NGC 3184 in Ursa Major (Hidden Treasure 52), NGC 3621 in Hydra (Hidden Treasure 57), and NGC 2903 in Leo (Hidden Treasure 51), which travel through space with NGC 3344 in the Leo Spur of Galaxies. NGC 3344 is the smallest of these systems, however, measuring some 40,000 light-years across (about half the size of the Milky Way) and shining with a total luminosity of 4 billion Suns.

NGC 3344 is an SABbc-type spiral galaxy. This type of galaxy will not be found in Edwin Hubble's original 1925 classification scheme of galaxy evolution. It is an elaboration found in the so-called *Revised Morphological Galaxy Classification System*. NGC 3344 is a mixed spiral system that lies somewhere between a barred spiral (SB) and an unbarred (SA) one. It has a very small, bright nucleus (bc) in a weak bar that's surrounded by an almost complete inner ring. As usual in tightly wound face-on spirals, this ring is the product of overlapping tightly wound opposite spiral arms that appear to be connected by dust lanes to the central region. This inner ring marks the beginning of the spiral structure and the ring is dominated by a young stellar population. The galaxy's two principal arms begin from this ring, and other knotty arms branch off from them. The brightness of the arms gets fainter the farther one looks from the center.

Seen in detailed images, the bar, not the bulge, dominates the galaxy's core. Close inspection of the inner ring and the initial spiral arms shows that the beginning parts of the spiral structure and its accompanying dust, in part, wrap tightly around the ring. This region is also dominated by a population of young stars, which is commonly seen in more typical barred spiral galaxies (see Hidden Treasure 51).

Curiously, where the main spiral structure stops in the outer halo, astronomers have detected a very faint pseudoring, which, surprisingly, is not centered on the nucleus. This outer pseudoring, like the inner ring, is populated by young stars, indicating that it has active star-forming regions. Typically, isolated spiral galaxies – those away from the gravitational tug of other systems, do not show such anomalies. Equally odd, the neutral hydrogen gas beyond this ring is warped. How did such complexities arise? In a 2000 *Astronomy and Astrophysics* paper, L. Verrdes-Montenegro (Instituto de Astrofisic de Andalucia) and his colleagues suggest that the galaxy recently cannibalized another system, which warped and modified its outer disk. The prominent spiral arm on the southeast side, for instance, exhibits a definite warp. And a galactic-merger event is the most likely cause. Such a scenario would explain the relative rarity of such curious features in an average late-type spiral. The warping, however, could also be due to a near miss by an unknown passing object, or some massive internal energetic activity.

Seeing this isolated galaxy on star charts, is like seeing a lone sailing ship waiting to be plundered. As Commodore Downes, who ultimately took down the Malay pirates in the early nineteenth century, recounts in *The Pirates' Own Book: Authentic Narratives of the Most Celebrated Sea Robbers* (first published in 1924, and reprinted by Dover in 1993), the pirates "lie concealed under the land, until they find a fit object of opportunity. The time chosen is when a vessel runs aground, or is becalmed, in the interval between the land and sea breezes." NGC 3344 is one such becalmed target – drifting alone in the void, "sails down," waiting for the swift strike of your eye swooping in like a pirate.

Before you attack, you must first navigate the open waters that caress Leo's back. You can start at 2.6-magnitude Delta (δ) Leonis (Zosma) in the Lion's hindquarters. Just 6° to the northwest is 4th-magnitude 54 Leonis, which forms the southern apex of an acute triangle with the 6th-magnitude stars 48 and 50 Leo Minoris 45′ to the north. Use binoculars to confirm this, if you have to. Now, 54 Leonis also marks the eastern apex of a near 3°-wide equilateral triangle with 6th-magnitude 41 Leo Minoris and the slightly fainter star 40 Leo Minoris. NGC 3344 lies midway between these two stars. If you observe under a dark sky and can see these stars with the naked eye, all you have to do is place the crosshairs (or laser spot) of your finder midway between these stars and use low-power to sweep it up. On the other hand, if your telescope is on an equatorial mount, you could simply center 54 Leonis (which is a beautiful double star) and swing the telescope 3°

55

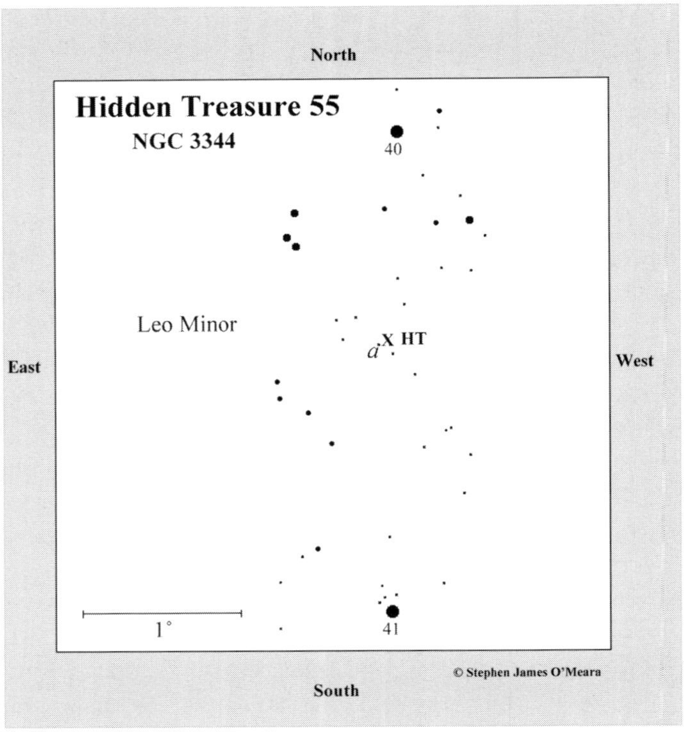

west to center the galaxy in a low-power eyepiece.

Under a dark sky the galaxy is obvious. The late Walter Scott Houston had no problem sweeping it up in his 4-inch from Connecticut. NGC 3344 is a full magnitude brighter than NGC 3184 (Hidden Treasure 52) in Ursa Major. And at 23×, it's immediately intriguing. The galaxy is about 5′ west and slightly north of a magnitude 10.5 star (*a*), which sits on the very flank of the galaxy's halo. A roughly 12.5-magnitude star lies midway (and a bit northwest) between it and the galaxy's bright nucleus. Beware: at high power, a 14th-magnitude star lies even closer to the nucleus to the southeast. The arms form a large halo of uniform milky light that envelopes all of these three suns.

At 72× and a relaxed gaze, the galaxy begins to lose its "milky" uniform luster. If you give this galaxy time, small telescope users may be able to make out some fine details. I find that using low and moderate powers are not good enough to see the faint and delicate spiral structure. Not that much of it can be seen in a small telescope. But when I use 101× and 182×, the dimmest parts of the galaxy disappear, leaving hyper fine hints of dense spiral arcs mingling with the field stars projected on the galaxy's face. The inner arms I saw (after much trial and tribulation) formed two arcs around the central bulge. Interestingly, when I study the bulge at moderate power, I see it oriented

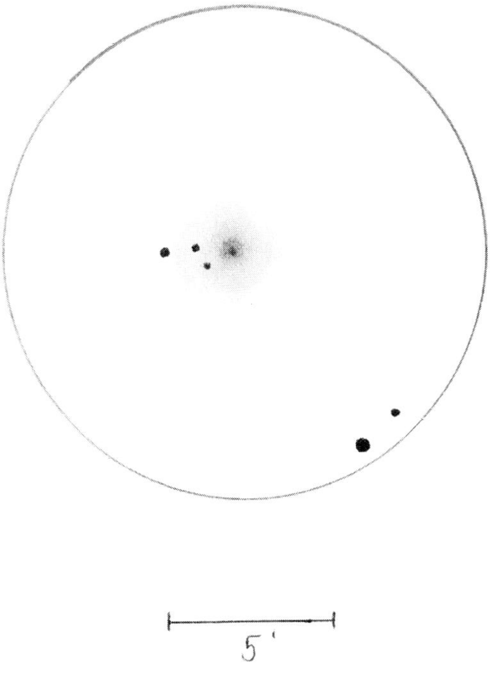

east–west, which is 90° from the direction of the bar (which I cannot see). I take this to be caused by intensity variations in the spiral arms which form a lopsided loop around the nucleus. Studying this galaxy is like peeling away the skin of an onion – it has to be done in layers. And your carving tool is magnification.

If you haven't already, don't forget to check out 54 Leo Minoris. It is a gorgeous double for a 4-inch, even when atmospheric seeing is not that great, it withstands magnification fairly well. When the atmosphere steadies, the pair is simply glorious, a gold primary and green secondary. Of this star, Adm. Smyth writes, in his illustrious way, "[a] neat double star just over the Lion's back, where it is preserved from the Lesser Lion by one of the map-maker's nooks." He saw a white primary and a gray secondary. He also notes that William Herschel, the star's discoverer, saw the same colors as himself, "in this instance."

Hidden Treasure 56

NGC 3521

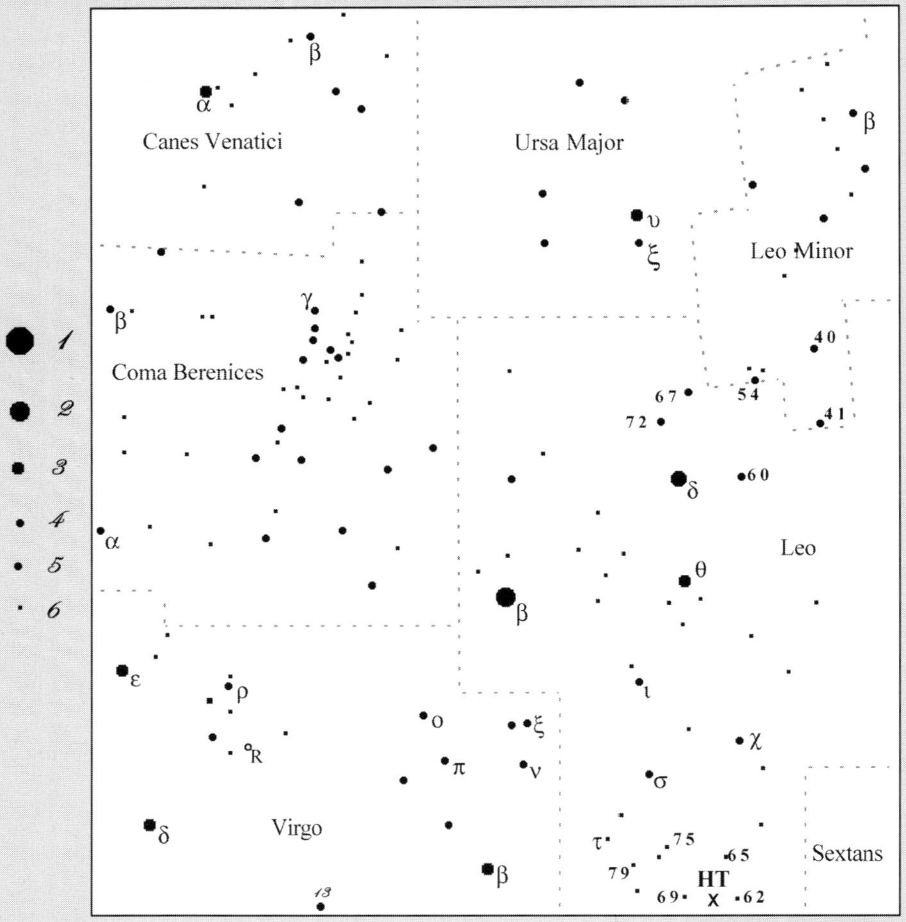

Tirion: Chart 13

Uranometria: Chart 236

56

NGC 3521
Type: Mixed Spiral Galaxy (SABbc)
Con: Leo

RA: $11^h\ 05.8^m$
Dec: $-00°\ 02'$
Mag: 9.1 (O'Meara); 9.0
Dim: $11.7' \times 6.5'$
SB: 13.6
Dist: 23 million light-years
Disc: William Herschel, 1784

W. HERSCHEL: [Observed February 22, 1784] Very bright, much extended in the direction of the meridian of the nebula, suddenly much brighter middle, 7′ or 8′ in length. (H I-13)

NGC: Considerably bright, considerably large, much extended toward roughly position angle 140°, very suddenly much brighter in the middle to a nucleus.

DEEP IN THE FAR RECESSES OF southern Leo, hidden from view due to the lack of bright stars nearby, lies the beautiful spiral galaxy NGC 3521 – one of the most neglected bright galaxies in the heavens. NGC 3521 is equal to, or nearly as bright as, many of the Messier galaxies in Leo, Virgo, and Coma Berenices; it is a full magnitude brighter than M108 in Ursa Major, and only 0.6-magnitude fainter than M82, also in Ursa Major. Yet few seek out NGC 3521. This galaxy, which lies on the Celestial Equator, also escaped the careful eyes of Messier, Mechain, and their comet-seeking contemporaries. Not until February 22, 1784, less than two months after William Herschel began his systematic search for nebulae with his 20-foot telescope on December 19, 1793, did light from NGC 3521 first fall directly on a human eye.

Herschel classed the object as a bright nebula. And Smyth after him noted that this "enormous mass of luminous matter is an outlier of the vast nebulous tract which appears to be posited nearly at right angles to the galaxy; but in regular occurrence." Today we know NGC 3521 is a spiral galaxy seen obliquely (29° from edge-on). It is a fairly large system, measuring 72,000 light-years in true physical extent and has a total mass of 150 billion Suns. It belongs to the Leo Spur of Galaxies and is receding from us at 804 kilometers per second.

Like NGC 2903 (Hidden Treasure 51) and NGC 3344 (Hidden Treasure 55), NGC 3521 is a mixed spiral system, having characteristics of both a barred spiral and normal spiral galaxy. In photographs, the galaxy shows a very small and very bright nucleus nestled in a hexagonal-shaped lens from which many knotty and filamentary spiral arms flee outward. The inner lens has a high surface brightness, but the thin spiral arms do not. In fact, the outer arms are defined more by the galaxy's dust lanes than they are by the luminous fragments that make up its tightly wound arms. The largest H-II regions in the disk have diameters of 4″.

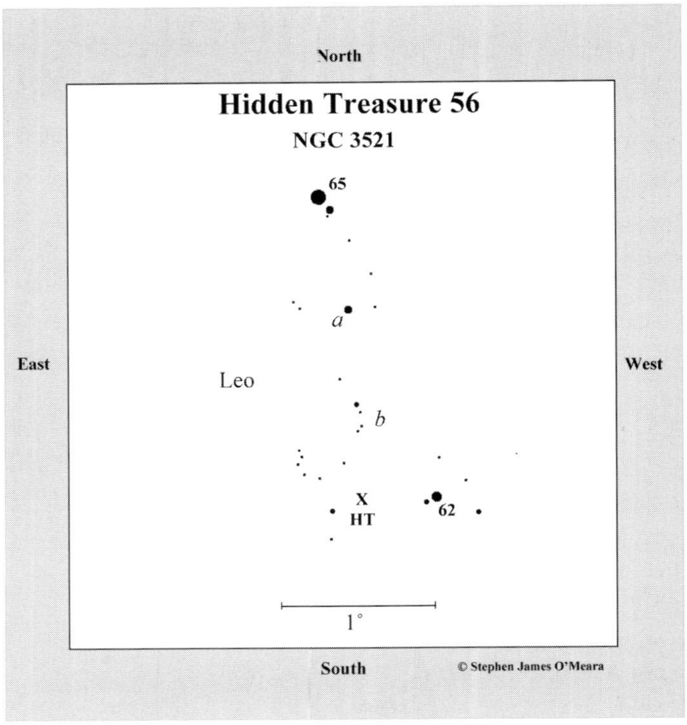

The view is reminiscent of the mottled segments seen in NGC 2841 (HT 49) in Ursa Major. Still, two distinct spiral arms can be traced. One, the eastern arm, is long and continuous, extending over 180° in azimuth; the other, the western arm, appears to split in two, leaving a continuous spiral arm over about 130° of azimuth. Deep photographs show a cylinder of light that is perpendicular to the plane of the galaxy and concentric with the center. Interestingly, the galaxy's disk appears to rotate at different rates. The inner disk rotational velocity increases steeply at 10″ (1,695 light-years) from the central region, but is then followed by a dip in rotational speed at 20″ (3,260 light-years). It then increases slowly until it reaches a broad maximum rotation at 1.5′ from the core.

Observations of the galaxy's central lens reveal two dusty structures that appear to slash across the nucleus in an X-like fashion; the feature is reminiscent of a similar structure observed in M51, the Whirlpool Galaxy in Ursa Major, though more observations are needed to confirm the feature's existence. It's possible that a large-scale starburst event may have occurred within the past billion years. The central lens also has a bar, which we see almost end-on. Astronomers have also detected the presence of a fast rotating gaseous disk in the nucleus, whose spectrum also shows strong and broad H-alpha absorption – a feature that could indicate the presence of a bright *A*-type star cluster, evidence again that a large-scale starburst may have occurred in the nucleus within the last billion years.

To find NGC 3521, all you have to do is locate the right rear foot of Leo, but few, if any know that Leo even has a right, rear foot. It is comprised of a 3°-wide polygon of five 5th- and 6th-magnitude stars: 62, 65, 69, 75, and 79 Leonis – about 15° due south of 3rd-magnitude Theta (θ) Leonis (Chertan).

Use your naked eye or binoculars to star hop to the polygon. Start at Theta Leonis and look $5\frac{1}{2}°$ to the southeast, where you'll find 4th-magnitude Iota (ι) Leonis. Another hop $4\frac{1}{2}°$ to the south-southwest will bring you to 4th-magnitude Sigma (σ) Leonis. And a hop 4° in the same direction will bring you to 5th-magnitude 75 Leonis, the northern tip of the polygon. You'll know if you have 75 Leonis because it has a 6th-magnitude companion about 30′ to the southeast. Just $2\frac{1}{2}°$ west of 75 Leonis is 5.5-magnitude 65 Leonis. Another short 2° sweep to the southwest will bring you to 6th-magnitude 62 Leonis, which is the star you should center in your telescope. NGC 3521 lies a little more than 30′ east and slightly south of 62 Leonis. You can also star hop to it from 65 Leonis. Center that star, which has a 7th-magnitude companion about 6′ to the southwest. About 45′ to the south-southwest is a roughly 6th-magnitude star (*a*). A 40′ hop south will bring you to a nice 10′-long arc of four roughly 10th-magnitude suns (*b*), which is oriented north–south. Your target is only 35′ south of that arc of stars.

The galaxy is visible in 7 × 50 binoculars and in my antique telescope it looks like a faint splotch of light 10′ northwest of an 8th-magnitude star. At 23× the galaxy looks like a spindle, oriented northwest–southeast, lying among a rich group of dim stars. It is quite easy to see, a small but pleasing sight, bright and obvious. It also has a slight cometary appearance, being a simple ellipse within an ellipse, with a gradual brightening toward the core. The core is condensed and its surrounding lens a bit mottled. I suspected one wide-flung spiral arm.

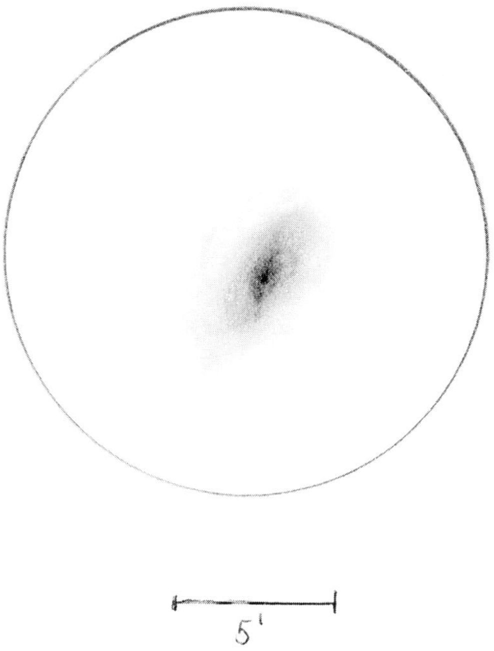

At 72×, the galaxy is a more complex sight, though all the details are delicate. The nucleus appears sharp and bright, and an oval coma surrounds it. The southeast side of the galaxy is sharply defined and has what appears to be an arm reaching out from the center, like a bar. The inner disk looks warped and mottled, though these features are subtle.

In his concluding remarks about NGC 3521, Smyth stands on a soap box and sings praises to William Herschel, who, through his great sky survey and discovery of innumerable nebulous objects, greatly advanced man's understanding of the universe:

Besides the more condensed masses, diffused nebulosity exists in an abundance which exceeds all imagination; and the indefatigable [Herschel] examined more than 150 square degrees of it. His conclusion is, that the high degree of rarefaction of the nebulous matter, should not be considered an obstacle to the theory of its finally being compressed into a body of the density of our Sun: for, supposing the nebula to be about 320 billions of miles distant, and its diameter subtending an angle of 10′, then must its magnitude exceed that of the Sun by more than 2 trillions of times! This presents magnitude and mass vast and inconceivable; and has staggered many a tyro . . . The developments which crown [Herschel] with imperishable fame, will for ages draw forth both practical and theoretical talent, so that his reasonings and conclusions on the condensation of nebulous matter into suns and planets, will be rigorously reviewed and tested; and there is no doubt but that future exertions will create progressive advances in means. Already has the Earl of Rosse . . . produced the most perfect telescope that ever was constructed; and he has now undertaken another of 6 feet aperture and 50 feet focus, with every prospect of attaining perfection. May diligent observation and faithful records follow, in the true Herschelian spirit of advancing sidereal astronomy.

When you turn your telescope to NGC 3521 and view the galaxy in all its subtle grandeur, see if some subsurface emotion does not suddenly well up inside you like magma in a volcano and cause you verbally to explode.

Hidden Treasure 57

NGC 3621

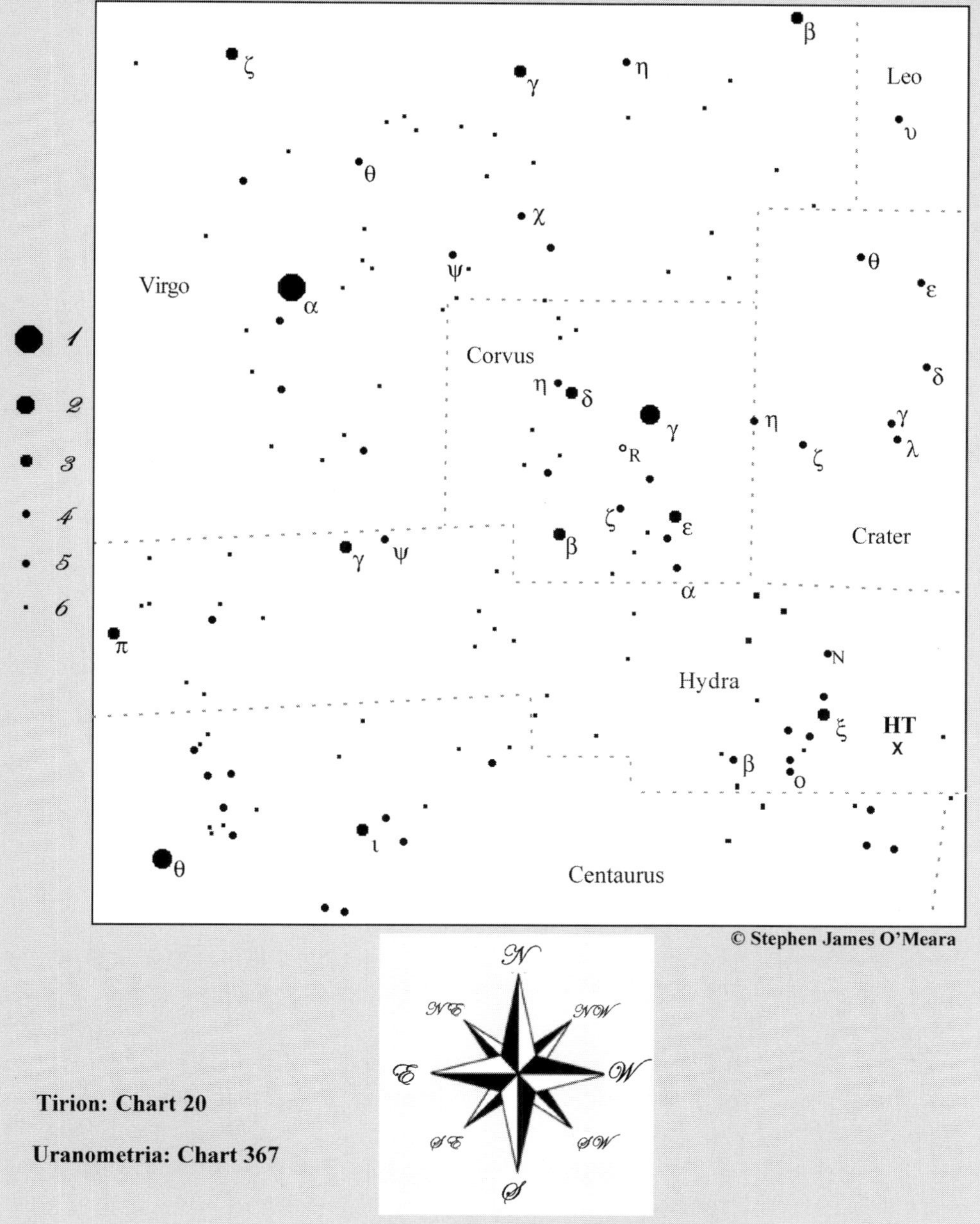

Tirion: Chart 20

Uranometria: Chart 367

57

Frame Galaxy
NGC 3621
Type: Mixed Spiral Galaxy (Sd)
Con: Hydra

RA: 11^h 18.3^m
Dec: −32° 49′
Mag: 8.5 (O'Meara); 8.9
Dim: 9.8′ × 4.6′
SB: 12.9
Dist: 22 million light-years
Disc: William Herschel, 1790

W. HERSCHEL: [Observed February 17, 1790] Considerably bright, extended, 70° in a direction from north preceding to south following, very gradually brighter middle, 7′ length, 4′ breadth, within a parallelogram. (H I-241)

NGC: Considerably bright, considerably large, extended toward position angle 160°, among 4 stars.

HYDRA IS HOST TO INNUMERABLE galaxies, many of which can be picked up in small amateur instruments. Its biggest deep-sky prizes, of course, are M83, a magnificent spiral galaxy, and NGC 3242 (Caldwell 59), the Ghost of Jupiter planetary nebula. Spiral galaxy NGC 3621 is another bright denizen of Hydra but very little attention has been paid to it. The galaxy is not included in many popular handbooks, though it is similar in size to M83 (albeit nearly $1\frac{1}{2}$ magnitudes fainter). In a 1993 *Astrophysical Journal* paper, NGC 3621 was referred to as "another comparatively poorly studied galaxy despite having a large angular size and abundance of H-II regions." Once again, the low southerly declination and the lack of bright stars nearby might explain the neglect.

In long-exposure photographs, NGC 3621 is a remarkable pinwheel just 25° from edge-on. The galaxy is large (93,000 light-years across) and has a total luminosity of 13 billion Suns. Its small, bright nucleus is nestled in a bright ellipse of spiral arms (cut by rifts of dust and dappled with numerous young star clusters) that gradually unwind into a fainter outer halo. One prominent arm extends southward from the small bulge; it carries some of this galaxy's most luminous H-II regions.

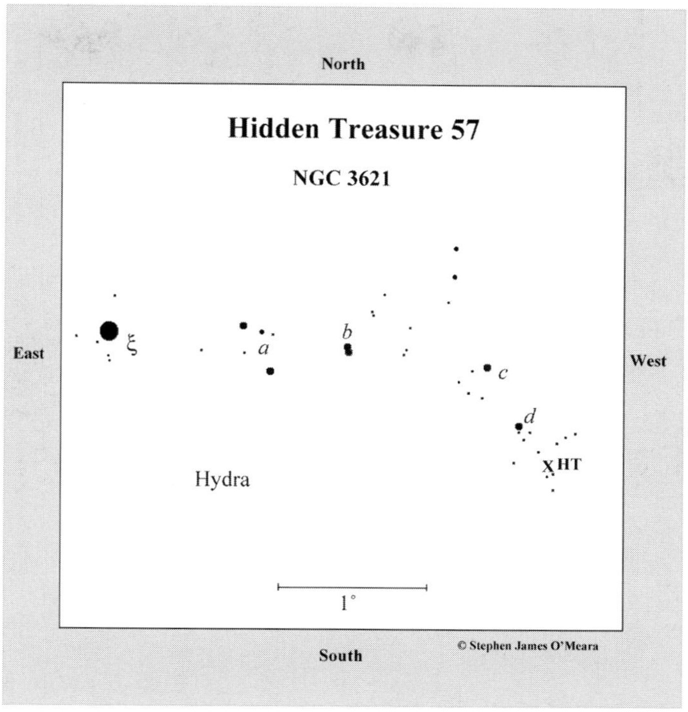

The spiral pattern of NGC 3621 is similar to the bright galaxy NGC 2403 (Caldwell 7) in Camelopardalis. H-II regions are present in the spiral arms and the largest have halo diameters of 4″.

NGC 3621, which belongs to the Leo Spur of Galaxies, was one of 18 galaxies observed as a part of the Hubble Space Telescope's Key Project on the Extragalactic Distance Scale, whose goal was to measure the Hubble constant to an accuracy of 10 percent (see Hidden Treasure 17). NGC 3621 contributed 69 Cepheid variable stars, with periods ranging from 9 to 60 days, to the project, which yielded a distance to NGC 3621 of 18.3 million to 22.8 million light-years. In 2001, astronomers announced that some of NGC 3621's brighter stars were also studied at the European Southern Observatory's very large telescope at Paranal Observatory in Chile. The scope resolved 10 supergiant stars, which were used as standard candles to help establish the galaxy's distance at 21.8 million light-years, which is well beyond the Local Galaxy Group. The universe's expansion is carrying it away at a speed of 730 kilometers per second in recession.

You'll find NGC 3621 only $3\frac{1}{4}°$ west and slightly south of magnitude 3.5 Xi (ξ) Hydrae. To locate it, though, from suburban sites at mid-northern locals, it's truly a treasure hunt. First locate 3rd-magnitude Epsilon (ε) Corvi. About 2° south and slightly west of Epsilon is 4th-magnitude Alpha (α) Corvi (Alchiba). Alpha Corvi is the easternmost star in a roughly 10°-wide, near-equilateral triangle with magnitude magnitude 4.5 Beta (β) Crateris (about 13° west-northwest) and Xi Hya (about 10° to the southwest). A fist held at arm's length will cover the area made by this triangle. You want to point your finder at Xi then sweep about 1° west-southwest, where you should encounter a 20′-wide pair of 8th-magnitude stars (*a*), which are oriented northeast–southwest. About 40′ to the west is a fine double star (*b*) with a magnitude 8.0 primary and magnitude 9.8 secondary to the south-southwest. Less than 1° to the west, and slightly south, is a solitary 7th-magnitude star (*c*). Just about $\frac{1}{2}°$ to the southwest is an 8th-magnitude star (*d*). NGC 3621 is almost equidistant from that star to the southwest.

57

NGC 3621 is the 617th object in James Dunlop's 1827 *Catalogue of Nebulae and Clusters of Stars in the Southern Hemisphere*. As seen through a 9-inch reflector at Parramatta, New South Wales, Dunlop describes it as "[a] very faint pretty large nebula, about 2′ broad and 4′ long, very faint at the edges. The brightest and most condensed part is near the south following extremity; a [faint] star is involved in the north preceding extremity, and there are two [faint] stars near the south extremity, but not involved."

From the dark skies of Hawaii, and I'm sure from other southerly locals in the continental USA, NGC 3621 is visible in 7 × 50 binoculars as a soft peach-like glow, even with the crescent Moon up. In my antique scope, it's a dim, blotted haze; the challenge in this star-poor region is to know exactly where to look. Telescopically, the galaxy appears as a big diffuse glow with a bright core and a faint, elliptical halo. Its spiral arms are faint and will challenge small-telescope users. Still, Ernest J. Hartung notes that the spiral system is "quite easy, though faint," in a 3-inch. He also found it "lying in a trapezium of four stars in good contrast with a scattered star field."

In the 4-inch at 23×, it's a fabulous sight with three dim stars forming a cap on the galaxy's southeast edge, which seem to pose as supernovae suspects. In fact, the galaxy looks more like an elongated globular cluster just starting to be resolved. Do not be fooled by the photographic dimensions given in the table. The galaxy is really segmented into two parts: a bright inner region that measures about 5′ × 3′ surrounded by a gentle outer halo, which is but a breath of moist air on a cold morning. That halo will vanish in small apertures with any increase in power.

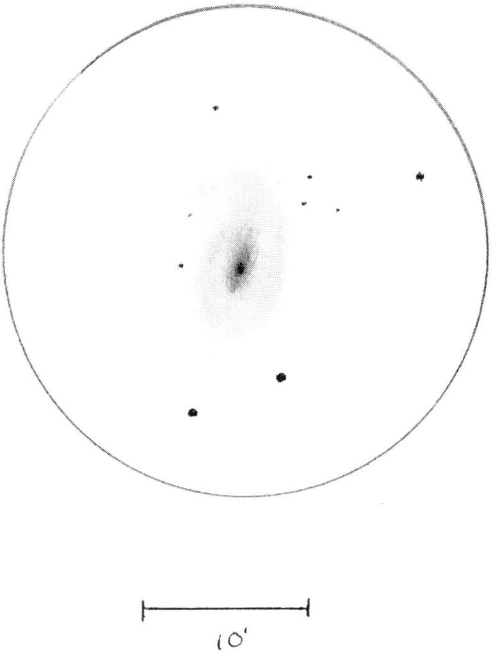

At 72×, the bright inner lens is fantastic. With any concentration it splinters into a veritable ornament of dim lights. The northeastern side is darker than the southwestern side, which is indicative of the dark and dappled veins of dust that riddle the galaxy in photographs. By moving my attention from my retina's periphery to its fovea, I can register several bright areas. First the nuclear region is a diffuse circular haze, like a fuzzy peach core. A bright nub of an arm extends to the south-southeast and, with time, I can follow two spiral arms that form a backwards S – one arm starts in the southeastern quadrant and loops around to the west and the other starts in the northwestern quadrant and loops around to the east. The southwest portion of the former spiral appears blotchy. A thin arm extends from the northeast side of the central peach pit and arcs toward the southwest.

With powers ranging from 101× to 168× the galaxy offers yet another layer of wonders. The core takes magnification particularly well, so do not hesitate to push it. At 168×, the core is lost in what appears to be a northwest–southeast oriented bar. Thin dark lanes of dust border the bar to the east and west and separate it from equally thin segments of inner spiral arms. Using a 10-inch Newtonian reflector under the dark rural skies of Potter Valley, California, Todd Hansen also notes the presence of these two lanes and arms. Two roughly 13th-magnitude stars occupy the northwest flank of the galaxy and a sprig of dappled haziness lies just to the east. In photographs it appears this region is a mixture of foreground stars and extragalactic wonder. David Levy of Arizona calls NGC 3621 the Frame Galaxy because it lies in a nice parallelogram of stars – like a picture in a frame. Note too that William Herschel called attention to this parallelogram in his description of the object in the table above.

Hidden Treasure 58

NGC 3628

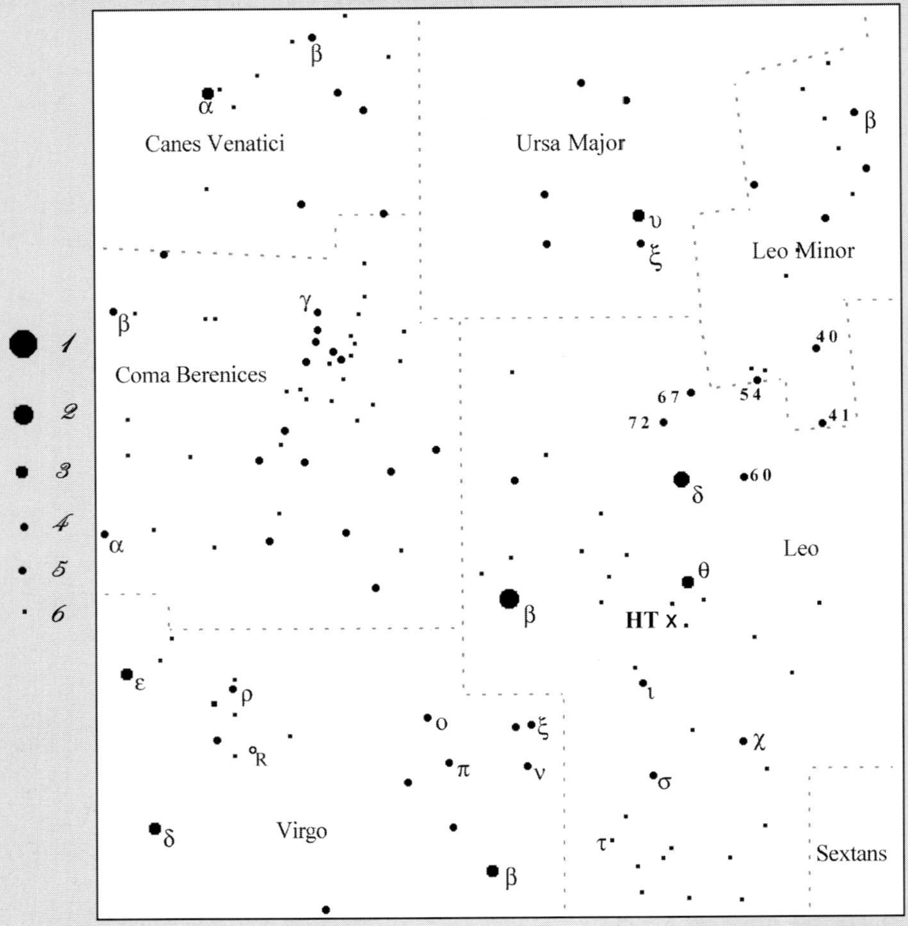

Tirion: Chart 13

Uranometria: Chart 191

58

King Hamlet's Ghost, Vanishing Galaxy
NGC 3628
Type: Peculiar Spiral Galaxy (SbP)
Con: Leo

RA: $11^h\ 20.3^m$
Dec: $+13°\ 35'$
Mag: 9.5 (O'Meara); 9.5
Dim: $14.8' \times 3.3'$
SB: 13.7
Dist: 25 million light-years
Disc: William Herschel, 1784

W. HERSCHEL: [Observed April 8, 1784] Bright, extended, almost in the direction of the parallel of declination, but a little in a direction north preceding south following, near 15′ in length. (H V-8)

NGC: Pretty bright, very large, very much extended toward position angle 102°.

JUST $1\frac{1}{2}°$ SOUTHEAST OF 3RD-MAGnitude Theta (θ) Leonis (Chertan), and 1° due east of 5th-magnitude 73 Leonis, is one of the sky's most celebrated galaxy triplets in the heavens – M65, M66, and NGC 3628, my "gift-to-you" hidden treasure. Normally, I would not have included this object, for the simple reason that it is part of this famous triplet, lying only 35′ due north of M66. Certainly anyone who has seen M65 and M66 has also seen NGC 3628. Nevertheless, just look at the comparison list in Appendix D; NGC 3628 is overwhelmingly popular among many observers.

That's probably because, unlike M65 and M66, we see NGC 3628 only 3° from edge-on. If there is one sight in the heavens that sets the imagination on fire, it's that of an edge-on spiral galaxy. Not only are bright ones rare, but they are exceptionally mysterious. They are the embodiment of all the wonder we feel when we look across the plane of our own Milky Way and try to fathom what wonders lie on the other side. Seeing these massive systems lying on their sides, obscured largely by dust, is nothing short of being handed a box that contains the Holy Grail but not having any way to open it. All that is

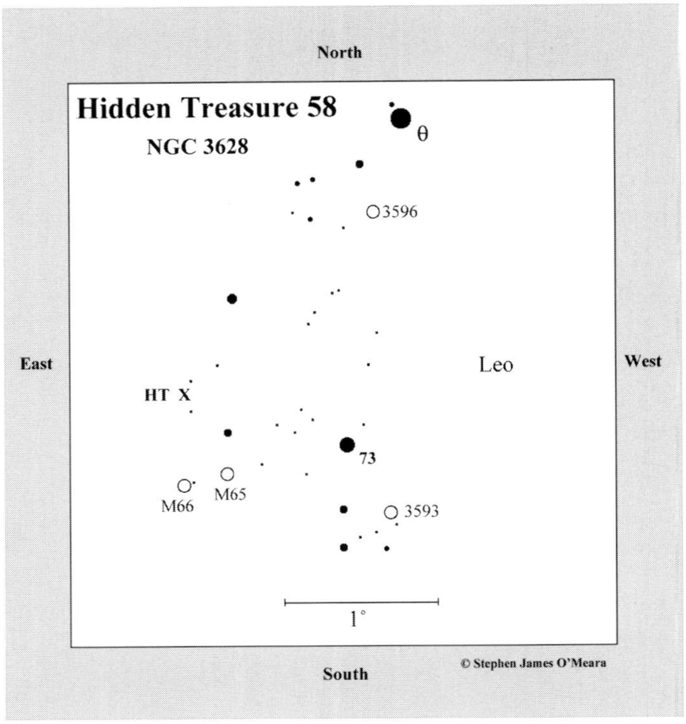

a few degrees to each other, and the ends of the spiral arms are obviously warped, with evidence of star formation. The galaxy's bright bulge also has an unusual "peanut" shape, which is most likely the result of gravitational interactions with M65 and M66 as the three galaxies orbit one another. Then again, one theory of "peanut" galaxies is that they are simply barred spirals seen edge-on. Otherwise, it's an eye-for-an-eye relationship, whereby the three members are simultaneously distorting each other over time. Deep charge coupled device (CCD) images have also revealed a 260-light-year-long stellar tail or plume extending to the south from NGC 3628 – another product of the tidal effects. This plume consists of clumpy star-formation regions that are roughly a hundred million years in age.

Radio observations reveal a roughly 4″-long string of a dozen components aligned with the major axis of the galaxy. These could be star-forming regions in the disk. Indeed, the Very Large Array, detected about 100 highly compact H-II regions near the center of the galaxy. These components are regarded as evidence for enhanced star formation in structures that are often produced in interacting systems. There is also a strong central radio source, which is possible evidence for a nuclear starburst (star formation taking place on an unusually large or rapid scale). Observations with the Nobeyama Millimeter Array have identified

precious to our understanding lies hidden from view.

The dust lying in the galaxy's plane was certainly dense enough to veil the object from the keen eyes of Pierre Mechain, who discovered M65 and M66 in 1780. And while NGC 3628 is only 0.5-magnitude fainter than M66, the galaxy is larger, so it has a lower surface brightness (13.7 vs. 12.5, respectively). Smyth pays no tribute to NGC 3628, nor does Webb. Ironically, NGC 3628 is the more fascinating member of the Leo triplet.

Like several of our hidden treasures, NGC 3628 belongs to the Leo Spur of Galaxies. It has a linear diameter of 78,000 light-years and a total mass of 135 billion Suns, so it is similar in physical size to NGC 3521 (Hidden Treasure 56). The galaxy is not without its peculiarities. For instance, it has two equatorial planes, marked by dark matter tilted four expanding shells and they appear to have arisen in a single starburst episode which has occurred in the last few million years.

The presence of a nuclear starburst has made NGC 3628 a popular target for X-ray observations. Einstein observations of the galaxy detected two X-ray sources, one of which was coincident with the nucleus, together with a bright X-ray outflow aligned along the minor axis of the galaxy. The central X-ray source lies about 1 kiloparsec from the nucleus and may be an intermediate-mass black hole candidate. This source has no known optical, infrared, or radio counterpart. It is also extremely variable. At a June 2001 American Astronomical Society meeting, David K. Strickland (John Hopkins University) notes that this source "is clearly another example of an intermediate luminosity X-ray object, very similar to the most-luminous X-ray source in M82 that has attracted so much attention."

The ROSAT X-ray satellite, launched in 1990, detected a 91,000 × 130,000 light-year halo of hot tenuous gas around NGC 3628. It is about 2 million K, with the most likely heating source being the shocks that result from a supernova-driven outflow wind from the circumnuclear starburst region. In many respects this soft X-ray halo resembles those of other nearby edge-on starburst galaxies like M82 and NGC 253 (Caldwell 65).

At 23× in the 4-inch, NGC 3628 is very obvious and very beautiful. It makes up in length what it lacks in width. First impression is that of a sharp needle in an area of haze: the ends are warped and broad and with just a little concentration I can see an obvious dust lane. The east end is most stubby. It's caged in a little house of moderately bright field stars.

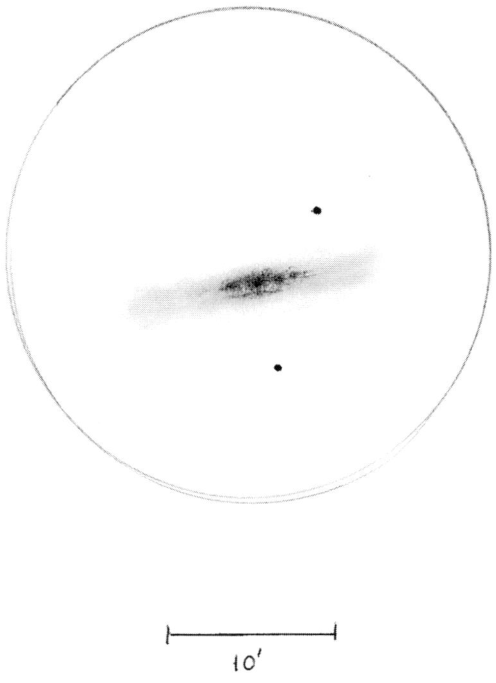

At 72× I can follow the warped dust lane throughout the body; the galaxy's bright regions are segmented here, and the west end is definitely fainter than the east end. Again the east end is also wider. The core is beaded and, with great concentration, tube tapping, and gentle breathing, I can see the dust lane against the core with scalloped edges. The galaxy is equally fascinating at 101×. The nuclear bulge looks like a comet, beaded and broken apart, thin and fine. The galaxy is difficult to observe if you want to see fine details in a small telescope. As Walter Scott Houston noted, "The galaxy spans half the Moon's diameter and can stretch clear across the field of a medium-power eyepiece." It requires time and patience and a willingness to be frustrated at peering into an eyepiece and seeing *almost* nothing . . . at first. As I reveal in *Deep-Sky Companions: The Messier*

Objects (in the essay on M65 and M66), I call NGC 3628 the "Vanishing Nebula," because with each increase of power, the galaxy blends more and more into the background of deep space, until it all but vanishes. This happens because the thick dust lane, which runs across the galaxy's entire length, overpowers the feeble light outlining it. So as you increase power, you magnify the dark lane, while spreading the faint light across a larger area of sky.

What's amazing about this galaxy is that the more you observe it, the more accustomed you become to seeing it. For that reason, I more recently nicknamed it King Hamlet's Ghost, in honor of the troubling spirit in Shakespeare's play *Hamlet, Prince of Denmark*. In Act I, Scene I, Officers Marcellus and Bernardo enter into the following conversation shortly before they see King Hamlet's ghost for a third time:

> MARCELLUS: What, has this thing appear'd again?
> BERNARDO: I have seen nothing.
> MARCELLUS: Horatio says 'tis but our fantasy,
> And will not let belief take hold of him
> Touching this dreaded sight, twice seen of us:
> Therefore I have entreated him along
> With us to watch the minutes of this night;
> That, if again this apparition come
> He may approve our eyes . . .

What we learn from this conversation is the importance of having patience with, and making repeat observations of, all things ghostly – and that includes challenging edge-on galaxies like NGC 3628. And, just as readers of *Hamlet* still debate whether Shakespeare intended the ghost of Hamlet's father to be real or imaginary, sky watchers still find the visibility of NGC 3628 and its details in small telescopes a mix of fantasy and fact. The answers to these questions are the same: it's simply a matter of whether seeing is believing. Only *you* can be the judge.

Hidden Treasure 59

NGC 4214

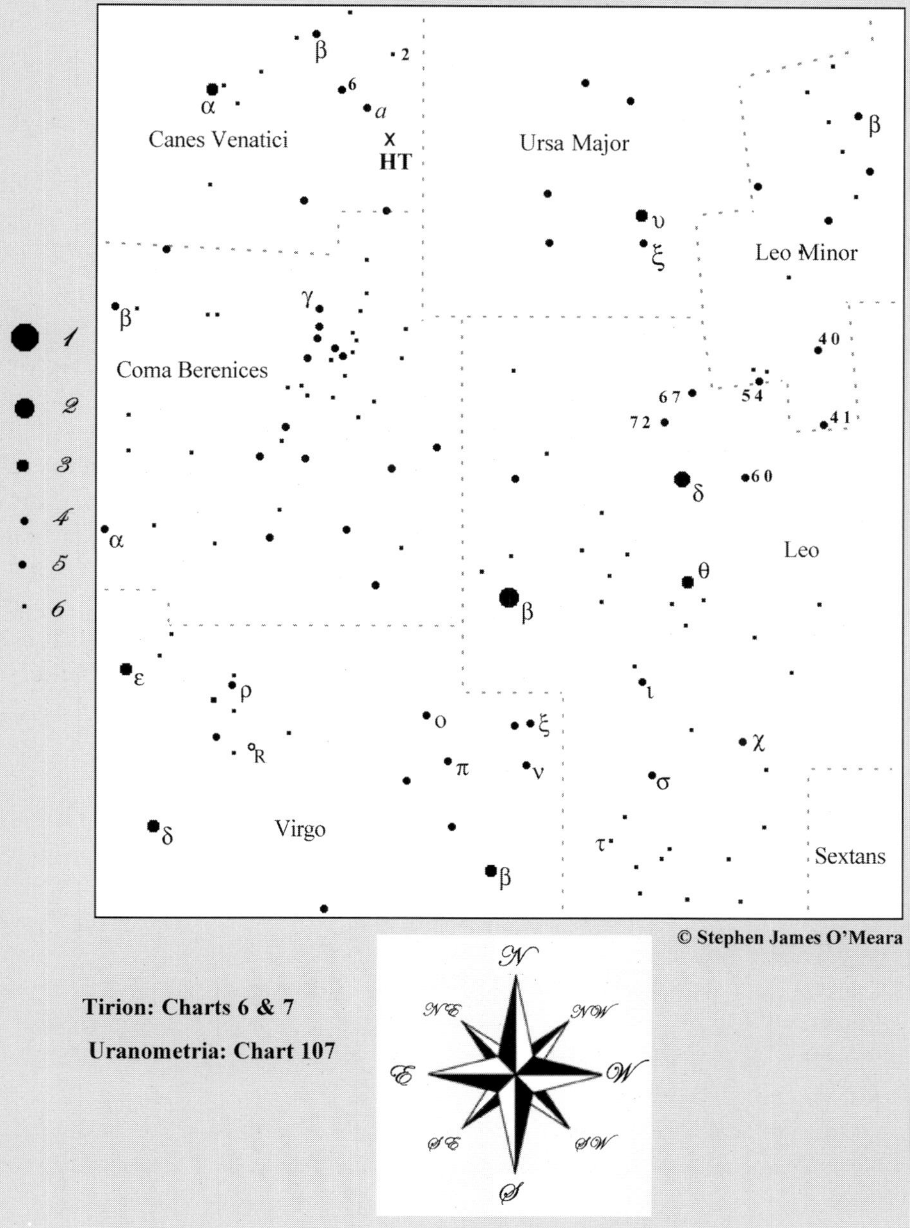

Tirion: Charts 6 & 7

Uranometria: Chart 107

59

NGC 4214 = NGC 4228
Type: Mixed Irregular Galaxy (IABm)
Con: Canes Venatici

RA: $12^h\ 15.6^m$
Dec: $+36°\ 20'$
Mag: 9.1 (O'Meara); 9.8
Dim: $9.6' \times 8.1'$
SB: 14.4
Dist: 11 million light-years
Disc: William Herschel, 1785

W. HERSCHEL: [Observed April 28, 1785] Considerably bright, considerably large, extended in a direction from north preceding, south following, with a brighter middle, 4′ in length and 3′ in breadth. (H I-95)

NGC: Considerably bright, considerably large, irregularly extended, binuclear.

IF WE COULD SOMEHOW REPOSITION the Large Magellanic Cloud (LMC) so that it is 11 million light-years from our Milky Way, it would appear very much like our next hidden treasure, NGC 4214 – a Magellanic-type irregular galaxy with the same linear extent (30,000 light-years). When William Herschel discovered this system in 1785, he had no idea what a chaotic wonder he had found. Not until the galaxy revealed itself to photographic emulsion did modern-day astronomers begin to fathom the galaxy's complex state.

When Edwin Hubble imaged the galaxy in 1955, he saw an object whose entire face resolved into individual stars and H-II regions. By the 1960s, astronomers were imaging a barlike core with dark markings and traces of spiral structure with highly resolved whorls. Gerard and Antoinette de Voucouleurs and Hal Corwin first classified it as an S-shaped Magellanic-type irregular in 1976. Today we recognize NGC 4214 as a metal-deficient Magellanic irregular galaxy with an inner blue starburst region and an older red disk. We see the galaxy 37° from face-on. It belongs to the Coma–Sculptor Cloud of Galaxies and has a total mass of 4 billion Suns – about half that of the LMC. It has a complex and interesting structure consisting of two distinct emission regions – each breaking up into several smaller

clumps – in a north–south S shape; the western end of the northern region of the S is an almost semi-circular arc.

In 1983 John Huchra, Margaret Geller (Harvard-Smithsonian Center for Astrophysics) and their colleagues made a pioneering study of the system in ultraviolet, visible, and infrared light. They found that strong winds, typical of *O*-type stars, are responsible for about 50 percent of the ultraviolet emission; they found few evolved supergiants. These and other observations suggest that a relatively recent burst of star formation has occurred within the galaxy, which may be the result of a merger or strong interaction with a companion.

During the *Astro-2* Spacelab mission in March 1995, the Ultraviolet Imaging Telescope (UIT) investigated NGC 4214 in far-ultraviolet light, which can trace recent massive star formation. The UIT data confirmed the presence of a starbursting core and identified several strong far-ultraviolet components, including a luminous central knot; an inner region (16,000 light-years across) with about 15 resolved sources, and a population of fainter knots extending to the edge of the optical disk. The data support Huchra and Geller's suspicion that the massive star formation in NGC 4214's core is the result of an interaction, possibly a tidal encounter, with one or more dwarf companions.

In 1997 NGC 4214 was imaged with the Hubble Space Telescope's Wide Field Planetary Camera 2. The image shows an amorphous bloody-yellow mess – like a yellow amoeba that's been hit by a truck. What the image actually shows is a highly resolved view of the star formation across the galaxy's face. Filigreed clouds of glowing gas surround at least six bright blue star clusters – the youngest of which are probably less than 10 million years old – in the galaxy's southern flank. The fast "stellar winds" flowing away from these young, hot objects attain speeds of thousands of miles per second and literally blow bubbles in the surrounding gas. Over millions of years, the bubbles increase in size as the stars inside them grow older. The activity is so great in this galaxy that it is sometimes referred to as a Wolf-Rayet Galaxy. These bubbles are due to the fact that most massive stars in a starburst spend a large fraction of their lives buried inside their original molecular cloud. We are seeing them just as they are emerging from this envelope.

The most spectacular feature in the Hubble picture, though, lies near the center of NGC 4214. As John MacKenty (Space Telescope Science Institute) explains, this object

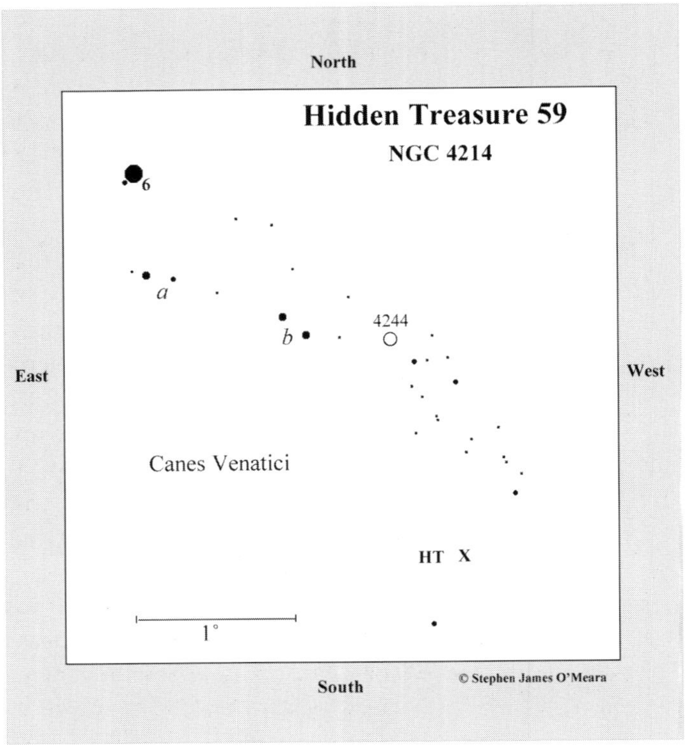

gradually fade away. The entire galaxy is swathed in a faint aura of older Suns, showing us that episodes of star birth have been occurring in NGC 4214 for billions of years.

This magnificent 9th-magnitude object lies about 6° southwest of 4th-magnitude Beta (β) Canum Venaticorum. To star hop to it, first use your eye or binoculars to locate 6 Canum Venaticorum, a 5th-magnitude star nearly 3° to the southwest. Center that star in your telescope. About 40′ south, and slightly west, is a pair of 8th-magnitude suns (*a*). Now swing the telescope about 1° to the southwest, where you will encounter an unequal pair of 7th-magnitude suns (*b*); the 10th-magnitude, edge-on galaxy NGC 4244 (Caldwell 26) lies just 30′ west of that pair of stars. NGC 4214 is a generous $1\frac{1}{2}$° sweep further to the southwest. Seen together in a wide-field telescope, NGC 4244 and 4214 look somewhat similar in shape and orientation to M81 and M82, but with diminished apparent size and brightness.

is a cluster of hundreds of massive blue stars, each of them more than 10,000 times brighter than our own Sun. A vast heart-shaped bubble, inflated by the combined stellar winds and radiation pressure, surrounds the cluster. The expansion of the bubble is augmented as the most massive stars in the center reach the ends of their lives and explode as supernovae. (Paul Wild discovered a 10th-magnitude supernova 84″ east and 216″ south of the nucleus in 1954, and HST detected a supernova remnant embedded in one of the two large H-II complexes.) Deprived of gas, this central cluster will be unable to form further new stars, and its luminous stars will continue to go supernova and disappear. Elsewhere in the galaxy, however, gas will start to collapse and form yet another new generation of stars, even as the clusters visible today

Larry Mitchell of Houston, Texas, notes that NGC 4214 is another object where historical confusion reigns. It is listed as both NGC 4214 and NGC 4228. When Herschel discovered the object on April 28, 1785, he wrote that it was considerably bright and large with some elongation and a bright middle. His son John, however, listed it as #1146 in his catalog, and then as #1157 at a slightly different position; John Herschel's

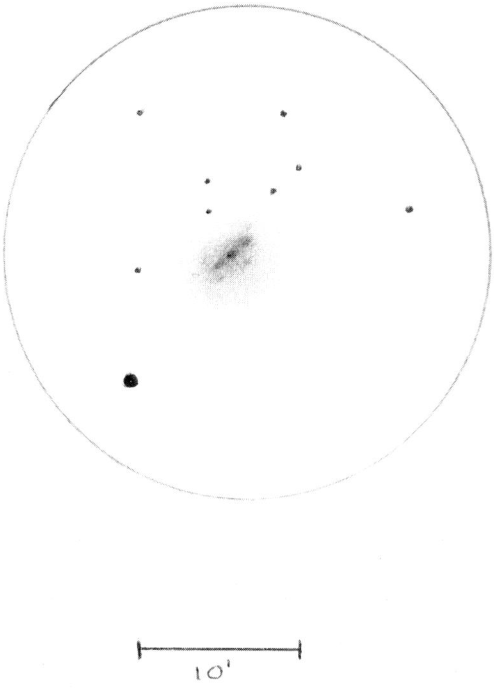

1157 later became NGC 4228. "William Herschel never reported observing this object," Mitchell says, "nor did Heinrich Ludwig d'Arrest (1822–1875) who looked for it in vain. John's description of NGC 4228 was 'very faint, large, round with a gradually brighter middle,' which almost sounds like a different object? Since nothing else is nearby, John must have made a positional mistake."

In the 4-inch at 23×, NGC 4214 appears as a bright oval disk, about 5′ northwest of a roughly 12th-magnitude star. It is ever so slightly oriented northwest–southeast. At first, the disk may appear uniform in brightness, but with averted vision it gets progressively brighter toward the middle, though not to a sharp stellar point. The inner core is simply a bright oval of light within a fainter one. At 72×, the inner oval seems mottled. With time and concentration I can make out a dual core – two fuzzy pearls of light. Smyth could not make these out, noting that, "despite all my coaxing, I was unable to see the two remarkable nuclei, so beautifully figured by [Herschel]." Clearly, then, despite Herschel's catalog description in the table above, he was the first to break up the galaxy's components before the advent of photography. Equally interesting, Luginbuhl and Skiff, saw only a "diffuse central pip" with a 10-inch telescope.

At 101×, the twin nuclei are flanked to the southeast by an even fainter glow. These three features define the galaxy's major axis, which Luginbuhl and Skiff refer to as a "bar." The northeastern flank of the inner lens is slightly brighter than the southwestern side, which appears as a thin brow around the region. The overall beauty of the galaxy in small telescopes begins to be lost at this power, though careful study of the outer lens does reveal its boxy shape. The object must be a marvel in larger instruments. As Ron Buta describes in the Webb Society's *Deep-Sky Handbook* (Volume 4), NGC 4214 through a 30-inch telescope appears "[b]right, mottled and fan-shaped with a wisp on one end which ends in a double knot; six very faint very small knots line the major axis, including what appears to be a conspicuous nucleus."

Hidden Treasure 60

NGC 4216

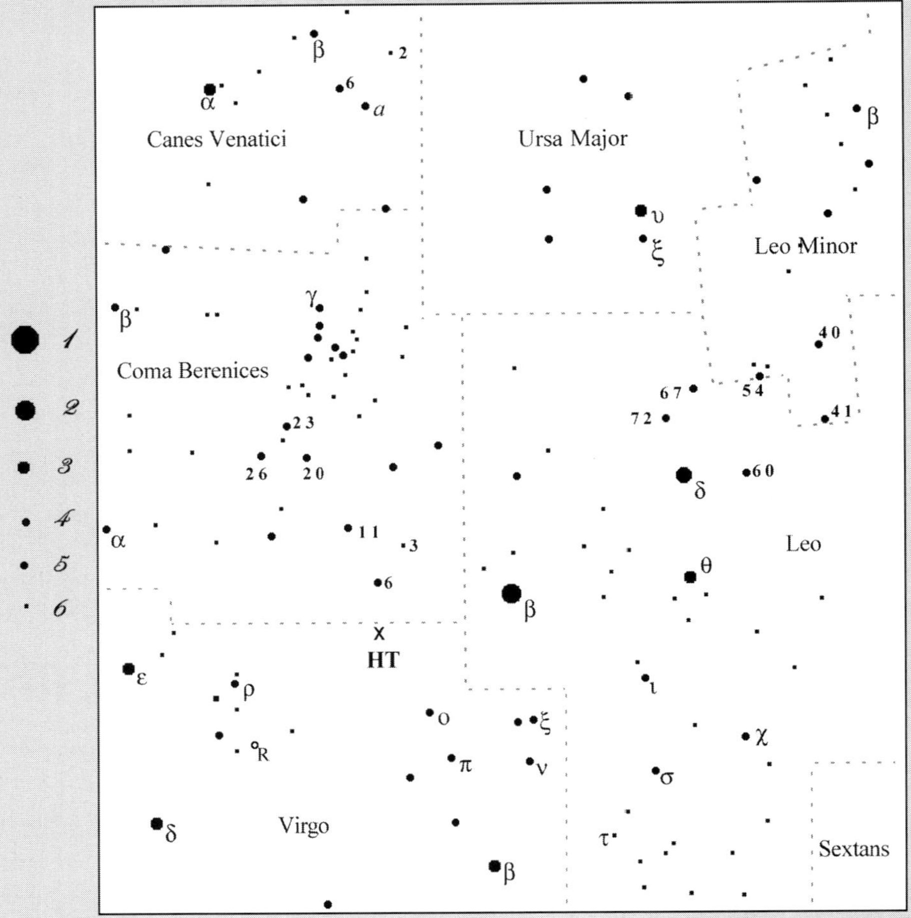

Tirion: Chart 13

Uranometria: Chart 193

60

Silver Streak Galaxy, Weaver's Shuttle Galaxy
NGC 4216
Type: Mixed Spiral Galaxy (SABb)
Con: Virgo

RA: 12^h 15.9^m
Dec: +13° 09′
Mag: 10.0 (O'Meara); 10.0
Dim: 7.9′ × 1.7′
SB: 12.6
Dist: 55 million light-years
Disc: William Herschel, 1784

W. HERSCHEL: [Observed April 17, 1784] Bright, very much extended, very bright in the middle, 9′ or 10′ in length. (H I-35)

NGC: Very bright, very large, very much extended toward position angle 17°, suddenly brighter in the middle to a nucleus.

NGC 4216 IS A PICTURE-PERFECT galaxy in Virgo. It lies only $1\frac{1}{2}°$ south of M98 and M99 in Coma Berenices and some 2° west of the very heart of the Great Coma–Virgo Cluster of Galaxies. Compared to other bright galaxies in the region, NGC 4216 stands apart, being inclined to our line of sight by just 1°. But that slim viewing angle is sufficient enough to let us glimpse, in stunning clarity, its bright core, nuclear bulge, luminous arms, and spiral dust patterns. It was among the first spiral "nebulae" photographed with the 36-inch Crossley reflector at Lick Observatory at the turn of the century. Lick Observatory's second director James Keeler (1857–1990), had in 1898 focused his attention on the study of Herschel's nebulae. Keeler believed that by comparing photographs of a spiral nebula over time, it might be possible to sense its direction of rotation. Heber Curtis (1872–1942), who joined the staff at Lick Observatory after receiving his Ph.D. in astronomy from the University of Virginia in 1902, carried on Keeler's work in 1914, noting that it was important because "a knowledge of the proper motions or of any rotational movements which these bodies may have would be of great value in investigations as to the size and distance of the nebulae, and therefore as to their place in the structure of the visible universe." By 1943 Edwin Hubble was using a negative image of NGC 4216 to illustrate a discussion

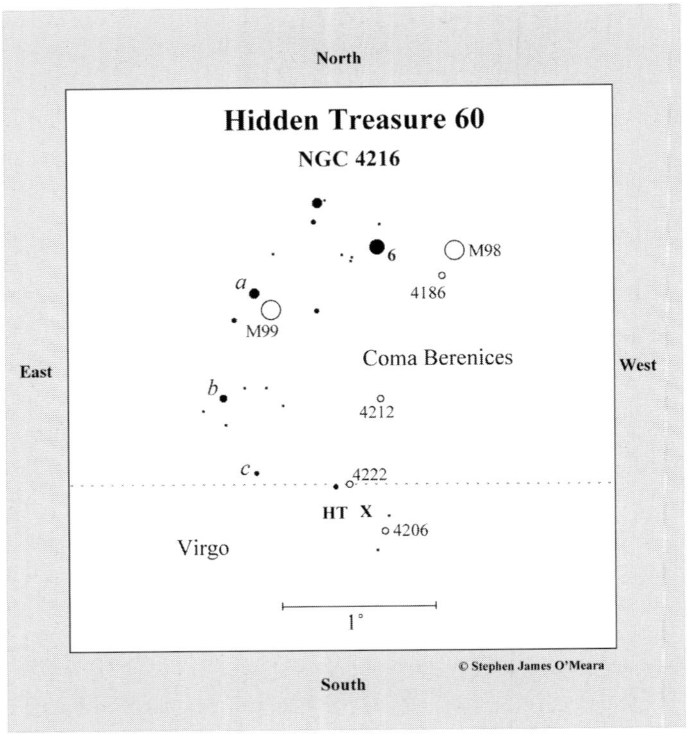

on the opening of spiral arms relative to the sense of rotation.

Today, images of NGC 4216 are often used in textbooks to illustrate the classic components of a typical, middle-aged Sb spiral galaxy. The galaxy displays a small, extremely bright nucleus partially hidden by dark lanes. A fuzzy halo of light, as soft and transparent as angels' breath marks the position of the galaxy's central bulge. Knotty, regular arms swirl away from this core in luminous concentric rings lined with very strong dark lanes. The outer regions of the disk are very dusty. The intensity of the nucleus and the symmetry of the rings gives the galaxy a look of quiet calamity.

A peculiar brightening some 50″ north-northeast of the nucleus, at the very lip of the central lens, could indicate the presence of a bar. As early as 1979, John Kormendy (University of Texas) found that bars in lens systems usually fill the lens in one dimension. If true, the position of NGC 4216's bright spot at the extreme of the lens argues that NGC 4216 is an edge-on bar.

In a 1999 *Astronomical Journal* paper, O. K. Sil'chenko (Sternberg Astronomical Institute, Moscow) and his colleagues obtained high-resolution spectra of NGC 4216 with the Soviet 6-meter reflector and found a distinct metal-rich nucleus, whose age appears to be different from that of the surrounding bulge. Normally, these two galactic components are of the same age. But NGC 4216's nucleus is about twice as young as its bulge. It is thought that the age of the nucleus is 8 to 12 billion years, while the bulge of the galaxy is older by a factor of 1.5 to 2. The astronomers also found spectral evidence for starburst activity which may form a circumnuclear ring in the plane of the galaxy's disk. If NGC 4216 is like other mixed spiral galaxies (see Hidden Treasure 51), the small bar in the lens may be triggering star formation in the nucleus, keeping it "young."

To find this beautiful system, first locate 5th-magnitude 6 Comae Berenices, which is about $6\frac{1}{2}°$ due east of Beta (β) Leonis (Denebola), the Lion's Tail. NGC 4216 lies $1\frac{3}{4}°$ due south of 6 Comae. You could also start with the famous 10th-magnitude galaxy M99, which is only 50′ southeast of 6 Comae and 10′ southwest of a magnitude

6.5 star (*a*). A 40′ hop south-southeast of M99 to bring you to a 7th-magnitude star (*b*). A little more than 30′ to the southwest will bring you to a roughly 9th-magnitude star (*c*). NGC 4216 lies almost 50′ to the west-southwest.

In the 4-inch at 23×, the galaxy is very bright and beautiful – quite a simple yet stunning sight. And though it is obvious in a small telescope under a dark sky, it is also small. Look for a bright, 5′-long silver streak, oriented north-northeast–south-southwest. With averted vision, the edges sharpen to a knife edge. The northern half of the lens also looks slightly brighter than the southern half. Admiral Smyth called it a "long pale-white nebula." He also called it a "very curious object, in shape resembling a weaver's shuttle." In keeping with the times, I suppose, the Rev. T. W. Webb seconded Smyth's opinion, noting that the object has a "long shuttle-like nucleus." For these reasons, I call the galaxy the Weaver's Shuttle.

NGC 4216 is all the more stunning at higher powers. At 72×, a bright core – tack sharp and extremely luminous – burns through the soft and smooth spindle, the ends of which gradually taper to a point, or so it seems. And at 101×, the brilliant nucleus lies at the center of a white egg-shaped core beyond which extends the oblique and dynamic lens. Smyth makes an interesting comment about the visibility of the galaxy's nucleus as the eyes work to absorb its photons: "and the centre exhibits a palpable nucleus, which in my instrument brightens at intervals, as the eye rallies." As Smyth's observing eye weaved back and forth like a shuttle across the length of the galaxy, the nucleus flared to prominence, then dimmed, in accordance to where its light focused on his retina. If you spend some time surveying the galaxy's length at 101×, see too if the major axis doesn't seem to suddenly and occasionally fluoresce – like a pulsating neon tube light. This may be due to the presence of a dust lane along the major axis, which the eye intermittently sees. Observers have spied this lane clearly in 12-inch and larger telescopes.

By the way, if for any reason you find yourself unimpressed with the appearance of this system, or somehow feel that it's an insignificant wonder, consider that in physical size and mass (linear diameter = 91,000 light-years; total mass of 190 billion Suns), NGC 4216 is comparable to our own Milky Way system. So as you stare at this near-edge-on galaxy, and perhaps shrug your shoulders at its dimness and simplicity, imagine how an alien observer from another galaxy might feel seeing our Milky Way 55 million light-years distant through a small telescope. It is a failing of human nature that we can so easily brush aside anything that does not immediately appeal to our visual senses. Our inability to comprehend fully all that we see is one of our greatest weaknesses.

In fact, NGC 4216 is the brightest of three edge-on systems within a 25′ field of view. The other two galaxies lie beyond the "obvious" in our range of vision. Shining at magnitude 13.3, the exactly edge-on galaxy NGC 4222 lies 12′ to the northeast, just over the border in Coma Berenices. And 11′ to the southwest lies NGC 4206, a magnitude 12.2 galaxy in Virgo with a small and very bright nucleus.

In light of the profound nature of all these dim and distant systems in the Coma–Virgo Cloud of Galaxies, it is almost nonsensical to look out into the vastness of space with our telescopes and not contemplate at least the possibility of life elsewhere in the universe. That curiosity is what sets us aside from other forms of life on Earth. It is what makes us distinct beings, though not necessarily unique in a cosmic sense.

In his 1963 book, *The View from a Distant Star,* Harlow Shapley said we can shun such possibilities because we are "dumb," congenitally dumb. "Our failure to comprehend the universe," he says, "is that we have been and still are bedeviled by a natural and persisting anthropocentrism. Correctives for our vanities are provided by modern science, but we suffer relapses and return to believing that we are somehow important and supremely powerful and understanding. Of course we are not."

In the words of Salvador Dali, "The universe is a slight thing compared with the amplitude of a brow painted by Raphael."

Hidden Treasure 61
NGC 4361

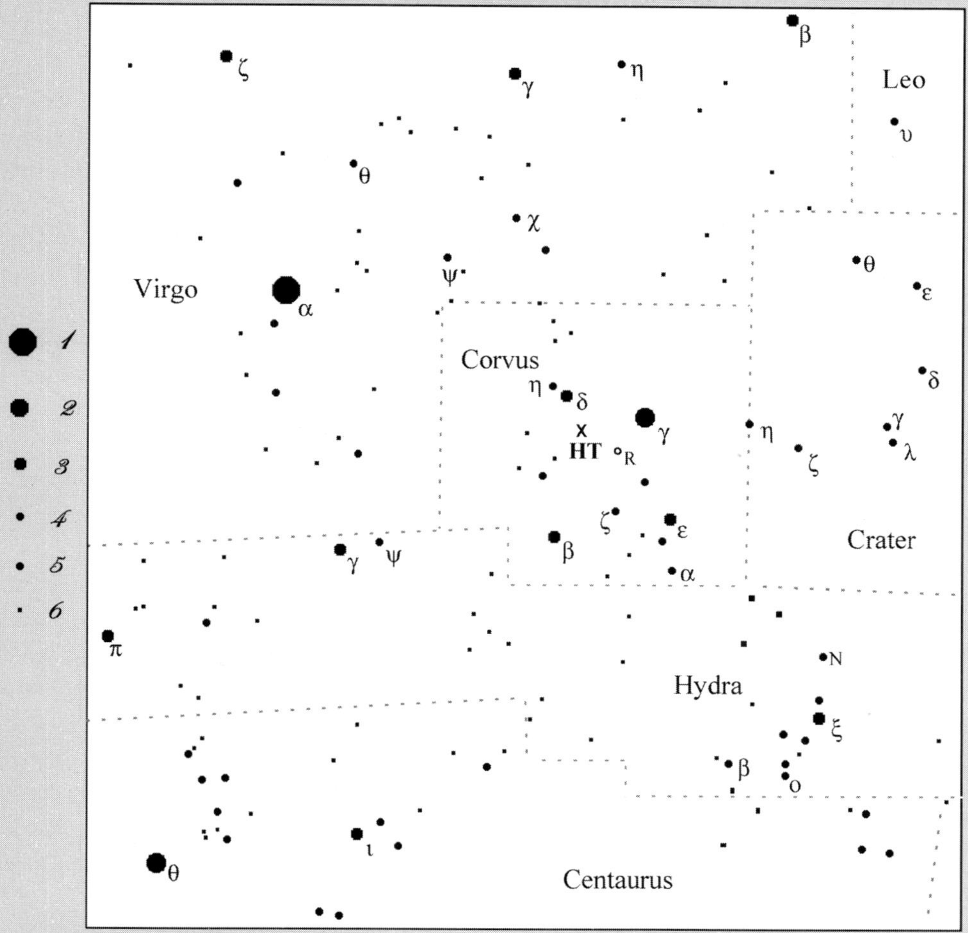

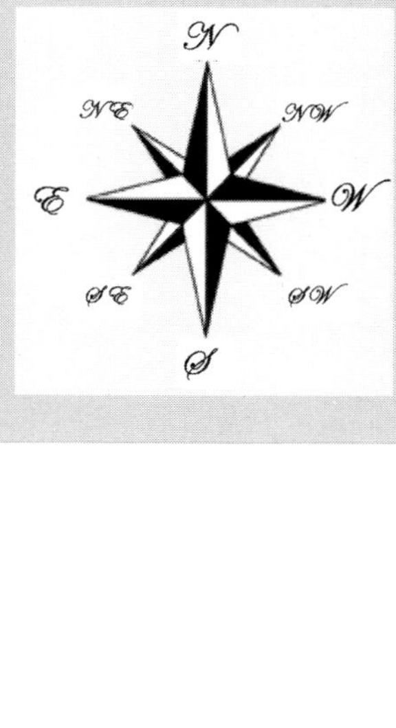

Tirion: Chart 21

Uranometria: Chart 328

61

NGC 4361
Type: Planetary Nebula
Con: Corvus

RA: $12^h\ 24^m\ 30.8^s$
Dec: $-18°\ 47'\ 05''$
Mag: 10.2 (O'Meara); 10.9
Dim: $1.9' \times 1.9'$
Dist: 3,900 light-years
Disc: William Herschel, 1785

W. HERSCHEL: [Observed February 7, 1785] Very bright, pretty large, irregularly round, bright middle, like a nucleus. (H I-65)

NGC: Very bright, large, round, very suddenly much brighter in the middle to a nucleus, [mottled].

CORVUS, THE CROW, IS A SOUTHERN constellation, whose four 3rd-magnitude stars form a prominent keystone pattern some 10° southwest of brilliant Spica in Virgo. To deep-sky enthusiasts, the constellation is most noted for its pair of interacting galaxies, NGC 4038–NGC 4039 (Caldwell 60 and 61), the famous Ringtail or Antennae Galaxy. But its brightest deep-sky object, and our hidden treasure (NGC 4361), is a very large planetary with one of the highest excitation states known.

When William Herschel discovered this peculiar object in 1785, he did not catalog it in his class of planetary nebulae (he did not know what these objects were), but as a bright nebula. His decision is understandable. Even in modern black-and-white images, the planetary, if quickly glanced, can be confused with a face-on, barred spiral galaxy – two symmetrical arms centered on a bright nucleus and central ring; adding to the confusion, the nebula lies on the southern outskirts of the Great Virgo Cluster of Galaxies. But there is no confusing the object in detailed color images, which show a 13th-magnitude central star surrounded by a fluorescent pink ring with a turbulent aquamarine outer shell. Here is the aged core of a red-giant star that has blown off its outer atmospheric layers across 2 light-years of space.

In a 1999 *Monthly Notices* of the Royal Astronomical Society, Spanish astronomer R. Vazquez (Institute of Astrophysics, Granada) and his colleagues describe the overall structure of NC 4361 as a fast, expanding halo being overrun by an even faster central bipolar outflow. In a 2001 *Astronomical Journal* paper, however,

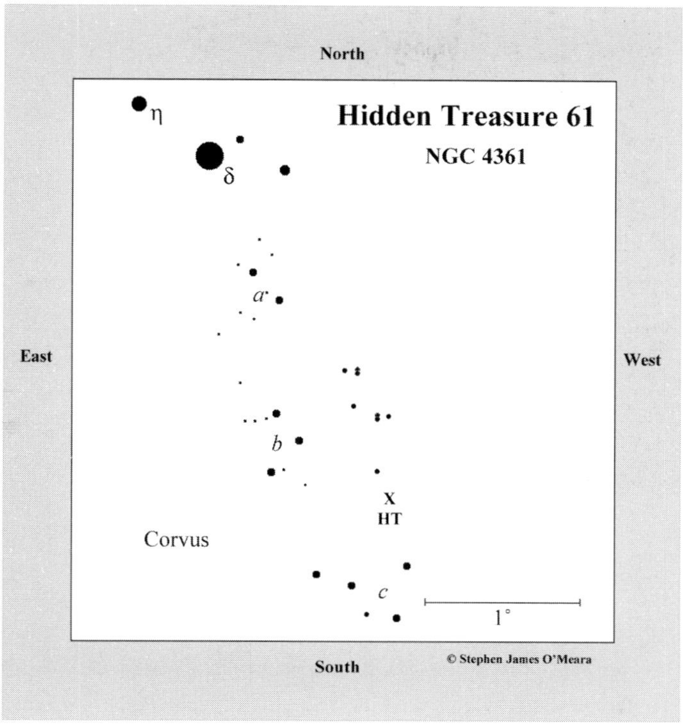

Indian astronomers C. Muthu and B. G. Anandarao (Physical Research Laboratory, Navrangpura, Ahmedabad, India) add another symmetrical axis, arguing that NGC 4361 has quadrupolar flows – most likely produced by a precessing equatorial disk. The interaction of these symmetrical outflows with the halo has, in the words of Vazquez, "already produced notorious signs of deformation in the [halo]." Careful inspection of the planetary's halo reveals it to have a complex curdled appearance, as if some unseen hand has stirred this brew of tenuous vapors with a stick. In a sense it has. Shocks produced by the interactions between the polar outflows and the halo cannot only disturb the nebula's appearance but also heat the nebula to the observed levels. Vazquez and his colleagues inspected Hubble Space Telescope (HST) images of the planetary but detected no binary core.

NGC 4361 was the first object the late Walter Scott Houston used to turn to in the Crow, and it is unanimously popular as one of the finest non-Messier non-objects of its class (see Appendix D). It lies in the north central part of the keystone (also known as the Sail), and marks the southern apex of a near-right triangle with Delta (δ) Corvi (Algorab) and Gamma (γ) Corvi (Gienah). To find it, center Delta Corvi, then look a little less than 1° southwest for a pair of 8th-magnitude stars that are part of a little line of 8th- and 9th magnitude suns (*a*). Another 1° hop to the south will bring you to a 20′-wide isosceles triangle of 7th-magnitude suns (*b*). NGC 4361 is about 50′ southwest of that triangle, about 30′ north of a 40′-wide Y-shaped asterism of 8th-magnitude suns (*c*). Under a dark sky I could just suspect it in 7 × 50 binoculars, which appears to be the case with other careful observers. It can be seen without trouble in my antique telescope.

At 23× in the 4-inch, the nebula is easily spotted as a small, round, glow, about 2′ wide. But with a good boost in power (I found 182× worked best in my small telescope), the nebula appears as an almost uniform glow around a 13th-magnitude star. Take the time, however, to survey this halo with averted vision. Be sure to breathe regularly and to take short breaks away from the telescope. When I do this, I can just make out the nebula's spiral inner structure, which

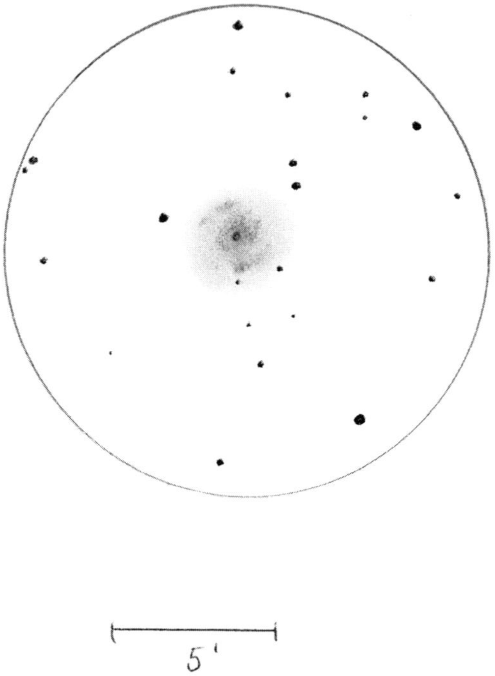

appears at first only as dim extensions, or lobes, oriented northeast–southwest. But if you persist, the spiral pattern becomes "obvious." Fainter lobes can also be seen oriented northwest–southeast, giving the nebula a cloverleaf appearance.

The most difficult aspect of observing this planetary with a small telescope is the low contrast between the arms and the surrounding halo. If you have difficulty, try lowering the magnification. At one point I did pump the power up to 303×, but the nebula became quite extended and dim. And while the same details did show, they were fantastically elusive; I had to fight to see them with averted vision, even then, the features seemed almost supernatural – a real eye strainer.

So it's best to study this object with a variety of magnifications over different nights. The more you get acquainted with this object, the more you'll see, and the easier the task will become to see the finer details. Overall, aside from the central star, the most obvious feature is a knot on the south-southwest arm, which forms the end of a "bar" of material that eventually sweeps round to the west. Another knot on the opposite side, but farther out from the central star, is similarly shaped; but its material sweeps from the northeast to the east-northeast. That's where it touches a fan of material that forms the eastern clover leaf. A dimmer clover leaf, the faintest of them all, lies to the west. The overall shape of the nebula is indeed round, which is best seen at low powers. But at high powers, this planetary for all intents and purposes looks like a barred spiral galaxy. There is no mistaking the similarity. The lobes of each clover are not of a uniform light, they appear mottled, and they are not evenly lit. It is quite the feathery jumble of whirling gases, a most complex sight, and perhaps one of the most visually bizarre planetaries visible in small telescopes.

By the way, Delta Corvi is a beautiful double star. At 23× the yellow primary and burnt-sienna secondary look like a sun, like our own, with a Jupiter-sized planet.

Hidden Treasure 62

Melotte 111

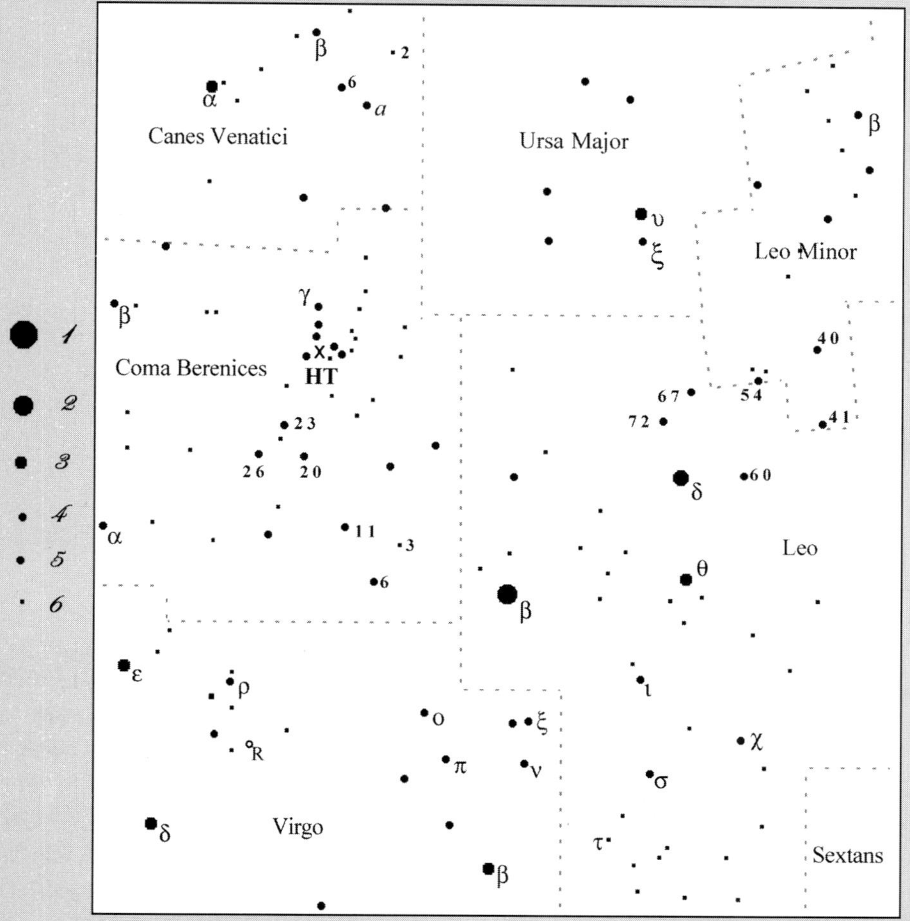

© Stephen James O'Meara

Tirion: Chart 7

Uranometria: Chart 148

62

Coma Berenices Cluster, Berenice's Hair, Ariadne's Hair, Thisbe's Veil, Cobweb Cluster, Black Bart's Hair, Flying Witch Melotte 111
Type: Open Cluster
Con: Coma Berenices

RA: $12^h\ 25.1^m$
Dec: $+26^\circ\ 07'$
Mag: 1.6 (O'Meara); 1.8
Diam: 300.0′
Dist: 288 light-years
Disc: Known since antiquity; first alluded to by Eratosthenes; Ptolemy cataloged it in his Almagest

HERSCHEL: None.

NGC: None.

ON COOL SPRING EVENINGS, WHEN Leo is leaping high across the meridian, and Arcturus in Bootes burns brilliantly above the beast's lower shanks, another constellation, dim but glittering, fills the void between them. Large and inconspicuous, Coma Berenices (Berenice's Hair) is a 5°-wide spangle of seven prominent, and twice that many fainter, naked-eye suns that glitter like dew on a cobweb under the cold light of the Moon. More than a fill-the-gap constellation, Coma Berenices is a marvel to behold in binoculars and rich-field telescopes under clear, dark skies. While the constellation is continually probed and prodded for its telescopic wonders – especially the brightest members of the Coma Berenices Cluster of Galaxies, which includes such splendors as M64 (the Black-Eye Galaxy) and the edge-on wonder NGC 4565 (Caldwell 38) – few observers recognize the constellation for what it truly is: one of the largest and brightest open star clusters in the night sky, one that has touched the hearts of poets over the centuries.

Conon of Samos (~280 BC–220 BC), court astronomer to Ptolemy III of Egypt, created the constellation in response to a theft. The story begins with King Ptolemy heading off to war to avenge the murder of his sister in Syria. Fearful of the outcome, Queen Berenice II, Ptolemy's lovely sister and wife, asks the goddess Aphrodite for help; Berenice vows that she will cut off her long amber locks if her husband returns

safely. Weeks pass in restless worry for the queen, but her beloved king returns victorious, and the nation rejoices. But the king is not amused when he learns of his wife's vow to Aphrodite. To his eye, nothing is more fair than Berenice's hair; it is also her pride, the joy of the nation, and the inspiration for poets. Alas Berenice fulfills her vow. With great ceremony, her beautiful tresses are clipped from her shapely head and carried to the Temple of Aprhodite. There, the priests lay the locks on an altar and offer them as a gift of thanks to the goddess of love. But the king's disappointment turns to outrage when he discovers the next morning that the locks have been stolen. The King summons the priest and would have shouted an "off-with-their-heads" command had it not been for the quick thinking of Conon. The court astronomer tells the king that he knows where the locks are being kept and encourages him to wait until nightfall. That night, as the ashen glow of twilight starts to fade, the king and queen and all the nation's people gather in the court, Conon looks at the nervous priests, gives them a wink, then points to the heavens, beyond the Tail of Leo. "Look," Conon suddenly cries, "Dost thou not see the clustered curls of they queen, too beautiful for a single temple to possess, placed there by the gods for all the world to see? Look. They glitter like a woven net, as golden as they were on Berenice's head."

Needless to say, from that time on the constellation has been known as Coma Berenices and people have sung its praises. Among the first was the great Roman poet Callimachus, who honored Conan's ingenuity in a poem entitled *Berenice's Lock.*

A more modern poem, *The Rape of the Lock*, by Alexander Pope (1688–1744), draws a parallel to the story of Berenice's hair. Pope's poem deals allegorically with the alleged theft of another lock – that of Arabella Fermor's hair, which her friend (and then potential husband), the Lord Petre, had cut off as a prank some time before March 21, 1712. Like Berenice's hair, Pope sees Fermor's hair (which he disguised as "Belinda's hair" in the poem) fleeing into the sky: "A sudden Star, it shot thro' liquid Air, / And drew behind a radiant Trail of Hair. / Not Berenice's Locks first rose so bright, / The heav'ns bespangling with dishevel'd light." Pope explained the reason for writing the poem: "The stealing of Miss Belle Fermor's hair, was taken too seriously, and caused an estrangement between

the constellation as a sheaf of wheat in the Herdsman's hand, while others depicted it in the Virgin's hands. Several cultures saw it as a laurel or wreath. In ancient China, the stars in and around Coma Berenices – including the tail of Leo, the torso of Virgo, and the stars of Canes Venatici – were a scene at the royal court. In the center of the court was the royal prince, T'ai-tsze, surrounded by all his dignitaries. Regardless of the confusion, we have to thank the great Danish astronomer Tycho Brahe (1546–1601) for first recording its position and fixing it as a distinct constellation in his 1602 catalog. Johann Bayer adopted Tycho's usage of Coma Berenices in his 1603 *Uranometria*.

the two families, though they had lived so long in great friendship before. A common acquaintance and well-wisher to both [John Caryll], desired me to write a poem to make a jest of it, and laugh them together again. It was with this view that I wrote *The Rape of the Lock*."

The stars of Coma Berenices have not always belonged to Berenice. In the third-century BC, Eratosthenes first alluded to the grouping as the Lock of Ariadne (or Ariadne's Hair), which he says is "near the tail of the Lion." Ptolemy also recognized the group as a "lock" of hair, but this hair belonged to Leo, the Lion, and was not recorded as one of his original 48 constellations. Indeed Arabian astronomers long knew it as either Al Halbah or Al Dafirah, which refer to either "The Coarse Hair" or "The Tuft" (the bushy tip) of the Lion's Tail. In his *Cycle of Celestial Objects*, Adm. William Henry Smyth writes, "Old Thomas Hill, in his *Schoole of Skil*, 1599, calls these sacred tresses by the homely designation of Berenice's Bush; and there has been a name still homelier." Bayer depicts

In his book *The New Patterns in the Sky: Myths and Legends of the Stars*, Julius D. W. Staal shares an old Babylonian story, related by Ovid, that tells how the stars of Coma Berenices came to be. Two lovers, Pyramus and Thisbe, decided to meet in secret one evening at a spring near the edge of the forest. Thisbe arrived first, but she was frightened away by a lion carrying its dead prey. As she ran, she lost her veil, which fluttered past the lion, who angrily snatched at it, staining the veil with blood. When Pyramus arrived at the scene, the lion had gone, but Pyramus immediately recognized Thisbe's veil. Seeing the blood and the mauled carcass nearby, Pyramus feared the worst. Blaming himself for the tragedy, he kissed the blood-stained veil, drew his sword, then plunged it deep into his heart. His limp body fell at the foot of a mulberry tree, whose roots thirstily absorbed the oozing blood, which caused its fruits to turn dark red. Unable to find help, Thisbe returned to the forest only to find her beloved Pyramus dead. Grasping Pyramus' sword, Thisbe plunged it deep into her side then swiftly joined her lover. To honor the pair for their devotion, Jupiter put Thisbe's veil among the stars, where it still "flutters today in the stars of Coma Berenices, wafted by an eternal cosmic breeze."

John Landseer (1769–1852), an English writer on archeological astronomy, says that, based on the study of an Assyrian symbolic monument, the seven brightest stars Coma Berenices were seen as a Dove; and since the first of these stars rise with the last bright star of Argus (the Ship [Noah's Ark]), as well as rising simultaneously with the hands of Bootes, the Herdsman, Landseer believes that Coma Berenices, not the Pleiades, is better associated with the tale of the Biblical flood as written in the stars – this, he argues, would solve the mystery of the missing Pleiad.

In my youth, I preferred to see Coma Berenices as a spider's web stretched across the void between Canes Venatici, Leo, Virgo, and Bootes. Although I grew up in what is now light-polluted Cambridge, Massachusetts, this sparkling array of dim suns was certainly bright enough to snare my imagination; such a view of it would not be possible today from my old backyard (unless there was a power failure on the East Coast); but at least I have those sentimental memories, which I can now relive from my new home. Garret Serviss also imagined a cobweb here, writing, "One might think the old woman of the nursery rhyme who went to sweep the cobwebs out of the sky had skipped this corner, or else that its delicate beauty had preserved it even from her housewifely instinct."

While Coma Berenices has long been noticed as a conspicuous gathering of stars, its true nature was not revealed until 1938 when R. J. Trumpler identified 37 stars as true cluster members. But P. J. Melotte had already listed it as the 111th object in his 1915 *Catalog of Star Clusters*, which is why it retains the catalog name of Melotte 111.

Melotte 111 is very young, having an estimated age between 400 and 600 million years – about the time when the first winged insects appeared on Earth. For a cluster that is so close, it is relatively dim. It is also relatively poor in stars, containing some 270 members ranging from magnitude 5 to 10.5; that's about a quarter of the number of stars found in the Pleiades. No fainter members have been detected. As R. J. Garcia Lopez (Astrophysical Institute of the Canaries) and his colleagues suggest in a 2000 *Astronomy and Astrophysics* paper, the Coma Berenices Cluster may not have enough mass to retain its less massive original stars, such as red dwarfs, which have long since escaped into space. The mass of the cluster is probably under 100 solar masses, and its density is about one star per 10 parsecs. The members we see today are arranged in a roughly triangular shape that spans 10 Moon diameters in extent (22.5 light-years). The cluster contains no giant stars, though the brightest members are just beginning to evolve toward the giant stage. The cluster is intrinsically only slightly brighter than the Sun, meaning that if we could place the Sun at the distance of the Coma Berenices Cluster, it would shine with the brilliance of a 1st-magnitude star.

A sweep of the cluster with binoculars or a rich-field telescope reveals very few

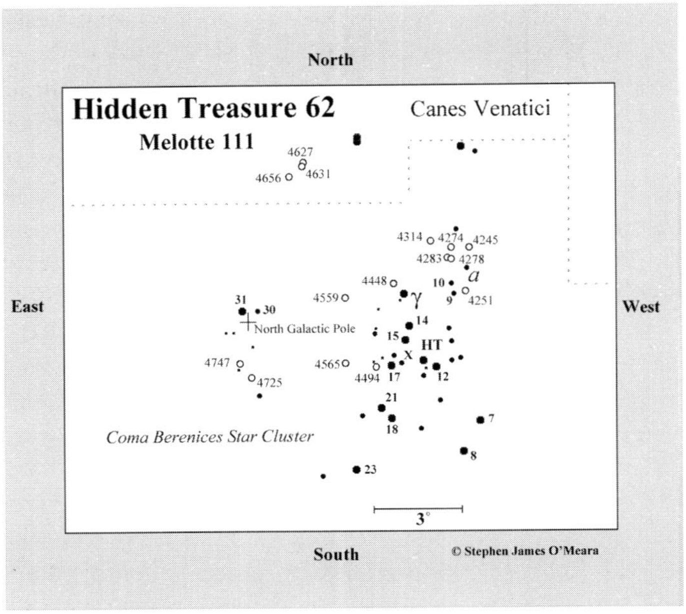

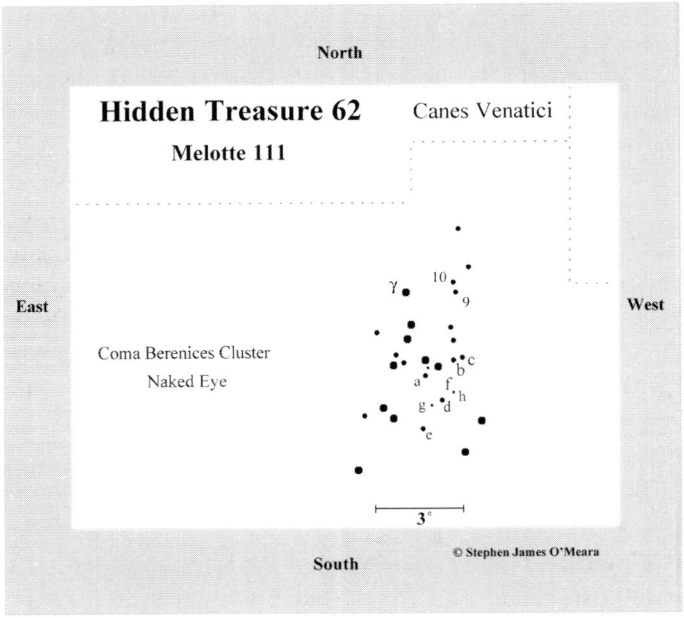

close doubles. The most prominent one is 17 Comae, but 16 Comae has four dim companions that make an attractive sight, like a dangling earring with a large central diamond. The entire cluster is fractured into mini asterisms that seem to float in the darkness like letters in alphabet soup. More than 100 stars can be seen with a casual glance in the triangular portion formed by 12, 14, and 17, the region around and Gamma Comae. But it's amazing how utterly "clean" the cluster looks through the 4-inch at 23×, just a scrim of stars beyond which lies a black and empty stage. In a local sense, this is simply due to an uncommon perspective; unlike most open clusters, the Coma Berenices Cluster is as far as one can get away from the stellar pandemonium of the Milky Way's disk. In other words, it lies only about 5° west of the North Galactic Pole – the point where our galaxy's axis of rotation intersects the north celestial sphere. In a universal sense, this background emptiness is an illusion, for the region is on the western fringe of the great Coma Berenices Cluster of Galaxies, a turbulent array of more than 3,000 galaxies (mostly ellipticals and lenticulars), 280 million light-years distant. It is the nearest massive cluster of galaxies and is scattered across 20 million light-years of space.

Ten reasonably bright galaxies in the Coma–Sculptor Cloud of Galaxies are within reach of a 4-inch telescope under a dark sky.

They all lie within 2° of the Coma Berenices Star Cluster: NGC 4225, 4251, 4274, 4278, 4283, 4314, 4448, 4494, 4559 (Caldwell 36), and 4565 (Caldwell 38). To hunt them down, start at Gamma Comae. (Note: do not be turned away by the magnitudes, because many of these galaxies have a bright central core.) You'll find 11th-magnitude NGC 4448 a little less than 30′ northeast of Gamma Com. A star hop $1\frac{1}{2}°$ west of Gamma Com will bring you to the 6th- and 7th-magnitude stars, 9 and 10 Comae, respectively; the 11th-magnitude galaxy NGC 4251 lies less than 20′ due west of 9 Comae. About 40′ northwest of 10 Comae is a 6th-magnitude star (labeled *a*). Another hop 40′ to the northeast will bring the magnitude 10.2 galaxy NGC 4278 into view, with the tiny 12th-magnitude galaxy NGC 4283 (with a very bright center) less than 5′ to the northeast. Center this galaxy pair and look 20′ to the north for the magnitude 10.4 galaxy NGC 4274. About 30′ due west is magnitude 11.4 NGC 4245 (very small and bright nucleus). Swing back to NGC 4274 and look about 40′ to the northeast, where you'll find magnitude 10.6 NGC 4314. If you return to Gamma Comae and sweep 2° east-southeast, you will encounter 10th-magnitude NGC 4559. Next, center the beautiful double star 17 Comae; just $\frac{1}{2}°$ east-southeast is 10th-magnitude NGC 4494. A little more than 1° to the east-northeast will bring you to the final object, the stunning magnitude 9.6 edge-on galaxy NGC 4565.

After admiring the cluster with binoculars or a rich-field telescope, put all your equipment aside, grab a chair, or lie on a blanket and see how many dim stars you can detect in the cluster without optical aid using the chart on page 312. The cluster contains nine stars between magnitude 6.0 and 7.4 with separations ranging from 12″ to 31″. The following magnitudes were derived from Hipparcos satellite data:

Star	Mag.	Star	Mag.
a	–	*f*	6.7
b	6.2	*g*	7.4
c	6.4	*h*	7.3
d	6.2	9	6.4
e	6.0	10	6.6

Melotte 111 is extraordinarily close. Using astrometric data from the Hipparcos satellite, astronomers have determined the cluster's distance to be 288 light-years. This makes it one of the nearest open star clusters, only the Ursa Major Cluster and the Hyades are even closer. If you look up at it on the eve of September 12, 2046, the light reaching your eyes would have left the cluster the night Charles Messier independently discovered M1, the Crab Nebula – the object that inspired him to create his now famous catalog of deep-sky objects.

If you see Coma Berenices in 2010, you will be seeing light that left the star in the year when the last legal execution for witchcraft happened: In Protestant, Scotland, an elderly woman was burned at the stake, after being accused of turning her daughter into a pony and riding her to a witch's sabbat. You can see the "witch" fly today in the stars of Coma Berenices: Gamma Comae is the witch's head, the triangular form of the main cluster is the witch's body. The stars 14, 16, 17, and 21 Comae form the witch's wind-blown cape, and the stars 18, 23, and 26 form the witch's broom, which she clutches in her hand (12 Comae).

Finally, if you see the cluster on February 10, 2010, the light will have left the group on the day that the infamous pirate Bartholomew Roberts (Black Bart) went to Davy Jones's Locker. On that fateful day, Roberts was aboard his flagship, the *Royal Fortune,* battling captain Chaloner Ogle's company of *HMS Swallow* off Cape Lopez, Gabon. Roberts was killed when grapeshot fired from a canon struck him in the neck. As told in *The Pirates' Own Book: Authentic Narratives of the Most Celebrated Sea Robbers,* "Roberts . . . made a gallant figure at the time of the engagement, being dressed in a rich crimson damask waistcoat and breeches, a red feather in his hat, a gold chain round his neck, with a diamond cross hanging to it, a sword in his hand, and two pair of pistols hanging at the end of a silk sling flung over his shoulders, according to the custom of the pirates." By the way, some historians argue that Roberts, who had a full head of dark curly locks, might have been a woman, perhaps Anne Bonny after she escaped the gallows. The stars of Coma Berenices, then, could also represent the hair of that mysterious pirate.

Hidden Treasure 63

NGC 4490

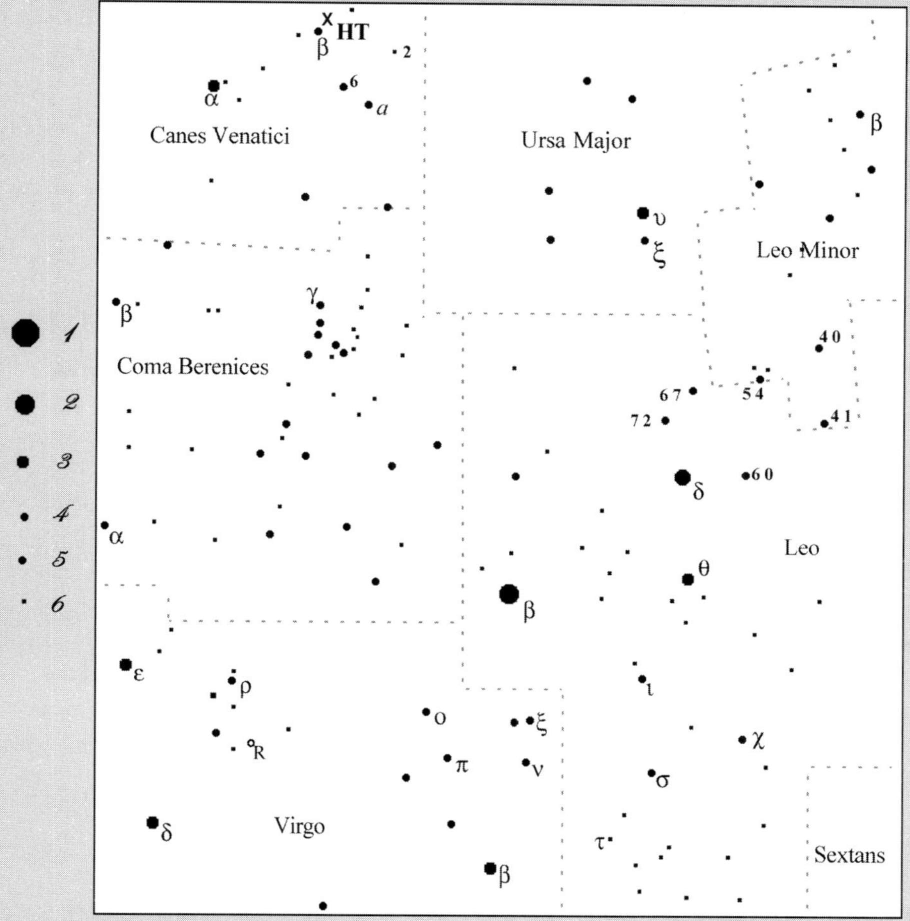

Tirion: Chart 7

Uranometria: Chart 75

63

63

Cocoon Galaxy
NGC 4490
Type: Peculiar Barred Spiral Galaxy (SBdP)
Con: Canes Venatici

RA: 12^h 30.6^m
Dec: +41° 39′
Mag: 9.5 (O'Meara); 9.8
Dim: 5.6′ × 2.8′
SB: 13.0
Dist: 25 million light-years
Disc: William Herschel, 1788

W. HERSCHEL: [Observed January 14, 1788] Very bright, very large and irregularly extended. The northern member [NGC 4485] is bright, pretty small, an irregular figure and is 1′ distant. (H I-198)

NGC: Very bright, very large, much extended toward position angle 130°, [mottled], south following of 2.

CANES VENATICI HARBORS A WEALTH of extragalactic treasures, and NGC 4490 is one of the more prominent ones. This peculiar galaxy lies a mere 40′ northwest of 4th-magnitude Beta (β) Canes Venaticorum (Chara), and it is an easy target for small telescope users. NGC 4490 is a late-type spiral that forms an interacting pair with NGC 4485, which is two magnitudes fainter and only about 3.5′ to the north. Together the pair constitute Arp 269.

NGC 4490's discoverer, William Herschel, treated NGC 4490 and NGC 4485 as one object with two parts, which he cataloged as H I-197 (i.e. NGC 4485) and H I-198 (i.e. NGC 4490). But NGC 4490 is our hidden treasure, since NGC 4485 is two magnitudes fainter. But one cannot look at NGC 4490 without at least trying to see NGC 4485, so both objects are discussed here.

NGC 4490 and NGC 4485 are remarkable in that they are imbedded in a very large, symmetrical envelope of neutral hydrogen gas that, apparently, has little to do with the tidal interactions shared by these two galaxies. Instead, as M. S. Clemens, P. Alexander, and Dave A. Green suggest in a 1999 *Monthly Notices* of the Royal Astronomical Society, this cloud can be explained by the energy input to the interstellar medium by supernovae. Only a few percent of the mechanical energy output of each supernova is required.

In a 2002 *Monthly Notices* of the Royal Astronomical Society, UK astronomers Marcel Clemens and Paul Alexander (Cavendish Laboratory, Cambridge) report finding an outflow of hydrogen-alpha emission and a radio filament out to a distance of about 10,000 light-years. They interpret this as an outflow of ionized gas that is related to the neutral hydrogen envelope, which they traced out to a distance of about 100,000 light-years. The neutral hydrogen envelope then, they argue, can be explained if the entire disk of NGC 4490 has been undergoing a high, ongoing star formation at a constant rate for at least 100 million years. Combined with other evidence Clemens and his colleagues suggest that NGC 4490 is a young galaxy with an age of approximately 2 billion years, and that it has had an approximately constant star-formation rate during this period equal to the current rate of about 4.7 solar masses per year. They also note that the effects of the interaction between the two galaxies have not had sufficient time to affect the star-formation rate in NGC 4490 significantly, although NGC 4485, which is about eight times less massive, has been severely affected by the interaction. In fact, most star formation is occurring *between* the two galaxies. Radio observations show a ridge of neutral hydrogen gas extending from NGC 4490 toward NGC 4485; this feature coincides with a previously known chain of giant H-II regions.

One of the most striking results from the Infrared Astronomical Satellite (IRAS) survey was that interacting galaxies appear commonly to show significantly stronger far-infrared emission, presumably from star formation, than do similar non-interacting galaxies. Indeed, it has been suggested that "ultraluminous" galaxies and quasars might be the result of a collision between a pair of gas-rich galaxies.

Interestingly, the nineteenth-century British observer, Rev. T. W. Webb described NGC 4490 as a "[l]ong, easily resolvable" nebula. Indeed, the galaxy's size, shape, and brightness rival those of some of the more fashionable Messier galaxies in neighboring Coma Berenices, such as M88, M99, and M100. At a distance of 25 million light-years, however, NGC 4490 is half as far away as they are and, with a diameter of 35,500 light-years, less than half as big.

In high-resolution images taken with large telescopes, we see that NGC 4490 is inclined 65° from face-on and is a most fascinating system with a ring-tail core wrapped in a cocoon of stellar filigree and H-II regions. The galaxy's northwest end looks warped, and it is here that bizarre drips of starlight make a dogleg to the north, to yet another a pool of tortured nebulosity surrounding the galaxy's companion, NGC 4485. In these images NGC 4490 looks like an eviscerated galaxy – one whose entrails have been ghoulishly ripped from the carcass and tossed wildly asunder.

Both components show massive, robust star formation over their faces. Individual stars begin to resolve at about 18th magnitude in each galaxy, but the surface brightness is so high over most of their disks as to make the photometry of individual stars difficult. Also there is much dust over NGC 4490's face.

To find this enigmatic wonder, again, all you have to do is look 40′ northwest of 4th-magnitude Beta Canes Venaticorum

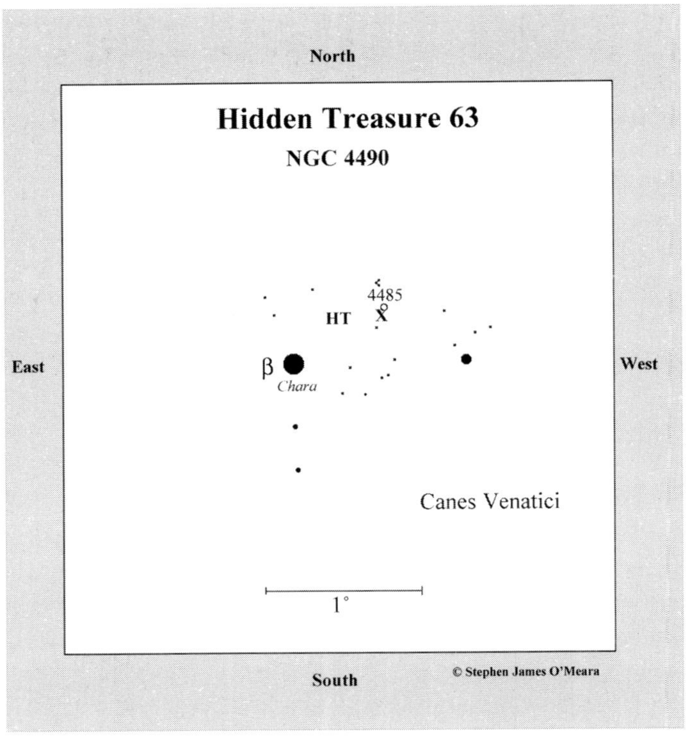

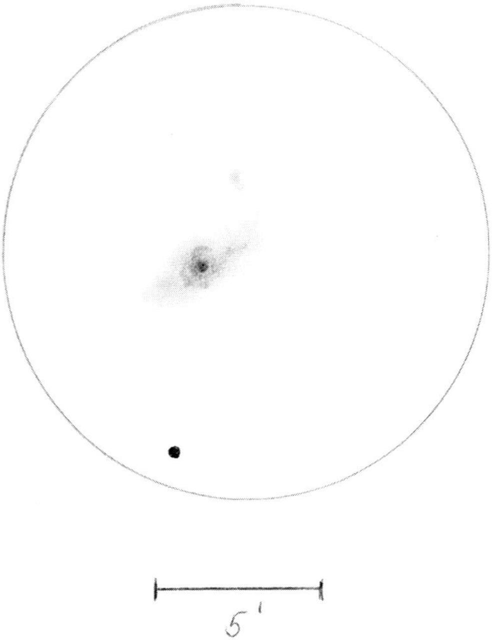

(Chara). From a dark sky, it is visible in 7 × 50 binoculars as a small dim glow. My antique telescope shows it as a larger, though ill-defined haze. Through the 4-inch at 23×, NGC 4490 is a flying-saucer-shaped lens of light with faint extensions. It is a very condensed, extremely distinct glow, so observers under suburban skies should have no problem with this one. With averted vision the galaxy looks mottled, especially if I give myself time. I can also see the dim companion, NGC 4485.

The companion really stands out at 72×. But it is NGC 4490 that demands attention. It is simply bright and beautiful. Look for a faint starlike core in a bright round core that's imbedded in a cocoon of slightly fainter light. With averted vision, the galaxy's inner lens fragments into quarters, with enhancements at the four cardinal directions. Increasing the power to 101×, transforms this cross of light into an extremely subtle S-shaped spiral pattern, but this is highly suggestive. Then again, this S-shaped pattern is real. As reported in Volume 4 of the Webb Society's *Deep-Sky Observer's Handbook,* Ron Buta used a 36-inch reflector to observe NGC 4490, noting that "in the centre a small, two-armed spiral pattern is suspected, apparently winding from a diffuse, non-circular central region; there is no nucleus within this area; the two arms are very short and fade rapidly into the surrounding nebulosity; the S-shape of the galaxy appears not to be related to this spiral structure." With averted vision and time, I can just see the northwestern segment of the galaxy curving toward NGC 4485 – a tidal warp.

The core is so bright that it takes magnification quite well. At 303×, the core appears as a bright star surrounded by waifs of light, each a nucleus within a curling wisp of nebulosity of what appears to be a part of the galaxy's spiral structure. In a fanciful sense, the nucleus seems to ignite with fairy dust. Luginbuhl and Skiff also noted this appearance saying, "there is a small central bulge with bright dots along the major axis near the center." They also noticed a dark band passing along the north side of the nucleus in a 12-inch telescope. Also in the 4-inch at 303×, NGC 4485 shows some minute structure, namely the otherwise circular spot becomes slightly elongated. With much attention I see a bright core and a dim patch to the south.

Hidden Treasure 64
IC 3568

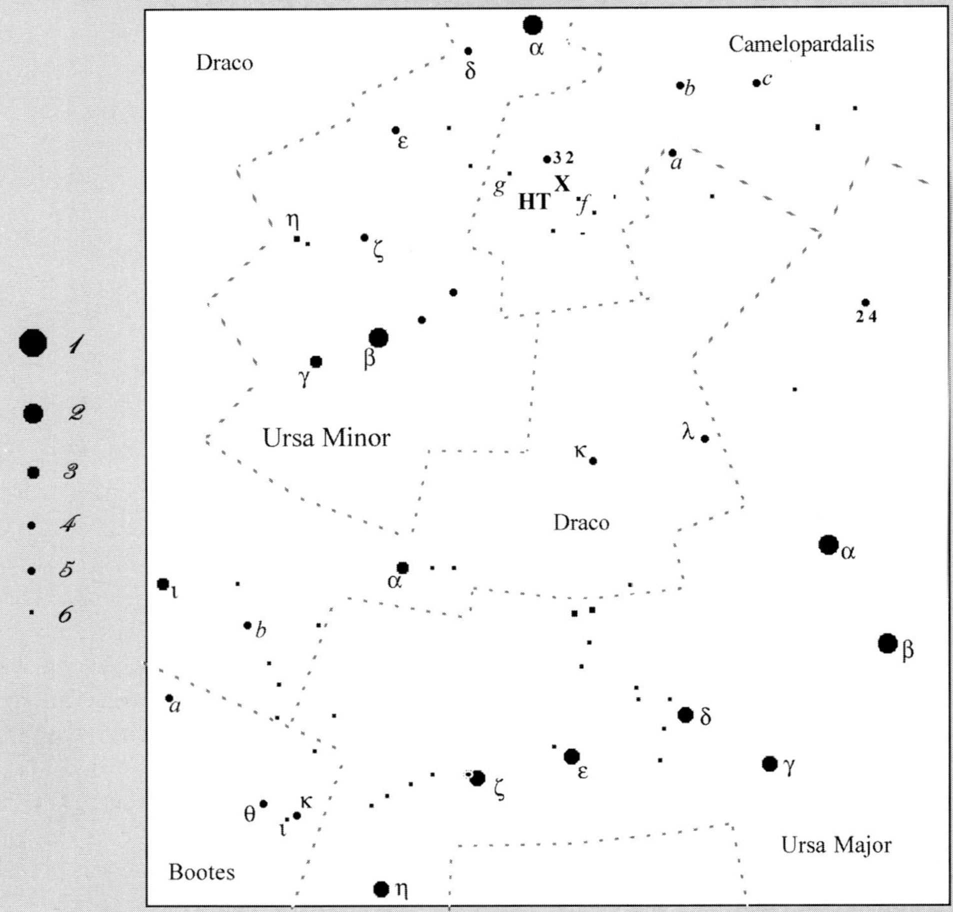

© Stephen James O'Meara

Tirion: Chart 2

Uranometria: Chart 9

64

Theoretician's Planetary, Sliced Lime Nebula
IC 3568
Type: Planetary Nebula
Con: Camelopardalis

RA: $12^h\ 33^m\ 06.9^s$
Dec: $+82°\ 33'\ 49''$
Mag: 10.3 (O'Meara); 10.6
Dim: $18'' \times 18''$
Dist: 3,300 light-years
Disc: Robert Grant Aitken, 1900

HERSCHEL: None.

IC: Planetary or nebulous star of magnitude 9.5; magnitude 13 star preceding 15″.

"THERE'S A SAYING AMONG PROSPECTORS: 'Go out looking for one thing, and that's all you'll ever find'." I love these words by legendary documentary filmmaker Robert Flaherty (1884–1951), because they apply to prospectors of all kinds, including us celestial treasure hunters. Consider the relatively recent discovery of McNeil's Nebula. On January 23, 2004, Jay McNeil of West Paducah, Kentucky, discovered a new nebula in Orion. He found the diminutive 1′-diameter nebula (which shines at 15th magnitude) on a charge coupled device (CCD) image of M78; he took the discovery image from his backyard with a 3-inch refractor. But as McNeil stresses, "It wasn't me, 'an imager,' that made the discovery. It was me, 'an observer.' It was 20 years behind the eyepiece. The new nebula was much brighter than any of the other objects I had observed visually around M78."

McNeil's discovery was not "accidental." It demanded a certain level of knowledge and visual perspicacity. Indeed, between the time McNeil's Nebula turned on (some time between October 28 and November 15, 2003 – owing to a flare up of its illuminating star) more than 20 astrophotographers had imaged M78 and its surroundings; but only McNeil's acumen noticed the new nebula.

That same visual acumen was at work when Robert Grant Aitken (1864–1951) discovered our next hidden treasure, the planetary nebula IC 3568. Aitken, who worked at Lick Observatory from 1895 to 1935, specialized in observations of double stars, discovering 3,100 new pairs. He also measured positions of comets, and that's how he came to discover his one and only nebula. On the night of August 31, 1900, Aitken was using the observatory's 12-Clark-Refractor to examine Comet Borrelly–Brooks, when he noticed that the star BD +83° 357′ in Camelopardalis was surrounded by a small circular nebula. He confirmed the nebula's appearance with the 36-inch refractor the next night. In 1918, Heber Curtis described its photographic appearance as a "perfectly round disk fading out a little at the edges." Direct photographs of it taken by R. Minkowski with the 200-inch reflector showed it to consist of a small bright ring with an outer envelope that faded away smoothly.

In a 1969 *Astrophysical Journal* article, Paul Lee (Louisiana State University), Lawrence Aller (University of California, Los Angeles), and their colleagues noted that the symmetrical appearance of this nebula makes it the closest approximation to the "oft-mentioned theoretician's planetary." In other words, as Bruce Balick (University of Washington, Seattle) explained in 1996, "If IC 3568 did not exist, it might have been created by theoreticians," thus the planetary's nickname. Ironically, in 1973, the object was misclassified as a compact galaxy (UGC 7731) in the 1973 *Uppsala General Catalogue.*

No matter, to this day IC 3568 is considered one of the most circularly symmetric of all planetary nebulae. It is also one of the morphologically simplest planetaries in the sky. The nebula, which measures 0.3 light-years across in true physical extent, was imaged in stunning clarity by the Hubble Space Telescope. Here we see the nebula's magnitude 12.3 central star surrounded by a bright inner sphere with smooth edges and a complex internal structure – namely a series of symmetrical linear features radiating outward from the central star. A very smooth, faint halo surrounds this bright region. It appears slightly brighter in the middle, and gradually fainter toward the edges. It is expanding at about 10 kilometers per second. IC 3568's simple structure conforms well to the interactive wind theory (see Hidden Treasure 16).

Finding this "simple" planetary is, well, relatively simple. It lies only about 8° from Alpha (α) Ursae Minoris (Polaris), near the circular arc of the Little Dipper's handle. You need to first locate 6th-magnitude 32 Camelopardalis, which forms the western apex of a 6°-wide isosceles triangle with the 4th-magnitude stars Delta (δ) and Epsilon (ε) UMin in the Little Dipper's handle. It also marks the northern apex of a roughly 2°-wide acute triangle with the 6th-magnitude Stars *f* and *g*. The planetary lies 1° southwest of 32 Cam, midway between 32 Cam and Star *f*. It is also the northern "star" of two separated by a few minutes of arc.

At 23×, IC 3568 looks like a close 10th-magnitude double star. The planetary, again, is the north one of the pair. IC 3568 is extremely stellar at low power. If you believe you see a fuzziness, it might be because the planetary and its neighboring star are so close together, it gives the false impression

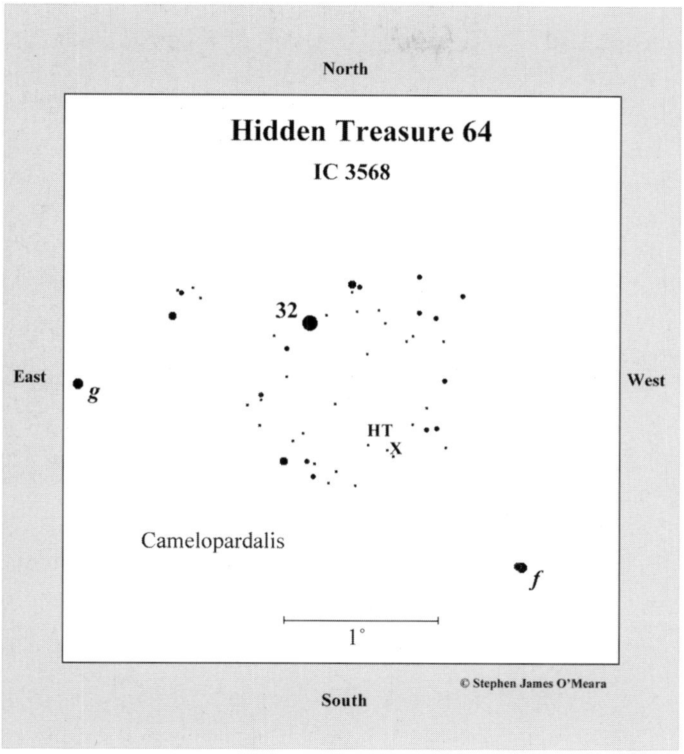

10″ diameter across, and the outer halo is nearly 20″ across. These are small dimensions; M57, the Ring Nebula in Lyra, is about four times as large. For this reason, and to quote Walter Scott Houston, "it is easily passed over in a hasty sweep with low magnification," so you do have to be careful. "Once spotted," he continues, "it resembles a ghostly green Mars at an average opposition distance."

Although IC 3568 is very small, it has a very high surface brightness – thanks to the bright inner core, which is what we first detect in our telescopes. It is one of the most stellar planetaries of a halo of light. In other words, IC 3568 may look fuzzy at low power for the same reason the double star M40 in Ursa Major looks fuzzy – it's an illusion. You really need magnification to see this planetary as a planetary. To confirm the planetary, I go directly to 101×; at this power IC 3568 appears as a very obvious swollen star, especially with averted vision. It also has an aqua tint. If I then drop the power to 72×, I can still detect the planetary with averted vision; it looks slightly bloated.

At 182× some detail is suspected – dim structures, perhaps shells. One very steady night, I increased the magnification to 504× and saw the planetary clearly with its very simple three-tier structure: a bright central star, bright inner halo, and diffuse outer halo. The inner halo measures roughly

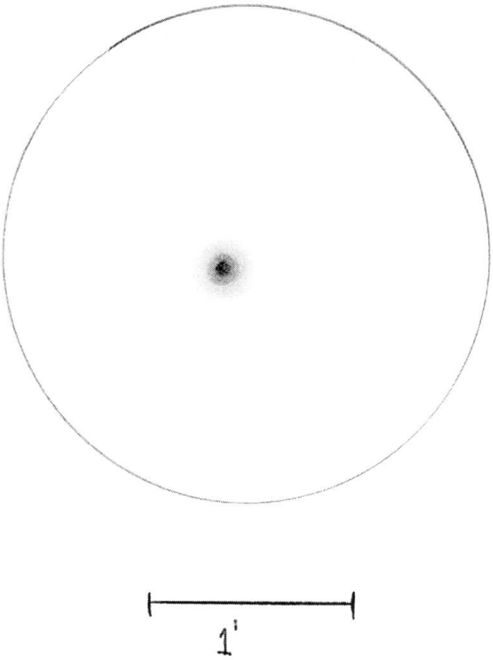

visible in small telescopes. How does it appear in larger telescopes? French amateur Yann Pothier (Chateau Renard Observatory) observed IC 3568 with a 17.5-inch f/4.5 reflector at 400×, describing it as "small, round and bright [with] fuzzy edges and brighter center. Ultra-high-contrast and oxygen-III filters give a good contrast gain and hydrogen-beta is of no use, except . . . that it renders the nebula more homogeneous (maybe by enhancing slightly the outer parts)." He also saw a magnitude 14 star just south of the nebula's west edge and a magnitude 11 star 1.5′ to the south. So while small-telescope users have to push their scopes to the limit of achievable magnification, they can, with patience, see just about as much detail as someone using a much larger telescope, though the observers with larger scopes get a much more relaxed view of this simple beauty.

Hidden Treasure 65

NGC 4526

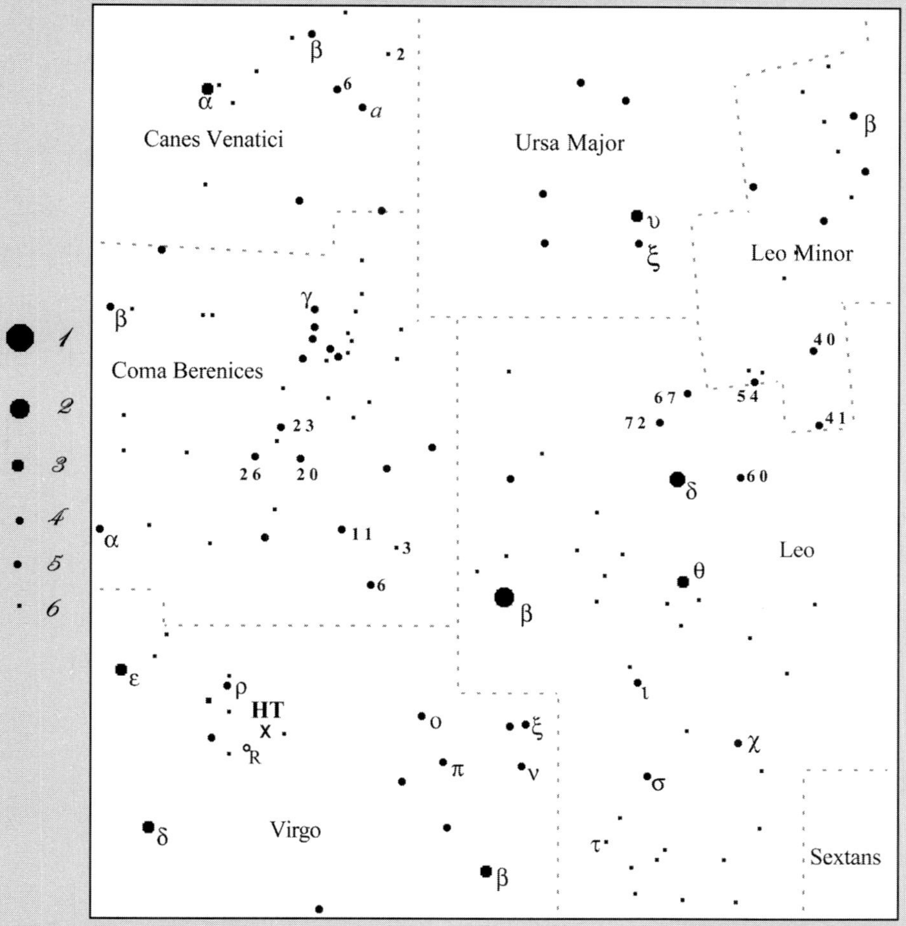

© Stephen James O'Meara

Tirion: Charts 13 & 14

Uranometria: Charts 193 & 194

65

Hairy Eyebrow Galaxy
NGC 4526
Type: Mixed Lenticular Galaxy (SAB0)
Con: Virgo

RA: 12^h 34.0^m
Dec: $+07^\circ$ 42′
Mag: 9.9 (O'Meara); 9.7
Dim: 7.4′ × 2.7′
SB: 12.8
Dist: 40 million light-years
Disc: William Herschel, 1784

W. HERSCHEL: [Observed April 18, 1784] Bright, very large, much extended, much brighter in the middle. (H I-31 = H I-38)

NGC: Very bright, very large, much extended toward roughly position angle 120°, pretty suddenly much brighter in the middle, between 2 magnitude 7 stars.

NGC 4526 IS ONE OF THE COUNTLESS galaxies swarming like bees around a disturbed hive between the Lion's Tail and the Virgin's waist. Seen projected on the sky, it is part of the Southern Extension (a subcluster of the great Virgo Cloud of Galaxies) and lies very close to the magnificent elliptical system M49. Indeed in his *Nearby Galaxies Catalog*, R. Brent Tully shows NGC 4526 and M49 being at the same distance: 55 million light-years. Recent studies, however, suggest that NGC 4526 may be a foreground system.

In a 1999 *Astronomy and Astrophysics* paper, German astronomers Georg Drenkhan (Sternwarte der Universitat Bonn) and Ton Richtier (Max-Planck-Institute for Astrophysics) describe how they used Hubble Space Telescope (HST) data to identify globular clusters within the galaxy. By calibrating and comparing the luminosities of these clusters with others of known distances, the team determined that the clusters in NGC 4526 lie about 40 million light-years distant, placing the host galaxy a substantial 15 million light-years closer to us than the Virgo Cloud.

This finding follows the 1994 discovery of a supernova eruption in NGC 4526. Supernova 1994d was a bright Type Ia event 9″ west and 7″ north of the galaxy's nucleus. Type Ia supernovae are believed to occur in a binary star system, where a star of an undetermined nature is accreting matter onto a white dwarf. So much mass piles up on the dwarf that its core reaches a critical density of 2×10^9 grams per cubic centimeter – enough to result in an uncontrolled fusion of carbon and oxygen, thus detonating the star. Supernova 1994d erupted near the edge of the central disk, just beyond or north of its prominent dust lanes. This region is dominated by diffuse starlight, which is quite red. The color implies the presence of *K*-type stars, presumably giants. It is possible then that the white dwarf companion that contributed to Supernova 1994d was such a *K*-type giant.

At the time of discovery (March 7, 1994, Universal Time) Supernova 1994d shined at magnitude 15.2. It achieved a maximum brightness of magnitude 11.8 and was captured in outburst with the HST; since Type Ia supernovae are the best standard candles in the universe, observing the event's rise and fall in brightness, as well as its color, can help astronomers to plumb the depths of the universe. If all Type Ia supernovae have the same intrinsic brightness, then the brighter a supernova appears, the closer it must be, and vice versa. Observations of several other extragalactic Type Ia supernovae have proven the case. But the initial studies of Supernova 1994d showed it to be anonymously bright for a Type Ia event. Drenkhan and Richtier argue that's because astronomers used the fact that the galaxy is a member of the Virgo Cluster. When Drenkhan and Richtier analyzed the light curve of Supernova 1994d using their newly derived distance of 40 million light-years, they found it to be a bright but normal Type Ia supernova. Based on their studies, I have adopted a distance of 40 mil-

lion light-years for NGC 4625 in the table above.

Observations near the location of SN1994d in 1995 with the William Herschel Telescope at La Palma, and 21-centimeter spectra obtained with the 100-meter Effelberg Radiotelescope, indicate calcium II and sodium absorption produced in interstellar gas in this region. This is the first detection of interstellar absorption originating in a galaxy of early morphological type. The gas appears to have near-solar abundances with very low dust content and is consistent with galactic models in which hot gas is expelled into the outer galactic halo and then subsequently cools.

In images taken by Edwin Hubble in the 1960s, the galaxy appears as a fuzzy elliptical object with a boxy bulge (slightly asymmetric), a diffuse central lens with a hint of spiral structure and a whisker of darkness along its major axis. Observations with the 60-inch telescope at Mt. Wilson show the presence of an internal ring of dust that is buried deep within the main body of the galaxy. This internal structure is an obscuration of material rather than a bright spiral arm of the galaxy. The outer disk is nothing but a highly amorphous ellipse.

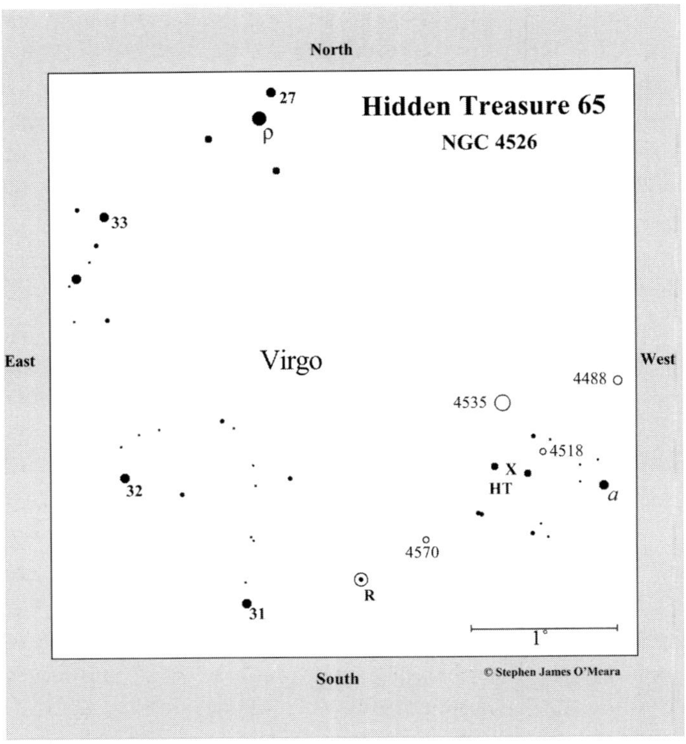

New images from the HST reveal that boxy inner lens as an amazingly dusty spiral with an inner bar seen 16° from edge-on. The dense dust forms a bushy brow (like a hairy eyebrow, and hence my name for the galaxy) on the near side of the lens, while a diffuse inner lens obscures much of the obscuring matter on the more distant flank. The core of the galaxy shines with a pale misty blue light, which forms a semi-transparent dome around a small but eerily bright and diffuse nucleus. While the inner dust lane is parallel to the major axis, extending out to about 20″ from the nucleus, the overall appearance of the dust is that of a slightly warped disk, as if someone had placed a frisbee on a hot stove and it started to warp. If we accept a distance of 40 million light-years, NGC 4526 diminishes in linear extent from 94,200 light-years to about 68,500 light-years.

To find NGC 4526, start at 3rd-magnitude Epsilon (ϵ) Virginis. Nearly 5° to its west-southwest is 5th-magnitude Rho (ρ) Vir. (Rho Vir is easy to identify, since it lies inside a 30′-wide near-equilateral triangle of 6th- and 7th-magnitude suns.) Now use your naked eye or binoculars to star hop along a 3°-wide C-shaped asterism formed by the 5th- and 6th-magnitude stars 33, 32, and 31 Virginis, the variable R Virginis – a Mira-type star that ranges in brightness from magnitude 6.1 to 12.1 every 146 days – and Star *a*. Center Star *a* in your telescope. The 8th-magnitude galaxy M49 should be 35′ to the northwest, and a little more than 30′ to the east-northeast will be a pair of 7th-magnitude stars separated by 15′. NGC 4526 lies midway between these two 7th-magnitude suns.

Under a dark sky, NGC 4526 is visible in 7 × 50 binoculars with averted vision. I see it best if I look at a point midway between M49 and NGC 4526 but focus my attention on NGC 4526. It's easier to see in my antique telescope, which better separates the two 7th-magnitude stars bracketing the galaxy. At 23× in the 4-inch, the galaxy is a very condensed fuzzy glow between the two studlike 7th-magnitude stars. You should immediately see a bright core with a starlike nucleus. Rather than the boxy appearance displayed in photographs, I see the bulge at low power as being very round.

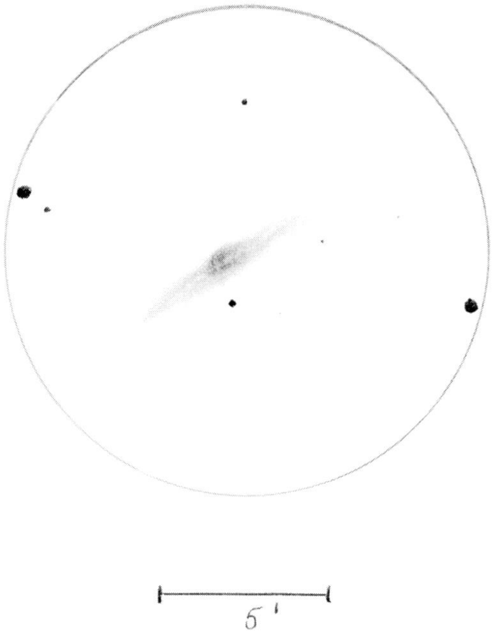

With averted vision I can see the extended lens oriented west-northwest–east-southeast; if I use direct vision, the dim ellipse vanishes. Increasing the power to 72× enhances the central lens, causing the dim elliptical outer regions to look like two spheres, one on either side of the central lens. The view is reminiscent of Galileo's first impression of Saturn with its nearly edge-on rings in 1610: "Saturn," he wrote, "not one alone, but is composed of three, which almost touch one another and never move or change with respect to one another . . . the middle one is about three times the size of the lateral ones."

A magnitude 12.5 star lies just few minutes due south of the nucleus. At 101×, I can begin to see what appears to be hints of a dust lane or lanes. It seems to run the length of the galaxy (though this may be an illusion). The northwest end of the galaxy is somewhat twisted – either that or the dust lane is twisted.

Aside from M49, three other galaxies are visible nearby: the 10th-magnitude spiral NGC 4535 is 30′ due north of NGC 4526, while the 12th-magnitude barred spiral galaxy NGC 4488 lies about 1° to the northwest. More challenging is NGC 4518 A+B, a pair of tiny interacting galaxies; although their combined light shines around 13th magnitude, the pair can be seen as a single object in a 4-inch telescope with magnification.

By the way, you should periodically check NGC 4526 for supernovae. Aside from Supernova 1994d, this galaxy hosted Supernova 1969e, a 16th-magnitude object at discovery, which appeared 11″ west and 27″ south of the nucleus.

Hidden Treasure 66

NGC 4605

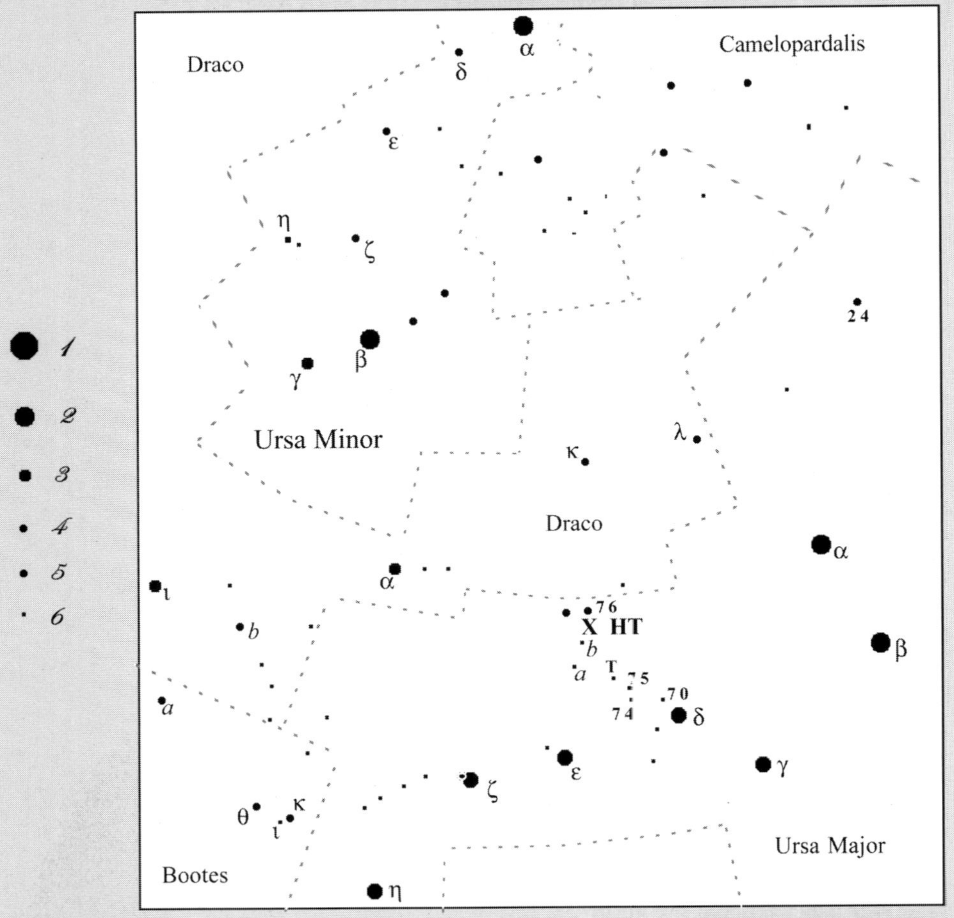

© Stephen James O'Meara

Tirion: Chart 2

Uranometria: Charts 25 & 26

66

Faberge Egg, Frankenstein Galaxy
NGC 4605
Type: Dwarf Barred Spiral Galaxy (SBc Pec)
Con: Ursa Major

RA: $12^h\ 40.0^m$
Dec: $+61°\ 36'$
Mag: 10.3 (O'Meara); 10.3
Dim: $5.7' \times 2.5'$
SB: 13.1
Dist: 13 million light-years
Disc: William Herschel, 1790

W. HERSCHEL: [Observed March 19, 1790] Extremely bright, extended in the direction of the parallel of declination, 5′ in length, all over equally bright except just on the edges. (H I-254)

NGC: Very bright, very large, much extended toward roughly position angle 120°, pretty suddenly much brighter in the middle, between 2 magnitude 7 stars.

NGC 4605 IS A PECULIAR DWARF galaxy seen only 21° from edge-on. It is situated about $5\frac{1}{2}°$ northeast of Delta (δ) Ursae Majoris (Megrez) in the Big Dipper's Bowl. The galaxy received some incidental recognition after January 1996, when the public first saw the now-famous Hubble Deep Field – a 10-day-long Hubble Space Telescope exposure revealing at least 1,500 new galaxies, some as faint as 30th magnitude, in an area of sky only 2′ across. The Deep Field lies only about 42′ to the northwest of NGC 4605, which often appeared in photographs and finder charts that were published to help amateurs locate the tiny sky patch scrutinized by Hubble.

Despite NGC 4605's proximity to the Dipper's Bowl, amateurs do not commonly target the galaxy, probably because it has tough competition from the better known Ursa Major galaxies M51, M101, and M108. Yet NGC 4605 rivals M108 in brightness, apparent size, shape, and level of detail it presents at the eyepiece.

But modern amateurs are not the only ones who failed to appreciate this hidden

treasure. The nineteenth-century observers Rev. Webb and Adm. William Henry Smyth do not mention it in their popular works – *Celestial Objects for Common Telescopes* and *Cycle of Celestial Objects*, respectively. Even *Sky & Telescope*'s late Deep-Sky Wonders columnist, Walter Scott Houston, did not spend much time discussing it, even though he said it was "surprisingly easy in average skies" and "obvious" in 65-mm binoculars.

Finding NGC 4605 is not difficult. Houston noted that it lies nearly on the extension of a line joining Gamma (γ) and Delta (δ) Ursae Majoris. If you want to star hop to NGC 4605 from Delta, it's quite an adventure, requiring several 1° hops. First head for 5th-magnitude 70 UMa, about 1° to the northeast. You'll want to pause here, because $\frac{1}{4}$° further on in the same direction is the famous double star Winnecke 4 (M40). Next hop about 1° to the east-northeast to 5th-magnitude 74 UMa, then 20′ north to 6th-magnitude 75 UMa. A little more than 1° to the northeast is the red Mira-type variable T UMa, which varies in brightness from magnitude 6.8 to 13.4 with a period of 267 days; if T UMa is at minimum you'll have to look $\frac{1}{2}$° further to the east-northeast for an arc of three 8th- to 9th-magnitude suns separated by 20′ and oriented roughly north–south. From this arc, move $1\frac{1}{4}$° to the east-northeast to a solitary 6th-magnitude star (*a*). It's time to change direction. From Star *a,* move 1° to the northwest to another 6th-magnitude star (*b*). Note that just 7′ to its southeast is yet another Mira-type variable: S UMa, which varies in brightness from magnitude 7.7 to 12.4 every 226 days. NGC 4605 is only 35′ northwest of Star *b.*

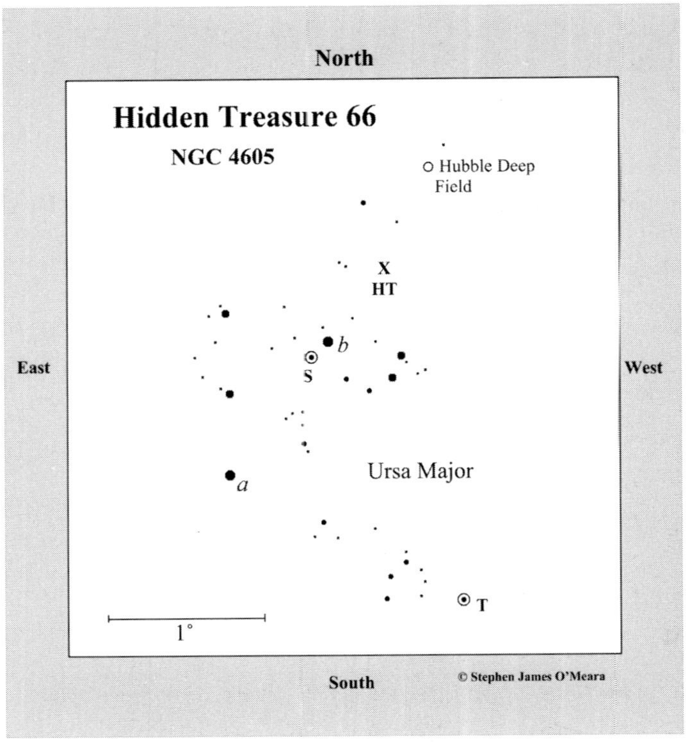

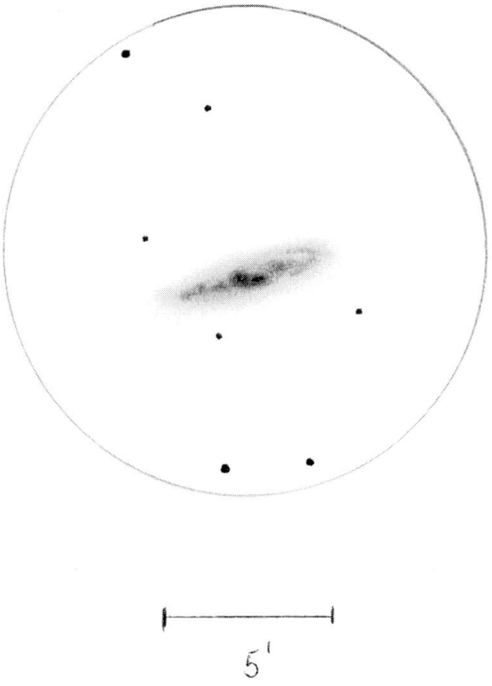

If you are not up for such a long adventure and observe under a dark sky, try eyeballing Star *a* first, or find it in binoculars, then center it in your telescope. Note that Star *a* forms the northern apex of a near equilateral triangle with Delta and Epsilon (ε) UMa. Once you center Star *a*, just follow the directions above to NGC 4605, which requires only two simple hops.

Luginbuhl and Skiff note that NGC 4605 is just visible in a $2\frac{1}{2}$-inch telescope as a small, slightly elongated patch. Under dark skies, NGC 4605 is visible in 7 × 50 binoculars; seeing it requires a little patience and time before the galaxy pops into view. My antique telescope shows it as a slightly fuzzy "egg" oriented northwest–southeast. Through the 4-inch at 23×, NGC 4605 (whose true linear diameter is a modest 22,000 light-years) appears very bright but small. With averted vision the disk is clearly oval. With direct vision, only the bright core emerges from the sky background. If I gently sweep the galaxy back and forth across my retina and fovea, the core becomes more elongated, very slender in parts, and looks suspiciously beaded.

At 72×, the galaxy sports a warped bar, oriented northwest–southeast, surrounded by a spindle of light. Its disk displays a mottled texture and has a large and obvious patch of dust in its southeastern quadrant. Long-exposure photographs reveal the northwestern side to be visually thicker but of a lower surface brightness; it is this side of the galaxy that shows a multiple arm pattern, which appears to be one spiral arm opening up. The images also reveal a larger envelope surrounding the spindle, several H-II regions, and what appears to be a tiny galaxy protruding from NGC 4605's southeastern tip. Try as I may, I could not see this tiny pip of fuzzy light.

You must take your time with this galaxy because it has a wealth of fine details. With its slender waves of light, and star-studded face, it looks like a translucent Faberge Egg with enameling. Peter Carl Faberge included *guilloche,* a surface treatment that could make waves and striations in the design. He also used round cut precious gems – sapphires, rubies, and emeralds for decoration. Now look at NGC 4605 with that in mind. When I study the galaxy with magnifications varying between 72× and 101×, and spend several minutes viewing with each change in magnification, I can trace three main components to the galaxy. First, the galaxy's "core" is a nebulous stellar knot that appears to be northwest of center; it burns forth like candlelight seen through gauze. Second, the northwest section of the galaxy is a beaded whisker of light. And third, the southeast section of the galaxy looks like an attached miniature of barred spiral galaxy with poorly organized structure; there appears to be a central hub with one spiral arm extending to the northwest before it curls to the north, and another longer and broader arm extending to the southeast before it curls to the south. Seen in this way, the galaxy looks like a mix between NGC 247 (Caldwell 62), a barred spiral galaxy in Cetus, and M82, an irregular galaxy in Ursa Major. In fact, considering the galaxy looks like it has been badly pieced together with different body parts, I've also coined it the Frankenstein Galaxy.

At 168×, the galaxy is still a marvel. The northwest whisker has a wiggle to it, just where three dim beads form a little arc near the core. The circular core region itself breaks down into finer details at this power. Aside from the central fuzzy bead, a dimmer one can be seen to the northwest, where it is separated from the whisker by a dark gap. The core is also fanned and serrated to the north, as if a multitude of supernovae had gone off in concert, and blown matter asunder. This fan, it turns out, is the north extension of the spiral arm mentioned earlier. A dark wedge separates the core from the rest of the southeastern component, which appears as a tiny fleck of light with a long tail that gracefully curls to the south before heading northwest. The disk is dark in between. A study in 1998 revealed no nuclear component that was visible.

Now for the final curve ball. Although NGC 4605 is located in Ursa Major, it belongs to the Coma–Sculptor Cloud of Galaxies, which includes another dwarf, NGC 404 (Hidden Treasure 5). NGC 4605 is receding from us at 140 kilometers per second.

Hidden Treasure 67

NGC 4656

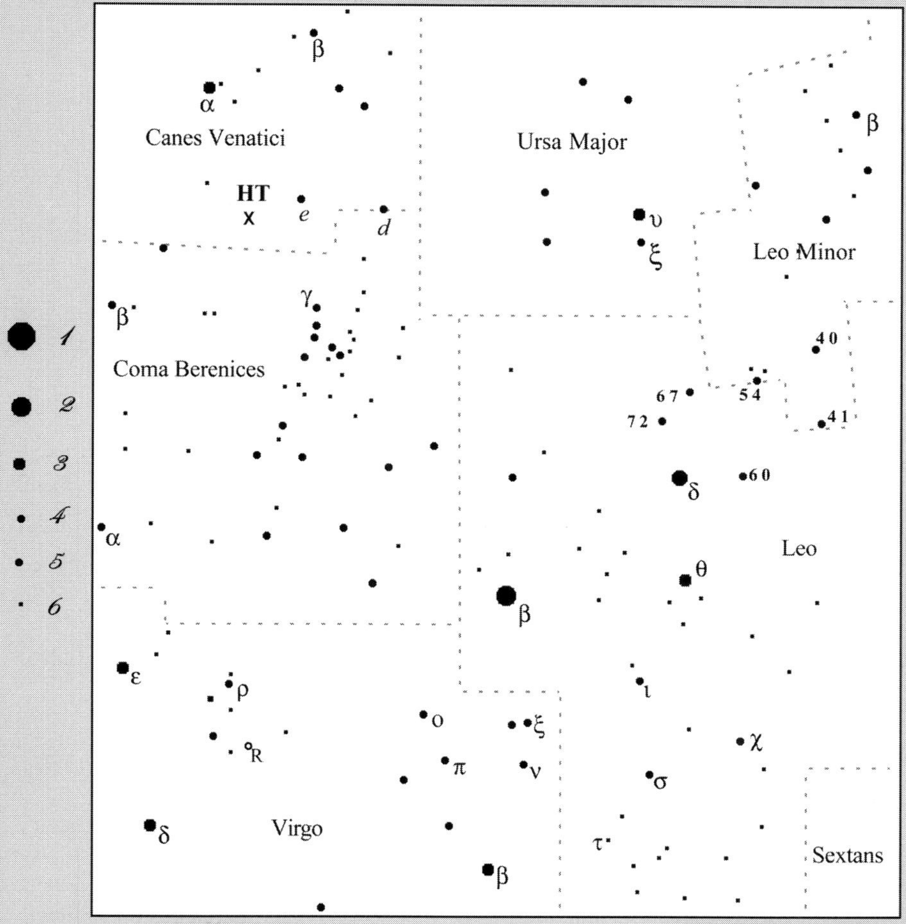

© Stephen James O'Meara

Tirion: Chart 7

Uranometria: Chart 108

67

"Messier's" Hockey Stick, The Hook, The Hummingbird
NGC 4656
Type: Barred Spiral Galaxy (SBm Pec)
Con: Canes Venatici

RA: 12^h 44.0^m
Dec: $+32°$ $10'$
Mag: 10.5
Dim: $18.8' \times 3.2'$
SB: 14.8
Dist: 23.5 million light-years
Disc: William Herschel, 1787

W. HERSCHEL: [Observed March 17, 1787]. Two. The following considerably bright. Extended, much brighter in the middle. The north pretty bright, extended, south preceding north following. Both joined and form the letter S. (H I-176 & 177)

NGC: Remarkable, pretty bright, large, very much extended toward position angle 34°, south preceding of two.

LARRY MITCHELL OF THE HOUSTON Astronomical Society first introduced me to NGC 4656 during the 1998 Texas Star Party. He calls it the Hockey Stick Galaxy, while his observing compatriot Barbara Wilson refers to it as The Hook. After seeing this galaxy through their large reflectors, I understand both monikers. But I have made one revision. I have since added the name Messier to Larry's Hockey Stick. Charles Messier had nothing to do with the discovery of NGC 4656 but he did discover deep-sky objects. No, the name Messier in this case refers to Canadian-born hockey star Mark Messier, who, among other triumphs, led the New York Rangers to win the 1994 Stanley Cup Championship. Messier also just happens to have a very appropriate last name to be tagged onto a celestial hockey stick.

NGC 4656 is an irregular barred spiral galaxy, similar in type to the Small Magellanic Cloud in Tucana and NGC 55 (Caldwell 72) in Sculptor. We see NGC 4656 merely 10° from edge-on, and it has a warped

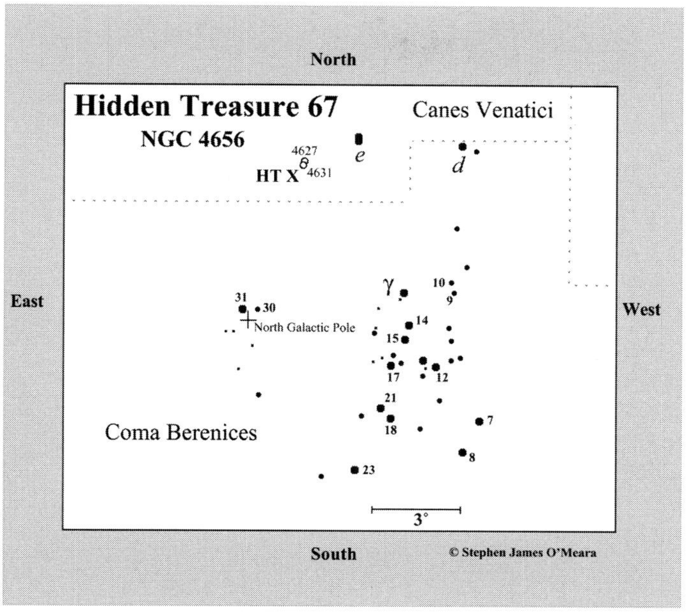

appendage on the northeastern end. Radio observations have also shown a possible supernova remnant akin to the Crab Nebula (M1) located near the southern edge of the galaxy, just off the edge of a giant molecular cloud.

NGC 4656 is a large system, with a physical diameter of 91,000 light-years. It belongs to the Coma–Sculptor Cloud of Galaxies and is receding from us at about 650 kilometers per second. A glance at just about any photograph of the galaxy reveals it to be highly asymmetrical in brightness, with the northeast half much brighter than the southwest half. In long-exposure images the galaxy resolves into a smattering of stars and H-II regions, the brightest of which lie approximately 18″ west of the apparent nucleus. Indeed, the galaxy is very active in star formation, but with most of the current activity concentrated to the nuclear region and the warped appendage.

Known as NGC 4657, the appendage juts out at right angles to the main shaft like the hooked blade of a hockey stick. Although William Herschel considered NGC 4657 a separate object, it is not. In fact, it is a part of NGC 4656. The most likely explanation for NGC 4656's distorted appearance is a tidal interaction between it and NGC 4631, which is virtually at the same distance. In his 1994 *Carnegie Atlas of Galaxies,* Allan R. Sandage writes that the pair's atomic-hydrogen gas has provided evidence of a tidal encounter between the two galaxies. The two are 145,000 light-years apart on the plane of the sky. Wilson likens these interacting spirals to "two ships nearly crashing in the night, both listing from the encounter, with [NGC] 4656 definitely the loser, but [NGC] 4631 . . . not unscathed."

Given the pirate theme of this book, I would add that, in images, NGC 4656 appears to be the victim of pirate attack. The ship's plundered hull has been blasted asunder. The dim southwestern segment of NGC 4656 is the ship's sunken bow seen through the shallow waters of some treacherous reef. Its battered stern remains helpless high above the waterline. Frothy waves splash against the grounded keel, which we see as the galaxy's turbulent nuclear region.

NGC 4656 is situated $\frac{1}{2}^{\circ}$ southeast of NGC 4631 (Caldwell 32). Since NGC 4631 is brighter than NGC 4656 by more than a magnitude, you should set your sights first for NGC 4631. You'll find it about halfway down, and a little southeast of, an imaginary line from Alpha^{1,2} ($\alpha^{1,2}$) Canum

try placing Star *e* just outside the field to the north, then sweeping two fields east. NGC 4631 should slip right in from the east side of your field of view. Again, NGC 4656 is just $\frac{1}{2}^\circ$ to the southeast of NGC 4631.

On clear and transparent evenings, NGC 4656 is visible in the 4-inch at 23× as a thin shaft of light. Observers' impressions of it vary, revealing that its visibility is sensitive to atmospheric conditions and light pollution. While Brian Skiff and Christian Luginbuhl say it is "barely visible" in a $2\frac{1}{2}$-inch telescope, Ernst Hartung, observing from Australia writes that it is "not suited" to small apertures. And while I agree that NGC 4656 will be a wash for even suburban observers, or that it might appear less-than-impressive if poorly positioned in a dark sky, observers

Venaticorum (Cor Caroli, a superb double star) to Gamma (γ) Coma Berenices. I simply use low power and a Tele Vue Qwik Point to aim at this imaginary location. But if your attempts prove unsuccessful, try starting from 4th-magnitude Gamma Com, the northernmost naked-eye star in the Coma Berenices Cluster. Next, look for two fifth-magnitude stars (*d* and *e*) that make an isosceles triangle about 5° to the north. The easternmost star (*e*) is a fine binocular double. NGC 4631 lies almost 2° southeast of Star *e*. If your have only a 1° field of view

under reasonably dark skies shouldn't have difficulty seeing it, especially since its nuclear region is brighter than NGC 4631's.

If you're having trouble pinpointing the galaxy, center NGC 4631 and look southeast of its eastern flank. There you should see a 30′-long chain of five 11th- to 12th-magnitude stars heading toward the southeast. NGC 4656 is immediately southwest of the southernmost star in the chain, oriented northeast–southwest. It is a dim glow at 23×, all the more so when seen next to NGC 4631 in the same field. But if you concentrate with averted vision, the galaxy does not appear uniformly lit. Like M82 in Ursa Major, it appears segmented, like three ghosts walking toward you hand in hand. At 72× and with averted vision, the galaxy's segments are more enhanced. The brightest segment appears just northeast of where it seems the nucleus should be. There is a slightly fainter "oval" of light adjacent to what may be considered its nuclear region; the galaxy is very faint on the southwestern side of the "nuclear region" and looks like a dim tentacle of light. In a fanciful way, the galaxy looks like a squid, and David Levy calls it the Hummingbird; indeed, in photographs, the galaxy looks like the body of a hummingbird hovering over an invisible flower. The hummingbird's beak then would be NGC 4657, the appendage that juts out at right angles to the main shaft like the hooked blade of a hockey stick. I could detect NGC 4657 in the 4-inch at 101× from the 12,000-foot level of Mauna Kea. Mitchell adds that the galaxy's largest H-II region has an angular diameter of 11 arcseconds. "This is large," he says, "and should be easily resolved in modest sized telescopes. Try a nebula filter."

Given its initial apparent faintness, NGC 4656 is truly a remarkable object once you sit down and begin to study it behind the eyepiece, even in a small telescope, under a dark sky. Taking the time to really study this object will help any observer sharpen his or her visual acuity. You just have to dedicate yourself to the task and don't give up. I would also recommend observing this galaxy over the course of several nights. I will guarantee that any observer who does, will walk away from the telescope more gratified each night.

If you are having any trouble believing that NGC 4656 is worth one's while, consider that it is only 0.5-magnitude fainter than M108 in Ursa Major and only 0.4-magnitude fainter than M98 in Coma Berenices. Besides, since NGC 4656 is in the same low-power field of view as NGC 4361, how can anyone *not* want to see it? There is little in the night sky that compares with the sight of two relatively bright, nearly edge-on galaxies in the same field of view. It is quite a remarkable and uncommon view. As Mitchell says, "NGC 4656 is one of my favorites. I never get tired of looking at this object and every time I see it I swear I see something in it I never have before – a truly fascinating stop in the heavens."

Hidden Treasure 68

NGC 4699

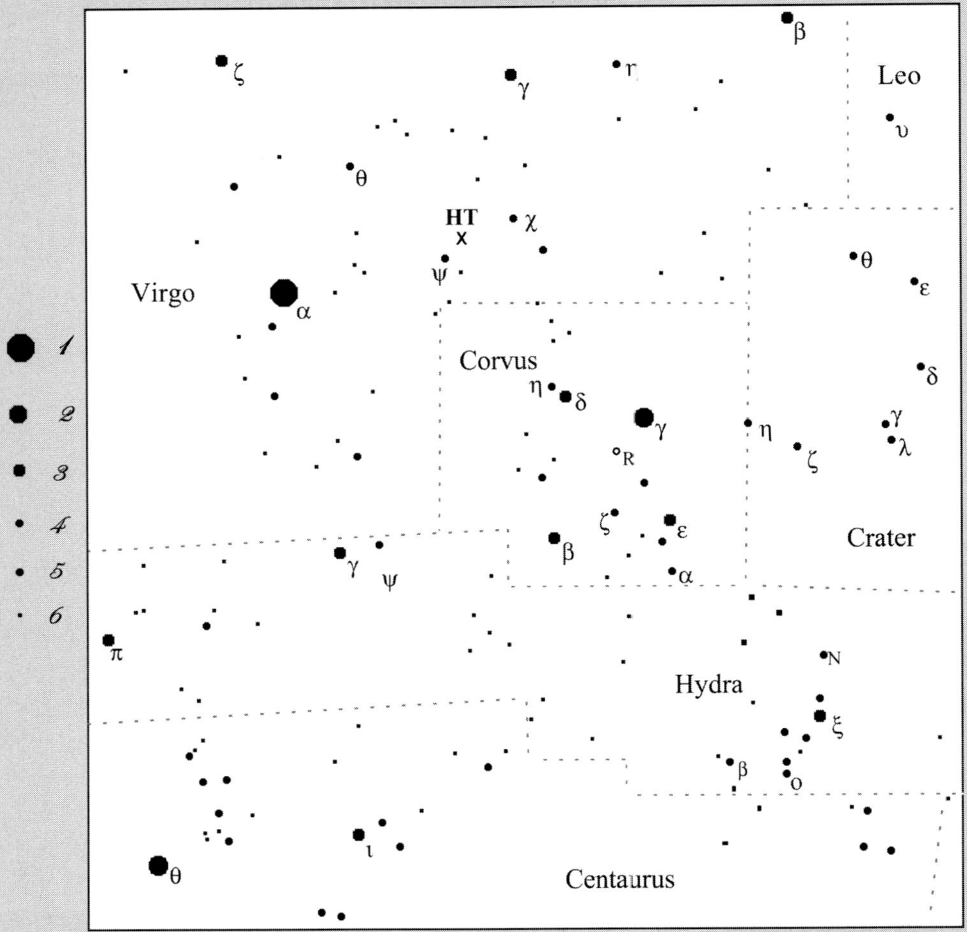

Tirion: Chart 14

Uranometria: Chart 284

68

Vinyl LP Galaxy
NGC 4699
Type: Mixed Spiral Galaxy (SABb)
Con: Virgo

RA: 12^h 49.0^m
Dec: -08° 40′
Mag: 9.5 (O'Meara); 9.5
Dim: 3.1′ × 2.5′
SB: 12.2
Dist: 84 million light-years
Disc: William Herschel, 1786

W. HERSCHEL: [Observed March 3, 1786] Very brilliant, irregularly round, very gradually much brighter middle. (H I-129)

NGC: Very bright, round, very much brighter in the middle resolved to a nucleus, [mottled].

NGC 4699 IS A SMALL BUT VERY bright multiple-arm, mixed spiral galaxy. It lies on the southern outskirts of the great Virgo Cluster, 9° west-northwest of brilliant Spica. "This object is a real sleeper," Larry Mitchell says. "It is located only a few degrees northeast of M104, one of the most observed galaxies in the sky. Yet few observers ever look at this object or even are aware of this gem's prime location . . . I know I wasn't, but I will look at it in the future, as I will never tire of looking at M104."

In form, it is somewhat reminiscent of NGC 2841 (Hidden Treasure 49), whose disk is fragmented into countless arcs of matter with no clear spiral structure. While NGC 4699's disk is broken into fragments, the structure is more uniform than that in NGC 2841, and a spiral structure can be detected. The spiral-arm fragments exist over the entire face of the galaxy, which is projected 41° from face-on. They are distributed differently in the bulge and disk.

In the central lens the spiral structure appears bright and tightly wound. It is also extremely intricate, consisting of highly fragmented, very thin spiral strands, which can be traced to within 12″ of the nucleus. The region is defined both by luminous fragments and by dust lanes, all of which can be traced to the center.

In the surrounding disk, the spiral structure is more apparent, especially since the arms are dimmer and looser. They appear to be unwinding gradually outward. The galaxy also has a large number of outer, low-surface-brightness spiral fragments in the region where the intensity of the bright bulge decreases sharply as it merges with the fainter disk. Close inspection of high-resolution images reveals an asymmetry in the spiral pattern on either side of the bulge, suggesting that dust is a significant component of the spiral pattern. The disk is also filled with H-II regions, more than 100 of which have been detected.

In short-exposures, the nucleus is a very small and very bright starlike object in a weak bar. However, the emission of the nucleus is very weak in this Seyfert-like galaxy. In a 1993 *Astrophysical Journal* paper, Gary Bower (University of Michigan) and his colleagues discuss their search for a supermassive black hole in the nucleus of NGC 4699 and other nearby galaxies. Quasars and active galactic nuclei are believed to be powered by accretion of matter onto a central supermassive black hole, which is required to have a mass of 10 million solar masses. If the host galaxy exhausts the matter capable of feeding the black hole, the energy output rapidly decreases leaving behind a dead quasar engine. Given NGC 4699's bright starlike core but weak nuclear emission, it's possible that such a quasar corpse exists here. Although Bower and his colleagues' results were inconclusive, the researchers do not rule out the presence of a 10-million-solar-mass black hole at the center of NGC 4699.

NGC 4699 was neglected by professional astronomers until the discovery of a 17th-magnitude Type II supernova on June 6, 1983, which flared to visibility 148″ east and 131″ north of the nucleus. Type II supernovae occur at the end of a massive star's lifetime, when its nuclear fuel is exhausted and the star is no longer supported by the release of nuclear energy. If the star's iron core is massive enough then it will collapse and become a supernova. Supernova 1983k, as it is called, was arguably the brightest Type II supernova with a plateau-shaped light curve – one that shows a rapid rise, followed by a broad dome that changes into a slow and gradual rate of decline. It was also the first Type II supernova for which high-quality spectra were obtained before it reached maximum light.

Virpi S. Niemela (Cerro Tololo Inter-American Observatory) and his colleagues believe that Supernova 1983k's light curve is evidence that the progenitor star was an exploding Wolf-Rayet star or a red supergiant, which had an extensive, pre-existing circumstellar shell. This suggestion was confirmed by E. Grasburg (Radio-Astrophysical Observatory) who explains the supernova's light curve by the interaction of a shock with matter in the extended circumstellar shell produced by a previous ejection from a star that was 9–14 solar masses. Grasburg also determined the distance to the supernova to be about 80 million light-years distant, which is comparable to Tully's determination in the table above. The galaxy then has a linear diameter of 73,000 light-years with a total luminosity of 78 billion Suns. NGC 4699, by the way, was also host to another supernova – 17th-magnitude Supernova 1948a – which appeared 46″ north of the galaxy's nucleus, so keep your eye on this galaxy for others.

To find NGC 4699, locate 1st-magnitude Alpha (α) Virginis (Spica). Spica marks the southeastern apex of a roughly 7°-wide near-equilateral triangle with the 5th-magnitude stars Theta (θ) and Psi (ψ) Virginis. You want to center Psi in your telescope. Now look for 8th-magnitude 39 Virginis 30′ to the north-northwest. NGC 4699 is a

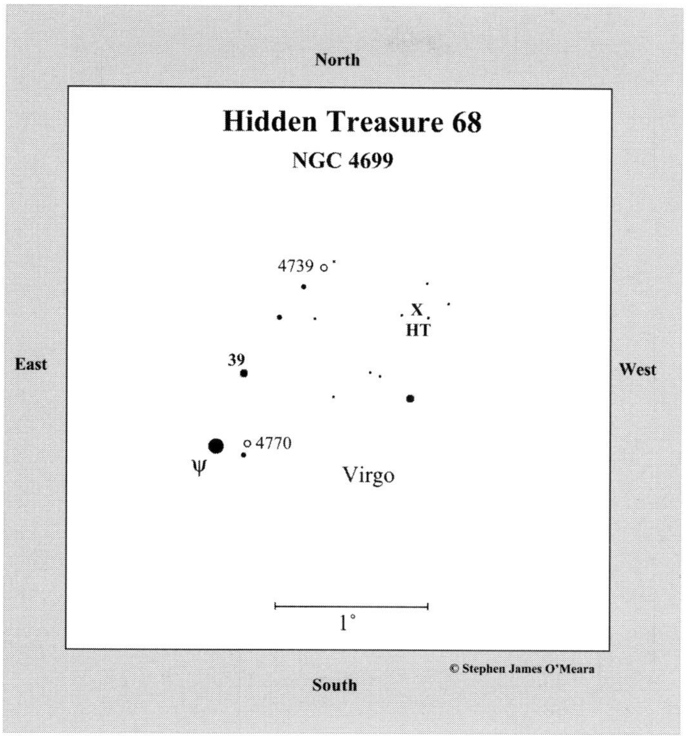

little more than 1° to the northwest of 39 Vir. Shining at magnitude 9.5, NGC 4699 equals M90 in brightness but it is about three times smaller, which probably explains why the object was overlooked prior to William Herschel's great sweeps of the heavens.

Under a dark sky, the galaxy can be seen in 7 × 50 binoculars as a star with a tiny halo. It looks much the same way in my antique telescope only a tad brighter. Still, had I been a comet hunter in the eighteenth century, I too would have passed over this largely stellar object. At 23× in the 4-inch, NGC 4699 shows a very intense and starlike core (the Seyfert-like nucleus) in a small but intense lens about 1′ in extent. This inner lens lies in a larger elliptical envelope 3′ in extent and oriented northeast–southwest. A 13th-magnitude star lies about 2′ to the south-southwest and a slightly fainter sun can be seen about the same distance to the east.

At 72×, the galaxy is a wonderful sight, showing much mottling. The Seyfert core is definitely elongated, indicative of a tiny bar that runs along the major axis of the central ellipse. The southern flank is slightly brighter than the northwestern one and looks curdled. Here are the multitude of spiral fragments hiding just beyond the limit of resolution. When I sweep my eye over the galaxy's tiny surface, I get the impression that someone is holding an old 78 LP record at a slight angle under the moonlight,

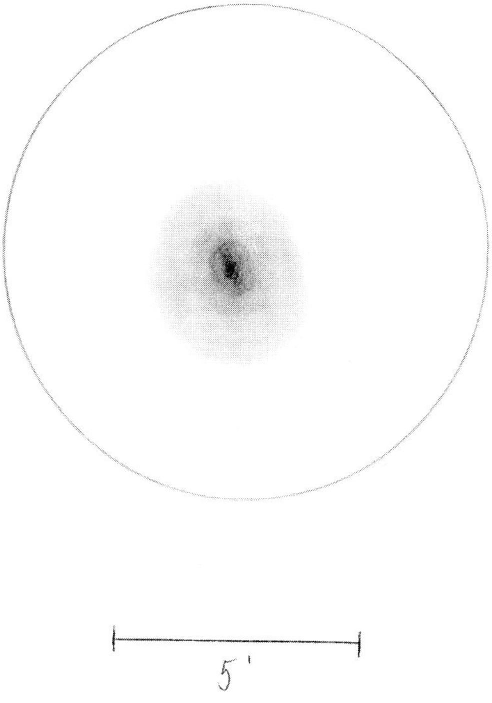

which is why I call it the Vinyl LP Galaxy – for those who remember owning such records. At 101×, the galaxy's dimmest sections in the 4-inch start to fade, appearing as an Irish Cross of enhancements at the northwest and southeast flanks as well as the northeast and southwest tips of the major axis. Once again, I get the impression of light shining off a shiny piece of vinyl.

NGC 4739, a magnitude 12.6 galaxy, lies about 40′ to the northeast of NGC 4699. I could see it as a pale oval glow. The magnitude 9.5 elliptical galaxy NGC 4697 (Caldwell 52) is almost 3° due north. And if you return to Psi Vir, see if you can't detect the tiny form of NGC 4770 a little more than 10′ due west. I could see its nucleus as a tiny star.

Hidden Treasure 69

NGC 4725

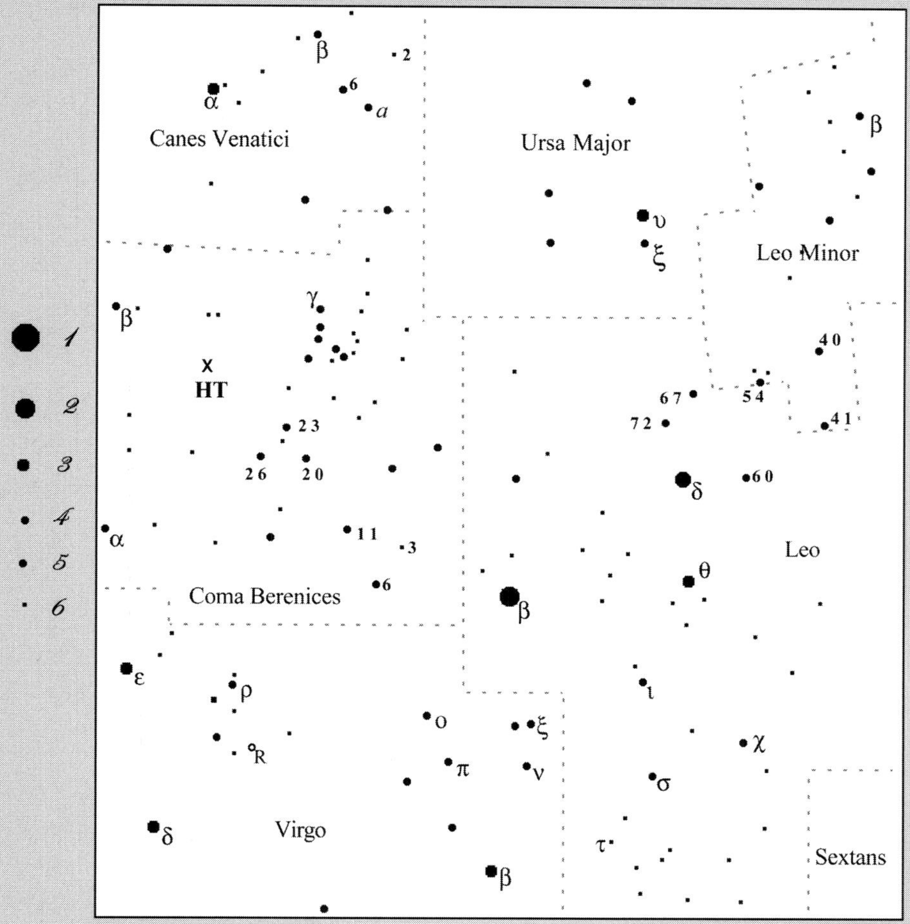

Tirion: Chart 7

Uranometria: Chart 149

69

NGC 4725
Type: Peculiar Mixed Spiral Galaxy (SABabP)
Con: Coma Berenices

RA: $12^h\ 50.4^m$
Dec: $+25°\ 30'$
Mag: 9.2 (O'Meara); 9.4
Dim: $10.5' \times 8.1'$
SB: 14.1
Dist: 42 million light-years
Disc: William Herschel, 1785

W. HERSCHEL: [Observed April 6, 1785] Considerably bright, irregularly round, small brighter middle, milky, 7′ or 8′ diameter. (H I-84)

NGC: Very bright, very large, extended, very gradually, then very suddenly very much brighter in the middle to an extremely bright nucleus.

HOW NGC 4725 ESCAPES NOTICE IS almost beyond imagining. Located about 5° east-southeast of Melotte 111 (Hidden Treasure 62), 2° south-southwest of the North Galactic Pole, and about 3° east-southeast of the stunning edge-on galaxy NGC 4565 (Caldwell 38), NGC 4725 is one of the classic hidden jewels of the night – especially if you consider that the only galaxy in Coma Berenices brighter than it is M64, located only 4° to the southeast. The fact is, the galaxy is like a leaf floating on the edge of a stream, it's far enough away from the rushing madness that it takes an act of serendipitous concentration to notice it – that is, of course, unless someone points it out to you, as I am now.

In photographs, NGC 4725 is a stunning example of a transition galaxy between a normal spiral and a barred spiral. If you look closely at the central lens, you will see that there is no well-defined bar. There is, however, a very small, extremely bright nucleus in a faint, smooth, broad bar with dark lanes. Also centered on the bar is the galaxy's most stunning feature – a high-surface-brightness spiral pattern that forms an almost complete ring inclined 43° from face-on; this feature is, in fact, one of the most complete near-rings of any galaxy known. The "ring" begins abruptly at the edge of the galaxy's amorphous central region – across which slice weak ripples of dust – and breaks into multiple, tightly wound spiral arms studded with rich star-forming regions. Therefore, the basic structure of NGC 4725 is closer to that of a normal spiral than that of a barred spiral galaxy.

On long-exposure images, two very faint external arms (generally smooth in texture with only hints of star-forming regions) loop almost completely around the bright central ring. These external arms are not on opposite sides of the nucleus but travel together, one lapped on the inside of the other. They are also branched. The principal one is an extension of a bright western inner arm, which forms part of the nearly complete inner ring. This arm branches into two, which then gradually unwind for about 200°. In the Hubble Space Telescope (HST) Extragalactic Distance Scale Key Project, the Cepheid distance of NGC 4725 was determined to be 42 million light-years, which is the distance I've adopted in the table above. If we accept that distance, then NGC 4725 is one of the more magnificent systems in the Coma–Sculptor Cloud of Galaxies, with a linear diameter of about 120,000 light-years and a total mass of 300 billion Suns. It is receding from the Sun with a velocity of 1,268 kilometers per second.

NGC 4725 is another great galaxy to watch for supernovae explosions. The first of three known events was discovered on May 5, 1940. At discovery, 1940b (a Type II supernova) shined at magnitude 12.8 and was 95″ east and 118″ north of the nucleus. A 15th-magnitude Type Ia supernova (1969h) erupted 18″ east and 10″ north of the nucleus in June 1969. The last known event occurred in December 1999, when a 19th-magnitude supernova (1999gs) appeared 3″ west and 105″ south of the nucleus. Astronomers used the HST to study the environment of Supernova 1940b, which occurred near the edge of what appeared to be two associations of bright, blue stars. The HST found that, indeed, the region contains many bright, young, blue stars, particularly in large *OB* associations; and several red stars, presumably red supergiants. The detected stars have ages that appear to range from 6 to 30 million years. Supernova 1940b may have been associated with the populations of young stars; if the supernova progenitor was a red supergiant, it may have had an age similar to the detected red supergiants, but could have been as young as 6 million years. The lack of a precise position for the supernova, however, prevented the researchers from determining the exact nature of the progenitor star.

To find this beautiful galaxy, first locate 4th-magnitude Gamma (γ) Comae, which marks the northern tip of Melotte 111, the Coma Berenices Cluster. Next use your naked eyes or binoculars to locate 31 and 30 Comae (5th- and 6th-magnitude), respectively, 5° to the east-southeast. Center 30 Comae in your telescope, then look a little more than 1° south-southwest for 7th-magnitude Star *a*. NGC 4725 is only about 1° south and slightly east of that star; another 7th-magnitude star (*b*) lies about 45′ to the southwest of the galaxy.

NGC 4725 is visible in 7 × 50 binoculars under a dark sky, and it appears as a soft and slightly elliptical glow in my antique telescope, like the core of M101. At 23× in the 4-inch, the galaxy is immediately obvious as a well-defined glow 20′ west of a nearly 20′-wide clustering of about a dozen 8th- to 12th-magnitude suns. The galaxy is also

69

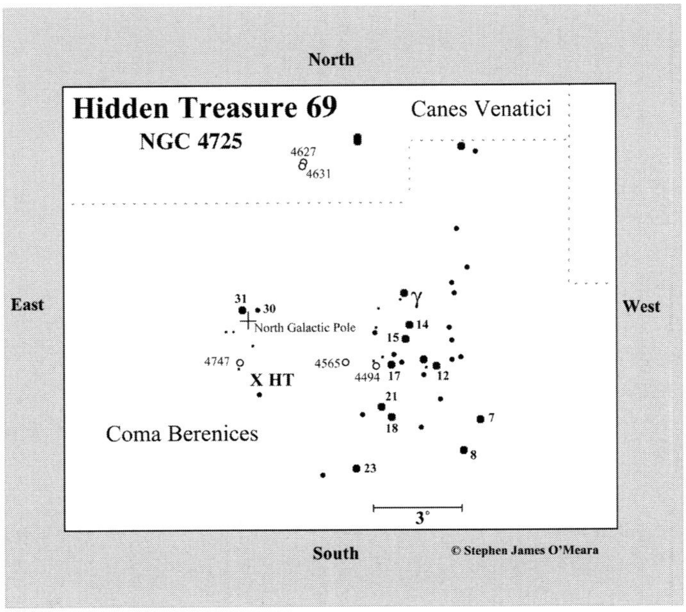

clearly elliptical, being oriented northeast–southwest, and very layered. The core is a brilliant star inside a subtle inner lens that rests inside a larger elliptical halo about 5′ across. The galaxy also lies atop a trapezoid of four 11th- and 12th-magnitude stars. Overall the galaxy has a low surface brightness; so, in my small telescope the more magnification I use, the less impressive the object appears. I find the most comfortable and enjoyable view to be at 72×, and I'm amazed at the subtle details that can be seen. First, the inner lens breaks down into what appears to be a bar oriented roughly north-northeast–south-southwest inside a sharp but broken S-shaped ring. A fainter, amorphous glow surrounds this inner lens, though I find it difficult to trace any definite shapes. The glow, however, does match the region of the galaxy's faint outer arms, which split and loop around the inner ring.

Interestingly, in a 12-inch telescope, Luginbuhl and Skiff could detect the nucleus but no core, as well as broad brightenings at the acute lens of the halo 2′ from the nucleus. Any enhancements I see are slightly askew from the ends, though this may be an illusion owing to dim starlight seen projected against the galaxy. I did see a roughly 13th-magnitude star close to the nucleus to the north-northeast and a fainter, perhaps 13.5-magnitude, star just to the west-northwest.

Who was the first person to detect NGC 4725's curious inner ring? None other than Lord Rosse, who in the mid 1800s observed this bright object with his great 72-inch Leviathan reflector at Birr Castle in Ireland. I had an opportunity to visit Birr Castle in 2003 and stand in the shadow of that newly renovated

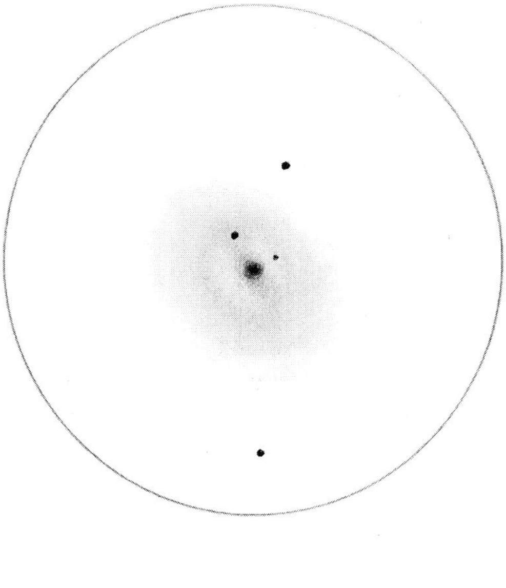

telescope.[1] The Leviathan is one of the great wonders of the astronomical world, a marvel of human ingenuity and technology. The 56-foot-long tube was too large to be in the Herschelian style, so Ross created his own temple to the stars to support it. He suspended the beast by chains and cables between two massive walls of masonry, each of which stand an impressive 56-feet high and are 72-feet long and 24-feet apart. As I stood before the mighty tube – a hollow column of dense wood held upright by a simple system of cables, pulleys, ropes, and winches; I could not help but feel as if some ancient and significant pillar of old was being raised for the first time from the hot sands of Egypt.

One night, at midnight, Tony O'Hanlon and Michael O'Connell of the local Shannonside Astronomy Club guided me up the 60-odd narrow, wooden steps that switchback up the steep western wall to the upper observing platform. As I pressed my body against the wooden railing of this movable platform – the only object separating me from the telescope – I realized (with a dash of dizziness) just how precipitous observing conditions were in Rosse's day. One misstep would send the observer down a sheer drop of 60 feet into a deep and black abyss. It also occurred to me that, given the telescope's limited access to altitude and azimuth – and considering that the telescope faces south – to discover the spiral nature of M51, Lord Rosse had to be observing with the telescope near its maximum height – a very precipitous position indeed. His observation of NGC 4725's ring would have been much more relaxing. By the way, Lord Rosse not only noticed that NGC 4725 has an inner ring, but he also said that it was an "incomplete" oval.

Rosse was a man dedicated to discovery, the importance of which he outlined in part of his address given to the 1854 meeting of the British Association for the Advancement of Science: "The love of truth; the pleasure the mind feels in overcoming difficulties; the satisfaction of contributing to the general store of knowledge; the engrossing nature of a pursuit so exalted as that of diving into the wonders of creation; all these are very powerful incentives to exertion."

Lord Rosse eventually used this telescope to find more than 60 of these "spiral nebulae."

[1] From the early 1900s to the mid 1990s, the Leviathan of Parsonstown was gradually subsiding into a state of "dignified decay," but, as David Bell (Shannonside Astronomy Club) explains, in 1996 Brendan Parsons, the Seventh Earl, set in motion a program of refurbishment aimed at breathing new life into this important artifact of Ireland's astronomical heritage. Over the next two years, the tube was recrafted to the original specifications, despite the lack of drawings and the sparsity of the Third Earl's notes. Much of the credit for rescuing William Parson's original vision from the mists of history must lie with a handful of people: people like Michael Tubridy, the Project Engineer, Owen McCarthy and his team, who took care of design considerations, and Bridget Roden, Project Manager. The patronage of various prominent Irish business people, as well as the support of several Irish government agencies, in addition to European financial aid, allowed the present Earl's vision to become a reality. As Fred Watson writes in his 2004 book *Stargazer: The Life and Times of the Telescope,* the telescope now has a new aluminum mirror and modern hydraulic mechanisms to move it. "Thanks to the vision and energy of William Brendan Parsons, Seventh Earl of Rosse, the telescope now represents an exciting fusion of old and new technology, a shining example of the best of both worlds."

Hidden Treasure 70
NGC 5102

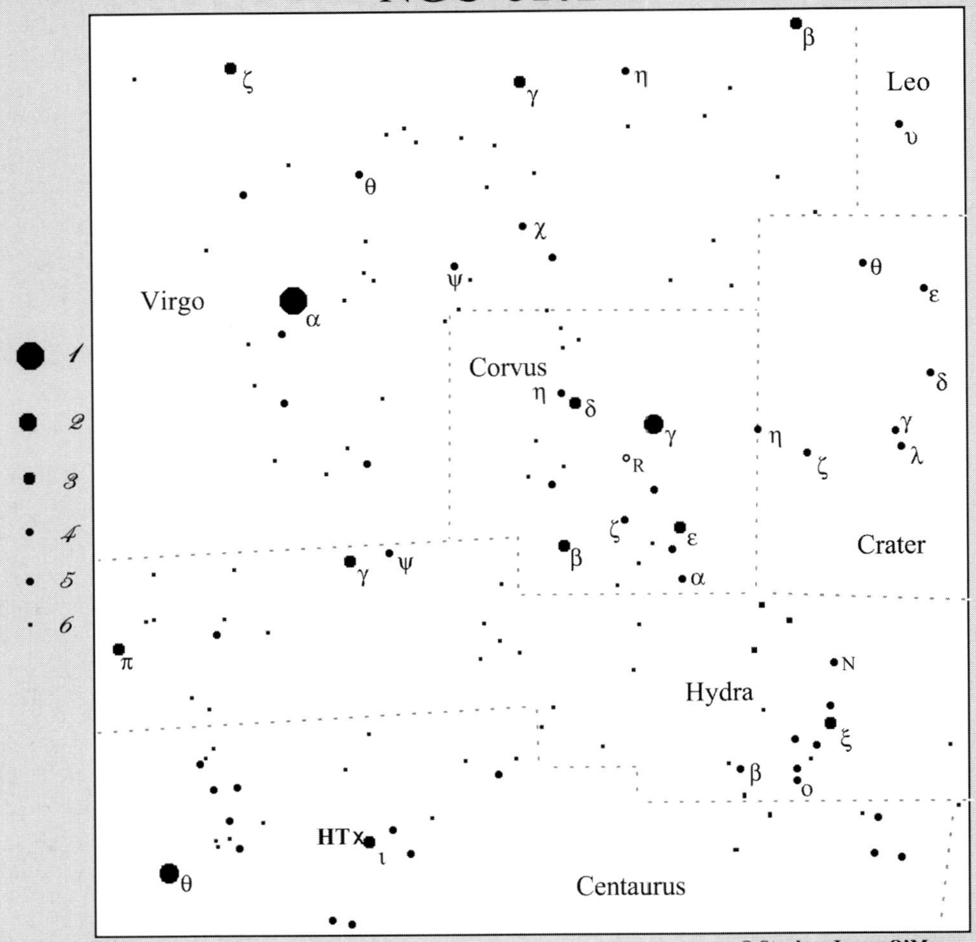

© Stephen James O'Meara

Tirion: Chart 21

Uranometria: Chart 370

70

Iota's Ghost
NGC 5102
Type: Lenticular Galaxy (S0)
Con: Centaurus

RA: $13^h\ 22.0^m$
Dec: $-36°\ 38'$
Mag: 9.3 (O'Meara); 8.8
Dim: $8.3' \times 3.5'$
SB: 12.6
Dist: 11 million light-years
Disc. John Herschel, 1835

J. HERSCHEL: Bright, round, pretty large, pretty suddenly much brighter in the middle to a star. (h 3492)

NGC: Very bright, pretty small, round, suddenly, very much brighter in the middle.

NGC 5102 IS A FANTASTIC LENTICULAR galaxy just 20′ east-northeast of 3rd-magnitude Iota (ι) Centauri. The galaxy hides in the bright "skirt" of its stellar "companion," though moderate to high powers can easily remove the star from the field of view. NGC 5102 is more easily spied than NGC 404 (Hidden Treasure 5) in Andromeda, being both brighter and farther from its neighboring star, and it makes a perfect hidden treasure for Southern Hemisphere observers. But Northern Hemisphere observers should note that NGC 5102 is only a little less than 2° further south than the open cluster M7 in Scorpius, which is the most southerly object in Messier's catalog. It is also more than 10° further north than Omega (ω) Centauri, which is essentially on the same parallel; Omega Centauri has been glimpsed from southern Canada. In fact, NGC 5102 is $\frac{1}{2}°$ further north than Lambda (λ) Scorpii (Schaula) the northern stinger star at the end of Scorpius's tail. So do not let the fact that NGC 5102 resides in Centaurus prevent you from attempting to see it on some crystal clear and transparent spring evening; NGC 5102 culminates on June 1st.

Interestingly, NGC 5102 is not mentioned in most observing handbooks and is not acknowledged in Ernst Hartung's *Astronomical Objects for Southern Telescopes.* Yet the galaxy's size, apparent magnitude, and surface brightness are comparable to those of M32, the elliptical companion to the Great Andromeda Spiral. This makes NGC 5102 a binocular object, albeit a challenging one because of its proximity to Iota

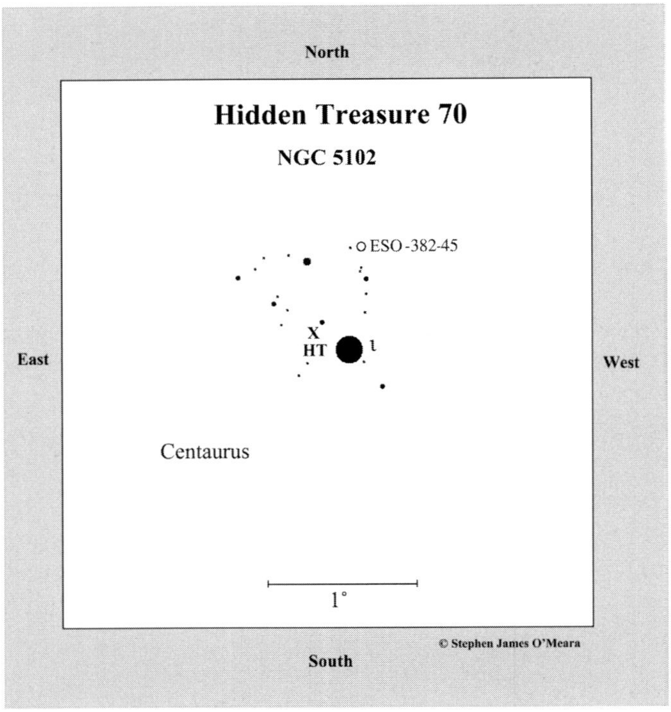

observations gathered in the 1980s, this facet has led astronomers to suggest that a burst of star formation occurred in the galaxy about 400 million years ago.

I first noticed the galaxy during a comet sweep across Centaurus with the 4-inch at 23×. It was a nagging glow just beyond the glare of Iota Centauri. I was ready to dismiss it as a ghost image of that star, but the star and the galaxy maintained their relative spatial relationship no matter how I moved the telescope. At low power, the galaxy displays a tack-sharp nucleus in a bright

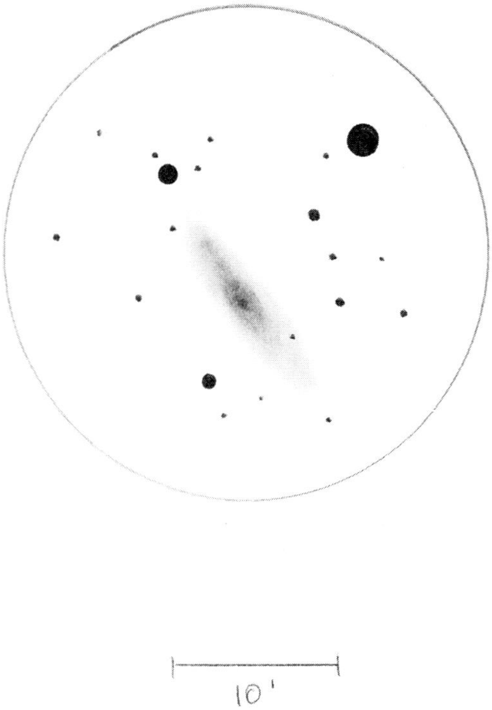

Centauri. Interestingly, Arizona amateur Steve Coe (Saguaro Astronomy Club) writes that, through a 17.5-inch f/4.5 Dobsonian reflector, "[t]his galaxy reminds me of a miniature version of M31 in Andromeda."

NGC 5102 belongs to the Coma–Sculptor Cloud of Galaxies and is regarded as a classical example of a Type S01 Lenticular. It spans 24,000 light-years and is receding from us at 460 kilometers per second. We see the galaxy inclined 19° from edge-on. Long-exposure photographs show that the lens may be dappled with dust, and one can try to imagine other dusty forms in the outer disk as well. A 1993 *Astrophysical Journal Supplement* series article classifies NGC 5102 as a very blue low-luminosity galaxy. In addition to an underlying population of old, red, metal-poor stars, the galaxy contains an abundance of hot, young, metal-rich stars. Together with infrared and obvious saucer of light, which is oriented northeast–southwest; a larger, uniform oval glow surrounds both these features. Use averted vision to see the full extent of this elongated glow. Over the years, I have found that the eye likes to elongate galaxies, especially if there is a bright star in line with the galaxy's major axis. And this is certainly the case with NGC 5102, which has a roughly 11th-magnitude star just beyond its northeast flank.

At 72×, the galaxy still has a very stellar core and an inner and outer lens, but sweep your eye carefully across the galaxy's length. Do you see any mottling? If you do, the reason may be due to a sprinkling of dim foreground stars that come in and out of view. Higher powers are required to make sense out of these field stars. Most notable is a triangle of 13th-magnitude stars "capping" the galaxy's southwestern flank and an equally dim sun on the galaxy's northeast flank. High power also reveals that the inner lens tapers to a point at the northeast end, while the southwestern edge appears rounded, like an arc of light with two knots: one to the west-southwest and one to the south. All these details are extremely tentative and seeing them requires you changing back and forth between medium and high powers.

If you live under a dark southern local, and you want to test the limits of your telescope, try to see the tiny 14th-magnitude galaxy ESO 382–45, only 40′ due north of Iota Centauri. It is just west of a magnitude 10.8 star and looks like a piece of lint in the moonlight.

Hidden Treasure 71

NGC 5281

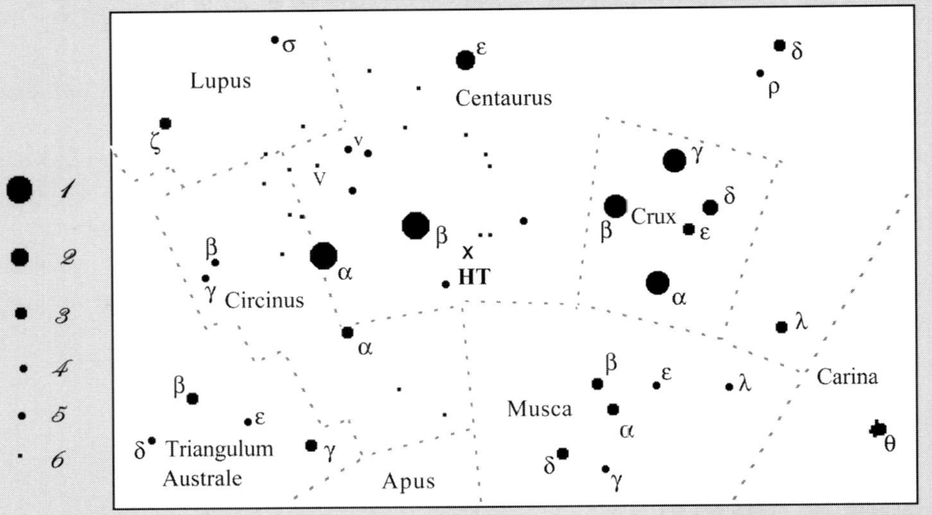

Tirion: Chart 25

Uranometria: Charts 451 & 452

71

Little Scorpion Cluster
NGC 5281
Type: Open Cluster
Con: Centaurus

RA: 13^h 46.6^m
Dec: $-62°$ 55′
Mag: 6.1 (O'Meara); 5.9
Diam: 8.0′
Dist: 4,200 light-years
Disc: Abbe Nicolas Louis de Lacaille, listed in his 1755 catalog

J. HERSCHEL: Small, compact irregularly round one 8th-magnitude star, and 15 or 20 smaller in a knot. No. 1 in Sweep 578 [NGC 5269] is an outlier of it. (h 3531)

NGC: Cluster, bright, small, pretty compressed, irregularly round, stars magnitude 10 to 12.

NGC 5281 IS A BRIGHT AND BEAUTIFUL open cluster in the dense southern Milky Way, just $3\frac{1}{4}°$ southwest of Beta (β) Centauri (Hadar), the 10th brightest star in the sky. Abbe Nicolas Louis de Lacaille (1713–1762) discovered this cluster while surveying the night sky over the Cape of Good Hope in South Africa, from April 19, 1751, to March 8, 1753. Through his $\frac{1}{2}$-inch f/50 telescope, the object appeared as a "small, confused patch."

Interestingly, he lists it as a Class I object ("nebulae without stars"). I say that's interesting because when I look at it with my antique telescope, I see it at first as an uncharacteristically large 6th-magnitude "star" that, with even the slightest bit of concentration, appears as three very prominent stars in a slight curve – like tiny jewels – surrounded by an irregular glow. On the other hand, NGC 5316, an equally bright but larger cluster lies just $1\frac{1}{4}°$ to the northeast; unlike NGC 5281, NGC 5316 does not resolve into starlight with the antique telescope but remains a diffuse glow. Is it possible that Lacaille got confused in this rich Milky Way region? Probably not. In his paper "On the nebulous stars of the southern sky," which appeared in the 1755 *Memoires de l'Academie Royale des Sciences,* Lacaille describes his Class I objects as being "no more than a whitish, ill-defined area, more or less luminous and of a very irregular shape: these patches are quite similar to the

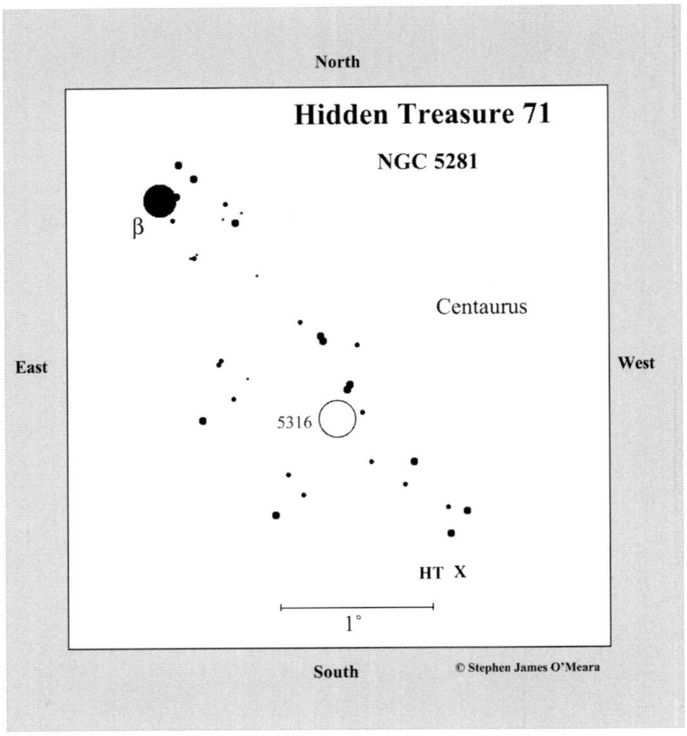

key, lift the lid of the chest, and see what lies within.

When I concentrate on NGC 5281 with low power, the first thing I notice is a bejeweled crescent of suns, oriented northeast–southwest consisting of stars of 8th magnitude and fainter. Two arms jut out from the northeast end, one to the north, one to the east. A fainter string of stars extends from the southwest end of the curve and gradually curls away to the west. With 72× and 101×, the cluster looks like a little scorpion with claws and a raised tail. And like the Milky Way around it, the cluster has all manner of geometrical patterns: long

nuclei of faint, tail-less comets." And that is just how NGC 5281 appears in my antique telescope.

Under dark skies, NGC 5281 is visible to the naked eye as a 6th-magnitude star, even when it is very low in the sky. The object appears swollen and ill defined in 7 × 50 binoculars – it does indeed look like the nuclear region of a tailless comet. Through the 4-inch at 23×, the cluster is a tiny package of light in the choppy seas of the Milky Way, which appears to be frothing haphazardly here and there with "windblown" clumps of dim starlight. What's interesting, because of its compactness, once I lock my eye onto NGC 5281, the cluster has the power to draw my attention away from everything else around it. It is as if I have found a floating treasure chest and want to do nothing else but use my telescope as a

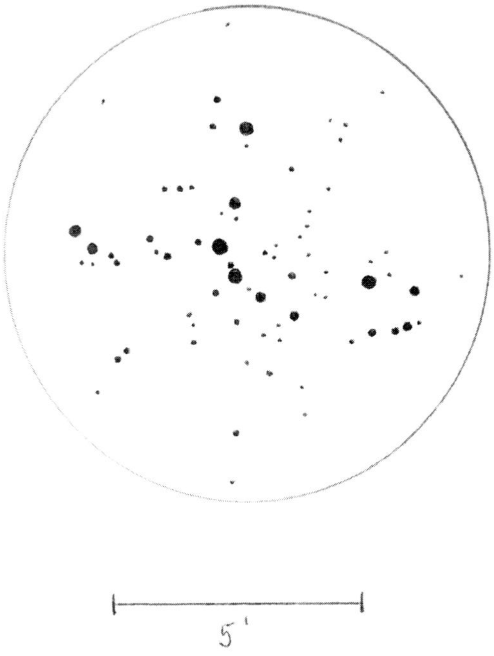

lines of stars, patches, tiny agglomerations of dim suns, arcs, diamonds, triangles. Despite the richness of the field, only 40 stars belong to the cluster. The remainder are background stars.

In larger telescopes, the main form of the cluster remains the same. For instance, when James Dunlop observed it from Paramatta, New South Wales in 1826 with a 9-inch f/12 telescope, he wrote, "This is a curved line of [faint] stars, about $1\frac{1}{2}'$ long, with a star of the 7th mag in the north extremity; a group of extremely [faint] stars on the preceding side of the crescent, and a multitude of very [faint] stars extended preceding and following." The only difference appears to be the sudden magnificence of the background Milky Way and the intensity of the stars' colors. As Ernst J. Hartung observed with his 12-inch reflector, NGC 5281 is a "beautiful scattered cluster of fairly bright stars merging into a fine field and concentrated at the centre in a pattern of two crossing curved lines of brighter stars, yellow, bluish, white and orange . . . on a clear dark night this is a most lovely field."

In true physical extent, the cluster is relatively small, measuring only 10 light-years across. In a 2001 *Astronomy and Astrophysics* paper, German astronomer J. Sanner (Sternwarte der Universität Bonn) and his colleagues present charge coupled device (CCD) photometry and proper motion studies of the three open star clusters including NGC 5281. The researchers found that NGC 5281 is a young open star cluster with an age of about 45 million years, making it about half as young as M45, the Pleiades.

Hidden Treasure 72

NGC 5363

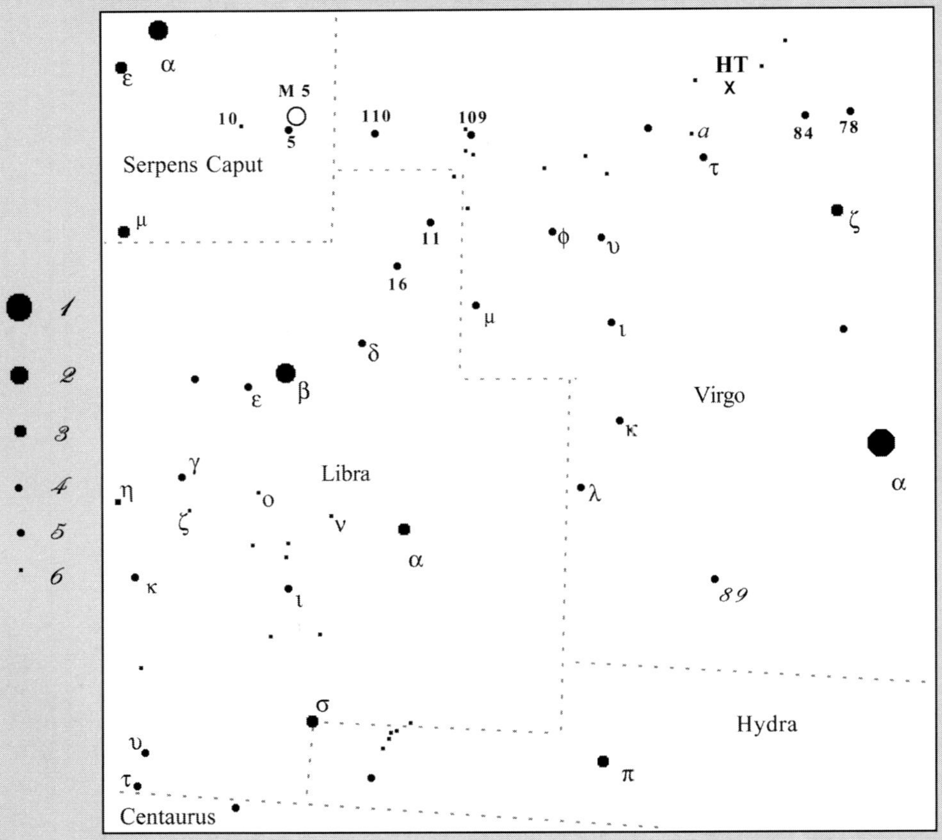

© Stephen James O'Meara

Tirion: Chart 14

Uranometria: Charts 196 & 241

72

NGC 5363
Type: Peculiar Galaxy (P)
Con: Virgo

RA: $13^h\ 56.1^m$
Dec: $+05°\ 15'$
Mag: 10.1 (O'Meara); 10.1
Dim: $4.7' \times 3.2'$
SB: 12.9
Dist: 73 million light-years
Disc: William Herschel, 1784

W. HERSCHEL: [Observed January 19, 1784] Very bright, pretty large, gradually much brighter middle. (H I-6)

NGC: Bright, pretty large, round, pretty suddenly brighter in the middle, magnitude 8 star north following.

NGC 5363 IS ONE OF THE MOST obscure, out-of-the-way galaxies in this catalog. It is also one of the most peculiar, if not, paradoxically, very simple in appearance. It belongs to the Virgo–Libra Cloud of Galaxies and hides in the midsection of the bleak and lonely corridor of space between 1st-magnitude Alpha (α) Virginis (Spica) and 0-magnitude Alpha (α) Bootes (Arcturus). Here is what I call the Dead Zone between the Great Coma–Virgo Cloud of Galaxies and the bejeweled hub of the Milky Way Galaxy. It is not surprising then that this object did not make it into anyone's "best" list, other than mine.

There is something about seemingly vast empty spaces that tends to "turn off" people's attention. Perhaps for the same reason that is why the Everglades National Park is one of the least visited of our nation's parks; endless tracts of swampland do little to assault the eye. Ironically, the Everglades is one of my favorite national parks, because, aside from the alligators, I enjoy the challenge of exploring its vastness and discovering the subtle riches within. I "discovered" NGC 5363 in a similar way. I found it while sweeping through that vast celestial marshland while comet hunting with the 4-inch at 23×. I was immediately taken by its small and bright form hiding in the glare of an 8.5-magnitude star. When I stopped to admire its beauty, my averted vision snapped up the slightly larger but slightly fainter glow of NGC 5364 just 15′ to the south – a double treasure. Actually, it turns out that NGC 5363 is the brightest member of a little flock of 7 galaxies within 1° of one another. And

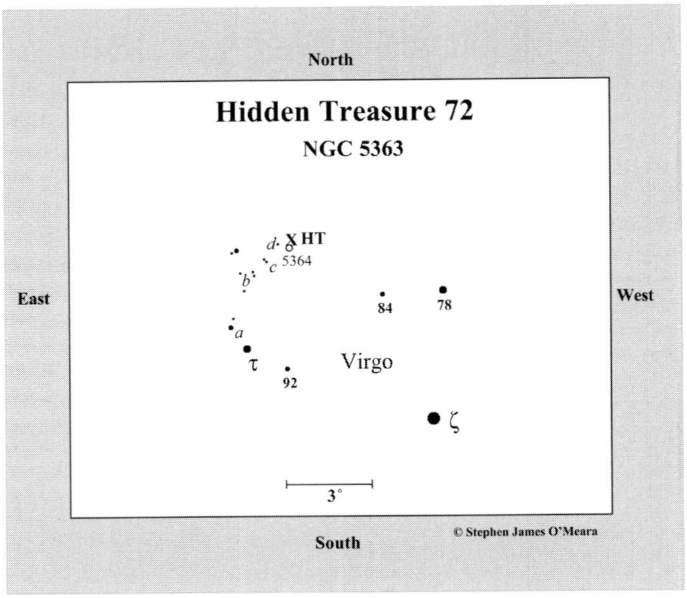

lies a little more than 25′ to the northwest. The galaxy is about 4′ southwest of that star.

I could not see NGC 5363 in 7 × 50 binoculars but I could see it in my antique telescope with time and averted vision; the difficulty, of course, is its proximity to the 8.5-magnitude star. At 23× in the 4-inch, the galaxy is immediately obvious, showing a sharp core and a halo that brightens gradually toward the middle. With little effort, I can see that it is elongated slightly northwest–southeast. As with my initial "discovery" of this object, as I focus on the galaxy, the ominous glow of NGC 5364 looms into view like a flaring comet; again, its amorphous glow and larger size makes it less

though I could not see it, the galaxy cluster Abell 1809 lies about 42′ to the west-southwest. Finding NGC 5363, then, was like a pirate finding a doubloon in the sand, which, with a little brushing aside of some grains, uncovered a chest full of riches.

NGC 5363 is as bright as M98 in Coma Berenices, though half the size, making it a more compact and easier target. To find it, you'll have to do some star hopping (exploring), so be patient, and dedicate yourself to the search. First locate Spica. A little more than 10° (a fist width) north-northwest is 3rd-magnitude Zeta (ζ) Virginis. Another substantial hop about 7° to the east-northeast brings you to 4th-magnitude Tau (τ) Virginis. A little more than 1° to the northeast is 6th-magnitude star (*a*). Star *a* marks the south-southeast end of a 2°-long line of four 7th- to 8th-magnitude suns. The star at the north-northwest end of the line (*b*) is a double. A short hop 40′ to the northwest takes you to a magnitude 7.6 star (*c*). The magnitude 8.5 star (*d*) near NGC 5363

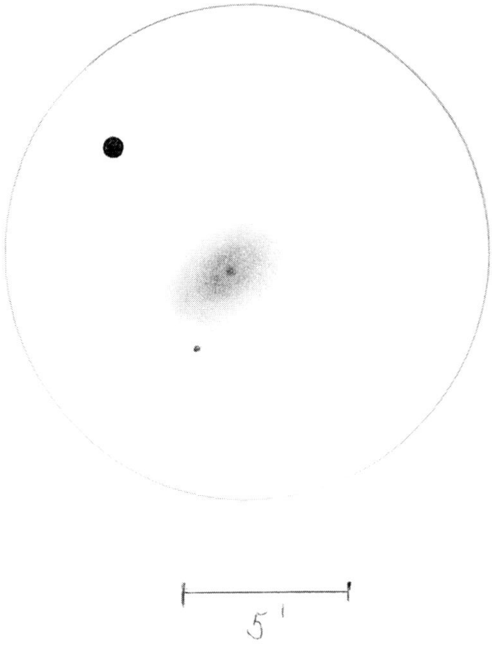

obvious in the 4-inch, though, with averted vision, it draws attention to itself.

At 72×, the core of NGC 5363 is mesmerizing. The nucleus is tack sharp – a star surrounded by a circular inner iris surrounded by a more diffuse and slightly elliptical lens. Its hypnotic nucleus is like that of a piercing eye, one that seems to probe into the depths of the soul. With imagination, I cannot help but feel that the brilliant nucleus is some sort of beacon, a message sent forth from a desperate civilization trying to make contact with other beings like itself. That said, it appears that this starlike nucleus is indeed a visual enigma, for a 12th-magnitude star is superimposed about 5″ southwest of the true (and equally bright) nucleus. The true nucleus is, in fact, a low-ionization nuclear emission-line region (LINER) – a type of gaseous region common in the centers of galaxies that have been shown to be a low-energy type of active galactic nuclei – like that of a Seyfert at its lowest level of activity.

The galaxy is classified as being either "irregular" or "peculiar" and has not been extensively studied. High-resolution images show the bright nucleus (and the superimposed star) with a narrow dust lane cutting across the major axis. Very Long Baseline Interferometry (VLBI) observations (in which the data from individual telescopes at different locations are sent to a central correlator and the signals combined) also confirmed the dark cloud in front of the core, as well as its neutral hydrogen profile whose emission characteristics are like those from the disk of an early-type galaxy. Although it is uncertain, NGC 5363 could be a lenticular galaxy with an anomalously large amount of dust. R. Brent Tully lists the object's inclination as 38° from edge-on. If we accept its distance of 73 million light-years, we find that it is a reasonably large galaxy with a linear diameter of 92,000 light-years and a total luminosity of 26 billion Suns. It is receding from us at about 1,126 kilometers per second.

Beware, several stars are superimposed on the face of this galaxy, confusing the view at higher powers. I could detect only one star, of about 13th magnitude, on the southwest flank. But those using larger telescopes could confuse any one of these field stars with a supernova.

For those adventurers with large visual shovels, you can go digging for the dimmer galaxies in the region. NGC 5338 (the brightest of the bunch) is a magnitude 12.4 barred spiral, a little more than 40′ due west and a bit south of NGC 5363. NGC 5356 shines at magnitude 12.6 about 18′ west-northwest of NGC 5363. NGC 5348 is a magnitude 13.1 barred spiral, 30′ due west and a tad south of NGC 5363. NGC 5360 is a magnitude 13.3 object about 8′ due west of NGC 5364. NGC 5373 is a magnitude 14.2 system 15′ due east of NGC 5363. Good luck!

Hidden Treasure 73

NGC 5662

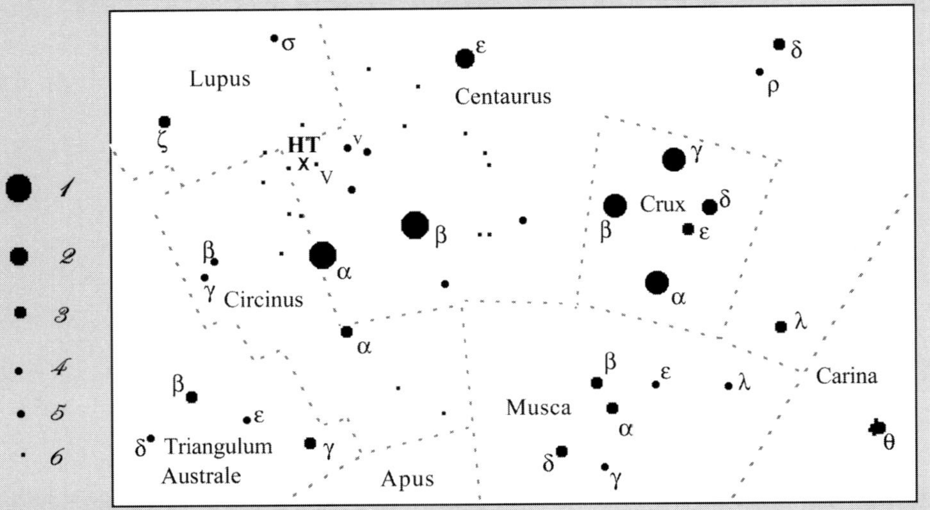

© Stephen James O'Meara

Tirion: Chart 25

Uranometria: Chart 431

73

Gypsy Moth Cluster
NGC 5662
Type: Open Cluster
Con: Centaurus

RA: 14^{h} 35.5^{m}
Dec: $-56°$ $40'$
Mag: 5.5
Diam: $30.0'$
Dist: 2,000 light-years
Disc: Abbe Nicolas Louis de Lacaille, listed in his 1755 catalog

J. HERSCHEL: Large, pretty brilliant, coarse, scattered cluster of Class VII which more than fills the field; 50 stars more or less 9th to 12th mag; chief star 7th mag, somewhat insulated, taken for place of cluster. (h 3573)

NGC: Cluster, large, pretty rich, little compressed, stars of magnitude 9 and fainter.

NGC 5662 IS A LOOSE CONCENTRATION of late *B*-and *A*-type stars lying in the rich Centaurus section of the Milky Way, a short jog north of Alpha (α) Centauri – the nearest star system to our Sun. The cluster's appearance is somewhat of a paradox. The cluster is moderately rich in stars, moderately bright, yet strangely detached. Its 280-odd members have, in fact, a most peculiar distribution. The "core" of the cluster is sparsely populated, though there is one obvious concentration of suns just north of it and a wider and more scattered concentration well to the south. This southern extension was not considered part of the cluster when Australian astronomer A. R. Hogg determined the cluster's diameter to be $12'$ in 1965. It's as if the cluster has been tugged on the northern and southern borders and suddenly ripped in two – a single double cluster of sorts. It's also possible that the bright Cepheid variable star, V Centauri (period = 5.494 days), situated about $35'$ southwest of the cluster's center is an outlying member of the cluster.

When the eighteenth-century celestial explorer Abbe Nicolas Louis de Lacaille discovered the cluster from the Cape of Good Hope with his $\frac{1}{2}$-inch $8\times$ telescope, he described it simply as "two stars in nebulosity." James Dunlop appears to be the first to note at least the extended nature of this

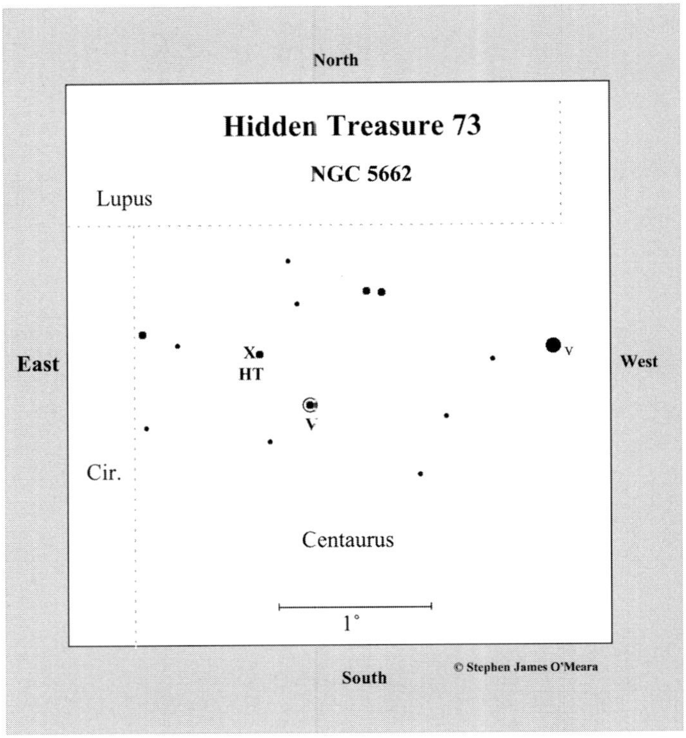

object. On July 10, 1826, he wrote, "A group of small stars of the 11th and 12th mag, with a multitude of minute stars mixt, extended [southwest] and [northeast.]" In one of his observations, John Herschel called it a "fine bright but not rich cluster" whose position is marked by the location of a "red star, the chief and centre of about 30 stars 9th to 13th magnitude. This red or high yellow star is 8th mag."

In a 1997 *Astronomy and Astrophysics Supplement* series paper, R. Sagar (Indian Institute of Astrophysics) and colleagues describe the color-magnitude diagrams for NGC 5662. Their data show that the cluster contains two red giants; one of them, a 7th-magnitude *K*5 star, is most likely the red star that Herschel spied. Sagar and his colleagues also determined that the Cepheid variable V Cen *is* a member of NGC 5662, so do not fail to include it in your view of this curious cluster. NGC 5662 has a total mass greater than 256 Suns, a true physical extent of 17.5 light-years, and a density of 0.5 stars per cubic parsec. Its age (80 million years) places it within the Pleiades age group. The cluster's metal content is nearly solar.

You'll find NGC 5662 is almost $4\frac{1}{2}°$ due north and slightly west of Alpha (α) Centauri (Rigil Kent) and about 3° due east of 4th-magnitude v Centauri, which forms the northern apex of a near-equilateral triangle with Alpha and Beta (β) Centauri. (Do not confuse V Centuari with the 7th-magnitude eclipsing binary star V Centauri 30′ to the southwest of NGC 5662.) It's an easy catch in 7 × 50 binoculars, appearing as a spindle of diffuse light the size of the full Moon. The "core" looks round. But this is the light of three stars that, with averted vision, form a V or triangle of stars. With averted vision, the cluster's northern and southern extensions separate from the core to form two triangular wings of unresolved starlight – base to base – one on either side of the central triangle: one wing extends to the north-northeast, the other to the south-southwest. V Centauri is very apparent as a bleeding star on the outskirts of the southern wing. In my antique telescope, NGC 5662 looks like Eta (η) Carinae seen with the naked eye. The central triangle of stars stands out prominently and the wings of starlight seem to

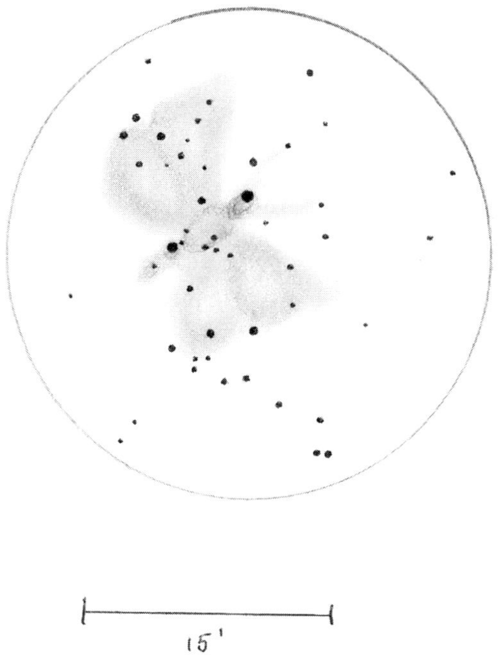

scintillate – like the powdery wings of a moth under a frost-covered porchlight.

The cluster is so large and scattered, and its suns rather dim (10th magnitude and fainter) that it simply looks best at low power. In the 4-inch at 23×, the crisp red central star burns forth in obvious splendor. Some 50 suns are clearly detected, though these are scattered and irregularly distributed along the cluster's length. At 72× they form pleasant groupings of three or more suns. It is quite a messy cluster, and it looks as if not only has someone tried to separate it into two parts but within these groups certain stars have gathered into different "teams." If you slightly defocus the view, the outskirts of the cluster have what appears to be loose spiral structure, giving the body of the cluster spidery legs.

I call this cluster the "gypsy moth" not only because it looks like one of those little critters with powdery wings but also to honor Etienne Leopold Trouvelot, the great astronomical artist, who spent time at Harvard College Observatory, making observations through the great 15-inch refractor and creating some of the most classic renderings of deep-sky objects. As Joseph Winlock, the observatory's third director, said, "[Trouvelot] combined in rare degree the qualities of an excellent observer with the skill of an accomplished artist." But Trouvelot also had an enterprising side to his character. And sometime between 1868–1869 he brought to North America some hardy silkworms to breed. Trouvelot wanted to develop a strain of silk moth that was resistant to disease as a part of an effort to begin a commercial silk industry. Alas, some of these insects escaped when a specimen jar fell from an open window at Trouvelot's home in Medford, Massachusetts. Later, more moths and caterpillars escaped from small populations growing on shrubs in Trouvelot's garden when high winds blew off protective netting. The artist notified townspeople about the accidents, but nobody thought the gypsy moth was a pest. *Wrong*!

The insects soon multiplied in a vacant lot next to Trouvelot's home. They then began to travel. Today, despite all control efforts since its introduction, the gypsy moth has persisted and has extended its range across state lines. Now the gypsy moth is the most important insect pest of forest and shade trees in eastern USA. The larval or caterpillar stage frequently strips entire trees and even forests of their leaves over wide areas. Trouvelot's gypsy moth turned out to be a true Pandora's box. Millions of dollars have been spent by federal, state, and local governments trying to exterminate them, yet the spread continues – which is why I have honored the southern skies with this pest. Besides, doesn't the cluster with all its holes and patches just look like the leaf of a deciduous tree being eaten by the gypsy moth caterpillar? *Gnaw, gnaw, gnaw*!

Hidden Treasure 74

NGC 5746

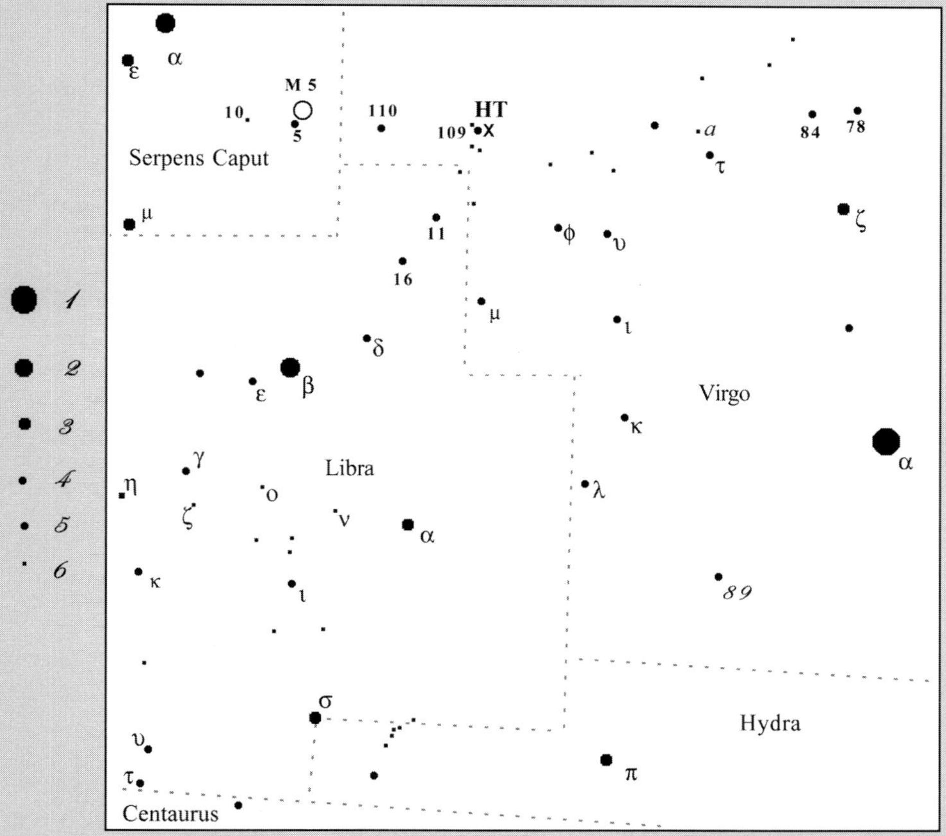

© Stephen James O'Meara

Tirion: Chart 14

Uranometria: Chart 243

74

Blade and Pearl Galaxy
NGC 5746
Type: Mixed Spiral Galaxy (SABb)
Con: Virgo

RA: $14^h\ 44.9^m$
Dec: $+01^\circ\ 57'$
Mag: 9.3 (O'Meara); 10.3
Dim: $8.1' \times 1.4'$
SB: 12.3
Dist: 96 million light-years
Disc: William Herschel, 1786

W. HERSCHEL: [Observed February 24, 1786] Extremely bright, much extended in the direction of the parallel of declination, bright nucleus, 8′ or 9′ in length. (H I-126)

NGC: Bright, large, very much extended toward position angle 170°, brighter then much bright to a nucleus.

LIKE NGC 5363 (HT 72), NGC 5746 IS a visual gem secreted away in that seemingly vast celestial wasteland between Virgo and Bootes. It lies about 12° southeast of NGC 5363, some 8° west of the brilliant globular cluster M5 in Serpens (Caput), and only 20′ west (and slightly north) of 4th-magnitude 109 Virginis; so it is truly hidden in the glare of a nearby star. But what a spectacle!

NGC 5746 is a massive edge-on system that belongs to the Virgo–Libra Cloud of Galaxies. It has a whopping linear diameter of 162,000 light-years and a total luminosity of 76 billion Suns. It is a shame that this object is often overlooked by amateurs since it is so easy to find. Larry Mitchell of Houston, Texas, says, "This object is one of my favorites and, yes, it is very much overlooked by amateurs. It is extremely easy to locate and does not require a star chart. It reminds me somewhat of a smaller version of NGC 4565." In fact, if you accept my visual magnitude estimate (9.3) of NGC 5746, it is just brighter than NGC 4565 (Caldwell 38) in Coma Berenices and NGC 891 (Caldwell 23) in Andromeda. Yet our hidden treasure is at least three times more distant than both those wonders.

NGC 5746 is an intermediate-type spiral with a very small, bright nucleus in a moderately small, though bright, peanut-shaped bulge. It is the brighter member of a non-physical pair with the normal spiral NGC 5740. We see NGC 5746's bulge just below the near-side dust lane as well as above it; for this to happen when the bulge is moderately small requires the line of sight to be within only a few degrees from exactly edge-on. A dark dust lane obscures most of the nucleus. One theory of peanut-shaped bulges is that they are bars seen end- or side-on. Indeed, spectroscopic observations of NGC 5746 suggest the presence of a bar seen partially side-on. The spiral pattern of bright and dark arms in the disk is well defined (save for the projection foreshortening). The arms appear regular and fairly smooth, although they have degree of clumpiness.

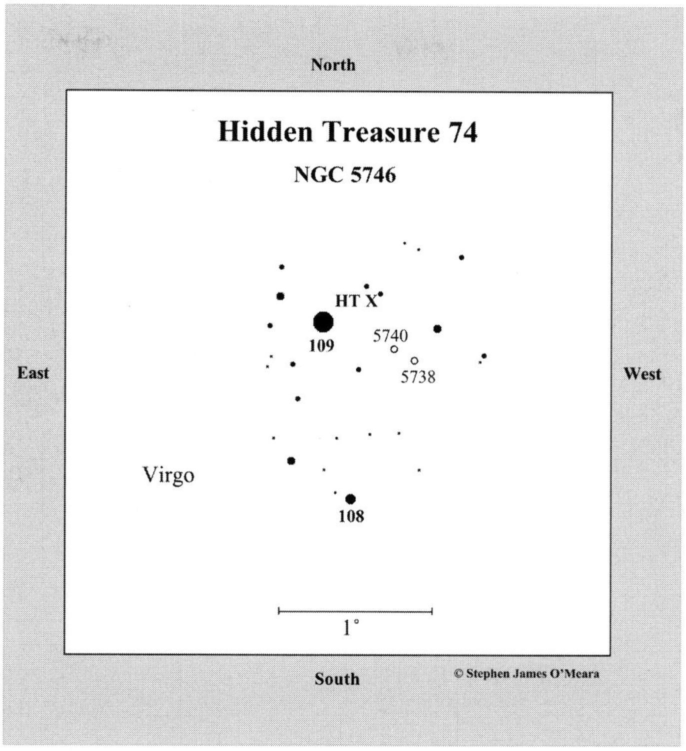

The arms of edge-on spiral galaxies are difficult to trace due to the orientation, but we do get an excellent view of the overall thickness of the disk. In NGC 5746, we get a good example of the galaxy's thickness and the absorbing dust that is present. By measuring and plotting the rotation curves (blue vs. redshift) of these edge-on galaxies, at increasing distances from the center, we obtain strong evidence that many galaxies contain large amounts of unseen material, which has been called "Dark Matter." In many galaxies, estimates of the amount of luminous and invisible matter show that the dark or invisible matter outweighs what we can see – sometimes by as much as a factor of ten.

NGC 5746 is similar in structure to NGC 5078 in Hydra, which is an interesting galaxy due to jets of gas that are being ejected from the disk. It is also similar to that of the Seyfert-like galaxy NGC 4699 (Hidden Treasure 68). And like NGC 4699, emission from the nucleus is very weak, which may be indicative of a large black hole that is currently not being fed with stellar mass. As Gary Bower (University of Michigan) and his colleagues state in a 1993 *Astrophysical Journal* paper, they cannot rule out the presence of a 10-million-solar-mass black hole at the center of NGC 5746. Another possibility is that a starburst episode has recently mostly shut down.

To find NGC 5746, all you have to do is locate 109 Virginis, which is nearly 18° north and slightly west of Alpha (α) Librae (Zubenelgenubi) or a little more than

13° northwest of Beta (β) Librae (Zubeneschamali). What's odd, is that to measure the discovery position of NGC 5746, William Herschel offset, not from 109 Virginis, but from 108 Virginis: "It precedes 108 Virg. of Flamsteed's Catalog by 0 min. 35 sec. in time and is 1° 15′ north of that star."

As with NGC 891 (Caldwell 23), NGC 4565 (Caldwell 38), and other edge-on systems, NGC 5746 has a low surface brightness. In the 4-inch at 23×, it all but fades from view with a direct glance, but just like the other edge-on galaxies, NGC 5746 swells into a fine sliver of light with averted vision; in fact, it is more apparent than NGC 891, appearing like a sharp needle with a bright bead balanced at the center like a pearl held steadily on a pirate's blade. With 72× and averted vision, the Seyfert nucleus is even more magnificent – a star in a thin wafer of light. The disk is not uniform, especially to the south, where there is a definite star or enhancement. With time I can also make out a waviness to the galaxy's bulge on the larger western side. Here is where light from the galaxy's disk, tilted ever so slightly toward our direction allows a glimpse of starlight populating the inner spiral.

In time, the dust lane is largely implied. The central pip is accompanied by two smaller (fainter) enhancements, which are probably tightly wound spiral structures seen nearly edge-on. At 101×, the galaxy remains visible but the contrast between it and the sky background diminishes greatly. Still, try studying the central lens at high power, to make out the finer details that waft in and out of view like luminous vapors. The dust lane is most prominent against the bulge and it is responsible for the waviness mentioned earlier. And the inner spiral enhancements look like bits of eyeliner, adding a spark of highlight. NGC 5746 hosted one supernova, Supernova 1983p, a Type Ia event that erupted to 13th magnitude 7″ east and 5″ south of the galaxy's nucleus.

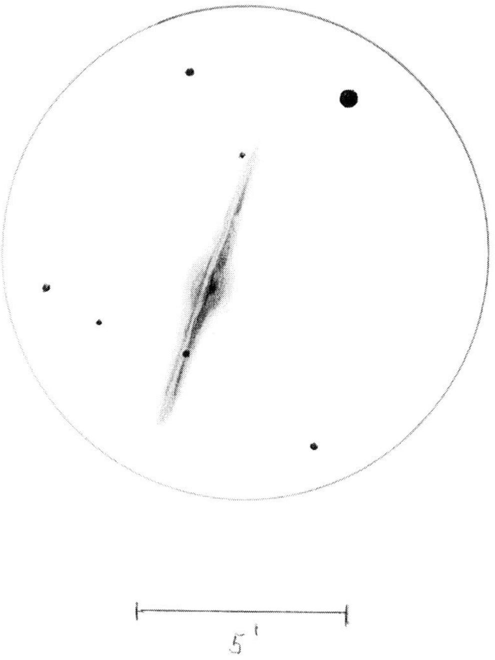

NGC 5746 has two companions nearby. NGC 5740 is a little spiral almost 20′ to the southwest; it shines at magnitude 11.9 and has a very bright nucleus. NGC 5738 is a 14th-magnitude pip about 25′ to the southwest. Do not be frightened by these magnitudes, they are not accurate. I could see both galaxies in my 4-inch, though it takes effort.

By the way, I forgot to tell you how I discovered NGC 5746. I'd like to say that I noticed it when Comet C/1996B2's (Hyakutake's) bulbous head skimmed 109 Virginis in March 1996, but the truth is, although I saw and plotted the comet during that approach, I failed to notice the galaxy. And while the late Walter Scott Houston mentioned it in his Deep-Sky Wonders column, it was really only in passing. I did not learn of, or take notice of, NGC 5746 until I started this project and saw it listed in the Saguaro Astronomy Club's "Best objects in the *New General Catalog*," and the Albuquerque Astronomical Society's "TAAS 200." The beauty of amateur astronomy is that we continually learn from each other.

75

Hidden Treasure 75
NGC 5866

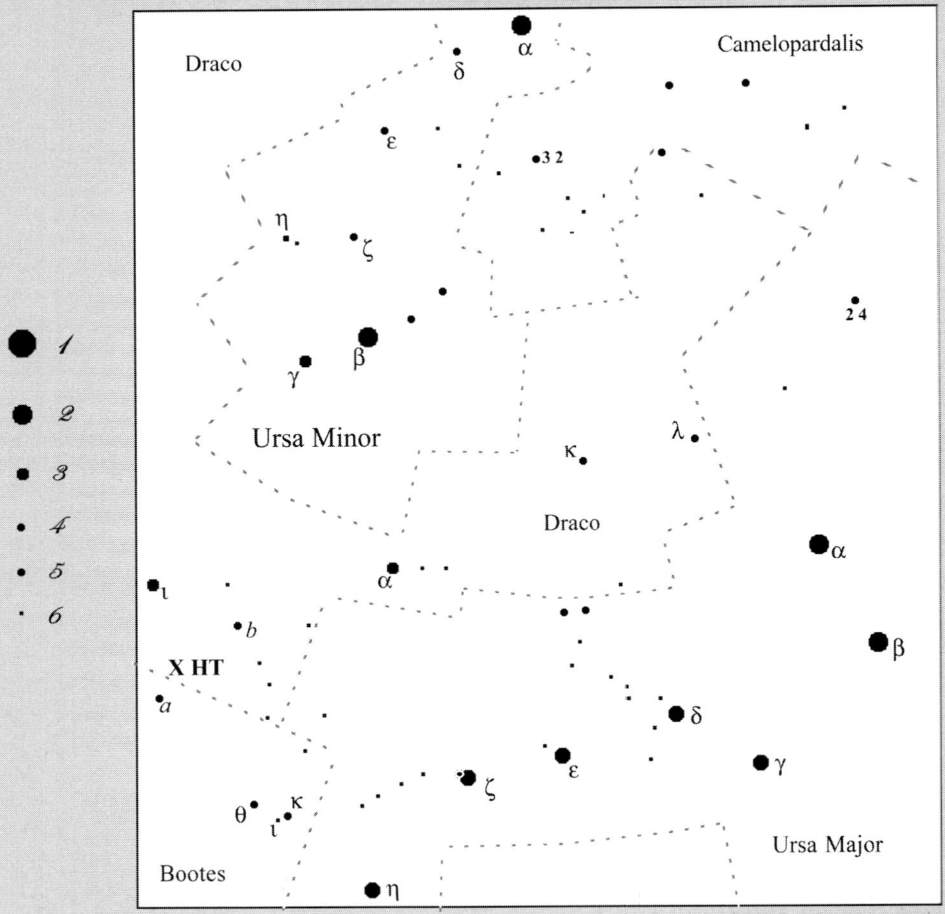

© Stephen James O'Meara

Tirion: Chart 2

Uranometria: Chart 50

75

Fool's Gold Galaxy
NGC 5866
Type: Lenticular Galaxy (S0)
Con: Draco

RA: $15^h\ 06.5^m$
Dec: $+55°\ 46'$
Mag: 9.8 (O'Meara); 9.9
Dim: $7.3' \times 3.5'$
SB: 13.1
Dist: 50 million light-years
Disc: William Herschel, 1788

W. HERSCHEL: [Observed May 5, 1788] Very bright, considerably large, extended, following 2nd star. (H I-215)

NGC: Very bright, considerably large, pretty much extended toward position angle 146°, gradually brighter in the middle.

THE UNASSUMING LENTICULAR GALAXY NGC 5866 in Draco, has long been at the center of a popular debate among deep-sky observers: whether it is M102 – one of the few ambiguous objects in Charles Messier's famous catalog of nebulae and clusters. The history surrounding this debate is worth repeating.

In Messier's catalog, the 102nd entry is described as:

> Nebula between the stars o [Omicron] Bootis and ι [Iota] Draconis. It is very faint. Close to it is a sixth-magnitude star.

Omicron Bootis and Iota Draconis are separated by nearly 40°, making them unlikely markers for even the most skilled observers of Messier's day. Thus, something is clearly amiss with the description. But what? To avoid the murky waters of speculation, let's look at the facts.

On April 13, 1781, Messier made the last observation of the objects included in his final supplement of the catalog. It was of M100, a "nebula without star" in Virgo, which Messier's contemporary, Pierre Mechain, had discovered the previous month. The supplement appeared in the French almanac *Connaissance des Temps* for 1784, which would be published in 1781. Mechain had supplied Messier with a number of new objects, and though Messier worked feverishly to observe all of them before his publishing deadline, time ran out and three objects went unseen. Still, Messier appended these objects as numbers 101,

102, and 103 to the supplement under the heading: "By M. Mechain, which M. Messier *has not yet seen,*" [emphasis mine]. Furthermore, Mechain had provided Messier with a measured position for only the 101st entry.

After the supplement was printed, Mechain noticed an error – entry 102 was not a new object, but rather a second observation of entry 101. After discussing the matter with Messier, Mechain drafted a letter of correction, which first appeared, in abbreviated form, in the *Histoire* of the Berlin Academy for 1782. A copy of the letter (dated May 6, 1783), was also sent to Johann Bernoulli in Berlin, who was editor of the *Berliner Astronomisches Jahrbuch.* Bernoulli translated the original letter into German and printed it in 1786:

> On page 267 of the *Connaissance des Temps* for 1784 M. Messier lists under No. 102 a nebula which I have discovered between omicron Bootis and iota Draconis. This is nothing but an error. This nebula is the same as the preceding No. 101. In the list of my nebulous stars communicated to him M. Messier was confused due to an error in the sky-chart.

Clearly, Mechain wasted little time in trying to undo the mistake. And his words solve the mystery "with complete certainty," according to Harvard historian Owen Gingerich. In his contribution to *The Messier Album* (Sky Publishing, 1978), Gingerich writes:

> In 1877 [Edwin] Holden included a reference to Mechain's letter in his useful *Index Catalogue of Books and Memoirs Relating to Nebulae and Clusters.* The correction is mentioned even more explicitly by [Guillaume] Bigourdan in his thorough analysis of old discoveries of nebulae in *Observations, 1907,* of the *Annales* of the Paris Observatory (published in 1917). Nevertheless, other investigators of Messier's catalogue seemed unaware of the correction until 1947, when Helen S. Hogg published Mechain's letter in the *Journal of the Royal Astronomical Society of Canada.*

Mystery solved? So it would seem. Yet despite these facts, many observers refuse to accept that M102 is a duplication of M101. They cling to the belief that M102 is a separate object. Such passion, however, is unfounded.

MURKY WATERS

The story of M102 can be briefly told: Mechain communicates to Messier a list of new objects. Messier confirms all but three of them before he submits his work for publication. In haste, Messier appends the three unconfirmed objects to his work and introduces an error. Mechain notices the mistake and publishes a letter of correction.

There is nothing mysterious about the story. It is well documented. The mystery stems from a popular theory based on a single assumption. The problem is that the theory introduces more mysteries than it attempts to solve. But before we look at that theory more closely, let me restate one very important fact: Messier never observed, nor did he attempt to observe, the 101st and 102nd objects in his supplement before it went to print. So there is no need to try to imagine what objects Messier might have seen in the area. The only person who saw anything was Mechain.

At the heart of the M102 theory is the error on Messier's sky chart. Many have speculated on this error and have come to the same conclusion, which was first introduced by William Henry Smyth in his 1844 *A Cycle of Celestial Objects.* It can be found in his discussion of M102:

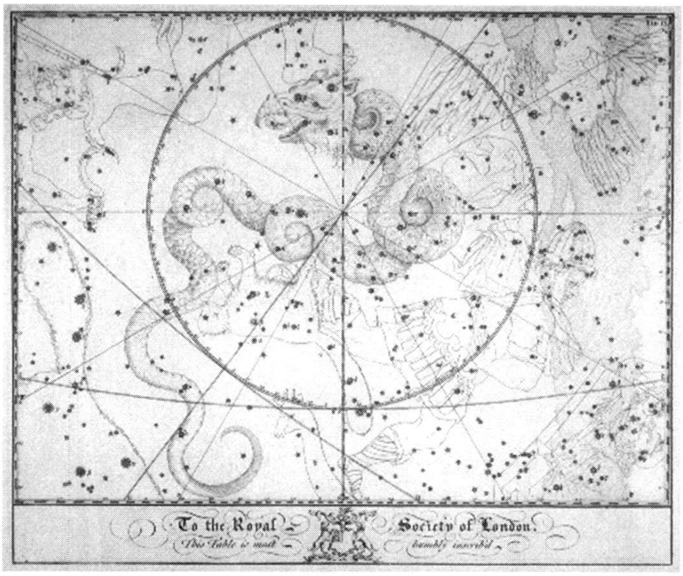

> A small but brightish nebula, on the belly of Draco, with four small stars spreading across the field, north of it. There may be a doubt as to whether this is the nebula discovered by Mechain in 1781, since Messier merely describes it as "very faint," and situated between o Bootis and ι Draconis. But there must be some mistake here; the one being on the herdsman's leg, and the other in the coil of the Dragon far above the head of Bootes, having 22° of declination and 44′ of time between them, a space full of all descriptions of celestial objects. But as the θ in the raised right hand of Bootes, if badly made, might be mistaken for an *omicron*, this is probably the object seen by Mechain, and [John Herschel's] 1910; it being the brightest nebula of five in that vicinity.

In addition to being the first to suggest that Messier confused the star Theta Bootis for Omicron Bootis, Smyth appears to be the first to suggest that M102 is NGC 5866 (which John Herschel listed as object 1909 [h 1909] in his catalog; Smyth confuses the issue by misidentifying this same object as h 1910), which is situated between Theta Bootis and Iota Draconis.

Smyth's solution is beautiful except for one fact: if Mechain had discovered NGC 5866, he was not aware of it. In the letter he sent to Bernoulli – the same one in which he corrects the error concerning M102 – Mechain included the descriptions of six nebulae he discovered in 1781, and NGC 5866 is not among them.

But mistaking Theta Bootis for Omicron Bootis, as Smyth suggests, can't be the only error, since M101 does not lie between Theta Bootis and Omicron Draconis. If Mechain had discovered a nebula between these stars, wouldn't he have told Bernoulli? As the late *Sky & Telescope* columnist Walter Scott Houston once quipped about Smyth's identification of M102, "one wrong assumption coupled with immaculate logic produced a can of worms."

Some have argued that Mechain's descriptions of M101 and M102 are suspiciously different, since he calls M101 a "nebula without star, very obscure and large" and M102 a "very faint" nebula near a "star of the 6th magnitude." The difference between a very obscure nebula and a very faint nebula is trivial. M101 is two magnitudes brighter than NGC 5866. M101 is a round glow nearly 30′ in diameter, while NGC 5866 is a mere 7′ × 3′ spindle of light. Through a small telescope, M101 is highly diffused, while NGC 5866 is highly condensed. The two objects are dramatically different in size and appearance.

SO WHAT IS M102?

Is there a way to use known facts to solve the mystery of M102? Yes. All we have to do is accept Mechain's words – there was an error on Messier's star chart; Messier was confused; and M102 is a second observation of M101.

Here is how Messier presented Mechain's description of M101 in the 1784 *Connaissance des Temps*:

> Nebula without a star, very obscure and extremely large, 6 to 7 minutes in diameter, between the left hand of Bootes and the tail of Ursa Major. Difficult to distinguish when the crosshairs are illuminated.

Messier also included a position determined by Mechain on March 27, 1781, that is quite accurate in declination but about $1\frac{3}{4}^\circ$ west of its true position. As reported on the Students For Exploration of Space website (http://www.seds.org/messier/m/m102d.html) – where, by the way, you will find an interesting opposing view to my argument – in his personal copy of the catalog printed in the *Connaissance des Temps* for 1784, Messier penned a position for M102 that falls almost midway between θ Bootis and ι Draconis. Just as Mechain had stated in his letter to Bernoulli, there was an error on Messier's star chart. Either θ Bootis was mislabeled o Bootis or, in his haste, Messier simply mistook θ for o.

Unfortunately this discovery does not solve the mystery of M102, because nothing exists at the penned position. The closest object is the 12th-magnitude galaxy NGC 5687, which was too faint for Mechain to see. NGC 5866 is 3° east-northeast of the penned position, and M101, on the other hand, is more than 6° west of it.

If M102 is indeed M101, and if θ Bootis was mislabeled on Messier's star chart (which makes perfect sense because θ is in the "left hand of Bootes"), then the only thing that does not make sense is the position of M102 that Messier penned in his personal copy of the *Connaissance des Temps*. Did Mechain mean that Messier was confused because he went in the wrong direction when he plotted Mechain's second observation?

Since θ Bootis is the brightest naked-eye star near M101, Messier likely would have used it as an offset star. Indeed, note that Messier does use "the left hand of Bootis [θ Bootis]" to help position M101. So I plotted Messier's penned position of M102 on a star atlas and stepped back to study the scene. I was immediately struck by how well this position mirrored M101's position with respect to θ. If Messier's penned position of M102 is plotted west of θ Bootis instead of east of the star, then it falls almost right on top of the modern position of M101.

Finally, is there any physical evidence that two positions of M101 actually exist? Yes, and these positions have been in our hands, or at least many bookshelves, for decades. *Sky & Telescope* magazine senior editor Dennis di Cicco noticed that in *The Messier Album* (Sky Publishing, 1978), in the facsimile of the Messier catalog (as Messier himself published in the *Connaissance des Temps*), there are two right-ascension positions for each object – one in time, the other in degrees, minutes, and seconds. Unlike all the other objects, the two positions given for M101 differ from one another. The 1781 position of M101 in time, when precessed to epoch 2000.0, is right ascension $13^h\ 51^m\ 33^s$; declination $+54^\circ\ 19'\ 09''$. That position places M101 $1\frac{3}{4}^\circ$ to the west of M101's modern position. It also places

M101 a mere 40′ from the 6th-magnitude star 86 Ursae Majoris; this, of course, would explain why Messier, in his description of what he thought was *M102,* writes, "Close to it is a sixth-magnitude star." On the other hand, the 1781 position of M101 recorded in degrees, minutes, and seconds, when precessed to epoch 2000.0 and converted to time, is $13^h\ 59^m\ 55^s$; declination $+54°\ 20'\ 25''$. The NASA Extragalactic Database position for M101 (epoch 2000.0) is $14^h\ 03^m\ 13^s$; declination $+54°\ 20'\ 57''$ – a difference of only 1′ from the *second,* more refined, position of M101 published in Messier's catalog.

In summary, if you accept my argument, M102 is simply a second, more refined, observation of M101. There is no reason, then, to be suspicious of Mechain's letter to Bernoulli; Mechain and Messier did not conspire to hide the truth – unless you also want to believe that Mechain and Messier were members of the Illuminati and M102's erroneous position is a clue to help its members locate the Holy Grail. No, Mechain's letter is full of truths. And while the theories that Smyth and others have put forth over the years have certainly been enticing, if not entertaining, close inspection renders them unlikely.

If you accept my argument, the mystery concerning NGC 5866 is over and the principle of simplicity has prevailed. Of course, that does not mean you shouldn't look at this marvelous object, which, undeniably, was discovered on May 5, 1788, by William Herschel. On the contrary, NGC 5866 is a remarkable object, especially through the high-quality telescopes so commonly used by amateurs today. We can see the object in ways that Messier and Mechain could only imagine. And that is why I have made it a hidden treasure.

THE MODERN VIEW

NGC 5866 is a remarkable nearly edge-on object (90,000 light-years in diameter) with a very bright ring and a dust lane – both of which are tilted against the equatorial plane of the elliptical halo – and contains an extended halo of globular clusters. Indeed, the outer envelope simulates an elliptical galaxy (which is generally known to be globular-cluster rich), but the lens shows a narrow ring of dark material surrounding the large nucleus. So it has a classic lenticular form.

The galaxy has an unusually rapid gas rotation curve with a measured maximum rate of 320 kilometers per second within about 3,000 light-years of the galactic center. This is much faster than would be expected based upon the light distribution of the bulge and the disk, and two possibilities have been proposed. One is that there is an unaccounted mass component found within the disk and the other possibility is a decoupled system located at the galactic center which is the result of a past capture or merging process.

NGC 5866 is also remarkable for being immersed in a massive molecular cloud. NGC 5866 has two wide companions (NGC 5907 and NGC 5879): NGC 5907 is a thin 10th-magnitude ellipse about $1\frac{1}{2}°$ to the northeast, and NGC 5879 is an 11th-magnitude ellipse about $1\frac{1}{4}°$ to the north-northeast. All three galaxies have very similar redshifts which suggests the galaxies form a wide physical group. They are located at approximately half the distance from each other that the Milky Way and M31 are, so hypothetical backyard observers living on a planet within the disk of any one of these three galaxies certainly would have two stunning naked-eye extragalactic wonders in their night skies to look upon and wonder.

Otherwise, NGC 5866 is relatively isolated, and there appears to be no evidence for any tidal distortions. So the existence of such a large gas shell, which is bright in infrared light, indicates that NGC 5866 may be entering an era of star formation fueled with gas donated by its aging stellar population. G. K. Kacprzak (New Mexico State University) and G. A. Welch (Saint Mary's University) presented evidence to support this contention at a 2003 American Astronomical Society meeting. After analyzing the dust and gas components of the interstellar medium around NGC 5866, the researchers concluded that NGC 5866's X-ray hot interstellar medium is likely being swept by perhaps a global wind powered by supernovae.

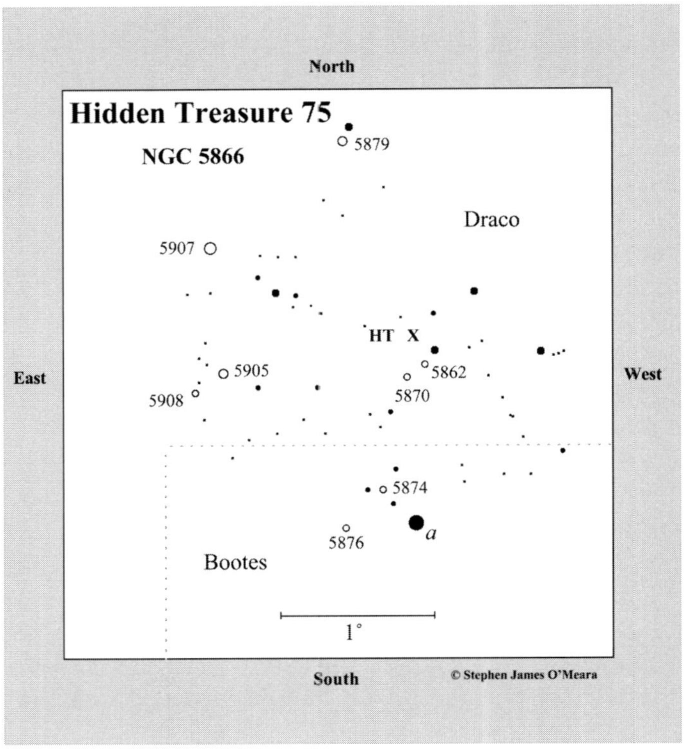

To find NGC 5866, first locate 3rd-magnitude Iota Draconis (Edasich). Now look for a 5th-magnitude star (*a*) just about 5° southwest. Use binoculars if you have to, but Star *a* should be easy to identify, since it is the brightest star in the region. NGC 5866 is just a little more than 1° due north of Star *a*, and 10′ northeast of a magnitude 7.5 sun. Once you get the field, try relaxing and seeing it with binoculars. I find that under a dark sky, NGC 5866 is *just* visible with 7 × 50 binoculars with effort. You have to know exactly where to look. The galaxy is easier to see in my antique telescope, but it is not something that calls attention to itself

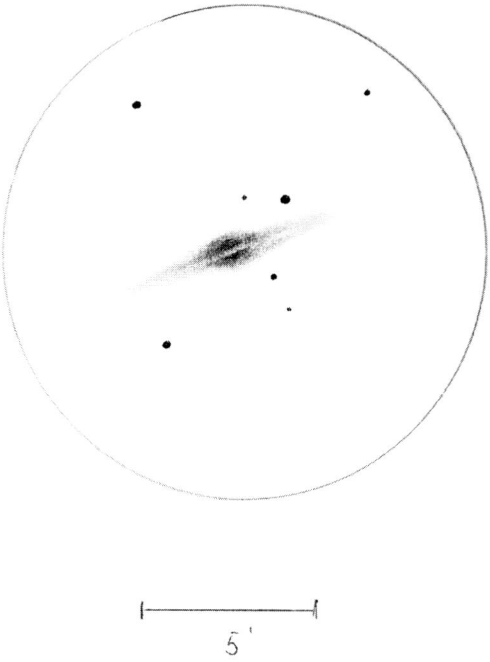

in a sweep. Again, you have to know exactly where to look to see it.

In the 4-inch at 23×, it is immediately obvious as a condensed elliptical glow with white winglike extensions. At this low power, the galaxy's light is almost combined with that of an 11.5-magnitude star about 1.5′ to the northwest. The galaxy, however, stands out much more brightly. Its core is very small (~2′) and very compact. The galaxy's spindle shape is nicely revealed at 72×. A number of fainter suns peek out from the background immediately surrounding the galaxy, adding a bit of dimension. It's a nice view at this power, and the lens looks extremely uniform and smooth. But an object as condensed as this one begs for higher powers. At 303×, the dust lane can be seen as a whisper of darkness – one that cleanly splits the central bulge in two. I needed extreme averted vision to see that thin sliver of darkness extend into the disk. But, in time, see if the dark matter just doesn't *pop* into view. The central bulge at high power is itself interesting – a prominent ellipse of light that gradually becomes more circular to a point. In the modern scopes of today, with their superb optics and light-gathering power, NGC 5866 is, overall, a much more dramatic object than either M99 or M100 – or M101, for that matter.

Hidden Treasure 76

NGC 5897

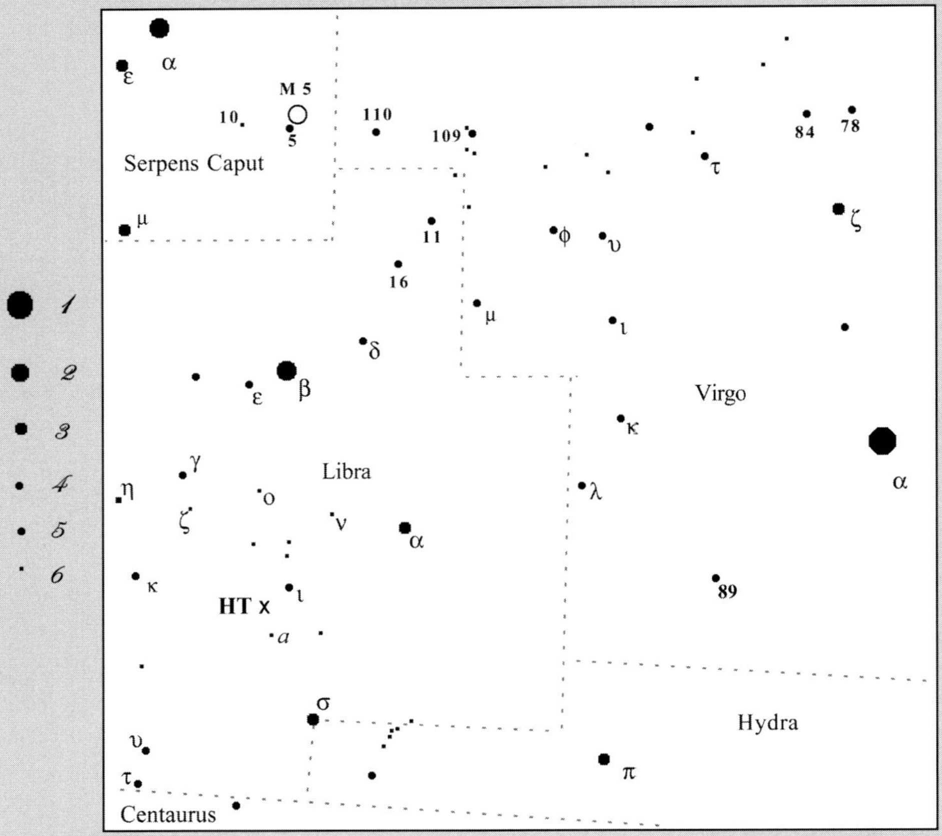

© Stephen James O'Meara

Tirion: Chart 21

Uranometria: Chart 334

76

Ghost Globular
NGC 5897
Type: Globular Cluster
Con: Libra

RA: 15^h 17.4^m
Dec: −21° 01′
Mag: 8.2 (O'Meara); 8.4
Diam: 11′
Dist: 40,400 light-years
Disc: William Herschel, 1785

W. HERSCHEL: [Observed March 10, 1785] A very close, compressed cluster of stars 8′ or 9′ in diameter, extremely rich, of an irregular round figure, a little extended. The stars are so small as hardly to be visible, and so accumulated in the middle as to look nebulous. (H VI-19 = VI-8?)

NGC: Globular cluster, pretty faint, large, very irregularly round, very gradually brighter in the middle, well resolved, clearly consisting of stars.

NGC 5897 IS AN UNUSUALLY LOOSELY structured globular cluster that haunts the dim and unassuming stars of Libra. William Herschel found the cluster on March 10, 1785, and logged it as the 19th object in his Class VI category of objects. It is possible that he observed it earlier, on April 25, 1784, when he logged his H VI-8 with an inaccurate position. He considered it an important discovery because, in his view, it was "one of the gradations from palpable congeries of stars . . . towards the distant nebulae." In his explorations of the nebulae, Herschel tried to connect the visual appearance of clusters and nebulae with some sort of physical reality. He believed that individual stars were closer to us than the nebulae. He believed that all nebulae were universes in the process of formation but seen at different distances. He believed that a nebula's degree of concentration was directly related to its state of evolution – highly evolved systems had high degrees of concentration and vice versa. NGC 5897 was a system that was intermediate in distance and in evolution.

Writing of this discovery in the *Philosophical Transactions* Herschel says:

> When I pursued these researches, I was in the situation of a natural philosopher who follows the various species of animals and insects from the height of their perfection down to the lowest ebb of life . . . In the same manner we pass through gentle steps from a coarse cluster of stars, such as the Pleiades, the Praesepe, the Milky Way, the cluster in the Crab, the nebula in Hercules, that near the preceding hip of Bootes, &c. &c, without any hesitation, till we find ourselves brought to such an object as the nebula in Orion, where we are still inclined to remain in the once adopted idea, of stars exceedingly remote and inconceivably crowded, as being the occasion of that remarkable appearance. It seems, therefore, to require a more dissimilar object to set us right again. A glance like that of the naturalist, who casts his eye from the perfect animal to the perfect vegetable, is wanting to remove the veil from the mind of the astronomer. The object I have mentioned above [NGC 5897], is the phenomenon that was wanting for this purpose. View, for instance, this cluster, and afterwards cast your eye on that cloudy star, and the result will be no less decisive than that of the naturalist we have alluded to. Our judgment, I may venture to say, will be, that *the nebulosity about the star is not of a starry nature.*

Herschel went on to call the cluster a "wondrous but difficult object." Admiral William Henry Smyth called it a "large compressed cluster of minute stars," saying "it is faint and pale, but owing to the fineness of the night, steadiness of gaze, and excellent action of the telescope, [it] was as well seen as so low and so awfully remote an object could be expected to be." In modern times, and seen from the Southern Hemisphere, Ernst Hartung simply calls it "irregularly round" and "conspicuous."

The cluster can be seen on clear, moonless nights as a large amorphous glow nearly one-third the apparent size of Omega (ω) Centauri (Caldwell 80). Look for what appears to be the detached head of a flaring comet nearly halfway between, and just south of, a line joining the wide double star Alpha^{1,2} ($\alpha^{1,2}$) Librae (Zubenelgenubi) and Pi (π) Scorpii. It may be helpful to use Beta (β), Delta (δ), and Pi Sco as the base of an arrowhead that points directly to Alpha^{1,2} Lib. Next, look for 4.5-magnitude Iota (ι) Lib, which lies about one-third of the way from Alpha Lib to Pi Sco in a direct line. If you

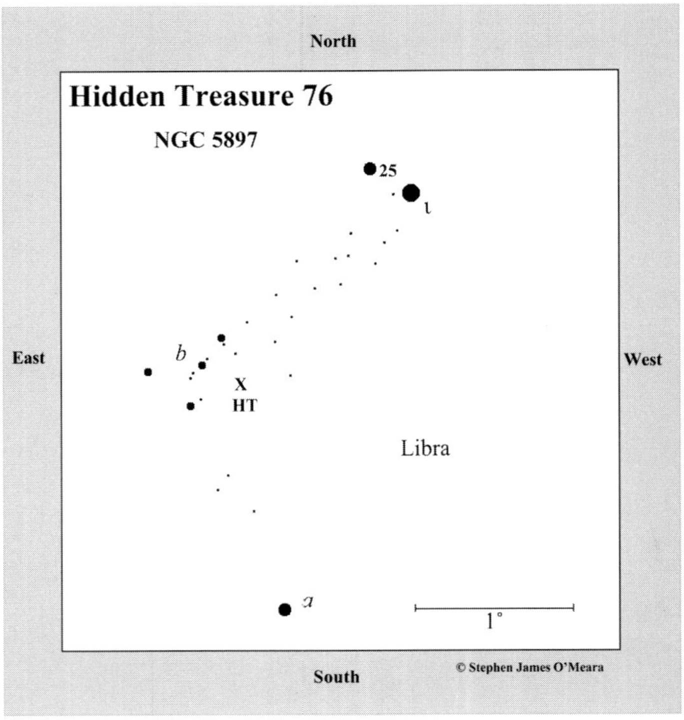

cannot see the star with your naked eye, use binoculars to locate it, then aim your finder at that location. There will be no mistaking Iota for two reasons: (1) it is the only star of that magnitude in the region; (2) 6th-magnitude 25 Lib lies only about 17′ to the northeast, so you will see a fine pair of bright stars in your binoculars or telescope. The globular is about $1\frac{3}{4}°$ southeast of 25 Lib and 15′ west of a 25′-long, Y-shaped asterism (*b*) comprised of four 8th-magnitude stars. If you cannot find the Y, use your binoculars to locate 25 Lib, look $2\frac{3}{4}°$ south-southeast for a solitary magnitude 5.5 star (*a*). The globular is about 30′ east of the midpoint between these two stars.

From a dark-sky site, NGC 5897 is visible in 7 × 50 binoculars. My antique scope shows it as a beautiful but dim cometary glow that is part of a Hershey's Kiss asterism, comprised of the Y-shaped asterism to the east and an 8th-magnitude star to the west. NGC 5897 is the chocolate inside the wrapper. At a glance in the 4-inch at 23×, the cluster is a perfect comet without a tail, reminiscent of M56 in Lyra. Indeed, the two globulars have a similar brightness and apparent size, though M56 is about 10,000 light-years closer. NGC 5897 is hypnotic at low power. The longer I look at it the more mysterious it looks. With averted vision, I see a central pip that swims in and out of view. Also, a multitude of phantom stars float across the cluster's periphery.

At 72× NGC 5897's oblate disk appears buffed and featureless, as if its stars had been scrubbed away with bleach. But with a little concentration I could resolve the glow into a patchwork of glittering hazes that lacks a strong central condensation. The cluster's brightest stars shine at magnitude 13.3, and they appear to form slight needles of light

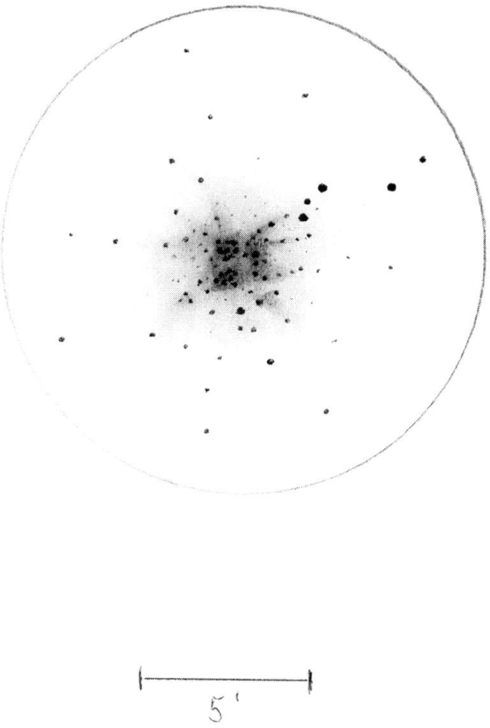

that border minute "holes" at the core. With concentration, these holes become patchy dark lanes cutting through it in all directions. With 101×, the cluster's core is an hourglass of starlight offset east of center. The core itself is dark. If the atmosphere is steady, try defocusing the cluster ever so slightly and see if you do not see the light of the globular caged by bars of darkness oriented north–south. When I do this, the dark core becomes the central bar. Now refocus the telescope and lower the magnification. This time, concentrate on just the outer halo. Does it look lily shaped? The brightest stars in the halo extend outward in weak and wavy arms, and with imagination look like wilting flower petals. An arc of foreground stars (magnitudes 11 to 13) illuminate the cluster's northeast border. If I were William Herschel, I would see four tiers of stars. The bright arc of stars would be closest. The cluster's hourglass shaped core would be on the next tier, followed by the sprinkling of dimmer suns in the halo, and finally the luminous "ether" in the background.

NGC 5897 resides in our galaxy's halo more than 40,400 light-years from our Sun and 23,800 light-years from the galactic center. I call it the Ghost Globular because it resembles a ghost image of the globular M55 in Sagittarius. Actually, NGC 5897 measures nearly 135 light-years across, making it 68 percent larger than M55. In size and morphology, then, NGC 5897 is more like NGC 288 (Hidden Treasure 4), which is only about 20 percent smaller. Yet, while NGC 5897 and NGC 288 are both loose clusters with significant blue straggler populations, the two clusters differ greatly in metallicity. NGC 5897's members each contain, on average, about $\frac{1}{50}$ to $\frac{1}{79}$ as much iron (per unit of hydrogen) as does the Sun; NGC 288's members each contain, on average, about $\frac{1}{13}$ to $\frac{1}{20}$ the solar abundance.

These numbers are significant because metal-poor globular clusters, like NGC 5897, probably formed at an early stage of the evolution of the galaxy, during the collapse of the protogalactic cloud. Metal-rich clusters, on the other hand, probably formed later, out of more metal-enriched gas. NGC 5897's metallicity is more similar to that of M3 in Canes Venatici, whose estimated age is 13 billion years (with an uncertainty of about 2 billion years). In a 1992 *Astronomical Journal* paper, Ata Sarajedini (Yale University) determined that NGC 5897 is about 2 billion years older than M3.

Hidden Treasure 77

NGC 5986

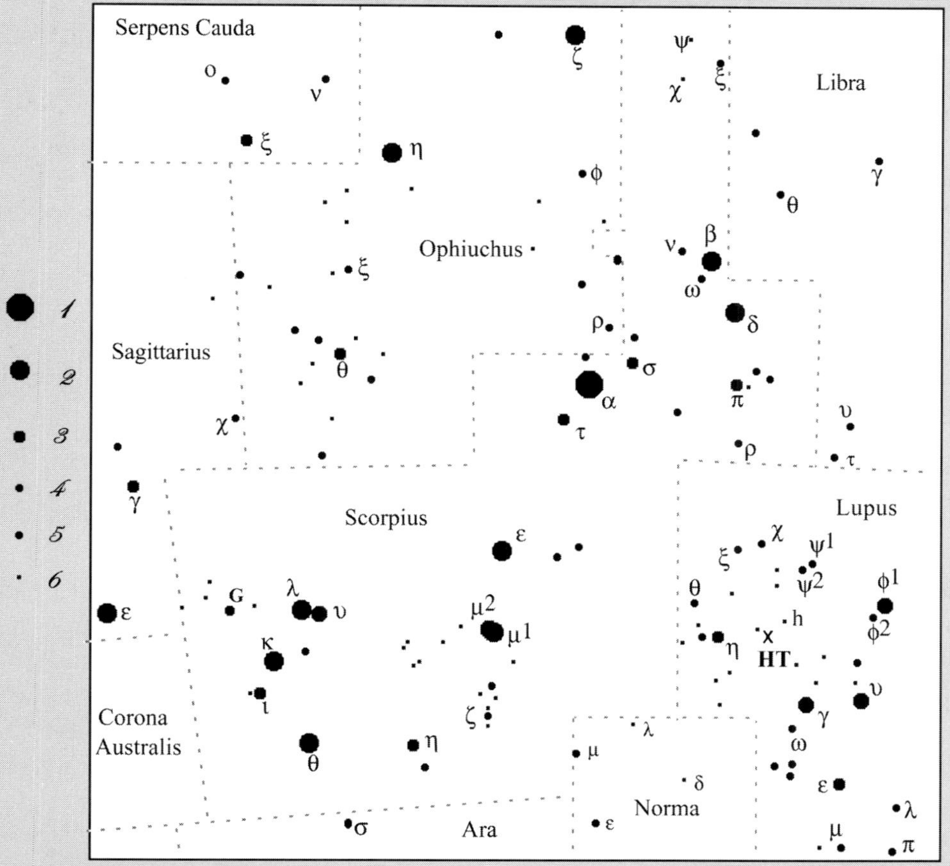

Tirion: Charts 21 & 22

Uranometria: Chart 374

77

NGC 5986
Type: Globular Cluster
Con: Lupus

RA: 15^h 46.1^m
Dec: $-37°$ $47'$
Mag: 7.8 (O'Meara); 7.6
Diam: $9.6'$
Dist: 34,000 light-years
Disc: James Dunlop, included in his 1827 catalog

NGC: Remarkable, globular cluster, very bright, large, round, very gradually brighter in the middle, stars from magnitude 13 to 15.

NGC 5986 IS A LITTLE-STUDIED, compact globular cluster lying in the Milky Way's inner halo 15,600 light-years from the galactic center. It is partly immersed in a rich swath of Milky Way that sweeps through northern Lupus, the Wolf. An ancient constellation, Lupus probably dates to the Akkadians of Mesopotamia (~2350 BC), who saw the stars in this region not as a wolf but as *Urbat,* the Beast of Death. Who, or what, that beast was remains unknown. The Greeks and Romans later identified it as a wild beast and, in the second century BC, Aratos referred to it in his *Phaenomena* as the sacrificial beast clutched in the Centaur's hand. The constellation did not become recognized as a Wolf until Renaissance times, apparently after an erroneous translation of *Al Fahd,* the Arabian name for either a lioness, leopard, or panther. And while the constellation has no clear mythology, some Greek scholars believe that the Wolf is the wicked King Lycaon of Arcadia. According to one version of the story, Zeus, disguised as a day laborer, descended to Earth to investigate some rumors of Lycaon's evil ways. When Zeus arrived at Lycaon's palace, the "laborer" was greeted with mock hospitality and treated to a feast that included one of his own sons as dessert. Outraged and disgusted, Zeus, disposed of his disguise and blasted Lycaon's palace with thunderbolts. The terrified king tried to escape Zeus's wrath by fleeing into some neighboring fields, but Zeus was quick to transform this savage beast into, well, a savage beast. Lycaon tried to cry out for help, but all he could do was howl, as his royal robes transformed into shaggy hair and his mouth filled with hideous fangs, which he used to ravage his own sheep with Lycaon barbarity.

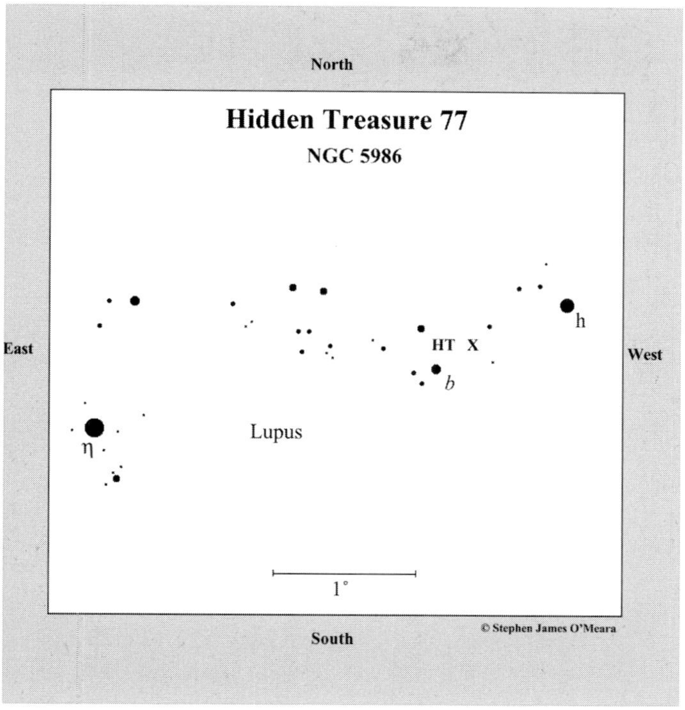

James Dunlop discovered this sheepish cluster on May 10, 1826, while surveying the southern skies from Parammata, New South Wales. He listed it as the 552nd object in his 1827 catalog of nebulae and star clusters. Through his 9-inch f/12 reflector, he described it as a "beautiful round pretty bright nebula . . . pretty well defined." While NGC 5986 is best viewed from southern locals, it should be a decent target for many Northern Hemisphere observers, since it is only 3° further south than the famous open star cluster M7 in Scorpius, and 30′ further south than Upsilon (υ) Scorpii, the westernmost star in the Scorpion's stinger.

To find it, simply draw a straight line 8° west-southwest from the stinger stars to the naked-eye double star Mu^1 (μ^1) and Mu^2 (μ^2) Sco in the Scorpion's Tail, then 10° further to 3rd-magnitude Eta (η) Lupi, which is the southwestern arm of an upside down Y-shaped asterism of 3rd- to 5th-magnitude suns. Now use binoculars to sweep $2\frac{1}{2}$° to the west-northwest of Eta Lupi where you'll find a magnitude 7.5 star (*b*) with a 5th-magnitude star (*h*) a little more than 1° to the northwest. NGC 5986 is a compact 7th-magnitude glow 20′ west-northwest of Star *b*. Under a dark sky, the cluster is easily spied in 7 × 50 binoculars. It's an absolutely beautiful object in the antique telescope, being a clearly defined cometlike glow. Had this object been just a little further north, Messier and his contemporaries would have certainly nabbed it for masquerading as a comet.

At 23× in the 4-inch, NGC 5986 displays a bright, moderately condensed core surrounded by an aura of dim suns. These stars, the brightest of which shine at 13th magnitude, flit about like milkweed seeds on a hot summer's day. At low power the cluster has a soft aquamarine hue, at higher powers it looks somewhat yellowed, like an eye with a cataract. In fact, depending on how you look at it, the cluster's total color can alternate between the two hues. This confusing view made sense after I saw a charge coupled device (CCD) image of NGC 5986 made with the 35-inch reflector atop Cerro Tololo in the Chilean Andes. It shows the cluster to be comprised of a fine mix of yellow and blue suns distributed equally throughout the cluster.

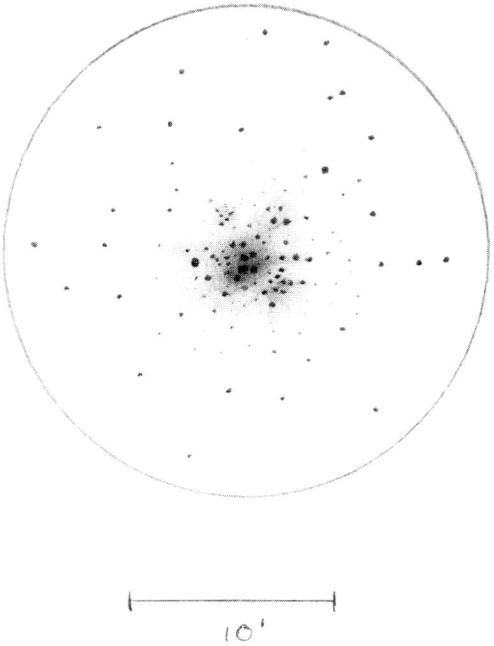

As an inner halo object, NGC 5986 has an intermediate metallicity with $\frac{1}{30}$ to $\frac{1}{49}$ as much iron (relative to hydrogen) as does our Sun. Two very bright and highly evolved Type *A–F* stars have also been identified in the cluster; these pulsating red giants shed matter in the form of a strong stellar wind and are probably nearing the planetary-nebula stage of their evolution. NGC 5986's color-magnitude diagram reveals a blue horizontal branch of stars – those of spectral Type *B*3 to *A*0 that have evolved past the red-giant stage and are now burning helium instead of hydrogen at their cores. The color-magnitude diagram is very similar to those of M2 or M13, which are similar in age, being around 13 or 14 billion years old. But, as William Harris says, "An age difference of ± 1 billion years is at the limits of distinguishability for present-day precision photometry. Any differences quoted smaller than that, I simply would not believe unless it is based on really exceptional data and really careful model fitting." The cluster also contains about half a dozen RR Lyrae variables, which yield a distance of about 36,500 light-years from the Sun, which is in good agreement with William Harris's distance to the cluster. NGC 5986's true physical diameter, then, is about 100 light-years – almost one-third the size of Omega (ω) Centauri, the biggest globular in our Milky Way galaxy.

At 72× in the 4-inch, I can see a number of stubby arms, like those of a starfish.

The cluster is punctuated on the eastern flank by a 12th-magnitude foreground star. Look also for two 13th-magnitude stars in the cluster's northeast quadrant; these are the two Type *A–F* stars mentioned above. NGC 5986 takes magnification well, so go as high as you want. I found 216× best for seeing the finer details in a 4-inch. The inner core measures about 2.5′ across and is very complex. Several stars, or groups of stars, punctuated a circular central patch, which has rays of stars extending to the south, southwest, and the west. The larger outer halo is fractured and frayed with long tufts of unresolved starlight that look like weeds growing out from the base of a circular rock; some are populated with bits of starlight, looking like ants in the grass. The most prominent tufts extend to the southwest and northwest. Can you see a dark, L-shaped rift in the cluster's halo on the southwest side? All these features make the cluster look disheveled, aged, and worn, like a weather-beaten flag flapping above a pirate's rig. It's quite a wonderful globular.

Hidden Treasure 78

NGC 6210

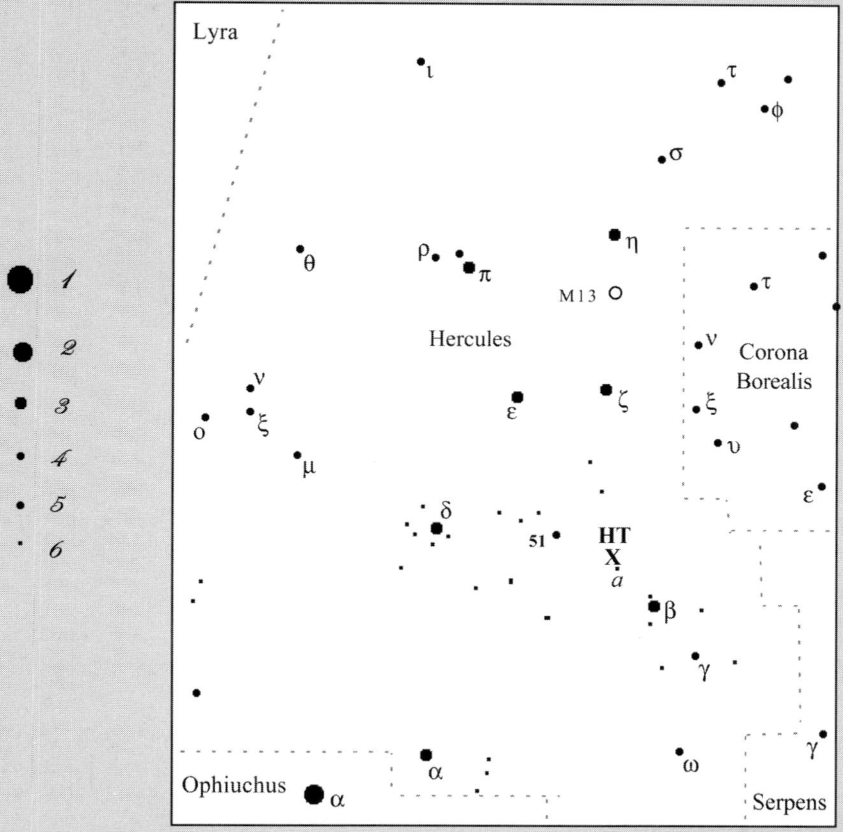

© Stephen James O'Meara

Tirion: Chart 8

Uranometria: Chart 156

78

78

Turtle Nebula
NGC 6210
Type: Planetary Nebula
Con: Hercules

RA: $16^h\ 44^m\ 29.8^s$
Dec: $+23°\ 48'\ 00''$
Mag: 8.4 (O'Meara); 8.8
Dim: $48'' \times 8''$
Dist: 3,600 light-years
Disc: Friedrich Georg Wilhelm von Struve, 1827

J. HERSCHEL: Struve's 5th nebula in the list at the end of the Dorpat Catal of D stars. Very bright, equal to a star of 8 or 8.9 magnitude, 8″ diameter, and of a uniform light, but with the edges boiling and ragged. A fine object like a star out of focus. Viewed between clouds. Struve's place. (h 1970)

NGC: Planetary nebula, very bright, very small, round, disc and border.

DURING HIS PRODIGIOUS CAREER, Wilhelm Struve (1793–1864) – the first of seven Struves in five generations to obtain a Ph.D. (or its equivalent) in astronomy, and the director of Dorpat and later of Pulkovo Observatory – was one of the greatest astronomers of all time. Among his many accomplishments, he founded Pulkovo Observatory and made fundamental contributions in studies of double stars, stellar parallax, galactic structure, astrometry, and geodessy. It was while he was searching for double stars between 1824–1826 with the Dorpat 9.6 f/17.8 Fraunhofer refractor that Struve discovered seven *NGC* objects. Prime among them is NGC 6210, a brilliant planetary in Hercules and our hidden treasure; it was the fifth object listed in the appendix of his 1827 *Catalogus novus generalis stellarum duplicium et multiplicium.* The other six *NGC* objects are as follows: NGC 629, an asterism in Cassiopeia; NGC 6572 (Hidden Treasure 90), a planetary nebula in Ophiuchus; NGC 6648, a double star in Draco; NGC 2202, a double star in Orion; NGC 6871, an open cluster in Cygnus; and NGC 6873, an asterism in Sagitta.

While Hercules harbors two Messier globular cluster (M13 and M92), it's hard to believe that a planetary nebula as bright, or brighter, than M57, the Ring Nebula in Lyra, went unnoticed by not only Messier, Mechain, and their contemporaries, but also by the great William Herschel. But it's true. Equally surprising is that Burnham does not spotlight it in his *Celestial Handbook.* Perhaps, for that reason, NGC 6210 has remained one of the best-kept secrets in Hercules. Even the dim members of the great galaxy cluster in Hercules get more attention these days than this bright planetary.

Professional astronomers have not shied away from this Herculean gem, and it has been the subject of many studies. Descriptions of the planetary's spectrum date to Royal Observatory director Richard Hugh Stoy's pioneering work in 1935, while the intensities of the principal lines were first measured by Lawrence Aller (University of California, Los Angeles) in 1941. Early photographic images show an amorphous shape with a concentration of light toward the center. And a comparison of calibrated spectroscopic plates taken between 1938 and 1970 showed no obvious intensity changes. Likewise, O. C. Wilson's radial velocity measurements in 1950 indicated nothing unusual about the object's internal motions. In a 1970 study, Aller notes that he and his colleagues could not detect any obvious small-scale knots or blobs. The central star, however, did show a relatively high-excitation spectrum; its temperature being probably near 60,000 K. Aller and his colleagues regarded the nebula as "smooth."

By the mid 1990s, technological changes allowed astronomers better resolution in spectral imaging. In a 1996 *Astronomical Journal* paper, for instance, Spanish astronomer J. P. Phillips and L. Cuestra (Astrophysical Institute of the Canaries) acquired low- and intermediate-resolution spectroscopy at 11 positions across the nucleus of NGC 6210, which enabled them to map variations in velocity, temperature, line excitation, and density over the object's core and halo. They discovered that temperatures reach a peak value of aboout 11,000 K close to the nucleus and decline by 24 percent toward the periphery of the core, some 7.5″ from the 13th-magnitude central star. The nebula shows a greater degree of variation in its density, being 70 percent denser toward the north than the south, with smaller variations elsewhere across its face. Indeed, long-exposure images of NGC 6210 reveal a circular halo 65″ in diameter, containing two pairs of armlike extensions on opposing sides of the nucleus. All these properties lead the investigators to believe that NGC 6210's nucleus, in fact, may be ejecting material along two, and possibly four, opposing directions, the shock of which interacts with the ionized shells of gas, creating all manner of complex details. The jet also diffuses and curves the farther

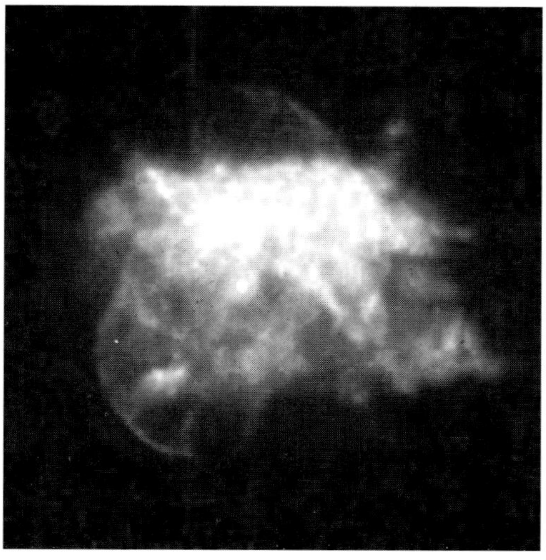

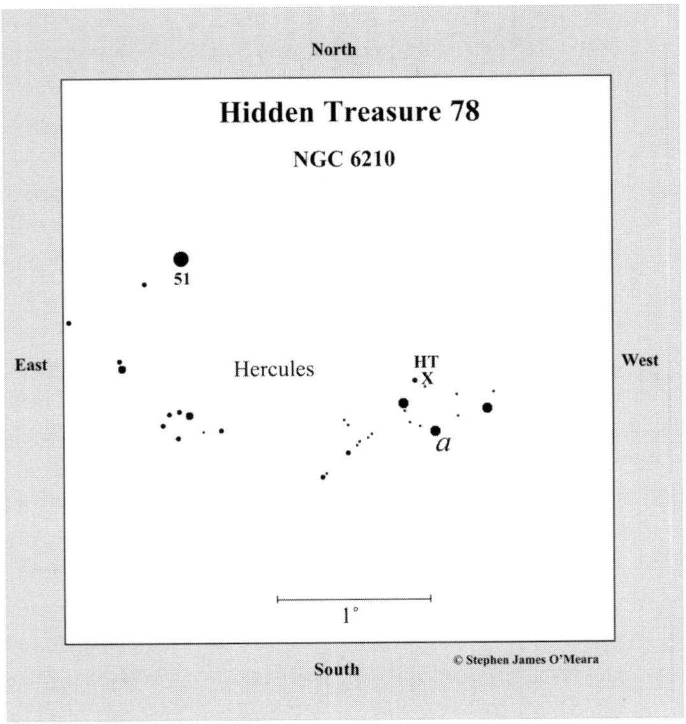

"seashell" is the nebula's central spirograph of ionized wonder. I think it looks more like a mid-air collision of two intergalactic warships shortly after impact. What the image really shows are the jets of hot gas streaming through holes in the nebula's older, cooler shell of gas. In true physical extent the nebula measures 0.8 × 0.1 light-years.

To find this treasure, first locate the 3rd-magntude stars Delta (δ) and pale yellow Beta (β) Herculis (Kornephoros), the shoulder stars of the mythological Strongman. Midway between these stars and a bit northwest is 5th-magnitude 51 Herculis; it marks the west end of a $2\frac{1}{2}°$-wide diamond (oriented east–west) of slightly dimmer suns. Now hop 2° to the southwest where you will find a roughly 30′-wide triangle of 7th- and 8th-magnitude suns, with the brightest member (Star *a*) forming the southern apex. NGC 6210 is 18′ north and slightly east of Star *a*; it is also 10′ northwest of the magnitude 7.3 star marking the triangle's eastern apex.

The nebula is easily visible in 7 × 50 binoculars as a roughly $8\frac{1}{2}$-magnitude stellar "spot." It is the same in the antique telescope only more prominent. In the 4-inch it is visible at a glance as a virtually stellar object. It is obvious, then, why Messier and his contemporaries passed over it. But since I have the advantage of knowing what this object is, I can concentrate on it for a while with averted vision. And when I do, it

away it is from the nucleus, which may indicate that the collimating beam is rotating – perhaps as the central star and an unseen companion revolve around a common center of mass.

The tide of public attention suddenly turned to NGC 6210 when the Hubble Space Telescope published its titillating image of it (seen above). While we have become accustomed to seeing planety nebulae in many fantastic shapes and forms – hourglass, butterfly, stingray, spheres, and eyes – the HST image of NGC 6210 added "another bizarre form to the rogues' gallery of planetary nebulae: a turtle swallowing a seashell." Indeed, as seen in hydrogen-alpha light, the bright inner shell of gas has a boxy, turtle-shell shape. Seen together with its four, flipper-like arms, the nebula does indeed resemble a turtle. The injested

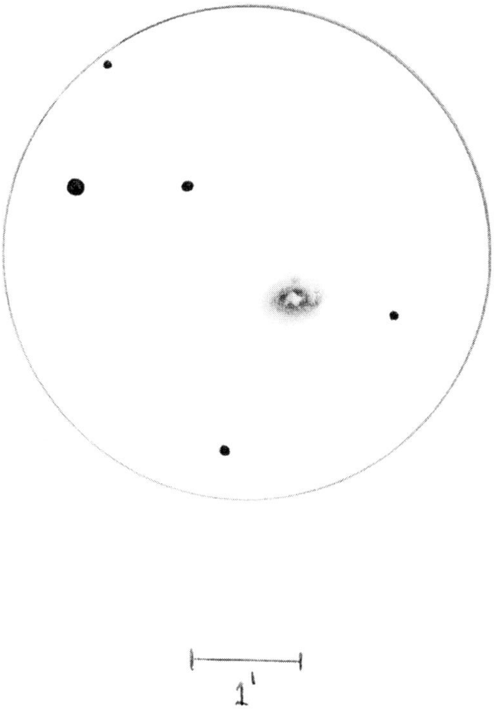

becomes apparent that it is not stellar, even at this low power. Its light has a softer glow than that of a star. It also has a slight aquamarine sheen. (The great Earl of Rosse saw it as an intense blue through his magnificent Leviathan).

These appearances become more apparent at 72×; though, it's important to note that, had I not known that this object is a planetary nebula, I probably would not have noticed. The Rev. Thomas W. Webb saw it as "small, not sharply defined; exactly like a star out of focus," which is how John Herschel described it. And here is the reason, I'm certain, that Herschel passed over it. A night of even mediocre atmospheric seeing will turn stars into slightly swollen disks, mimicing the appearance of NGC 6210.

Once again, the more time I spend with this very high-surface-brightness object, at 72× the more detail I see. There is a bright core wrapped in some very tight layers of haze. I also get the impression of a diffuse outer halo. Through a $5\frac{1}{2}$-inch telescope, Webb, too, says that he saw a glow around the out-of-focus-starlike core. The view improves only slightly at 101×, so I increase the power to 182× and can see the planetary elongated west-northwest–east-southeast. But even at this power the object has an extremely tight core, though it begins to look mottled and layered.

At high powers, it is a complex mix of knots, rings, and other amorphous shapes. So, my impression is a complex nebula with two ellipses perpendicular to one another, like two galaxies colliding – with one passing through the nucleus of the other. I get that impression with all manner of high powers. Another impression I cannot shake is that of a "drawing compass" of knots, with the focus being to the northeast; two other knots – one on either side of the "compass" at the ansae of the most elliptical parts of the annulus (to the northwest and southeast) complete this knot complex. I have to wonder if this complex is what Italian astronomer Angelo Secchi (1818–1878) observed when he thought he resolved the nebula with a power of 1,500×. The central complex of knots can be easily confused as a central star at low powers. The central star of the nebula is supposed to be magnitude 13.7 with possible variability. Anyway, that's about the brightness of the central star in M57, the Ring Nebula, which is difficult to see in a wide open annulus. There is no hope of seeing it in a 4-inch because the core or inner ring is too tight.

Smyth notes that "there are four stars in the field, of which that in the [southeast] quadrant is of the 6th magnitude and brightly reddish, affording a fair test of comparison with the pale blue nebula." He also notes that "it presents a visible disc of 8″ in diameter, at a distance probably equal to that of the star near it, the vastness of its dimensions is within the range of reasonable conjecture, however it may stagger the comprehension." You have to love it when our astronomical forefathers tried to use their minds to penetrate the two-dimensional sky and plumb the depths of space with their imaginations. That in itself staggers the comprehension.

Hidden Treasure 79

NGC 6242

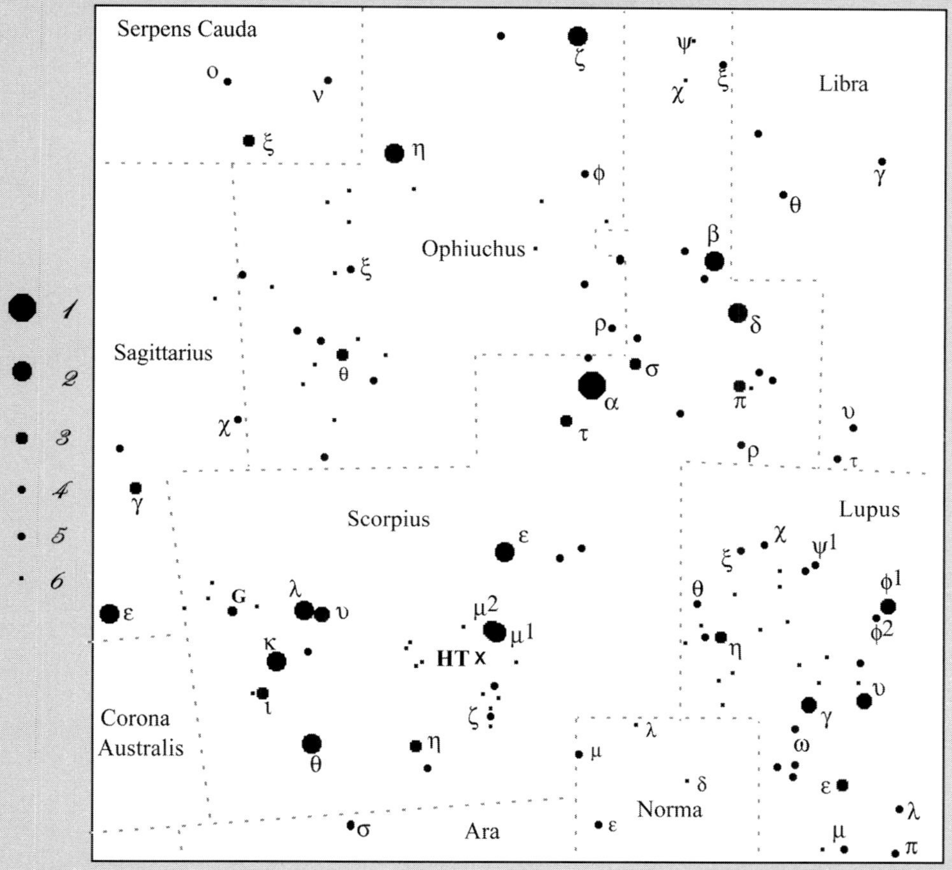

© Stephen James O'Meara

Tirion: Chart 22

Uranometria: Chart 407

79

NGC 6242
Type: Open Cluster
Con: Scorpius

RA: $16^h\ 55.5^m$
Dec: $-39^\circ\ 28'$
Mag: 6.2 (O'Meara); 6.4
Diam: 9.0′
Dist: 3,600 light-years
Disc: Abbe Nicolas Louis de Lacaille, listed in his 1755 catalog

J. HERSCHEL: Bright, large, rich, discrete, 12′, irregular figure, very little brighter in the middle, fine object; place of a red star 9th magnitude, rest 11th magnitude, white. (h 3654)

NGC: Cluster, bright, large, rich, stars from magnitude 8 to 11.

NGC 6242 IS ONE OF THOSE BRIGHT open clusters that, had it been anywhere else in the sky, it would have been a celestial showpiece. Unfortunately, it lies close to, but just far enough away from, one of the most dynamic sections of the Scorpius Milky Way – the Scorpius *OB*1 Association. This widely scattered family of young stars marks the location of a giant H-II region in the Sagittarius–Carina spiral arm. Its area includes such wonders as Zeta^{1,2} ($\zeta^{1,2}$) Scorpii and the most cometlike naked-eye spectacle in the heavens; the comet's head is open cluster NGC 6231 (Caldwell 76), while its tail is the elongated open cluster Trumpler 24. Trumpler 24 is flanked to the northeast by the diffuse nebula IC 4628, and to the west by the dark nebula Barnard 48. NGC 6242 lies about 1° north of the comet's "tail," which is far enough away to make it yet another stellar outcast. It is like a bit of celestial lint that we visually brush aside so that we can better appreciate the more prominent wonders.

NGC 6242 is yet another discovery by the Abbe Lacaille, during his pioneering telescopic survey of the southern sky. He listed it as the 10th Class I object (nebulae without stars) in his 1755 catalog of nebulous objects, though the only reason this cluster looked nebulous to him was because he used a $\frac{1}{2}$-inch diameter telescope and a power of 8×. Through that instrument, the object looked "like a faint patch, oval and elongated." Lacaille's instruments were not of the highest quality, and he did not spend much time observing these new objects. In fact, he writes, "[i]t is possible that each of these nebula [in Class I] may really be a faint comet: time did not allow me to decide, by searching the sky, to see if they remained always in the same place."

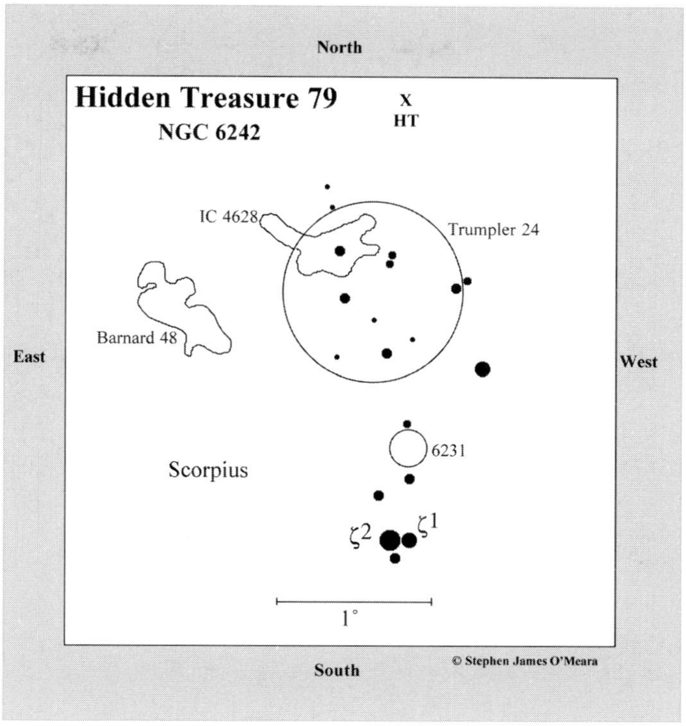

But we do have the leisure to take the time to investigate Lacaille's wonders. This particular one's declination is less than 5° further south than M7, the most southerly Messier object. The object is a little less than 3° due north and slightly east of Zeta[2] Sco. Once you know where to look, NGC 6242 is visible to the naked eye and is quite nice even in the twilight in my 7 × 50s. Under a dark sky, the cluster is partially resolved in binoculars. At first, it looks like a round haze, but with just a slight effort, its brightest stars pop right out and the core becomes or appears triangular. With my antique telescope, the cluster is also very nice, with a half dozen or so stars packed into a sparkling globe of fainter suns, which form a distinct haze.

In the 4-inch at 23×, the stars at the cluster's core form a distinct X, which lies on a carpet of faint suns whose ill-defined light shines like nebulosity. With averted vision this carpet sparkles as if bits of glass lie shattered in the weave, which is being illuminated by glancing light. A bright 7th-magnitude golden *K*5 star burns like a candle on the southeast end of the cluster. It is a foreground star, being 1,000 light-years closer than the cluster itself, which is about 2,500 light-years closer to us than NGC 6231 and Trumpler 24, which are in the Sco *OB*1 Association. All of these clusters, though, are members of the inner spiral arm next to the Sun. NGC 6242 is slightly dimmed by nearly half a magnitude by foreground dust, which is common with objects close to the galactic plane. At its distance of 3,600 light-years, NGC 6242 measures 9.5 light-years across. It contains only about 23 members, so much of what we see in the faint carpet are stars in the background Milky Way. David Malin of the Anglo Australian Observatory, notes that despite the foreground dust, astronomers have been able to estimate the age of the cluster by measuring the color and relative brightness of its stars. "The age turns out to be about 50 million years," Malin says, "which is quite mature for an open cluster of this type. By now, most of the brightest stars will have turned into supernovae and vanished, leaving this quite modest group

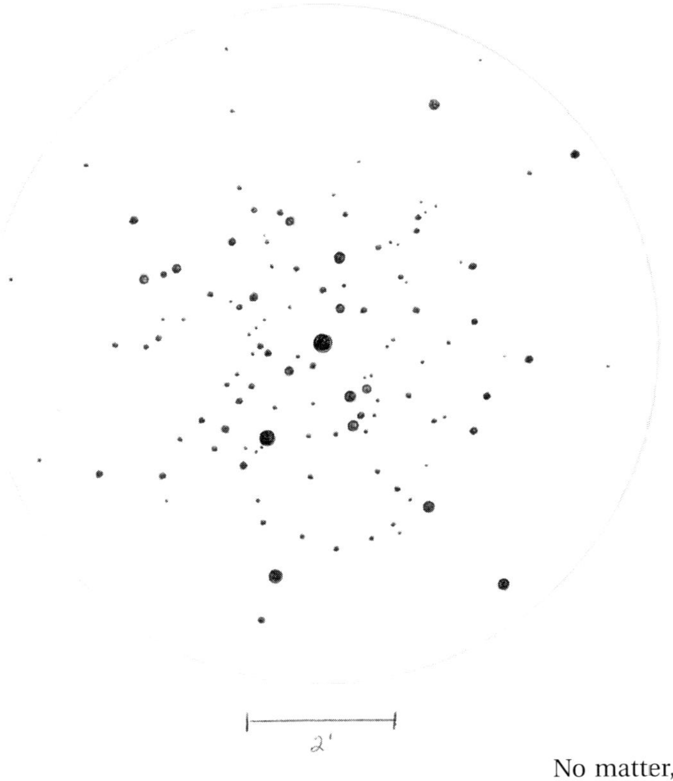

to compete with the many bright clusters of stars in this direction."

One mystery concerns John Herschel's second observation of NGC 6242 with the 18-inch reflector in South Africa. He reports seeing it as a large and rich cluster with "a red star 8–9th magnitude in centre." When Dunlop observed this object from Paramatta, New South Wales, before him in 1826 with his 9-inch f/12 telescope, he described the cluster as having "considerable compression of the stars towards the centre of the group," but noted no color. Herschel's third observation, however, revealed only "one [colorless] 7–8th magnitude [star] near middle."Color photographs of the cluster today fail to show any bright red member at the center. Could a variable star be lurking here?

Equally interesting, neither Herschel nor Dunlop refer to the 7th-magnitude golden star at the southeast end of the cluster. Of course, different people see the star's color differently. I see it as golden, Luginbuhl and Skiff record no color, and Hartung sees it as a "bright orange red." Steve Coe of Gundale, Arizona, observing with a 17.5-inch reflector at 100× does not single out the star or its color, as do not some other observers.

No matter, the cluster and its surrounds are, of course, glorious, especially at low power. In a way, it's hard to discern the extent of the cluster, whose arms radiate wildly in all directions. With 72×, the cluster is all patches and generous sweeps of starlight, like sprinkled glitter. On the eastern side of the cluster is a long dim banner of stars running north-northwest–south-southeast. It is bordered to the west by a parallel lane of relative darkness, which gives the cluster an asymmetrical appearance – being brighter on the western half and dimmer to the east. This may be due to the presence of intervening dust, where Archinal notes there is a dark nebula.

Hidden Treasure 80

NGC 6281

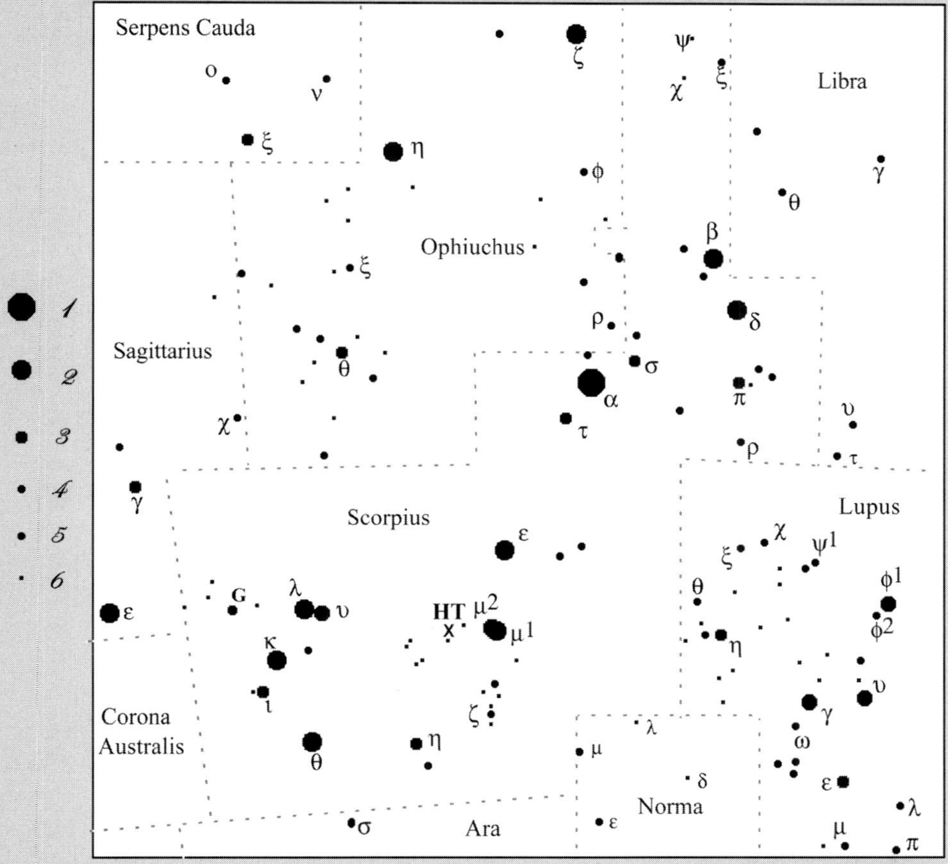

Tirion: Chart 22

Uranometria: Chart 376

80

Moth Wing Cluster
NGC 6281
Type: Open Cluster
Con: Scorpius

RA: $17^h\ 04.8^m$
Dec: $-37°\ 53'$
Mag: 5.8 (O'Meara); 5.4
Diam: 8.0′
Dist: 1,800 light-years
Disc: James Dunlop, included in his 1827 catalog

J. HERSCHEL: A pretty rich, large, pretty bright, cluster VII class, of loose stars ranging from magnitude 9 to 11, which fills $\frac{2}{3}$ of the field. (h 3664)

NGC: Cluster, large, pretty rich, little compressed, stars from magnitude 9 to 11.

THE TAIL OF SCORPIUS, THE SCORpion, is projected against the hub of our galaxy, a region littered with open clusters and asterisms. Some might argue that once you move beyond the Messier and Caldwell catalogs, open clusters are all the same – little puffs of concentrated starlight. But one, NGC 6281, stands out from the multitude as a fine target for small telescopes. In fact, at magnitude 5.4, NGC 6281, is the brightest non-Messier/non-Caldwell open cluster in Scorpius, though one hardly hears about it.

On June 5, 1826, James Dunlop swept up the cluster with a 9-inch reflector in New South Wales. He listed it as the 556th object in his 1827 catalog, calling it "[a] curiously curved line of pretty bright [faint] stars, with many very [faint] stars mixt." John Herschel later described it as both "pretty rich" and "pretty bright." Indeed, the cluster's surface brightness is quite high, making it a fine naked-eye target. NGC 6281 contains some 70 stars to 13th magnitude, with the brightest shining at magnitude 9.0.

NGC 6281 is a medium-age galactic star cluster (220 million years) with variable reddening across its face. It is the same age as M7 in Scorpius, but NGC 6281 is both smaller and dimmer than it in apparent size. Not only is NGC 6281 more than twice the distance of M7, it is more than four times smaller in true physical extent. NGC 6281's distance is more comparable to that of M6, which is only 200 light-years closer. But even M6 is twice the size of NGC 6281 in true physical extent. Indeed, NGC 6281 is quite

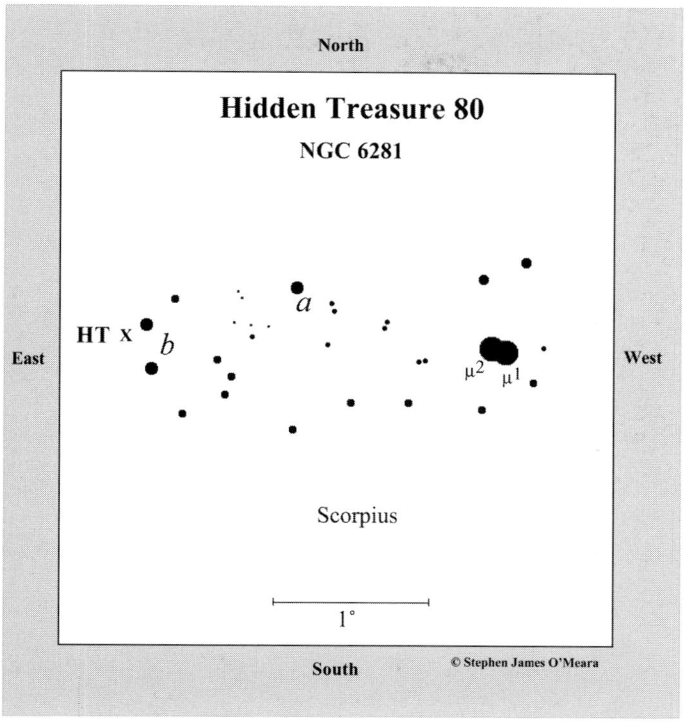

Although bright, NGC 6281 is also neglected, mainly because the cluster is near a host of other popular objects. It is $4\frac{1}{2}°$ north-northeast of NGC 6231 (Caldwell 76), 2° southwest of the Bug Nebula (Caldwell 69), 10° southwest of M7, and 8° northeast of NGC 6124 (Caldwell 75). At least NGC 6281 is conveniently located. It simply lies $2\frac{1}{2}°$ almost due east of $Mu^{1,2}$ ($\mu^{1,2}$) Scorpii, about one-third of the way toward the paired Stinger stars at the end of the Scorpion's tail. Try to locate it first in binoculars. From $Mu^{1,2}$ look for a solitary 6th-magnitude star (*a*) almost $1\frac{1}{2}°$ to the east-northeast. Next look just 1° to the southeast of Star *a*, for a pair of 6th-magnitude stars (*b*), which are oriented north–south and separated by 20′. NGC 6281 is just east of that pair of stars and essentially tangential to the northernmost star (HD153919) in the pair. If you are under a dark sky and can see to 6th magnitude, look for NGC 6281 with the unaided eye. The only difficulty should be trying to separate its light from that of HD153919's.

In the antique telescope, NGC 6281 is a beautiful object, appearing as a rosette of sparkling suns. Through the 4-inch, NGC 6281 is best seen at 23×, when its stars form a marvelous dipper of bright suns superimposed on a fog of Milky Way. The cluster's brightest star is a nice double star with a 9th-magnitude primary and a 10.8-magnitude companion; this double is the second star in the dipper's handle, which is comprised

tiny, spanning only 4.2 light-years of space, making it the smallest galactic star cluster (in true physical extent) in the Messier, Caldwell, or Hidden Treasure catalogs. It was once believed that a 6th-magnitude star (HD 153919) near the northwestern flank of NGC 6281 was physically related to the cluster, but this no longer appears to be the case. The star is very young, about 1 million years. Instead, A. Feinstein and J. C. Forte (Astronomical Observatory of the National University of La Plata, Argentina) suggest that HD 153919 may be a runaway star from NGC 6231 (Caldwell 76) – whose age is between 3.8 and 6.5 million years – because its direction of proper motion passes about 1.7° from the center of that cluster. HD 153919 is also an X-ray eclipsing binary star and an irregular optical variable.

it "a pretty cluster" for 60-mm telescopes.

Through the 4-inch, I counted some 40 suns to about 13th magnitude in an area of 15′, while another dozen or so stars form two prominent, though stringy, arms to the south and southwest; whether these arms belong to the cluster, I do not know. Brent Archinal and Steven Hynes list the cluster's diameter as only 8′ in apparent size, while several professional papers from the 1970s and 1980s include stars to 20′. Another fine double star occupies the center of the dipper's bowl. With imagination, it looks as if some fortunate prospector had scooped up two nuggets while panning for gold. The northwest side of the dipper forms the southern arc of a wider ellipse of roughly 11th- to 12th-magnitude suns; this arc is oriented northeast–southwest. Look carefully now for another ellipse of similarly bright suns tangential to this ellipse but oriented northwest–southeast. This second ellipse loops through the dipper's bowl. Seen together, the two faint ellipses look like moth wings.

of a straight line of five stars on the cluster's northwest side. The cluster's brightest stars are arranged in such a way that it inspires flights of fancy. With imagination, I can see the entire sprawl looking like a butcher wearing an apron and holding a meat clever. The late southern observer Ernst Hartung (1893–1979) sees my dipper as a "very beautiful . . . pyramidal group of stars; it contains several pairs and two bright orange stars, and the linear pattern is most striking." Christian Luginbuhl and Brian Skiff add that the cluster has "20 stars arranged over haze like Christmas tree lights," making

Before you move on, be sure to try for the magnitude 11.3 globular cluster NGC 6256, which lies 30′ to the north-northeast of 6th-magnitude Star *a*.

Hidden Treasure 81

NGC 6369

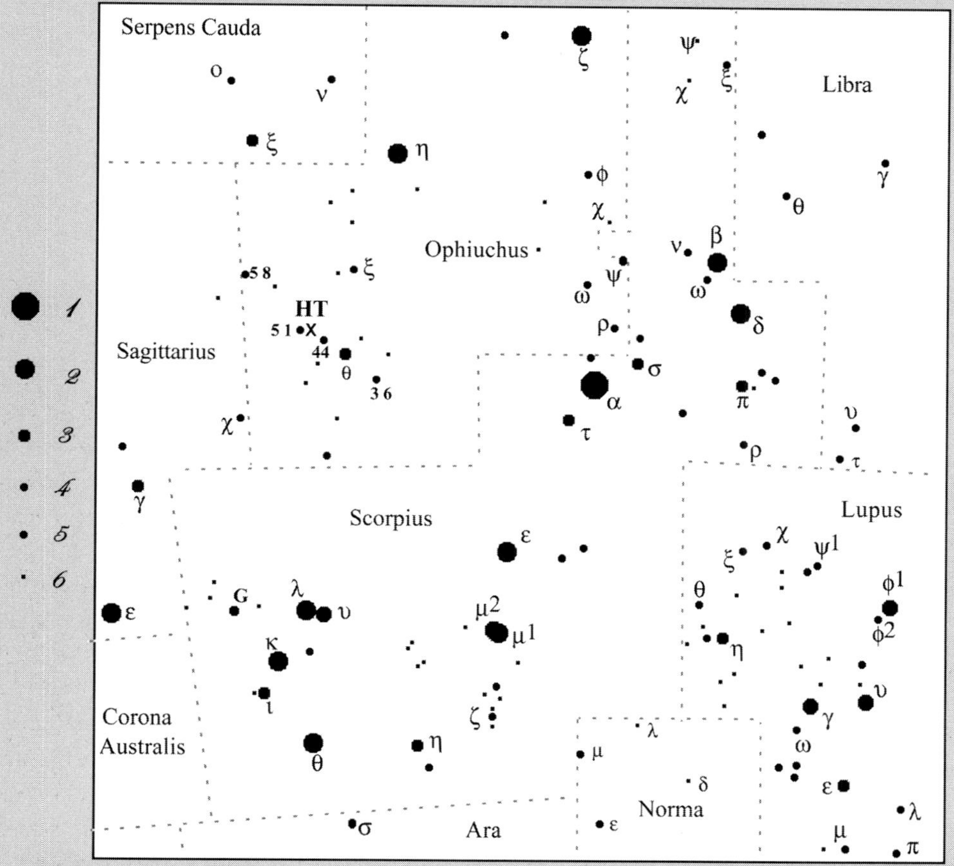

© Stephen James O'Meara

Tirion: Chart 22

Uranometria: Chart 338

81

Little Ghost Nebula, Ghost of Mars Nebula
NGC 6369
Type: Planetary Nebula
Con: Ophiuchus

RA: $17^h\ 29^m\ 20.7^s$
Dec: $-23°\ 45'\ 35''$
Mag: 11.4 (Skiff)
Dim: $58'' \times 34''$
Dist: Between ~2,000 and 5,000 light-years
Disc: William Herschel, 1784

W. HERSCHEL: [Observed May 21, 1784] Pretty bright, round, pretty well defined planetary disk, 30″ or 40″ in diameter. (H IV-11)

NGC: *Remarkable*, planetary nebula, pretty bright, small, round.

NGC 6369 IS A BEAUTIFUL RINGED planetary nebula in the rich Ophiuchus Milky Way. Actually, it lies in a significant black lagoon (Barnard's dark cloud, B77), just north of the famous Pipe Nebula (B78). Easily visible to the naked-eye, the Pipe Nebula is an enormous dark cloud, whose stem starts about 10° east of Alpha (α) Scorpii (Antares) and slices 7° through the celestial backwaters between the Scorpion's tail and the Serpent Bearer's feet. The Pipe's inner surface hugs a dense and conspicuous star cloud centered on 3rd-magnitude Theta (θ) Ophiuchi, "the observer's guide to the wonders of the Milky Way in this region," as Robert Burnham, Jr., called the star in his *Celestial Handbook*. One could spend an infinity exploring this dust-flecked complex of galactic wonder, which is probably why NGC 6369 is so elusive; it's like a tiny detail in a Georges Seurat painting seen from afar; it's easy for the eye and mind to overlook such an individual point of interest when the collective scene is so astonishing.

Another reason why this planetary is so overlooked is that the summer Milky Way abounds with popular Messier objects, including a wealth of globular clusters in and around Ophiuchus. There are also the famous planetaries M57 (the Ring Nebula), in Lyra, and M27 (the Dumbbell Nebula), in Vulpecula, to vie for one's attention, especially on short summer nights.

But if you're looking for a change of scenery, just turn your attention about 4° northeast of Theta Ophiuchi, to NGC 6369, more commonly known as the Little Ghost Nebula. Here is a star with a mass similar to that of our Sun near the end of its life. It has moved off the Main Sequence to become a red-giant star, but it has since blown off the outer layers of its atmosphere and hurled them into space. We see the result of that event as a glowing smoke ring, about 1 light-year in extent, surrounded by a thin and complex halo whose northeast and south-west edges bracket the ring like parentheses. If you fill in the annulus with imaginary light, the object will also look like Darth Vader's Death Star from the movie "Star Wars."

Burning inside that ring is a 16th-magnitude white-dwarf star about the size of the Earth, but whose matter is so dense that a tablespoonful of it would weigh as much as 15 tons. This star is sending out a flood of ultraviolet light which strips electrons off atoms in the smoke ring, causing it to fluoresce with a signature blue-green hue. Color images of the nebula also show that the outer fringes of the ring have a reddish-ocher sheen; this is where the ultraviolet light is less intense, and the ionization process is less advanced.

The nebula gained notoriety in 2002 when the Hubble Space Telescope (HST) imaged NGC 6369 with its Wide Field Planetary Camera 2 (shown here). The image reveals remarkable details of the ejection process that are not visible from ground-based telescopes. It was produced by combining images taken through filters that isolate light emitted by three different chemical elements with different degrees of ionization. The result shows an aquamarine smoke ring, whose color is the product of ionized oxygen atoms that have lost two electrons (blue) and from hydrogen atoms that have lost their single electrons (green).

The outer edge of the smoke ring is filigreed with golden-red strands of wispy nebulosity, like curled ribbons on a present; this color marks emission from nitrogen atoms that have lost only one electron.

If you look carefully at the HST image, you can see that the central star isn't at the center of the nebula. In reality, it is. It's just that we see the the central star obliquely through a cylinder of gas. Actually, in a 2004 *Astrophysical Journal* article, Hektor Monteiro (Cerro Tololo Inter-American Observatory) and his colleagues used the Cerro Tololo Inter-American Observatory 1.5-meter telescope to determine that NGC 6369's three-dimensional structure is more of a clumpy hourglass shape. Close inspection of the central star, which, they determined, has a mass of about 0.6 Sun, shows that the central star is closer to a dimmer edge of the nebula than to the opposing brighter edge. Although the brightest part of the nebula is roughly circular, the central star appears slightly off-center, and is on the side away from it.

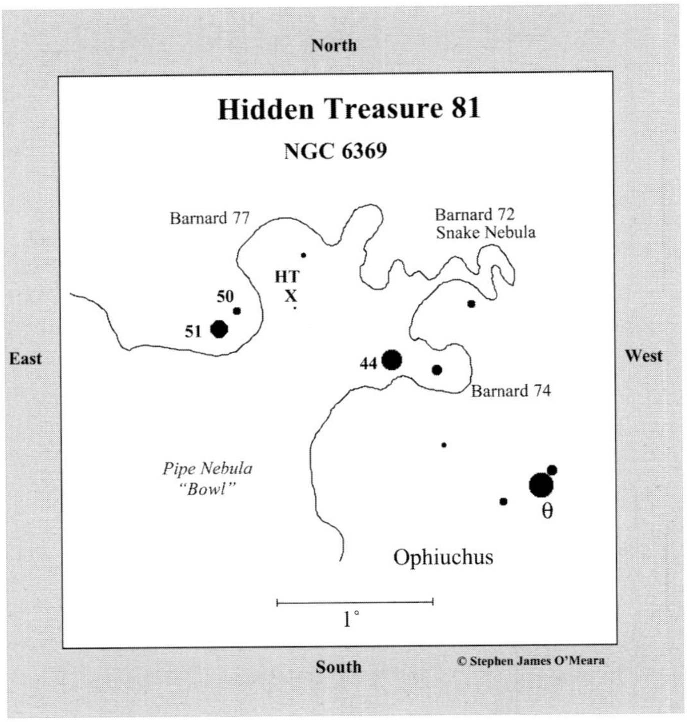

The HST also resolved in unprecedented detail prominences of golden-red gas blowing off the major axis of the ring. These are the vestiges of the first layers of the star to be ejected into space. This gas is expanding away from the central star at about 24 kilometers per second. After some 10,000 years, it will disperse and fade, like cigarette smoke in the air, until its particles blend with the dark background of interstellar space.

Not that you have to rush out and observe NGC 6369; it will be a Little Ghost for millions of years to come – until the white dwarf at the nebula's center cools off and winks out. But if you want to find NGC 6369, you have more than a ghost of a chance. First, locate Theta Ophiuchi. A little more than 1° to its northeast is 4th-magnitude 44 Ophiuchi. And a little more than 1° east and a little north of it you'll find 5th-magnitude 51 Ophiuchi. (Use your binoculars, if you must, to confirm this roughly $2\frac{1}{2}°$ arc of obvious stars.) You want to center 51 Ophiuchi, which is easy to confirm, since it has a 7th-magnitude companion (50 Oph) 10′ to the northwest. NGC 6369 is simply 30′ northwest of 51 Ophiuchi, or 20′ west-northwest of 6th-magnitude 50 Ophiuchi.

At magnitude 11.4, NGC 6369 is the faintest planetary nebula in the *Hidden Treasures* list. But that does not mean it is difficult to see. In fact, Walter Scott Houston said it is "within reach of all but the smallest telescopes." I too found that to be the case, since I could not detect it in 7 × 50 binoculars, nor in my antique telescope. But there is no mistaking its small form in the 4-inch at 23×. The nebula is projected against the blackness of B77, so, as its name implies, it appears like a translucent spirit materializing in the darkness. With any moderate magnification, the nebula is a perfect opal – smooth, round, and, well, opalescent. It is, as far as I can tell, one of the most perfect spheres imaginable in a small telescope. It gives me the pleasure of being able to see such a wonder and to imagine that I am truly seeing the disk of some gaseous body at a considerable distance. It brings to mind images of how William Herschel must have felt when he found the planet Uranus and then, later, these nebulous objects that mimicked so wonderfully that world's appearance.

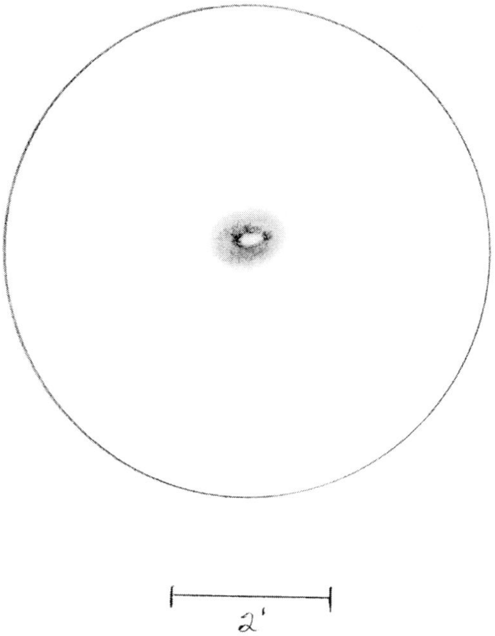

What I find remarkable is that Herschel failed to notice the object's annularity, which is a perfect smoke ring about the apparent diameter of Mars at a perihelic opposition; the nebula's dim and imperceptible outer envelope (at least in small telescopes) doubles the object's size. Luginbuhl and Skiff note that the annulus is clearly visible in a 6-inch telescope and measured the central hole at about 10″ in diameter. Perhaps Herschel's mind was on other things as he swept up this beautiful object. The ring didn't escape his son John's notice, who described it as an "annular nebula. Exactly round . . . well-terminated, but a very little cottony at the edge, and with a decided darkness in the middle . . . a beautiful specimen of the planetary annular class of nebulae."

Curiously, and of note, Ernst J. Hartung says that the "darker centre can just be made out with [an 8-inch]." I write "curiously," because in my 4-inch under significant magnification, the high-surface-brightness ring is obvious. Perhaps seeing the ring is a function of magnification and atmospheric seeing. In the 4-inch at 303× (under good observation conditions), the ring has a sharp interior that gradually dims outward. It's a stunning annulus that appears *ever so slightly* out of round with averted vision; the major axis being oriented northeast–southwest. After studying the nebula with various magnifications, I noted some variations in intensity along the annulus, namely, at the southwest ansa and along the northwestern rim. Interestingly, if you look closely at the HST image, the ring's northwest rim does indeed appear brighter.

As you peer into this nebulous wonder, imagine that it offers us a glimpse of our Sun some five billion years into the future, when it too will become a Little Ghost Nebula – one that will engulf the Earth and expand well beyond the orbit of Pluto. But that will not matter to us humans. We will be long gone; not as a race but as inhabitants of this Earth. By the time our Sun sheds its outer garments, we will have colonized new worlds. And with our telescopes set up in our new backyards, we will have front-row seats to an awesome spectacle: the dying gasps of our former Sun as it ends its life in the grandeur and glory that constitute the actions of a planetary nebula. Who knows, perhaps it will not be our first time.

Hidden Treasure 82

NGC 6400

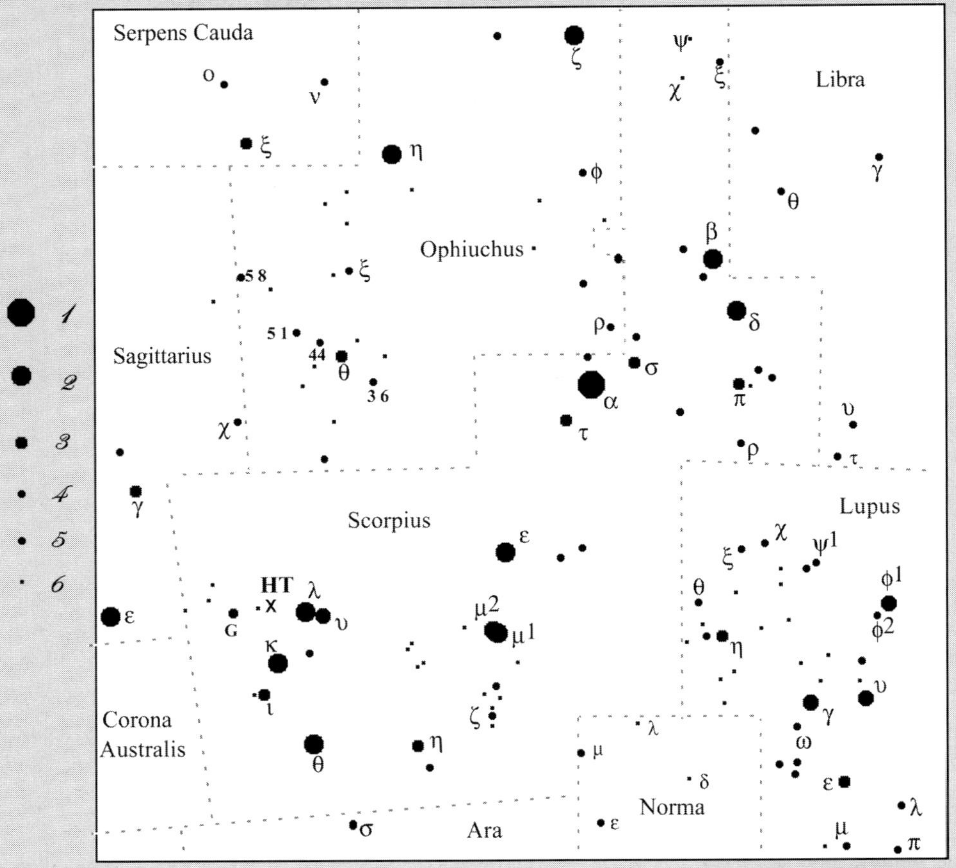

Tirion: Chart 22

Uranometria: Chart 376

82

Phantom Cluster, Long John Silver Cluster
NGC 6400
Type: Open Cluster
Con: Scorpius

RA: 17^h 40.2^m
Dec: $-36°$ $58'$
Mag: 8.3 (O'Meara); 8.8
Diam: 12.0′
Dist: –
Disc: James Dunlop, included in his 1827 catalog

J. HERSCHEL: Pretty rich, pretty large, irregularly round, 8′, stars of 9th and 10th magnitude. (h 3696)

NGC: Cluster, little compressed, stars [bright].

NGC 6400 IS NOT A BRIGHT CLUSTER, but it is one that certainly caught my eye. One night, while attending the May 1994 Texas Star Party, I swept the heavens for comets with a pair of 6-inch binoculars owned by Houston Astronomical Society member Larry Mitchell. As time passed, and my mind started to wander, my eyes suddenly focused on a most curious 8th-magnitude glow, round at first, then slightly elongated, as if it had a tail. My heart fluttered for a moment as the thought of a new comet knocked on the door of my mind. I paused to check with my naked eyes, just where I had been sweeping, which turned out to be immediately east of the Scorpion's Stinger stars – Lambda (λ) and Upsilon (υ) Scorpii. I could not recall ever having seen a deep-sky object in this present position. Then again, I was under dark skies using giant binoculars (how everything looked so different with binocular vision). Without wasting another moment, I grabbed the second edition of Tirion's *Sky Atlas 2000.0*. Now my heart pounded when I saw nothing plotted at that position. I just had to see the object with more power, to confirm it was truly cometary in nature and not some rich patch of the Milky Way. Larry had a 4-inch Tele Vue Genesis set up nearby, so I used it to sweep up this new object as I would from home. But the honeymoon didn't last. With one glance, I knew my "comet" was "just a cluster of 'little' stars."

The object turned out to be open cluster NGC 6400, which is plotted in the first edition of *Sky Atlas 2000.0*, but not in the second. Partly for this reason, I made

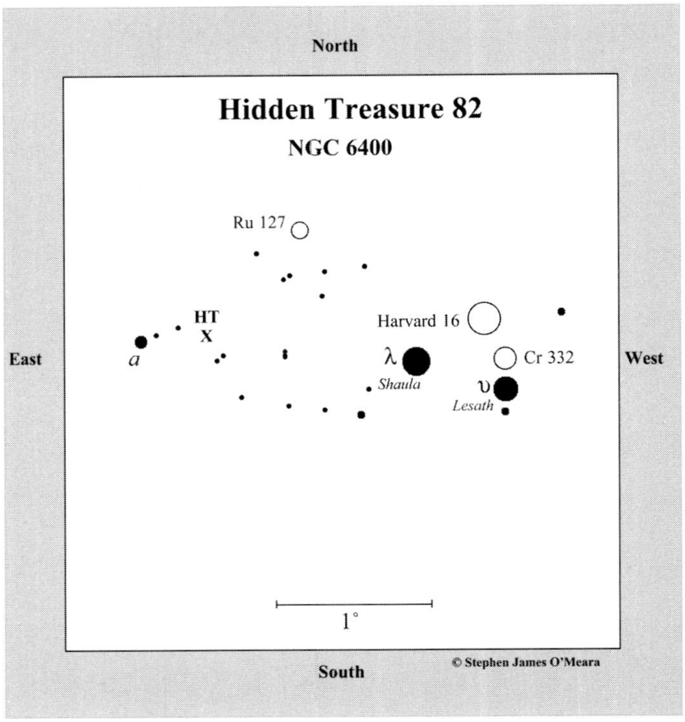

NGC 6400 a hidden treasure; perhaps like Charles Messier, I wanted to catalog an object that I believed would cause others to believe they have stumbled across a comet. The other reason is that I think it's a subtle but attractive sight. This particular cluster "speaks" to me like a lost puppy.

James Dunlop must have been equally curious when he discovered the object with his 9-inch telescope at Paramatta, New South Wales. He listed it as the 568th object in his 1827 catalog, calling it "a very faint cluster of very small stars, resembling faint nebula." I call it the "phantom cluster" not only because it vanished on the second edition of the *Sky Atlas 2000.0*, but because it at first looks more like a nebula than a cluster.

Now that you know it exists, scaring it up should be a snap. Though that depends on how you approach finding it. When Marsha Robinson of San Jose, California, first set out to find NGC 6400 with her 10-inch f/5.6 Dobsonian, she figured it would be easy. But she had some difficulty – until she realized it was in a direct line between G Scorpii and Lambda (λ) Scorpii (Shaula). "Then it was an easy kill," she said, calling it a "fairly large and lopsided [cluster] with a larger grouping of stars at one side."

The cluster should be an easy kill, at least from suburban or dark skies. It is a little less than $1\frac{1}{2}°$ due east and slightly north of Lambda Sco, almost midway between Lambda and G Sco. You should first set your aim, however, for magnitude 5.5 Star *a*, which is about $1\frac{1}{2}°$ west of G Scorpii. NGC 6400 is simply 25′ due west of Star *a*. The object under a dark sky is visible in 7 × 50 binoculars as a slightly elliptical glow, oriented roughly north–south. It also lies just 10′ north-northeast of what appears to be a single 3.5-magnitude star. Through the antique telescope, this star cleanly separates into two roughly 9th-magnitude stars (oriented west-northwest–east-southeast); the cluster rises like smoke from them. With effort, I can start to resolve some of the members, though they appear dim and elusive.

At 23× in the 4-inch, the near 3° field of view is absolutely spectacular. NGC 6400 marks the eastern end of a 2°-wide arc of open clusters that curves northward from Upsilon (υ) Sco (Lesath); these other clusters are, from west to east, 9th-magnitude Collinder 332, Harvard 16, and

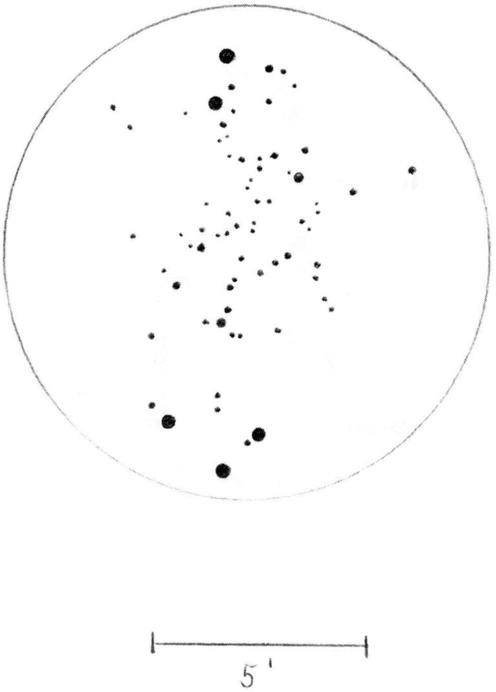

9th-magnitude Ruphrect 127. When I relax my gaze and let the entire scene sweep over me like the wind, I see the cluster as an insect trapped in a black web. This web is comprised of numerous threads of dark nebulosity, which is just a tiny part of the magnificent latticework of dust that drapes this inner spiral arm of our galaxy – one that is seen projected against the southern heartlands of the Milky Way. The dark nebulosity stands out even more if I slightly defocus the image, use averted vision, and relax my gaze.

When I concentrate solely on NGC 6400 at low power, which is almost a sacrifice, I can see it partially resolved. As Dunlop observed, it is "considerably congregated to the centre, irregular round figure." At 72×, all the cluster's details are clear. A thread of stars travels north-northwest from the double, until it branches into a little ellipse with a strong eastern border. Loose, swaggering arms extend to the southwest and east. All told, the cluster's 60-odd members, seen together with the 9th-magnitude pair to the southwest, present a distinct human form, which looks uncannily like a drunken pirate. And since the long spine of stars connects to a single leg, how could I not see this cluster as the most famous of all the pirates in classical literature – Long John Silver, that heartwarming character created by Robert Louis Stevenson in *Treasure Island.* There's even a tiny appendage of stars jutting east from the pirate's elliptical head to form a pipe, and three prominent stars extending west-northwest from the pirate's limp, western shoulder, which can be seen as none other than *Capt'n Flint,* Long John Silver's parrot, whose famous cries of "Pieces of eight! pieces of eight!" still rings in the ears of our youth.

With higher powers, the cluster reveals many fainter suns in no set manner. With imagination, these dimmer congregations of suns make the cluster look like an aerial view of a county fair, with all the stars its people – huddled here, crowded there, yet spread out among the fair ground. So the cluster is very nice for small telescopes and for people with fertile imaginations. Not much is added to the cluster in larger telescopes. Steve Coe of Glendale, Arizona, observing with a 17.5-inch f/4.5 reflector at 100× sees the cluster as "a long string of stars with several other members." And Auke Slotegraaf's (Director of the Astronomic Society of South Africa Deepsky Observing Section) view through a 15.5-inch f/9 Newtonian is of a "small open cluster of highly irregular shape lying to the north of an 8th magnitude star. The grouping is very irregular, the pretty bright stars allowing the eye to form many trails and curls; there are at least three star chains and one star loop easily visible. There are roughly two dozen 10th magnitude stars here."

NGC 6400 is not a blazing cluster but a simple one, it's a reminder that not all in the sky, as on Earth, has to be dynamic to be beautiful. Not every object in the sky has to be blazing to cause excitement. As Immanuel Kant says in his *Critique of Practical Reason,* "Two things fill the mind with ever new and increasing admiration and awe, the oftener and the more steadily we reflect on them: the starry heavens above and the moral law within. I have not to search for them . . . I see them before me and connect them directly with the consciousness of my existence."

Hidden Treasure 83

IC 4665

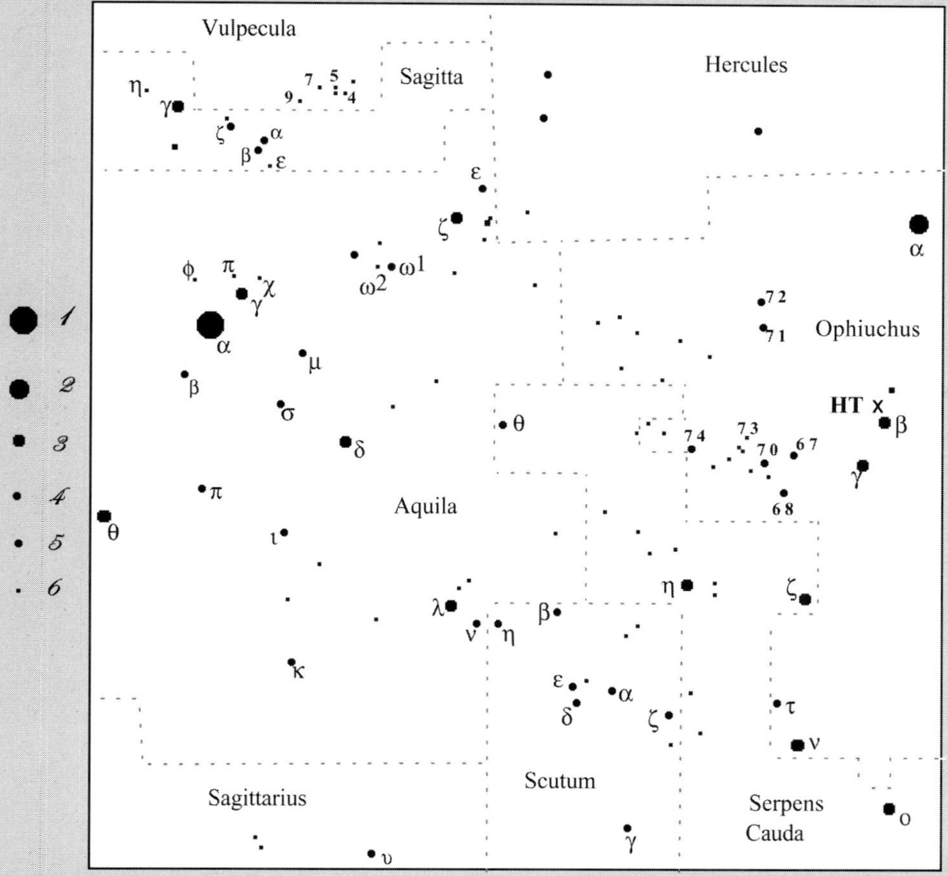

© Stephen James O'Meara

Tirion: Chart 15

Uranometria: Charts 203 & 248

83

Summer Beehive, Black Swallowtail Butterfly Cluster
IC 4665
Type: Open Cluster
Con: Ophiuchus

RA: 17^h 46.2^m
Dec: $+05°$ $43'$
Mag: 4.7 (O'Meara); 4.2
Diam: 70.0′
Dist: 1,100 light-years
Disc: Philippe Loys de Chéseaux, 1745–1746; most likely independently discovered by Caroline Herschel in July 1783

HERSCHEL: None.

IC: Cluster, coarse.

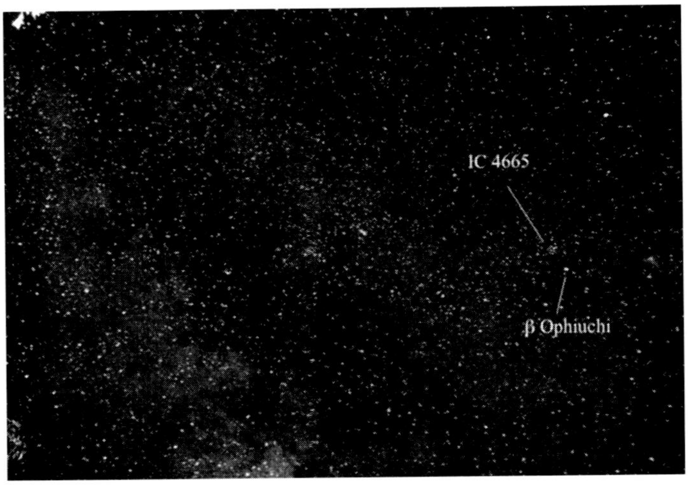

IC 4665 IS A REMARKABLE OPEN cluster that's bright enough to be a binocular marvel but just faint enough to avoid easy detection with the naked eye, which is one reason it is often neglected. Under a dark sky, IC 4665 is visible to the naked eye, but only once you know where to look – which is, simply, about $1\frac{1}{2}°$ northeast of 3rd-magnitude Beta (β) Ophiuchi (Cebalrai), the shoulder of the Serpent Holder.

IC 4665 is an enormous 4th-magnitude glow more than two Moon diameters across. So it can be very difficult to detect with the unaided eye from even decent suburban skies; I certainly never saw it that way from the city when I was growing up. In fact, I did not realize just how beautiful this object was until I moved to a dark sky and saw it in binoculars; here was a blizzard of dim starlight, one that Glenn Chaple, Jr., of Townsend, Massachusetts, calls the "Little Beehive." That's when I slapped my forehead for not having spent time looking for it before. But anyone of my age who used *Norton's Star Atlas* as his or her main deep-sky atlas will understand, because IC 4665 was not plotted on it. The only way we could have discovered it was to stumble upon it, as Chaple did.

The irony is that this big and bright cluster did not make it into Charles Messier's famous catalog of nebulae and clusters even though it was discovered more than a quarter century before Messier published his first list; the cluster was much too large and scattered to have been recognized by William Herschel in his grand survey of the sky. The *Second Index Catalogue* credits Solon Irving Bailey as the discoverer, who in his

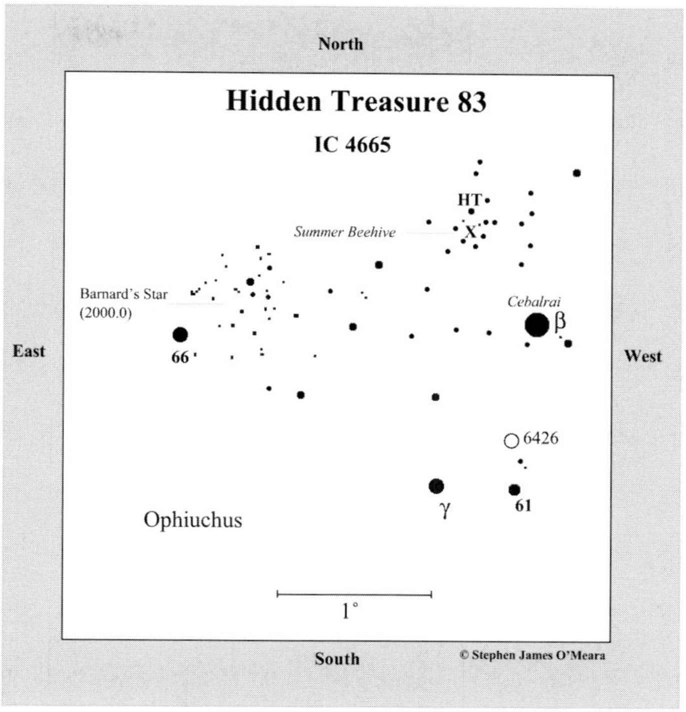

1908 *Catalogue of Bright Clusters and Nebulae,* lists 13 clusters not in the *NGC.* But the object's real discoverer appears to be the Swiss astronomer and mathematician Philippe Loys de Chéseaux (1718–1751), who published a list of 21 nebulous objects (compiled in 1745–1746), which, he says, "when seen in the telescope, are found to be simple clusters of stars." At least eight of these objects were previously unknown, and one of them, the second in his list, appears to be IC 4665.

I say "appears to be" because, while de Chéseaux's written description of the object and its position are fine ("Above the shoulder, β Ophuichi, a cluster of stars . . ."), he provides the precise position of the two brightest stars in the cluster but places them in a southern declination rather than a northern one. Caroline Herschel apparently did the same thing after she independently discovered it on July 31, 1783: "From beta Serpentarii towards [south], $1\frac{1}{3}$°, a cluster of stars" (see Appendix A). In her wide-field telescope, Beta Ophiuchi and IC 4665 would have been viewed with south oriented, "up." If we reverse the signs, IC 4665 is a direct hit for Caroline Herschel, though it is still 2° from de Chéseaux's position. No matter, the verbal descriptions by both de Chéseaux and Herschel are sufficient enough to give credit to both these great astronomers. Caroline had no way of knowing about de Chéseaux's discovery, although de Chéseaux presented his list to the French Academy of Sciences on August 6, 1746. Unfortunately, as Kenneth Glynn Jones notes in his 1975 book *The Search for the Nebulae,* "the list . . . does not otherwise appear to have been published and de Chéseaux' discoveries do not seem to have been known in detail to Messier." The list was "rescued from oblivion" by M. G. Bigourdan, who identified many of the objects in the *Annales* of the Paris Observatory for 1884.

While most of the Milky Way's open clusters are found in the plane of its spiral arms, IC 4665 is at a relatively high galactic latitude, almost 16° above the galactic plane. It is a young and nearby cluster, being only 1,100 light-years distant and about 35 million years young – younger than the Pleiades – about the time when two massive bolides impacted Earth – one at the Atlantic margin of North America around the area of

Chesapeake Bay, the other at Popigai in northern Siberia – perhaps triggering a rapid drop in global temperatures, of about 2 °C.

The location of IC 4665 relative to the galactic plane is suggestive of an environment for star formation that was distinct from that in other open clusters. Given IC 4665's uncharacterstically high galactic latitude, it would be unreasonable to expect that its chemical composition might differ from other nearby clusters closer to the galactic plane. Thus, astronomers have used the ROSAT satellite to investigate other young, nearby clusters to serve as a comparison. In a 1998 *Astronomical Society of the Pacific Conference Series* publication, M. S. Giampapa (National Optical Astronomy Observatories, National Solar Observatory) and his colleagues, found that the ROSAT data suggest that neither the formation environment nor the formation history of open clusters is reflected in the nature of stellar coronae, at least for clusters with solarlike metallicities.

The view of Beta Ophiuchi and IC 4665 under a dark sky with the unaided eye is uncanny, especially if you consider that the cluster is only about a magnitude fainter than the star, yet how much more brilliant Beta appears – these two objects would make an excellent demonstration of how a deep-sky object's size affects its visibility. IC 4665 is 70′ in angular extent, which translates to 22 light-years in true physical extent. The beauty of this cluster would have been more enhanced to the naked eye if it were slightly farther away.

Once I get the cluster in my sights, however, it's hard to let it go. The more time I spend with it, the more mottled the glow appears; the brightest stars in the cluster, after all, are near the limit of naked-eye visibility, and if you get two or more of these stars clumped together, the naked-eye haze should certainly break apart into tiny subsets of stellar hazes.

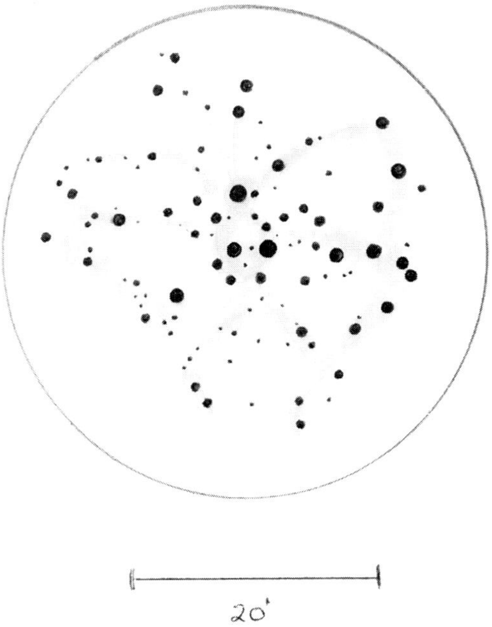

But like the Coathanger (Hidden Treasure 97), IC 4665 looks best in binoculars and rich-field telescopes. With 7 × 50 binoculars I could resolve the cluster remarkably well, with 15 or so stars forming a distinct irregular circle at the core with vast double overlapping wings attached to the west and east. The view is glorified at 23× in the 4-inch, which illuminates the clusters' 57-odd members plus numerous field stars in this outer expanse of the Milky Way. Two antennae stretch from the central rosette, and two taillike projections trailing southward from each wing, make the cluster appear like the most distinctive family of all familiar butterflies – the magnificent Swallowtails. In this case the Black Swallowtail found in practically all parts of North America south of Canada. This beautiful insect's black body and wings are lined with bright white spots, like the stars in IC 4665. It is also an appropriate name for this warm-weather cluster, since Black Swallowtails fly upon the wing almost any time in warm weather.

Finally, in his delightful book *Star-Hopping: For Backyard Astronomers,* Alan MacRobert notes that the cluster's inner group of bright stars spells the ragged word "HI" when southwest is up. "The greeting is not obvious in the eyepiece," he says, "until you notice it; then it looks shockingly impertinent. (Warning: anyone who sends this to the tabloids as proof of alien skywriting earns 100 years of bad karma.)"

Hidden Treasure 84

NGC 6445

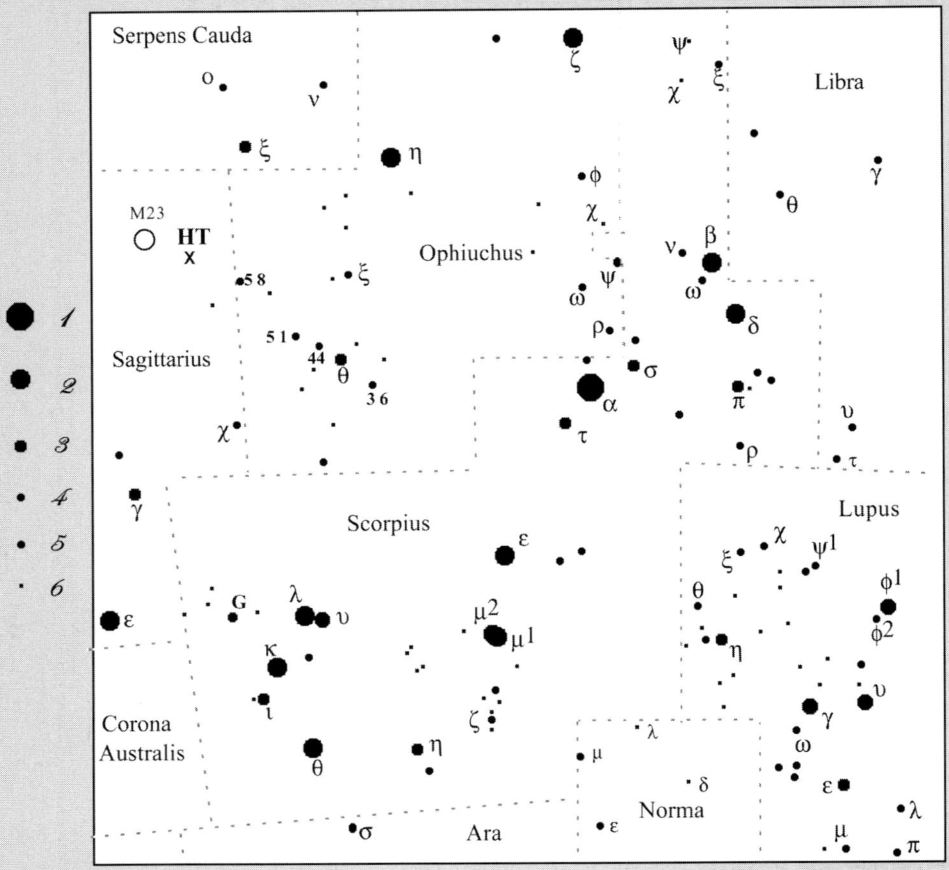

© Stephen James O'Meara

Tirion: Charts 15 & 22

Uranometria: Charts 338 & 339

84

Box Nebula
NGC 6445
Type: Planetary Nebula
Con: Sagittarius

RA: 17^h 49^m 15.1^s
Dec: −20° 00′ 34″
Mag: 11.2 (Skiff)
Dim: 3.1′ × 0.9′
Dist: 4,500 light-years
Disc: William Herschel, 1786

W. HERSCHEL: [Observed May 28, 1786] Pretty bright, small, irregularly faint. (H-II-586)

NGC: Pretty bright, pretty small, round, gradually brighter in the middle, [mottled], magnitude 15 star north preceding.

OF ALL THE DEEP-SKY OBJECTS IN the heavens, planetary nebulae are among the most exotic and photogenic. Their largely symmetrical shells take on all manner of colorful shapes and forms, including rings, butterflies, hourglasses, spheres, and eyelike ellipses. And like NGC 6210 (Hidden Treasure 78), the Turtle Nebula in Hercules, our next hidden treasure, NGC 6445, adds another dimension to this celestial zoo of expelled gas – namely, a box.

Astronomers once believed that the progenitor stars of planetary nebulae simply shed their outermost layer, producing a bubble of slowly expanding material that eventually dissipated into the background interstellar medium. "However," says Bruce Balick (University of Washington), "the process is far more interesting. Despite their individual differences in complexity, planetaries show overall symmetric structure." What process shapes these stars? Balick sees a donut of ejected material (assumed to have been expelled earlier) surrounding a white-dwarf nucleus with a fast-flowing wind. The wind flows freely until it rams into the disk along the equator. In other directions the outflowing stellar wind is less inhibited. A hot bubble of expanding gas is formed. "The nebula," Balick explains, "tends to grow slowest along the equator where the dense torus inhibits its growth, and fastest along the polar direction. The detailed models show that a crucial parameter in governing the evolving shapes of planetary nebulae is the original disk-to-polar density ratio."

Then there is the matter of illusion. As planetary nebula expert Sun Kwok explains in his book *Cosmic Butterflies*, "Maybe the different morphologies we see in planetary nebulae are an illusion. They all may have the same intrinsic structure after all . . . It is tempting to believe that such diverse morphologies are just different manifestations of a single, unified, basic three-dimensional structure."

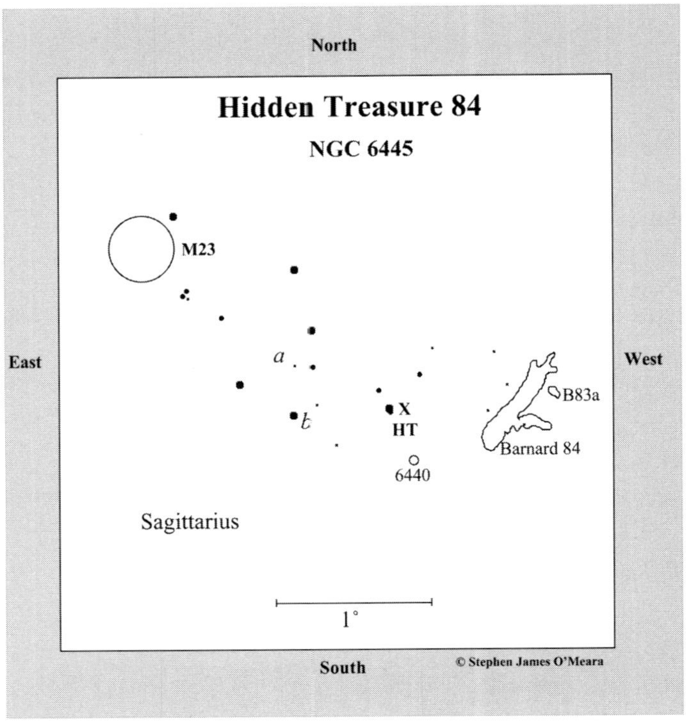

As early as 1968, Russian astronomer Gabriel Khromov and the Czech astronomer Lubos Kohoutek proposed that the different morphologies are an effect of projection, and that many of the observed morphologies can be explained by cylinders with open ends projected at different orientations in the sky. Today, however, astronomers believe the model of a planetary nebula is a bit more complex. The current thinking, as Kwok explains, is that while the brightest part of the nebula does have the shape of a short cylinder (a torus or donut), its polar extensions have the shape of fans or lobes. "The Torus," Kwok explains, "having higher densities, forces the outflow from the star to channel through the open ends. After breaking out, the outflow fans out to the side, giving the planetary nebulae a butterfly shape. Looking down the symmetry axis, the nebula looks like a ring with faint halos, exactly like the Ring Nebula. If we rotate this model to different orientations, then the projection to the plane of the sky will give rise to different morphologies."

Balick, however, believes that planetary nebulae begin their lives with round and spherical appearances and develop into increasingly asymmetric forms as they age. In this model, interacting winds play an important role in the shaping of the planetary. How does NGC 6445 fit into these models? Spanning 3′ × 1′ in apparent diameter (4 × 1 light-years in true physical extent), NGC 6445 is one of the largest known planetary nebulae, which also means it is among the oldest. It is also asymmetrical, whose brighter portion, as Heber Curtis (Lick Observatory) described in 1918, "is a very irregular square-shouldered ring . . . beyond which extended very faint ansae." It has also been likened to a "coral atoll" and "enclosed lagoon." Indeed at a 2004 American Astronomical Society meeting, Mexican astronomer R. Vazquez (Institute of Astronomy, UNAM) and colleagues announced that their studies of the planetary nebula NGC 6445 show it to be an asymmetrical bipolar nebula with a very bright central ring, opened lobes, and finer features.

NGC 6445 is easily spied in small telescopes. And P. Brennon of Regina, Canada, says that it is "easy to discern even in light summer skies." But few sky pirates seek it out, perhaps because it lies on the west bank of the rich Sagittarius Milky Way, where a flotilla of bright Messier objects are waiting to be visually plundered. Our hidden treasure lies only 2° southwest of the 5th-magnitude open cluster M23. To find it, first locate M23, which is about 5° west-southwest of M24, the Small Sagittarius Star Cloud. A little more than 1° to the southwest is a nearly 1°-wide asterism of five 7th- and 8th-magnitude suns (*a*) that resemble an upside down, crooked Y. If you center the southernmost star in that asterison (*b*), NGC 6445 will be about 40′ west and a tad north, just 5′ west of a magnitude 8.4 star.

The nebula is immediately obvious in the 4-inch at 23× as a non-stellar glow, which appears all the more obvious because it is so close to a bright star. In a nearly 3° field of view, the entire region is dappled with dimly glowing star clouds and veins of dark nebulosity. It is also graced with an obscure, 9th-magnitude globular cluster (NGC 6440) just 20′ to the south-southwest of our hidden treasure. In fact, in the 4-inch at 23×, M23, NGC 6440, and NGC 6445 fit in the same field of view. M23 lies 2,100 light-years distant, NGC 6445 lies 4,500 light-years distant, and NGC 6440 lies 24,000 light-years distant. Talk about an education in intergalactic distances. Use your mind and imagination and take a voyage through the cosmos.

At 72× in the 4-inch, the planetary looks annular; actually, it's more like two offset arcs of light. The northwest side is more difficult to observe because of the proximity of the 8th-magnitude star. Increasing magnification to 101× enhances the arcs and brings out some fainter stars, but higher power is really needed because of the proximity of the star to the nebula. The nebula takes power well, and I studied it with powers ranging from 252× to 504×, which breaks up the nebula into a mass of knotty enhancements surrounding an irregular hole. The Box Nebula reveals itself. The most prominent knots lie to the south and east, though these may be dim stars projected against the nebula in this rich field. The northwestern rim is fragmented into two sections, each with irregular enhancements, which could also be, in part, dim stars. I could not detect any outer loops. Do not hope to see the nebula's illuminating central star, which shines at a dim 19th magnitude.

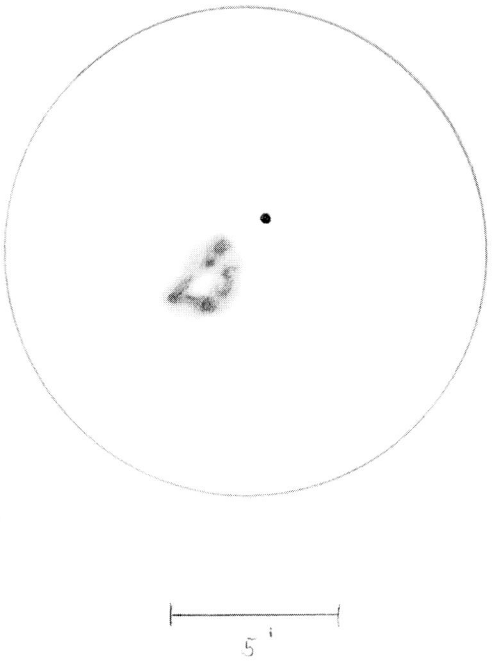

Hidden Treasure 85

NGC 6503

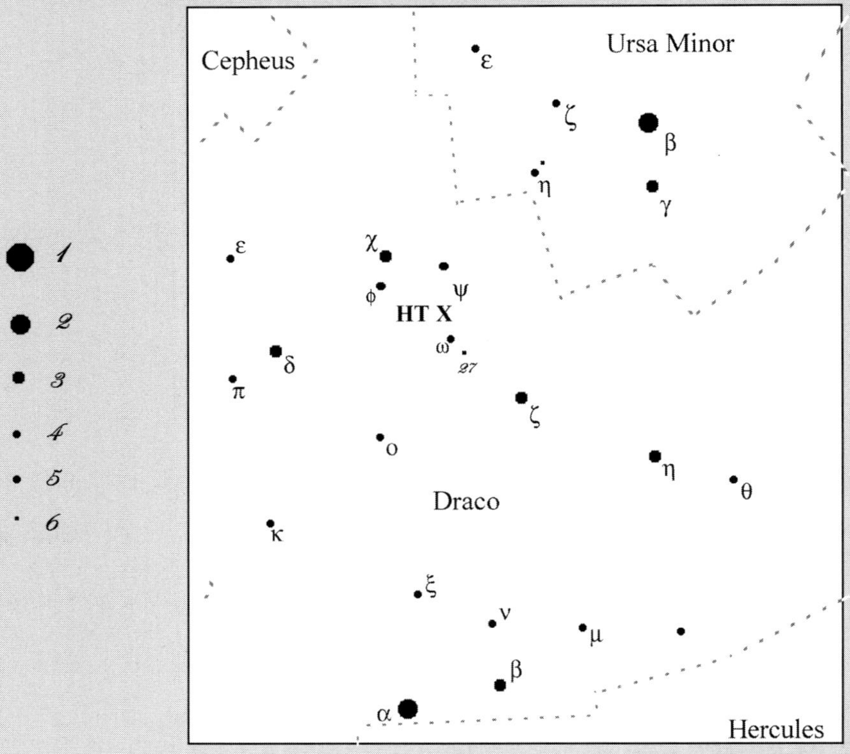

© Stephen James O'Meara

Tirion: Charts 2 & 3

Uranometria: Chart 30

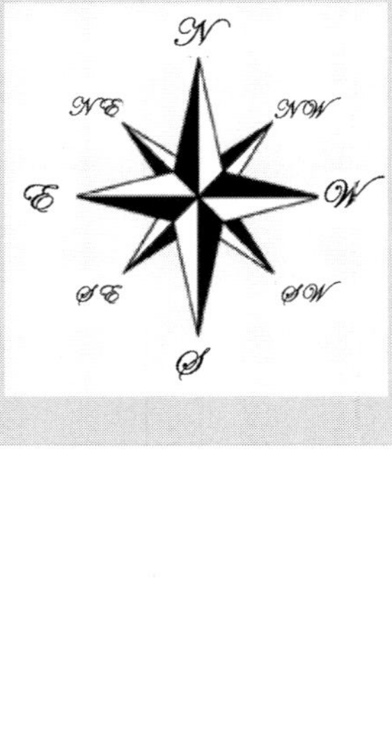

85

Lost-In-Space Galaxy
NGC 6503
Type: Spiral Galaxy (Scd)
Con: Draco

RA: 17^h 49.4^m
Dec: $+70°$ $09'$
Mag: 10.4 (O'Meara); 10.2
Dim: $5.9' \times 2.2'$
SB: 13.2
Dist: 17 million light-years
Disc: Georg-Friedrich Julius Arthur von Auwers, 1854

HERSCHEL: None.

NGC: Pretty faint, large, much extended, magnitude 9 star following 4′ (Auwers 37).

ON JULY 22, 1854, EIGHT YEARS before he joined the staff at Gotha Observatory, and seven years before he deduced the existence of a faint but massive companion to Alpha (α) Canis Minoris (Procyon), the budding and eclectic German astronomer Arthur von Auwers (1838–1915) discovered NGC 6503 in Draco with his own 2.6″ Fraunhofer refractor. At the time, he was a student at Göttingen University, as was Friedrich Winnecke (1835–1897), to whom he showed the new "nebula." The very next night, Winnecke would discover the open cluster NGC 6704 in Scutum with his personal 3-inch Merz refractor at Göttingen.[1] NGC 6503 is the 37th object in Auwer's list of new nebulae in *William Herschel's Verzeichnisse von Nebelflecken und Sternhaufen*, Königsberg 1862; he discovered only one other nebula in that list: NGC 4402, a galaxy in Virgo, which he found on March 5, 1862, with the 6.2″ Fraunhofer heliometer at Königsberg Observatory.

Today we know NGC 6503 to be a dwarf spiral galaxy, seen 16° from edge-on. It belongs to the Coma–Sculptor Cloud of Galaxies and is located at the edge of the Local Void – a huge (roughly 33 million light-years across), low-density hollow in the local universe between the Hercules, Coma, and Local Superclusters. The void is not completely empty, but contains small galaxy systems and poor clusters of galaxies. NGC 6503's diminutive size (about 30,000

[1] This discovery was more than fate or coincidence. Just after entering Göttingen University as a student of astronomy in December 1853, Winnecke discovered NGC 6791, an open cluster in Lyra, with his 3-inch refractor. He would also go on to discover his first of 10 comets on December 22, 1854. All told, Winnecke would discover nine new nebulae, including NGC 1398 (Hidden Treasure 19).

light-years in linear extent), therefore, is magnified by the fact that it is also one of the most spatially isolated galaxies in the local universe; it is the 873rd entry in I. D. Karachentsev's 1973 *Catalog of Isolated Galaxies.* Because of its isolation and neglect, I call it the "Lost-In-Space Galaxy."

In photographs, NGC 6503 displays a very small, very bright nucleus within a lens that contains bright, tightly wound spiral arms caked with dense star-forming regions. The spiral pattern is similar to that of the multiple-armed galaxy NGC 488 in Pisces, but is much later, and the arms are not as well defined. Individual stars and H-II regions resolve out on red-sensitive plates taken with the 200-inch telescope, but the resolution on the blue plates is not nearly as prominent.

In the mid 1970s, astronomers located a strong, compact radio source (a 17th-magnitude quasar) only about 10,000 light-years from the center of NGC 6503 and in the same plane. It was, at the time, believed to be a best-case possibility for a physical link between galaxies and quasars. In 1976, Halton Arp and his colleagues used a stacked plate photograph to show a filamentary emission feature extending from NGC 6503 in the general direction of the quasar (1749+70.1). But in a 1981 *Astronomy and Astrophysics* paper, Seth Shostak (Kapteyn Astronomical Observatory) and his colleagues used the Westerbork Synthesis Radio Telescope to show no real clear-cut association between the two objects.

NGC 6503 is also a Seyfert system – meaning the spectrum of its small, intensely bright, nucleus shows strong, broad emission lines. Actually, the nucleus of this galaxy has been classified as a miniature or "transition" Seyfert 2 nucleus – one that shows a broadening of all spectral lines, as opposed to a Seyfert 1 nucleus, which shows a broadening of only the hydrogen lines – those that can be associated with the highest gas densities. Either way, the spectral features suggest that the galaxy's core hosts large amounts of very hot, fast-moving gas, which is held in place by the gravitational pull of some large central mass. And while the nature of the central mass is uncertain, it could be a supermassive black hole. Indeed, high-resolution X-ray images by the now-defunct ROSAT show an extended and elongated source about 10″ from the galaxy's nucleus. The shape and energetics of the diffuse emission could be explained by a sort of intergalactic "hurricane" – a superwind blowing gas at speeds of thousands of kilometers per second away from the nucleus. A black hole could produce the enormous energy required to drive the jetlike wind. The Hubble Space Telescope observed stars along the galaxy's northwest edge and determined a distance of 17 million light-years, only 3 million light-years closer than R. Brent Tully's listing in his *Nearby Galaxies Catalog.*

The active galactic nucleus of NGC 6503, then, has been classed as a low ionization nuclear emission region (LINER). It is thought the LINER phenomenon may be the result of a starved monster supermassive black hole that is producing very little energy because it is receiving only a slow trickle of food in the form of infalling gas. Nearby galaxy searches have revealed strong evidence to suggest that the nuclei of several of the nearest galaxies may, in fact, contain supermassive black holes, with masses that are several millions to several tens of millions times that of the Sun.

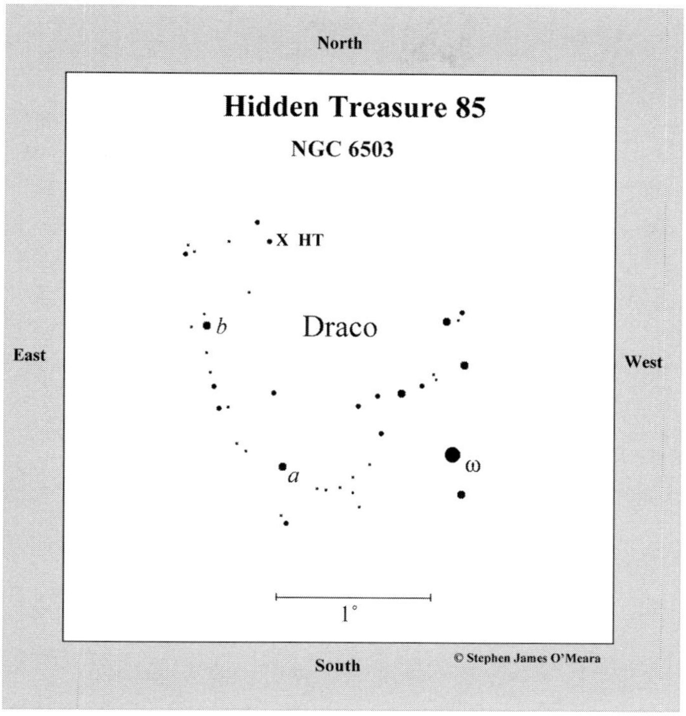

To visit this lonely and mysterious galaxy, first locate the trapezoidal Head of Draco, the Dragon – Beta (β), Gamma (γ), Xi (ξ), and Nu (ν) Draconis. Now you want to star hop with your naked eye along the Dragon's sinuous upper body. A little more than 10° (a fist) northwest of Nu Dra is 3rd-magnitude Eta (η) Draconis. About $6\frac{1}{2}$° to the northeast of Eta Dra is the equally bright Zeta (ζ) Draconis. Now look about 8° further to the northeast where you will find a roughly 3°-wide triangle of three $3\frac{1}{2}$- to $4\frac{1}{2}$-magnitude stars – Chi (χ), Phi (ϕ), and Psi (ψ) Draconis. Midway between Zeta and Phi Dra are two 5th-magnitude stars – Omega (ω) and 27 Draconis – separated by about 1° and oriented northeast–southwest, respectively. Center Omega in your telescope. Now look a little more than 1° to the east, and a tad south, where you will find a solitary 7th-magnitude star (*a*). Another roughly 1° hop to the northeast brings you to another 7th-magnitude sun (*b*). NGC 6503 is about 45′ northwest of Star *b*, just about 4′ west of a 9th-magnitude star.

Do not be fooled by NGC 6503's seemingly faint magnitude (10.4). The galaxy has a high surface brightness. Again, note that Auwers discovered it with a 2.6-inch refractor, and Skiff and Luginbuhl note that it is easy to see in a 6-cm refractor. I tried to see it in 7 × 50 binoculars but failed, though I believe it should be detectable in 10 × 50s, especially if they are mounted on a tripod. It is visible in my antique telescope with averted vision, looking like a faint streak of light – like the thin vapor trail of

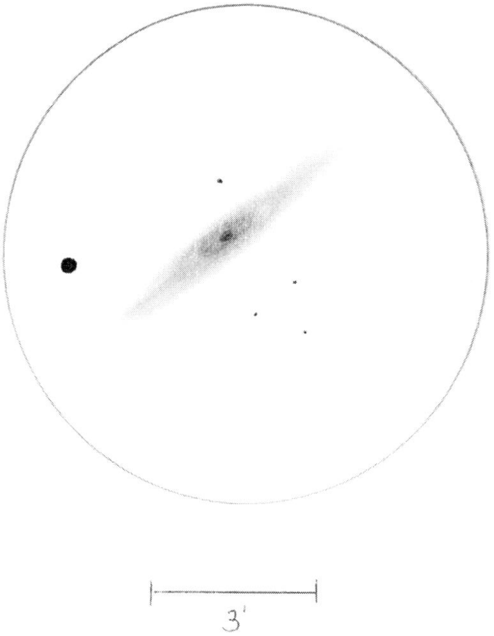

a quick and brief meteor. And that's the very impression I get at 23× in the 4-inch – a glowing meteor train oriented northwest–southeast.

This is a very nice galaxy even at 23×. I can see with averted vision a central pip (not obvious) just inside a bright lens that has fainter extensions to the north-west and southeast along the major axis. The surrounding field is very populated with starlight – an ironic illusion, given that the galaxy lies in the Local Void. NGC 6503 is even more stunning at 72×, which shows the lens breaking up into patches of dark and light streaks. The central pip is tack sharp – a tiny star in an elliptical mist. Continuing with the meteor metaphor, at this power, the galaxy looks like the train immediately after the head winks out and it scintillates with beads. The drawing shown here is a composite of views made with 23× to 168×. The knots and dust lanes are not at all difficult to see. The northwest end is brighter than the southeast end, but this may be an illusion due to the contrast with the neighboring 9th-magnitude star.

Hidden Treasure 86

NGC 6441

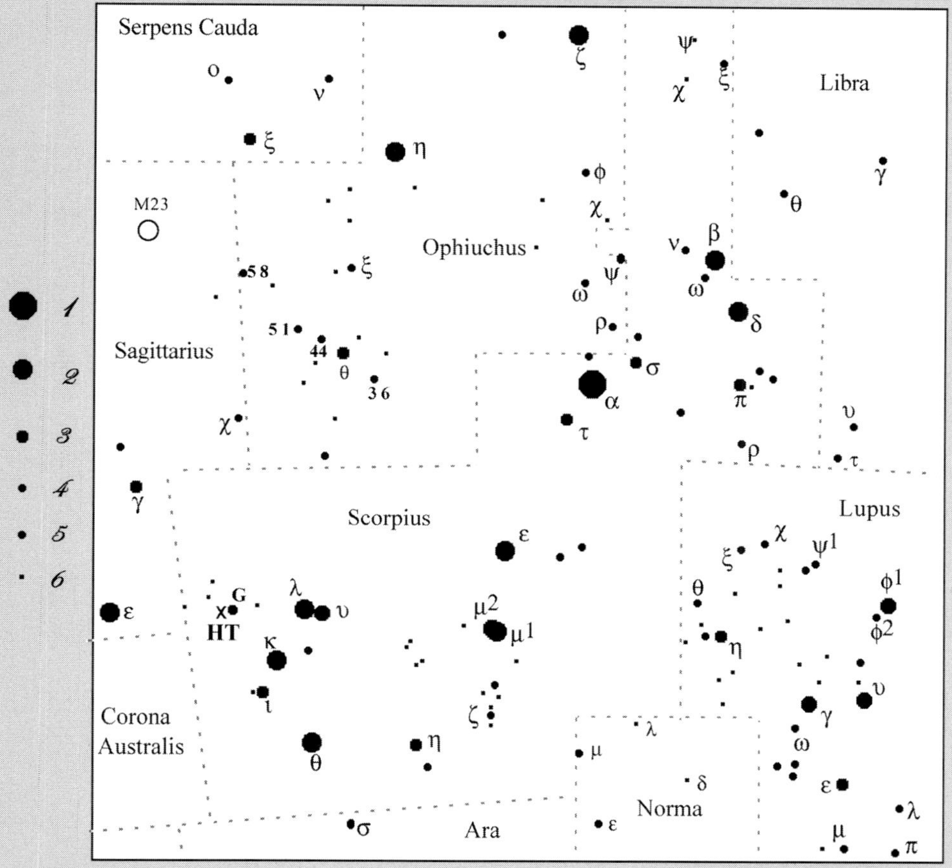

© Stephen James O'Meara

Tirion: Chart 22

Uranometria: Chart 377

86

86

Silver Nugget Cluster
NGC 6441
Type: Globular Cluster
Con: Scorpius

RA: $17^h\ 50.2^m$
Dec: $-37°\ 03'$
Mag: 7.2
Diam: 9.6′
Dist: 38,100 light-years
Disc: James Dunlop, included in his 1827 catalog

J. HERSCHEL: Globular, bright, round, very gradually brighter in the middle, up to a blaze. In field with Gamma Telescopii, and nearly on the same parallel; with left eye I barely see it resolved into stars of 18th or 20th magnitude. The whole ground of the heavens, for an immense extent, is thickly sown with such stars. A beautiful object. (h 3705)

NGC: Globular cluster, very bright, pretty large, round, very gradually much brighter in the middle, well resolved, clearly consisting of stars, 18 stars.

ANYONE SWEEPING HIS OR HER TELE-scope across Scorpius, the Scorpion, will find a rich assortment of deep-sky treasures buried in the stellar sands of the Milky Way. Globular cluster NGC 6441 is chief among them, but it is also one of the most obscure. That's because this jewel is lost in the glare of the 3rd-magnitude *K*-type giant star G Scorpii. The nineteenth-century astronomer James Dunlop discovered this tiny wonder while surveying the southern skies with a 9-inch reflector from Paramatta, New South Wales. He included it as the 557th object in his 1827 catalog of southern deep-sky objects, describing it as "a small well-defined rather bright nebula, about 20″ diameter." But even the smallest of today's telescopes will show it.

More than the globular itself, the beauty of NGC 6441 lies in its placement in the sky,

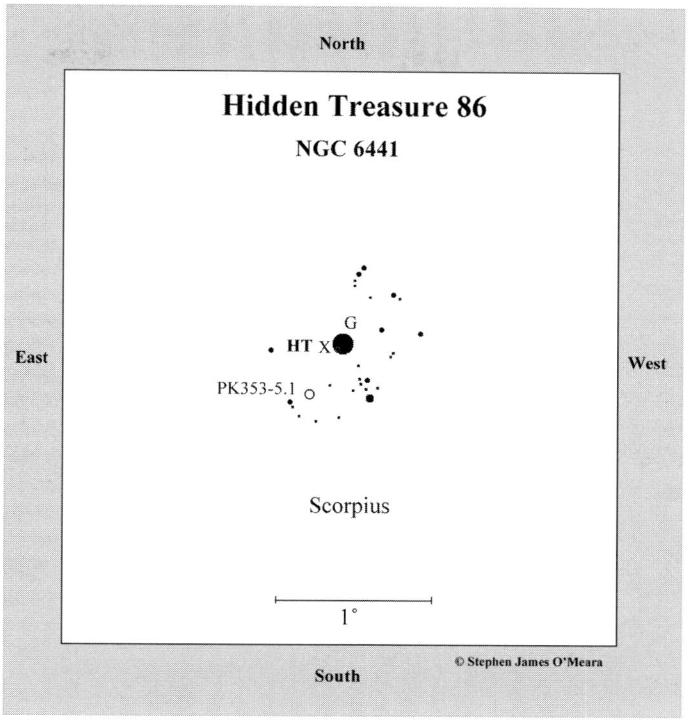

which through binoculars or rich-field telescopes is like opening a box of Whitman's Samplers on Valentine's Day. We see NGC 6441 projected against the southwest edge of the galactic bulge just east of the Scorpion's Tail, where colorful stars, clusters, and wide swaths of bright and dark nebulosity abound. In fact, the intervening dust in this region of sky is so dense that it diminishes NGC 6441's glow by 0.5 magnitude. If we could see this globular away from the galactic center in a relatively transparent region of sky, like Aquarius, it would teeter on the brink of naked-eye visibility. Indeed, NGC 6441 is one of the most intrinsically bright and massive globulars known. Without the effects of reddening, NGC 6441 would shine as brightly as M2 in Aquarius, which lies at about the same distance.

Unlike M2 (a metal-poor, halo object), NGC 6441 is a metal-rich, galactic-bulge globular. Its physical composition and nature more closely resemble that of NGC 6624 (HT 91), one of the most metal-rich globular clusters known in the galaxy. As with NGC 6624, NGC 6441's members contain, on average, about $\frac{1}{3}$ to $\frac{1}{4}$ as much iron (per unit of hydrogen) as does the Sun. If we presume that NGC 6441's age is similar to that of other galactic bulge globulars, it must be among the oldest – perhaps between 13 and 13.7 billion years old.

To find this hidden gem, first locate Lambda (λ) and Kappa (κ) Scorpii, at the end of the Scorpion's Tail. G Scorpii lies $2\frac{1}{2}°$ east-northeast of the midpoint between those two stars. NGC 6441 is only about 5′ east-southeast of G Sco; seen magnified, they look like gold and silver nuggets lying face up in the sand. I like to imagine them as part of the booty that Captain Kidd stole off the 400-ton ship *Quedah Merchant* on January 30, 1698. Before Kidd was arrested and hanged for this and other deeds, he allegedly stashed some treasure, which included gold and silver nuggets, on Gardiner's Island near New York (probably as security in case things went wrong); that action perpetuated the myth that pirates always bury their treasure. As David Cordingly writes in *Under the Black Flag*, "Although buried treasure has become a favorite theme in the pirate stories of fiction, there are very few documented examples of real pirates burying their loot."

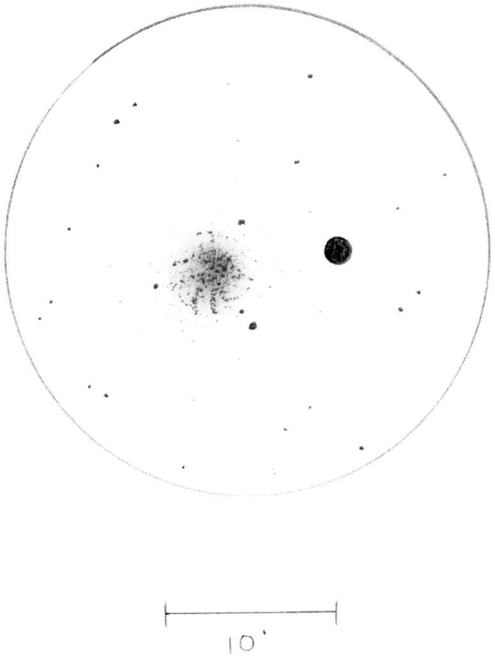

Rather than stash and save, pirates preferred to spend their plunder. But because Captain Kidd became one of the most famous pirates of history, "the matter of buried treasure," Cordingly says, "received more attention than it ever deserved." While modern-day treasure hunters still search for remnants of Kidd's stash on Gardiner's Island, we will turn our attention to the sky.

Using binoculars, challenge yourself visually to dig up the Silver Nugget globular cluster lurking ever so close to the golden glare of G Scorpii. Just hold the binoculars steady, look at G, then use averted vision. If you're a true pirate, it'll be hard to concentrate on this task, because so many other treasures surround it. There's the tug of the galaxy's hub, open clusters M6 and M7, and several other stellar gatherings nearby. No wonder NGC 6441 is so overlooked; there's so much to see around it.

In my antique telescope the cluster is a slightly swollen disk clearly separated from G Sco. And at 23× in the 4-inch, the globular is readily apparent as a tight, bright comet-like glow. A 10th-magnitude star shines only 1.5′ southwest of the cluster's center, and at least three other dimmer field stars surround the cluster even closer; all these stars add to the richness of the scene. With averted vision the cluster has a stellarlike core with an irregular halo. It's almost inconceivable to think that this tiny orb spans 107 light-years of space. At 72×, the cluster breaks down into three distinct layers: a dense nucleus, a narrow mantle, and a tenuous outer halo. The brightest stars in the cluster shine around magnitude 15.4, which are near the limit of detectability in a 4-inch. The cluster is so small and tight that it takes magnification well, so use as much as you're comfortable with. Even a slight increase in power, to 101×, will break up the cluster into a mottled haze, like wet cotton – most likely an appearance created by star clumping in such a small disk. Increase the magnification to 182×, and the core will no longer look stellar. Instead it appears as a central orb whose edges diffuse into an irregularly round midsection, which, itself, is surrounded by an "atomic" haze. A few stars or star clumps appear to be resolved in this outer envelope. Truly to resolve this cluster, however, you need a telescope capable of reaching magnitude 17.5.

The overall shape of the globular at high power is that of a face-on spiral galaxy. The eastern half of the globular is brighter than the western half. And the core is asymmetrical; it has a "C" shape that curls out from a central pip and is open to the south. Tiny patches of darkness can be seen throughout the cluster with averted vision, but none of them can be pinpointed exactly, they are so diaphanous. One prominent arm fans out from the globular's "mantle" to the east, then curves to the south. Another, thinner arm lies symmetrically opposite it to the west. Other weaker arms can be seen radiating in all directions. In fact, the impression I get is that the globular's arms are somewhat symmetrical, especially the two most obvious components, which look like the arms of a rotating water sprinkler.

The Hubble Space Telescope was used to survey the inner region of NGC 6441 for variable stars. A total of 57 new variable stars were found, including 38 RR Lyrae stars, six Cepheids, and 12 long-period variables. And in 1997 George H. Jacoby (Kitt Peak National Observatory) and L. Kellar Fullton (Space Telescope Science Institute) found a 5″-wide planetary nebula (JaFu 2) in NGC 6441 (right ascension $17^h\ 50^m\ 11^s$; declination −37° 03′ 27″. It lies 36,500 light-years from the Sun and is one of only four known planetary nebulae which are probably members of globular clusters. "Membership for the nebula in NGC 6441, though," Jacoby and his colleagues write in a 1997 *Astronomical Journal*, "is extremely likely based on common velocity, extinction, and proximity to the cluster center."

Before leaving the field, take the time to sweep your telescope across the surrounding Milky Way, which is filled with other hidden wonders, including open cluster NGC 6400 (Hidden Treasure 82), nearly 2° to the west-northwest. And about 25′ southeast of G Sco is another planetary nebula – PK 353–05.1; it's a 7″-wide object with no determined magnitude. If you have the capability to see G Sco and NGC 6441 in a 2° or larger field of view, spend some time sweeping back and forth from M7 to Lambda Scorpii, keeping your eyes peeled for veils of dust, which look like smoke from a fire (M7) blowing southwest toward the Stinger Stars. If you think you suspect the dust, change the sweeping pattern so that you are now moving northwest to southeast, against the grain. Its the visual equivalent of driving a car over washboard terrain. Enjoy the ride.

87

Hidden Treasure 87

Barnard's Star

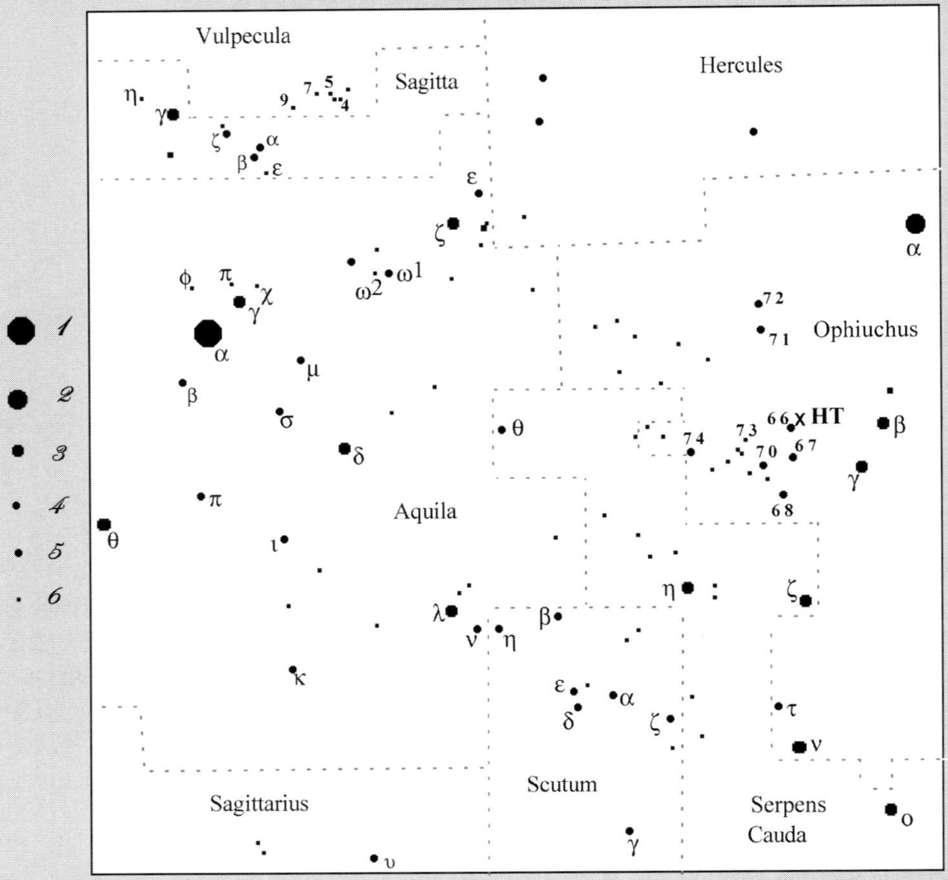

© Stephen James O'Meara

Tirion: Chart 15

Uranometria: Chart 204

87

Barnard's Star, Barnard's Runaway Star, Gilpin, O-bah-bahm-wawa-ge-zhe-go-qua
Type: High-Proper-Motion Star; Red-Dwarf Star (M3.8V)
Con: Ophiuchus

RA: 17^h 57.8^m (J1991.25)
Dec: +04° 40.1′ (J1991.25)
Mag: 9.5 (O'Meara); 9.5
Diam: –
Dist: 6 light-years
Disc: Edward Emerson Barnard, 1916

HERSCHEL: None.

NGC: None.

EDWARD EMERSON BARNARD, THE most celebrated pirate who ever "sailed" the celestial seas, scrawled his name across the heavens in swashbuckling fashion. A scan of the Milky Way will show his mark repeatedly; just look for the black flags of dark nebulosity, 349 of which bear Barnard's name. Barnard was a virtual "Blackbeard" when it came to astronomical discovery, meaning the man was a living legend. In Captain Johnson's *General History of the Pirates,* Blackbeard is described in celestial terms, as a man who "frightened America more than any comet that had appeared there a long time." While Barnard wasn't likely to frighten anyone, his name is attached to 14 of these "horrifying" apparitions. But it's the sheer number and magnitude of Barnard's other discoveries that really sends a shiver up the spine.

Barnard's name is associated with almost every type of deep-sky object visible in the sky today. The dwarf galaxy NGC 6822 (Caldwell 57) in Sagittarius is Barnard's Galaxy. The California Nebula near Xi (ξ) Persei, part of the Rosette Nebula in Monoceros, and Barnard's Loop (probably a supernova remnant) in Orion are all Barnard finds. He discovered Almathea (the fifth satellite of Jupiter), a great storm on Saturn, a naked-eye nova in Aquila, the binary nature of Beta^2 Capricorni, and the elusive gegenschein.

Between 1905 and 1921, Barnard accumulated 4,000 photographic plates of the Milky Way – each one being a rich trove of astronomical treasure. Anyone would be hard pressed to estimate their wealth. Today, we are still learning more about this man and his proficiencies as a photographer and an observer. For instance, in his monumental work *The Immortal Fire*

Within: The Life and Work of Edward Emerson Barnard, William Sheehan notes that Barnard had imaged Pluto in January 1921, nine years before Clyde Tombaugh discovered the world. These high notes are but part of a larger symphony of discovery that Barnard conducted throughout his life. "Known in his day as a man destined to be 'famous as long as fame shall last'," Sheehan writes, "few figures rivaled his renown, and today he continues to be remembered for the legendary keenness of his sight and his fantastic dedication to astronomy."

But there is one outstanding discovery by Barnard commonly neglected by visual observers: a rapidly moving star in the constellation Ophiuchus, known as Barnard's Runaway Star or, simply, Barnard's Star. It's funny that many mid-Northern Hemisphere observers lament the fact that, unless they travel, they cannot see the closest star system to our Sun: the Alpha Centauri system, with Proxima (4.22 light-years distant) being its closest member. It's funny because they have the second closest star to our Sun after the Alpha Centauri system within easy visual grasp. Lying just 6.0 light-years distant, Barnard's Star shines about two magnitudes brighter than Proxima and can be seen in a well-supported pair of binoculars.

Barnard discovered the star in 1916 while comparing plates he had taken in 1894 and 1916 with a "blink comparator" – an instrument that allows two plates of the same part of the sky to be viewed quickly and repeatedly in succession. If anything moves (like a comet or asteroid), or if anything flares to prominence or fades from view (like a variable star), the change is immediately apparent to the observer. But Barnard saw something unexpected – namely a star vanishing from one spot while another star popped into view a short distance away. After checking a 1904 plate of the same region, and then another taken in 1907, Barnard noticed that the appearing and vanishing stars moved chronologically in a straight line. After some puzzlement, he realized what he had found: a star with a high apparent angular rate of motion across the sky (high proper motion). Barnard's Runaway Star, as it was then called, was identified later on plates taken between 1888 and 1890 at Harvard College Observatory.

The path of Barnard's Star is now clear; it is moving almost due north (position angle 355.5°) at a rate of 1° every 351 years. It has the highest proper motion of any star. Typically, stars close to our Sun have proper motions of about 1″ per year. But Barnard's Star cuts across the rich stellar backdrop of the Ophiuchus Milky Way at a clip of 10.3″ per year. That's equivalent to the apparent diameter of Venus at full phase; in other words, if your telescope can resolve the disk of Venus, you can follow the annual movement of Barnard's Star. The star's apparent rapid motion across the sky is amplified by the fact that the star is approaching us at 140 kilometers per second, indicating a true space velocity of about 166 kilometers per second. The closest approach (3.9 light-years) occurs in 10,000 years. At that time, the star's proper motion will have increased to 25″ per year, and its brightness will have increased one hundred-fold.

Like Proxima Centauri, Barnard's Star is a red dwarf (M3.8V). It may be an old disk star that formed before the galaxy became enriched with heavy elements. It has less than 17 percent of the Sun's mass, 15 percent of its diameter, and $\frac{1}{2500}$ of its luminosity. If we were to replace our Sun with Barnard's Star, we'd be standing on a frozen

Earth, looking up at a very red Sun shining with an intensity only 100 times brighter than that of the full Moon.

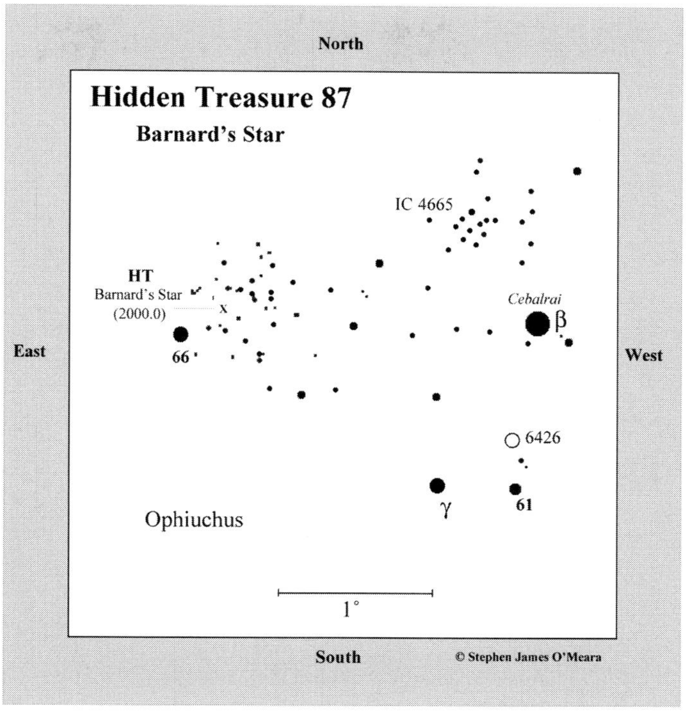

Astronomers have long suspected the presence of one, if not two, Jupiter-sized planets orbiting Barnard's Star. In the late 1960s, Peter van de Kamp (Sproul Observatory) measured a peculiar wobble in the star's motion across the sky. This wobble, he proposed, was due to a gravitational tug-of-war between the star and planet-sized companions. Although van de Kamp devoted most of his life to proving the existence of these invisible dancers, no one has yet been able to verify the claim. In 1995, the year van de Kamp passed away, George G. Gatewood (Allegheny Observatory) announced that while brown dwarfs exceeding Jupiter's mass by more than 10 times could not exist around Barnard's Star, planets having a mass smaller than Jupiter's may be present. To date, observations with the Hubble Space Telescope have failed to reveal a large Jupiter or brown dwarf-sized object to a limit of about 0.4 Jupiter mass. Although the idea of planets orbiting Barnard's Star is gradually fading from view, a final verdict in this case has not been reached.

To find this fascinating star, look $3\frac{1}{2}°$ east of Beta (β) Ophiuchi (Cebalrai) and 40′ northwest of 4.8-magnitude 66 Ophiuchi, the star marking the western horn of Taurus Poniatovii, the Polish Bull. Taurus Poniatovii – a now-defunct constellation created in 1777 by the Abbe Poczobut of Wilna, was named in honor of Stanislaus Poniatowski, then King of Poland. To see the Bull look for a close gathering of seven dim suns that form the letter V, one that resembles the Hyades in Taurus the Bull; the stars 77, 68, and 70 Ophiuchi, comprise the face of the Bull and the stars 66 and 73 the tips of the horns.

Look for a ruddy flame shining at magnitude 9.5. It is the middle and brightest of three stars – a magnitude 9.9 star 15′ to the northeast and magnitude 10.4 star to the southwest – forming a line 20′ long. The northeasternmost star points to a 15′ long fishhook of 8th- to 10th-magnitude stars (oriented northwest–southeast) 10′ away. Use the detailed chart shown here first to find Barnard's Star in a telescope. Once you know where exactly to look, it's a cinch in even the smallest of telescopes, including

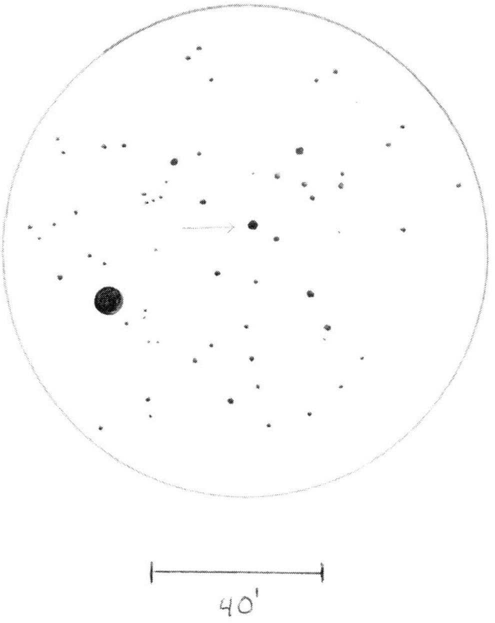

my antique scope. Don't forget to search for it in binoculars.

I have to wonder how many of us have seen Barnard's Star. I'll be the first to admit it. Although I've been observing the sky for about half a century and have long known about Barnard's Star, I did not hunt it down until June 2003 – after I had started this project. I've never been at a star party (and I've been to a lot), when a telescope barker shouted out for one and all to come and see this famous runaway star. Yet how easy it is to see, and how wonderfully educational an object it is. Once I saw it, I scolded myself for waiting so long. How much fun it would have been to have tracked this star since childhood.

Well, enjoy Barnard's Star while you can because it may already be around 10 billion years old; in another 40 billion years it will be on its way to becoming a black dwarf. By the way, did you notice the star's third nickname listed in the table above? After Barnard's discovery of the Runaway Star, the staff of Yerkes Observatory named it "Gilpin," after a character in a poem by William Cowper entitled *The Diverting History of John Gilpin*. Gilpin, a linendraper living north of London, is unable to restrain his horse and gallops 10 miles to Ware.

I've bestowed yet another poetical name on the star: "O-bah-bahm-wawa-ge-zhe-go-qua." You see, Henry Rowe Schoolcraft, a renowned historian, pioneering explorer, and geologist – who was also superintendent of Indian affairs for Michigan from 1836 to 1841– was married to Jane Johnston, a descendant of the Ojibway Indians of northern Michigan. Jane's Indian name is O-bah-bahm-wawa-ge-zhe-go-qua (The Woman of the Sound Which the Stars Make Rushing Through the Sky). If there is a star in the northern sky that should bear her Indian name, Barnard's Star is it.

Hidden Treasure 88

NGC 6520

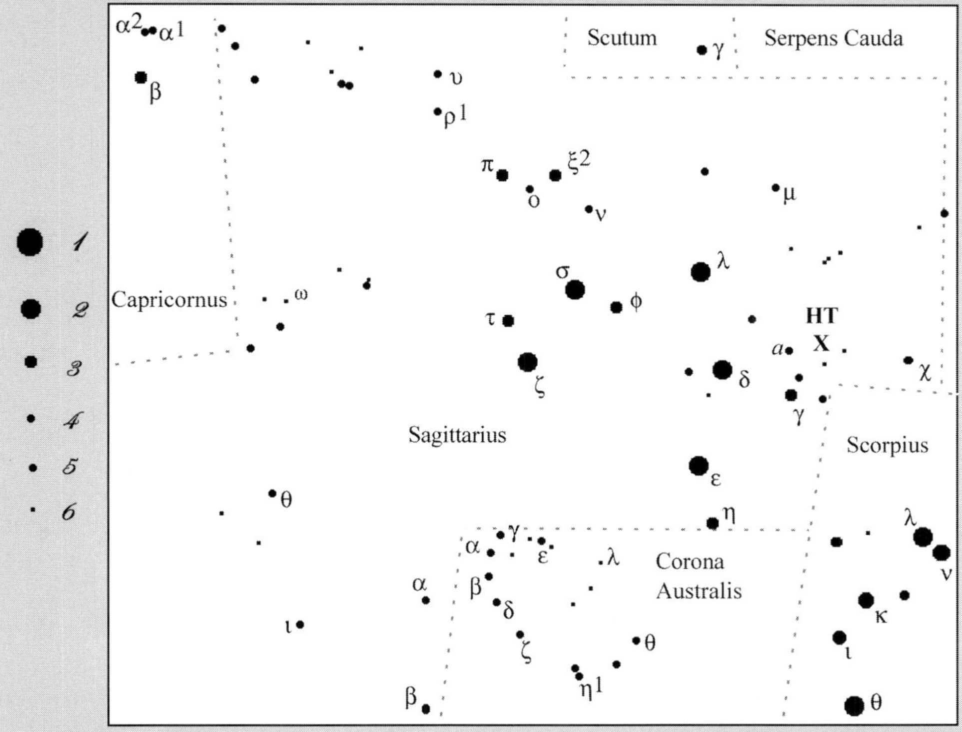

© Stephen James O'Meara

Tirion: Chart 22

Uranometria: Charts 339 & 377

88

Dead Man's Chest, Castaway Cluster
NGC 6520
Type: Open Cluster
Con: Sagittarius

RA: $18^h\ 03.4^m$
Dec: $-27°\ 53.5'$
Mag: 7.6 (O'Meara); 7.6
Diam: 5.0′
Dist: 5,300 light-years
Disc: William Herschel, 1784

W. HERSCHEL: [Observed May 24, 1784] A considerably rich, but pretty coarse, scattered cluster of stars, little more depressed in the middle. (H VII-7)

NGC: Cluster, pretty small, rich, little compressed, stars from magnitude 9 to 13.

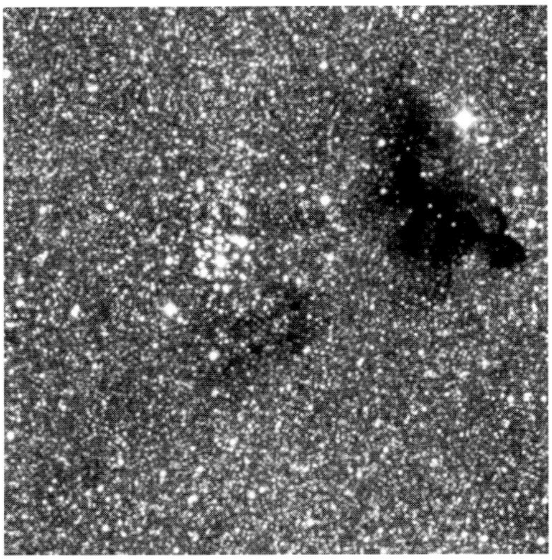

WHENEVER I LOOK AT THE TEAPOT of Sagittarius with the naked eye, my attention is inevitably drawn toward the northwest, where an archipelago of bright nebulae and star clusters line the Milky Way like so many islands in the sea. Vying for attention are the Lagoon Nebula (M8), the Trifid Nebula (M20), the Little Sagittarius Star Cloud (M24), and open clusters M23 and M25, to name but a few. But buried deep in the celestial sands of the Sagittarius Star Cloud, far from the maddening crowd of deep-sky objects mentioned above, is one of the sky's truly lost treasures – the open star cluster NGC 6520.

Located about $2\frac{1}{2}°$ north-northwest of 3rd-magnitude, Gamma (γ) Sagittarii (Al Nasl), the head of the Archer's arrow or the tip of the Teapot's spout, NGC 6520 lies in one of the most glorious fields of the Milky Way. Peppered with stars too numerous to imagine, veiled by dark mists and sheer luminous matter, the region is a virtual dreamscape of darkness and light set against a backdrop of endless wonder. So grand is the view around NGC 6520 that the tiny cluster looks like some lonely, uncharted isle in a tempestuous sea. For that reason, I call it the Castaway Cluster – in honor of Daniel Defoe's immortal hero *Robinson Crusoe*:

> September 30, 1659. I, poor miserable Robinson Crusoe, being shipwrecked, during a dreadful storm, in the offing, came on shore this dismal unfortunate island, which I called the Island of Despair.

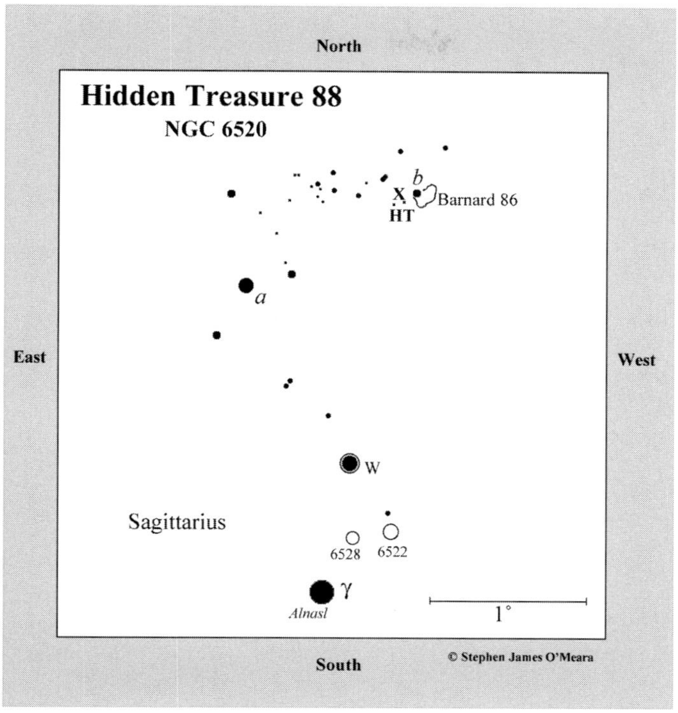

(2) Nearly 1° to the north-northwest is the bright Cepheid variable star W Sgr; it ranges in brightness from magnitude 4.3 to 5.1 every 7.6 days. (3) Just $1\frac{1}{4}°$ to the northeast of W Sgr is a magnitude 4.6 star (*a*). And (4) hop another $1\frac{1}{4}°$ to the northwest, where you'll find a solitary, magnitude 6.7 star (*b*) with a magnitude 7.6 companion about 7′ to the southeast; the magnitude 7.6 companion is the compressed glow of NGC 6520. There are no stars as bright as Star *b* in the region out to nearly 2°, so there is no mistaking it.

In 7 × 50 binoculars, NGC 6520 has a slightly fuzzy appearance, even under the light of a waxing quarter Moon. In the antique scope, the cluster looks elongated, like a little spit of starlight. I can resolve at least three stars. At 23× in the 4-inch, the cluster's six brightest stars lie along a northwest–southeast-trending line. It's as if the stars of Kemble's Cascade (Hidden Treasure 21) have been culled and compressed. The star in the middle of the line is a ruby among diamonds. With averted vision a circular halo of fainter suns surround the line like a crowd during a parade.

At 72× about 16 stars are obvious with others popping in and out of view. It's a very tight cluster, so high magnification is a must. Each boost in power is like digging deeper into a seaman's chest. After experimenting with various magnifications, I found the view best with a 5-mm Nagler and a 3× Barlow (303×). At this power, a stunning

Defoe was fascinated by pirates. Indeed, his hero, Crusoe, is modeled after Alexander Selkirk, a Scottish seaman marooned for four years on the uninhabited island of Juan Fernandez in 1704. Pirates used marooning as a way to settle differences. In Captain Johnson's 1724 *A General History of the Robberies and Murders of the Most Notorious Pyrates,* we find that the second of the eleven articles, drawn up and adopted by crew of different pirate captains, decrees that if any pirates "defrauded the company to the value of a dollar in plate, jewels, or money, marooning was their punishment." But in Selkirk's case, he was put ashore at his own request, after having a quarrel with Thomas Straddling, the ship's captain.

Finding marooned NGC 6520 is not difficult. In fact, try scouting it out with binoculars first. You can do it in four steps: (1) Start with 3rd-magnitude Gamma Sgr.

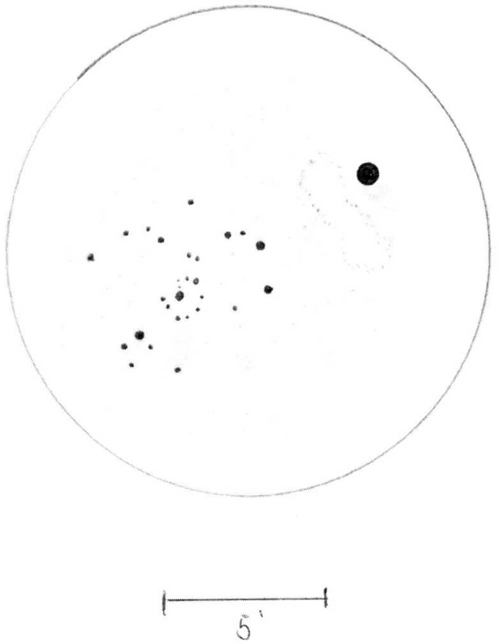

circlet of stars surrounds the red ruby heart of the cluster. One long, loose, arm flies off the circlet to the north, then curves around to the east. A tighter spiral swings from east to south, and two parallel arms jut out to the west. All told, I count only about 30 stars. Twice that number, though, are true members. The cluster, then, must be glorious in larger apertures.

The two brightest stars at the cluster's southern fringe shine with a tangerine hue. Together with the ruby heart of the cluster and the blood-red magnitude 6.7 star to the west, it's apparent that we are seeing two ages of stars juxtaposed. The 60 or so hot blue stars that truly belong to NGC 6520 lie only about 5,300 light-years distant and have an estimated age of 800 million years. In true physical extent they span about 8 light-years of space. Shining through these young suns are a handful of red giants, billions of years old. Most likely, they are distant field stars that belong to our galaxy's ancient hub, into which we are peering.

I also call this cluster the Dead Man's Chest because the red magnitude 6.7 star is part of a chain of four stars in a slight arc. It looks as if a bleeding man is lying prostrate next to a chest filled with precious jewels. There also appears to be a hint of glowing nebulosity, extremely tenuous, surrounding the four stars that make up the bleeding man; it's as if the dead man's spirit is leaving his body.

There's an added treat for visual observers. Wedged between NGC 6520 and the magnitude 6.7 star is one of the sky's blackest nebulae – Barnard 86. Edward Emerson Barnard discovered this "small triangular hole in the Milky Way, as black as midnight," on July 17, 1883. At the time he was sweeping for comets from his home in Nashville, Tennessee. As William Sheehan explains in his biography of the great man, *The Immortal Fire Within*, Barnard later described it as "a most remarkable small inky black hole in a crowded part of the Milky Way . . . with a bright orange star on its [north preceding] border and a beautiful little cluster [NGC 6520] following."

Six years later, on August 1, 1889, Barnard took his first successful photograph of the Milky Way. It was of the same region of Sagittarius, because it possessed "the most intricate and complex structure of any portion of the Milky Way above our horizon." Here is how he described that plate:

> This remarkable picture shows the cloud-like forms like waves of spray. A . . . curving lane [of darkness] runs from the lower left-hand portion of the picture . . . and curves gracefully

> upwards to the place of *Jupiter*. It is singularly like the stem of a great leaf. At the middle of the picture it is seen to pass behind some of the clouds of stars and emerge beyond, showing us clearly which part of the Milky Way at that point is nearest to us. Imagination may aid one, but it looks as if the lines of the cloud-forms, and of the stars and vacancies, all run more or less concentric with this extensive lane . . . The black hole is seen slightly to the left of the center, with the small cluster [NGC 6520] as a white spot close to the right of it.

If you have the chance to explore this region with a wide field of view, be sure to circumnavigate NGC 6520 with averted vision. The cluster, you may see, lies at the intersection of a much larger "X" of dark lanes. Like those old maps of buried treasure then . . . X marks the spot.

Hidden Treasure 89

NGC 6544

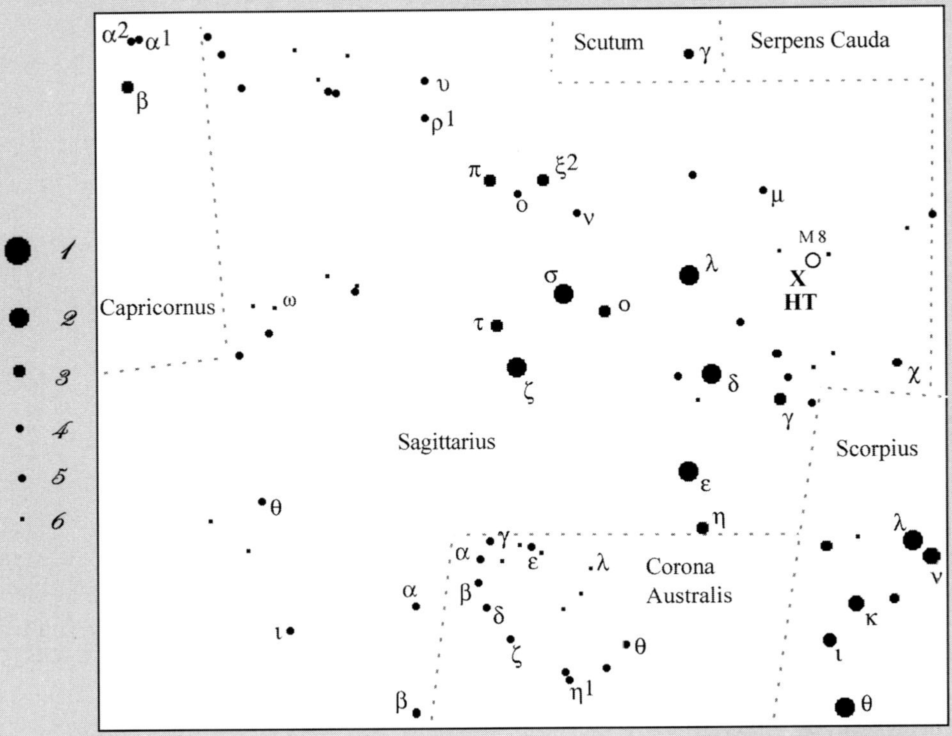

© Stephen James O'Meara

Tirion: Chart 22

Uranometria: Chart 339

89

Starfish Cluster
NGC 6544
Type: Globular Cluster
Con: Sagittarius

RA: $18^h\ 07.3^m$
Dec: $-25°\ 00'$
Mag: 7.5 (O'Meara); 7.5 (Skiff); 7.8 (Harris)
Diam: 4.6′
Dist: 8,800 light-years
Disc: William Herschel, 1784

W. HERSCHEL: [Observed May 22, 1784] Pretty bright, pretty large, irregularly round, resolvable. (H II-197)

NGC: Considerably faint, pretty large, irregularly round, resolved.

IT'S SO CLOSE, BUT YET SO FAR. Globular cluster NGC 6544 is one of those easy to see objects that's all but lost in the crowd of bright clusters and nebulosities that populate the Milky Way's hub around M8, the renowned Lagoon Nebula, which has been called by some the Orion Nebula of the South. So powerful is the sight of M8 that it's hard to draw our attention away from its splendor. Once we do, we usually set sail $1\frac{1}{2}°$ to the north, where we are certain to find yet another visual beauty – M20, the clover-like glow of the Trifid Nebula. And when we are finished marveling at M20, it's off to the bright open cluster M21, just 30′ to the northeast. In fact, M8 marks the southern end of a brilliant chain of Messier nebulae and open clusters that follow the galactic equator up to M16 in Serpens. Add the numerous *NGC* nebulae and clusters that infiltrate the spaces between these Messier objects, and it's a wonder we can break free from this chain at all. Then add to that the rest of Sagittarius, which contains 14 Messier objects worth scouting, and it is little wonder that lesser known *NGC* objects are neglected.

But NGC 6544 is one worth visiting, if only for brightness and position. The globular cluster is a fine binocular object, a tiny raft of light about 1° southeast of the visually weathered shores of M8. I wish I could see a raise of hands of how many have even bothered to tap their telescope tubes in its direction. I can honestly say that I hadn't – until only recently, when I swept it up during a comet hunt. That little pitter-patter my heart felt at that moment may have put me over the edge and coerced me to include this little gem in this catalog. NGC 6544 is

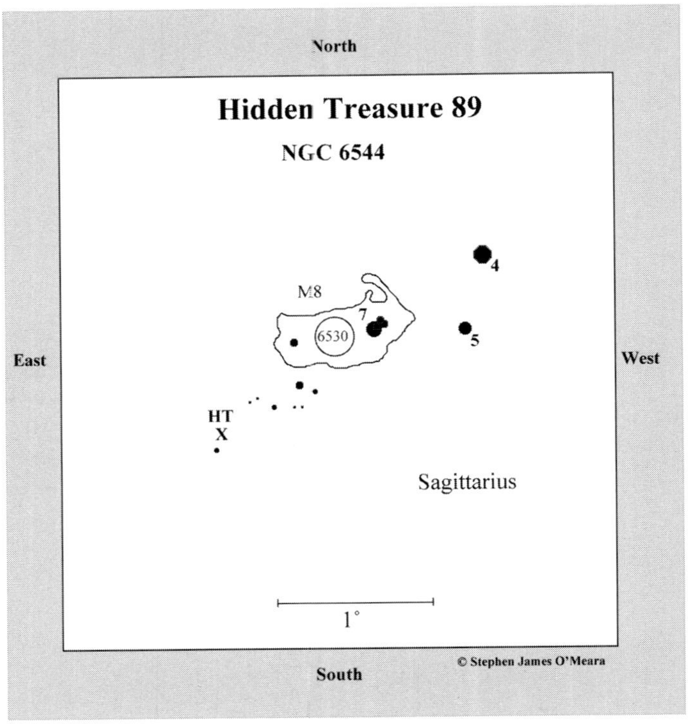

a magnificently modest object, magnifying the concept that lost pearls can hide in the shadows of the sky's grandest objects. Like an arrowhead lying on the lip of the Grand Canyon, NGC 6544 has its own story to tell.

We see NGC 6544 only about 6° from the galactic center, where intervening dust dims its light by 0.7 magnitude. Fortunately, it is relatively close, lying 8,800 light-years from the Sun and 17,300 light-years from the galactic center. It would have been a truly glorious object in our skies had it not been so small in true physical extent (12 light-years across). Compare NGC 6544 with globular cluster M22, which is only 10,000 light-years from the Sun but 96 light-years across, and you will appreciate the dynamical range of our galaxy's senior citizens.

In 7 × 50 binoculars and the antique scope, the globular is a fairly round but compact glow with a bright, starlike nucleus. Indeed, NGC 6544's core measures only 3′ in diameter and is the cluster's most prominent feature. It appears that the intense gravitational pull of so many stars in such a small volume seems to be causing stars to converge on the cluster's core like bees swarming to their hive. Such a collapse may be the process responsible for creating such exotic objects as X-ray pulsars, blue stragglers, and, possibly (though doubtfully), black holes. "A black hole has been suggested to exist at the center of M15, and also G1 in M31," Canadian astrophysicist William Harris (MacMaster University) notes, "but these are still just suggestions, and there are other more mundane and less exciting ways to interpret the measurements. Black holes less than about 500 solar masses at the centers of clusters would be virtually impossible to detect with present techniques."

Scott M. Ransom (Harvard University) has found a 3 millisecond binary pulsar with a 102 minute orbital period and an approximately 10 Jupiter-mass companion. Martha Hazen (Harvard-Smithsonian Center for Astrophysics) searched this cluster for variable stars and found two within the tidal radius of the cluster and two outside. One of the variables within the tidal radius is probably an eclipsing binary, the other is a RR Lyrae type that may be a cluster member. The two field variables are both RR Lyrae types.

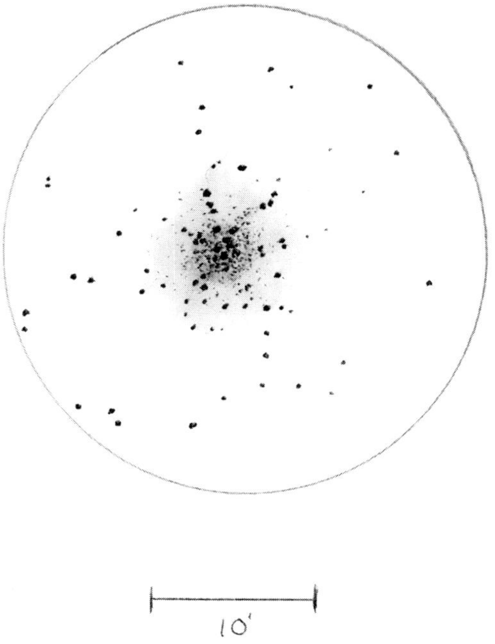

At 23× in the 4-inch, the globular appears as a large faint glow with a bright and highly condensed core, like the nucleus of an old comet surrounded by the faint and tenuous glow of its halo. An 8th-magnitude star lies 6.5′ south-southeast, a chain of three faint suns borders the cluster immediately to the east-southeast, and connects with a dim ellipse of suns that curves to the south before looping back to the cluster on the west side, where it connects to a solitary 12th-magnitude field star projected against the southwestern part of the cluster's halo.

Since the cluster is so small and tight, it begs for high power. At 101×, the cluster is very mottled. The cluster's brightest stars (12.8 magnitude) form a northwest–southeast-trending line across the core. A dimmer arm extends from the southeast end of that line and connects to the solitary 12th-magnitude star. So with averted vision, the brightest portions of the globular have a sideways V-shaped pattern. The cluster takes power well; I find the best views between 182× and 303×. With a horizontal branch magnitude of 15.0, observers with even modest-sized telescopes should be able to resolve many of the cluster's fainter stars – more so than in some of the dimmer Messier globulars.

At higher power the core becomes a stellar point surrounded by a nearly complete, though highly mottled, ring of star clumps. The ring is open to the southwest where a very dim arm extends to the 12th-magnitude star. A faint arm extends to the northeast before gently curving north. And fragmentary clumps of stars can be seen to the northwest and east. Overall, the brightest stars form a starfish-shaped pattern, so I call it the Starfish Cluster. Despite some claims that "if you've seen one globular, you've seen them all," no two globulars are alike. In a 12-inch scope, Brian Skiff and Christian Luginbuhl say that NGC 6544 is a "pretty sight at low power, sparkling throughout, with its bright stars near the center looking like a nucleus."

Hidden Treasure 90

NGC 6572

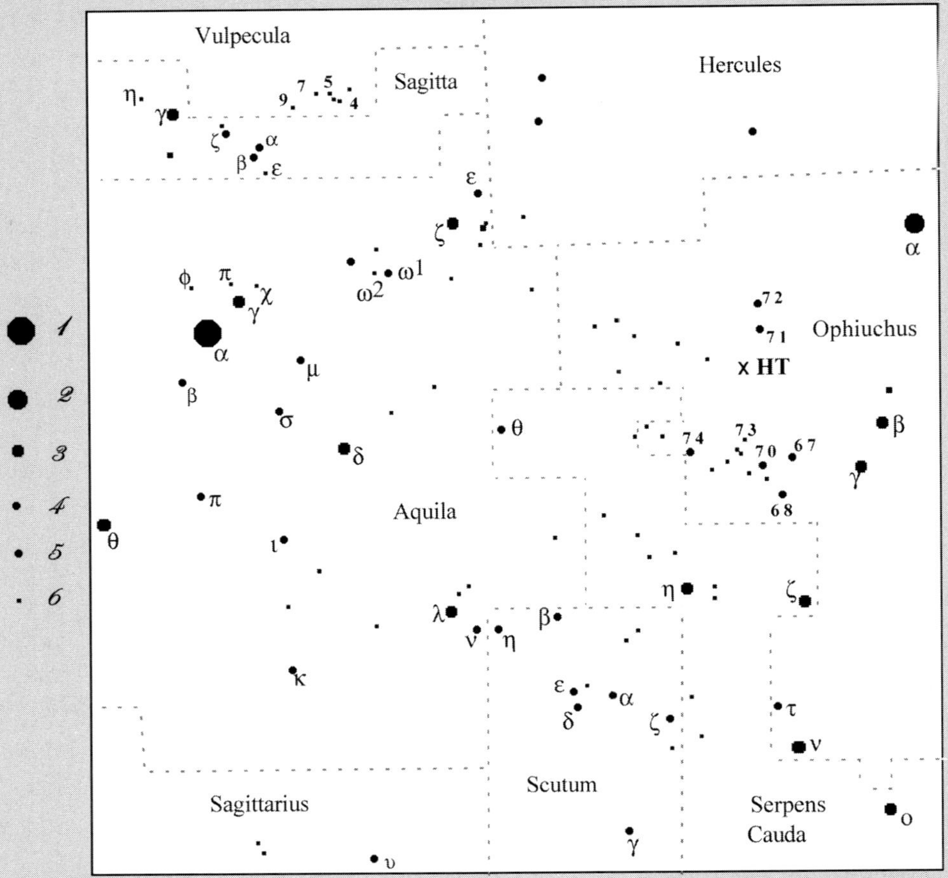

Tirion: Chart 15

Uranometria: Chart 204

90

Emerald Eye Planetary, Planet Krypton Nebula
NGC 6572
Type: Planetary Nebula
Con: Ophiuchus

RA: 18^h 12^m 06.6^s
Dec: +06° 51′ 13″
Mag: 7.3 (O'Meara); 8.1
Dim: 16″ × 13″
Dist: 4,800 light-years
Disc: Friedrich Georg Wilhelm von Struve, 1925

HERSCHEL: None.

NGC: Planetary nebula, very bright, very small, round, little hazy.

NGC 6572 IS THE 13TH WONDER OF Ophiuchus – the others being the seven bright Messier globular clusters (M9, M10, M12, M14, M19, M62, and M107) and four hidden treasures (Barnard's Star, IC 4665, NGC 6369, and NGC 6633). And what a wonder! Wilhelm von Struve, who discovered NGC 6572 in 1925 with the 9.6-inch Fraunhofer refractor at Dorpat Observatory (see also Hidden Treasure 78), described it as one of the most curious objects in the heavens – "it being a star surrounded by bright green ellipse of fuzzy light." In early photographic images, the nebula was classified as an "atmosphere" type – meaning that the nebula's light decreases uniformly outward from a bright core, like the atmosphere around a planet.

In 1925, not much was known about the nature of planetary nebulae. In 1922 Edwin Hubble had noticed that larger planetary nebulae have brighter central stars, and concluded that the nebulae must extract their energy from the star. A few years prior, Heber Curtis (Lick Observatory) compared his photographs of planetary nebulae with models of suspended fluids viewed by transmitted light, and concluded that nebulae showing a pronounced ring effect can be treated as homogeneous shells of various forms. Of course, we now know that the morphology of planetary nebulae are much more elaborate (see Hidden Treasure 84).

NGC 6572 is a compact planetary of very high surface brightness; not only is it 1.5 magnitudes brighter than M57 (the Ring Nebula) in Lyra, but also some five times smaller. In photographs, the nebula's structure appears simple: a bright ring surrounded by an amorphous halo with no filaments. Viewed in three dimensions, the nebula is most likely a thin cylinder of light,

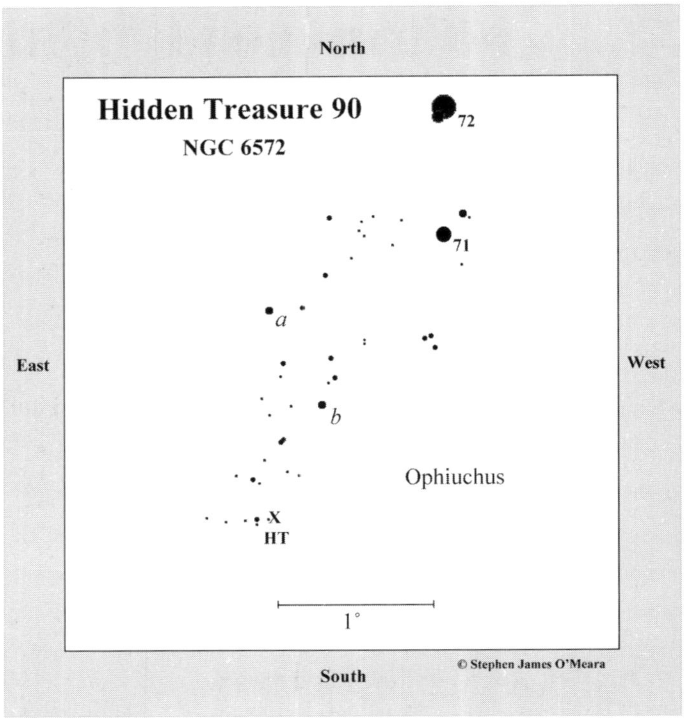

or a prolate ellipsoid, 0.4 × 0.3 light-years across, girdled by a dense torus of dust. Ultraviolet observations show it to be a medium excitation object; it also appears bright at radio wavelengths, indicating high density, and, therefore, youth.

In a 1994 paper published in the *Monthly Notices* of the Royal Astronomical Society, Seik Hyung and Lawrence H. Aller (University of California, Los Angeles) and Walter A. Fieibelman (NASA-Goddard Space Flight Center) determined that the central white dwarf has a mass of about 0.6 Sun; therefore, the progenitor star must have had a mass of about 1 Sun. They estimate the nebula is about 2,600 years young – about the time when the Maya were inventing (and gobbling) chocolate and the first skin transplants were being performed by Hindu surgeons. The nebula is expanding away from the central star at about 16 kilometers per second – about the speed that an average meteor travels in Earth's atmosphere.

To find this hidden treasure, turn your gaze to Alpha (α) Ophiuchi (Rasalhague) and Beta (β) Ophiuchi (Cebalrai). About 7° to the east is a pair of roughly 4th-magnitude suns – 71 and 72 Ophiuchi. Now look $1\frac{1}{4}°$ to the southeast for two 7th-magnitude stars (*a* and *b*), which are separated by 40′ and oriented northeast–southwest. NGC 6572 is about 50′ southeast of the southernmost 7th-magnitude star. It lies about 4′ due west of a magnitude 9.5 star. But the object is brilliant in 7 × 50 binoculars as a magnitude 7.3 star; the antique telescope shows the same appearance. The planetary remains stellar at 23× in the 4-inch, but as with NGC 6210 (Hidden Treasure 78) in Hercules, the more time I spend with it at this power, the more I suspect that I can tell it is "different" from the surrounding stars, not having the same visual sharpness as a star. But that's only with keen averted vision; even then it swells only slightly.

At 72×, it is a most mystifying sight. It shows a central star surrounded by an elliptical halo oriented north–south. But the central "star" shows a very tiny disk. This disk is what I believe is mistaken for the 10th-magnitude central star; the real central star shines at magnitude 13.6. That central disk reminds me of the planet Neptune passing through the Kuiper belt of comets – if you can imagine that. One night I wrote, "I also

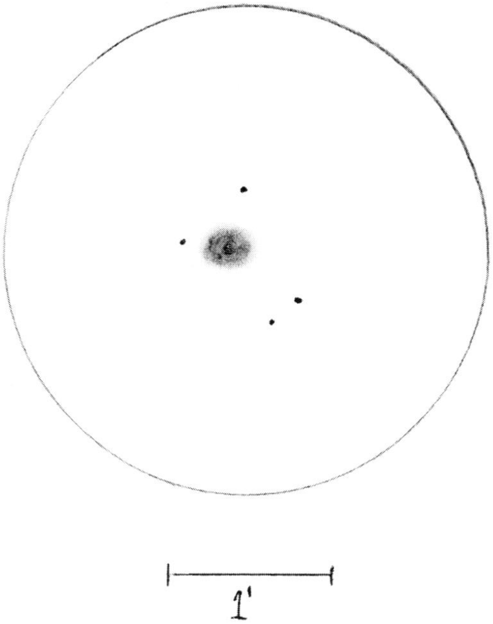

suspect an extremely dim elliptical outer envelope. It is as if a high cirrus cloud has passed overhead. But I cannot make other stars in the field wear this same shirt. If the night weren't as clear as it is, I might doubt this shirt's existence. But when I boost the power to 101×, it appears that this fuzz may be an optical illusion caused by dim stars abutting the nebula."

When I moved on to 182×, then to 303×, I really began to "see things." I still couldn't shake the thought that I could detect outer nebulosity. But this was only an illusion. "NGC 6572 is probably too young and compact of a planetary nebula to have much of a 'true' extended halo," says Jay McNeil, the visual guru of planetary nebulae. "Sure, there's definitely some tenuous material that was flung out pretty far from the main nebula by the progenitor star while it was still a red giant. On the other hand, NGC 6572 is considered to be 'optically-thick.' This means that nearly all of the radiation emitted from the current central star is being absorbed by the surrounding dense shell that forms the high surface-brightness disk that we see (hence its wonderful efficiency at producing the vivid blue-green color via the absorption of such ultraviolet radiation and reemitting it at the OIII lines). Generally, it is only once that this 'denser' material has thinned out some that enough of the radiation can escape the inner shell in order to illuminate the really faint outer stuff. On the Digital Sky Survey, I see that there are several stars of about the same magnitude within a 1′ radius of the nebula. I do agree that at lower powers under excellent transparency, these would likely form an illusive 'fuzziness' around the brighter nebula."

McNeil's visual observations with a 16-inch reflector at 586× under very steady seeing conditions showed a very compact and high-surface-brightness elliptical disk, evenly illuminated, elongated slightly north–south, and appearing about 12″ in greatest dimension. This "inner" disk was surrounded by a much fainter diffuse "halo" oriented in the same direction and extending to maybe 20″ at greatest. The brighter portion of the nebula was vividly greenish-blue in color, especially at powers below 200×. At higher powers, the nebula's apparent brightness and color faded. "However, the nebula's 13th-magnitude central star," he says, "can be easily spotted toward the object's center."

There is much confusion over the visible nature of the planetary's central star. It has been seen as 10th magnitude and as faint as 14th. Houston said, "such are the problems when trying to estimate a star's magnitude when it appears against a bright background." There are several reports of the star not being seen with 10-inch instruments. If you search for it, remember to use the highest magnification possible. Doing so dims the appearance of the nebula and increases the contrast between the star and its background. Hartung noted that this remarkable planetary nebula "shows a small disk 10″ across and in the prism the strong image is followed by another less bright on the violet side, with evidence of a third still farther along; these come from ionized oxygen and hydrogen. There is also a long thin spectrum streak, stronger towards the red, which is that of the otherwise invisible central star."

At 303× in the 4-inch, there appears to be a core with two arcs on either side, and a bead to the east. The dense inner nebula has a greenish tinge. Walter Scott Houston noted that his "old 10-inch reflector showed the vivid green color of this object with any power more than 50×." He also noted that older observers see NGC 6572 as green, while younger ones tend to call it "vivid blue." Admittedly, nothing I've seen or read, compares with the view of NGC 6572 through Larry Mitchell's 36-inch reflector. Here was an incredible neon green beacon – of a color and brilliance so powerful that I could see it with sunglasses on. The view made me feel like Superman looking at a piece of kryptonite – a fragment of the exploded planet Krypton. Krypton was the home world of the fictitious Marvel comic hero Superman – until it exploded into fragments as the result of a cataclysmic chain reaction originating at the planet's core. It was a planet of stunning beauty with innumerable visual treasures, including jewel mountains and a volcano that erupted gold instead of lava.

It is fitting that the neon beauty of NGC 6572, a planetary nebula, being the product of eruptive processes at the central star's core, be seen as the ultimate demise of the planet Krypton. As Superman's father, Jor-El, warned the inhabitants of Krypton before the end, "[t]he core of Krypton is composed of a substance called uranium, which, for untold ages, has been setting up a cycle of chain-impulses, building in power every moment. Soon, very soon, every atom of Krypton will explode in one final terrible blast. Krypton is one gigantic atomic bomb." Alas, the end came in one final cataclysmic eruption and the once mighty planet Krypton exploded into stardust. So the next time you look at NGC 6572, use your imagination and behold its green shell as the ultimate demise of Superman's home planet.

Hidden Treasure 91

NGC 6624

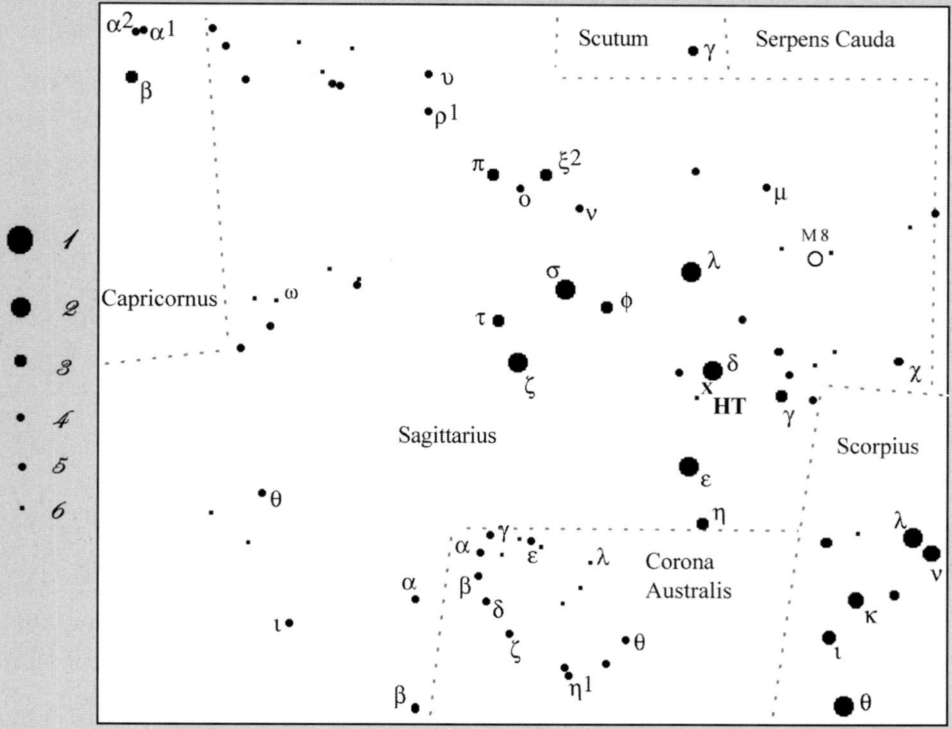

Tirion: Chart 22

Uranometria: Charts 377 & 378

91

NGC 6624
Type: Globular Cluster
Con: Sagittarius

RA: $18^h\ 23.7^m$
Dec: $-30°\ 22'$
Mag: 7.8 (O'Meara); 7.6
Diam: 8.8′
Dist: 25,700 light-years
Disc: William Herschel, 1784

W. HERSCHEL: [Observed June 24, 1784] Considerably large, round, very bright middle, milky. (H I-50)

NGC: Globular cluster, very bright, pretty large, round, well resolved, clearly consists of stars, 16 stars.

GLOBULAR CLUSTER NGC 6624 IS A subtle beauty basking in the glow of Delta (δ) Sagittarii (Kaus Media) – an orange 3rd-magnitude *K*-type giant star 45′ northwest of the Teapot's Spout. Despite the globular's brightness and proximity to a well-known star, hardly anyone looks at it, at least not in the Northern Hemisphere. Perhaps a generation of observers were put off by Robert Burnham, Jr.'s, remark in his *Celestial Handbook* that NGC 6624 is "not very impressive visually." At least Ernst Hartung noted in his *Astronomical Objects for Southern Telescopes* that it is "easily picked up as a hazy spot" with a 3-inch telescope. In fact, NGC 6624 is just as bright as the much sought after Sagittarius globulars M54, M69, and M70, and it is a full magnitude brighter than M75, also in Sagittarius.

Although NGC 6624 is seen against the central hub of our galaxy, which presents professional astronomers with the problems of interstellar reddening and field-star contamination, the cluster is well studied. The aged cluster has been literally dissected by astronomers using the Hubble Space Telescope (HST), who have determined that NGC 6624 has resided in the galactic bulge for nearly the last 14 billion years, making it another extremely aged globular cluster, like M69 in Sagittarius and the magnificent southern globular cluster 47 Tucanae (Caldwell 106). It is also among the most metal-rich globulars known; its members each contain, on average, about $\frac{1}{2}$ to $\frac{1}{3}$ as much iron (per unit of hydrogen) as does the Sun.

While globular clusters contain some of the oldest stars known in the galaxy, they

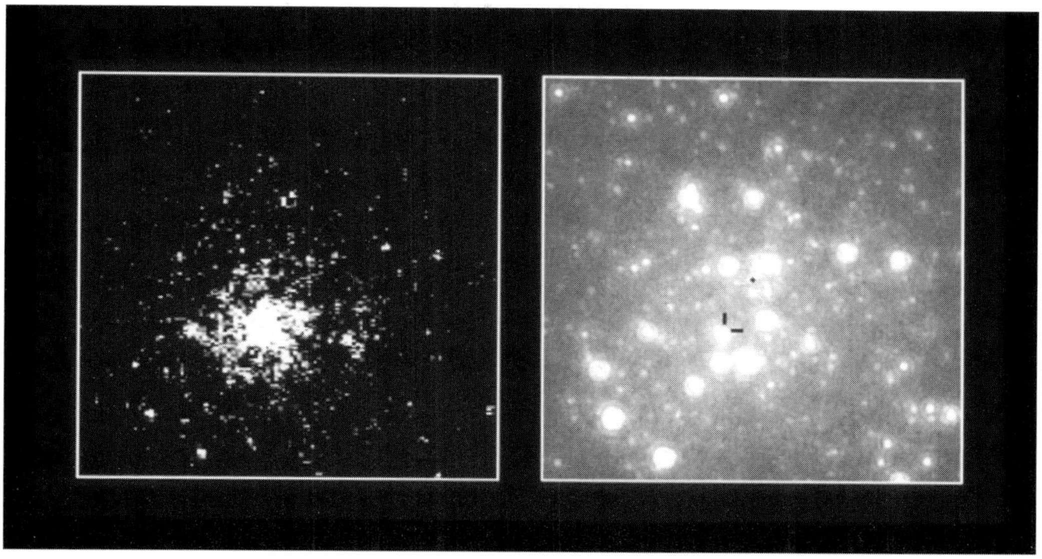

can also contain stars that are bluer and brighter than other cluster members. These enigmatic stars, called blue stragglers, are most likely formed either when the stars in a double-star system slowly merge, or when two unrelated stars collide. The HST detected at least ten candidate blue stragglers within five arcseconds of NGC 6624's center. The HST also found evidence for an unusual double-star system at the very heart of the cluster – a neutron star (a city-sized collapsed core of some ancient supernova explosion) and a white-dwarf companion (a planet-sized remnant of a burned out Sun-like star), which orbit one another every eleven minutes; it is the fastest binary system known. The stars are only 161,000 kilometers apart, or less than half the distance between the Earth and Moon.

Intense X-rays are being emitted from the system. No doubt, the fierce gravitational pull of the neutron star is ripping away gas from the white dwarf, which is slowly being cannibalized. The stolen gas forms a swirling disk around the neutron star, which heats up to temperatures greater than 100,000 °F. A steady stream of X-rays is then emitted as the gas falls onto the star. The infall of matter also supplies fuel for sporadic, intense X-ray bursts. These occur when enough gas accumulates on the surface of the neutron star to ignite spontaneously, creating an enormous nuclear fusion explosion. For a few seconds, the outburst abruptly outshines the usual steady X-ray emission.

To find this beautiful object, just center Delta Sagittarii in your telescope, then move slowly 45′ to the southeast. Try first in binoculars. I find that with 7 × 50s, NGC 6624 is visible as a starlike object, while my antique telescope reveals it as a slightly bloated disk. In the 4-inch at 23×, the cluster remains tiny and bright; it is easy to sweep over if you move the telescope too fast, so take your time scanning. The globular's core is highly condensed, but it becomes more globelike with averted vision. If you still have trouble spotting it in a sweep, just move to the globular's location, then use averted vision

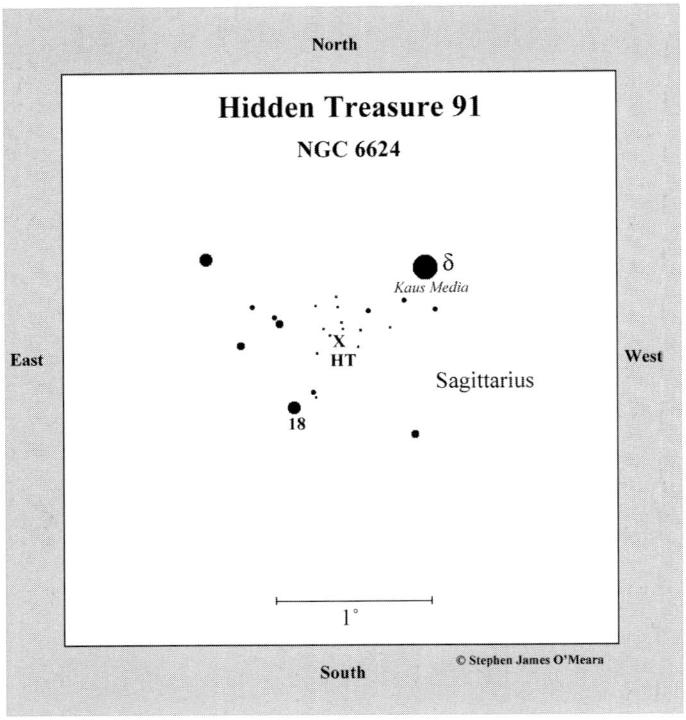

At 101× and averted vision, the core is tight and mottled, and looks like curdled milk. Sharp rays of starlight pierce the core like an arrow from the northeast to the southwest. Three wispy arms extend to the northwest. Look also for two stubby extensions: one to the southeast, which spirals toward the west, and another to the south. The outer halo is sprinkled with dim suns. Seen together, all these wispy arms make the cluster look like a spider web being stretched by an unseen hand. All the arms become much more apparent with 182×.

to scan the field for a starlike object surrounded by a tenuous haze. You'll find NGC 6624 at the southwest end of a 7′-wide chain of three "stars"; the other two stars shine around 10th magnitude, so they are reasonably dimmer. Then, immediately put in a higher power.

At 72×, some field stars scan be seen projected against the cluster's outer halo, especially to the west, northwest, and east. I see little color, except a pasty yellow gray, like that of nicotine-stained teeth. Take the time to scan the cluster carefully with averted vision, and see if it does not start to break up into mottled bits. The brightest stars in the cluster shine at 14th magnitude, so the combined light of clumped members should pop in and out of view, especially in the outer halo. The cluster is so condensed that it takes power well.

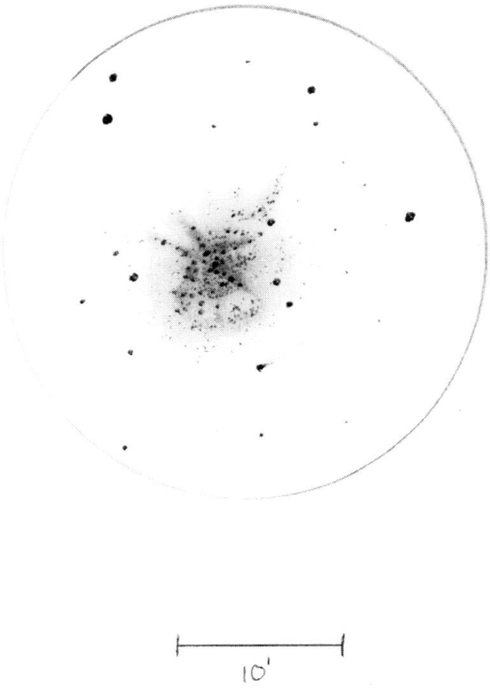

The southern edge of the cluster's outer envelope forms a discreet arc that seems to spiral to the west. A tighter arc – one closer to the nuclear region – lies to the north. You could spend hours trying to ferret out the secrets of this dense globular with a small telescope. The more time you spend with it, the more you will see – like a tiny spike of stars flowing gracefully to the north-northwest, another to the south; a sprinkling of suns here, a clump of suns there. All these details waver in and out of view, depending on how you move your eye.

One night, I was able to follow NGC 6624's dense core into bright twilight. At first I used 303×, then, as dawn approached, I switched to 168×. The core is so tight I had no problem staying with it, resolving it, while stars disappeared from the sky. Alas, there's only so much time you can give to a globular before you go mad with delight or frustration, because the patterns you see keep changing as you strive to resolve more and more stars. The view must be more astounding in larger telescopes. For instance, Hartung says that a 12-inch telescope will resolve this cluster into "faint stars scattering away in diffuse edges."

By the way, NGC 6624 is more than 12,000 light-years closer than NGC 6441 (Hidden Treasure 86), which is 0.6-magnitude fainter. That's because it is a physically smaller system. NGC 6624 spans 66 light-years of space, while NGC 6441's size is about 60 percent larger.

Hidden Treasures 92 & 93

NGC 6633 & IC 4756

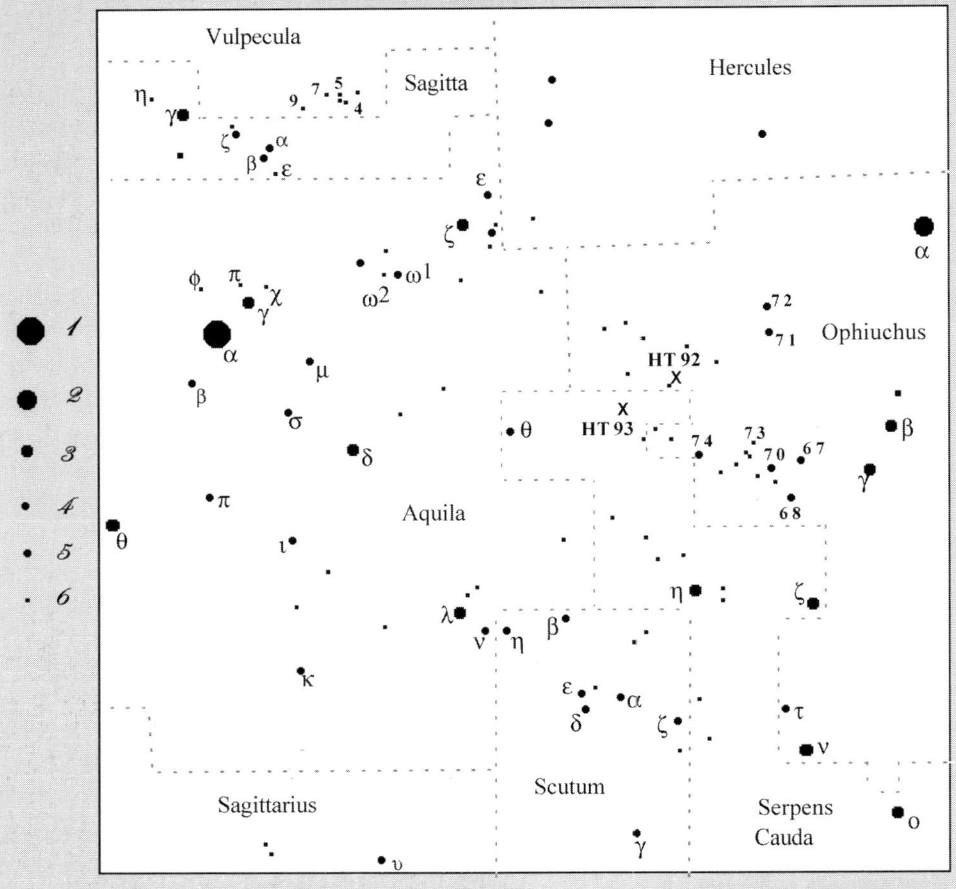

© Stephen James O'Meara

Tirion: Charts 15 & 16

Uranometria: Charts 205 & 250

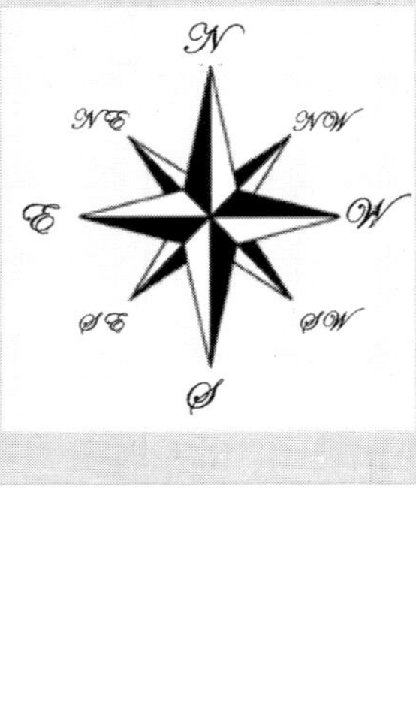

92

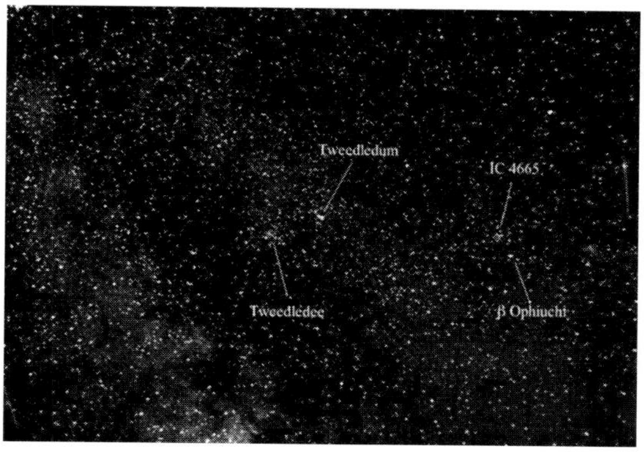

Tweedledum Cluster, Captain Hook Cluster, Wasp-Waist Cluster
NGC 6633
Type: Open Cluster
Con: Ophiuchus

RA: $18^h\ 27.2^m$
Dec: $+06°\ 30'$
Mag: 4.3 (O'Meara); 4.6
Diam: 20.0′
Dist: 1,000 light-years
Disc: Jean Philippe Loys de Chéseaux, 1745–46; independently discovered by Caroline Herschel in 1783

W. HERSCHEL: [Observed July 30, 1788] A cluster of coarsely scattered large stars. [Caroline Herschel] 1783. (H VIII-72)

NGC: Cluster, little compressed, stars [bright].

93

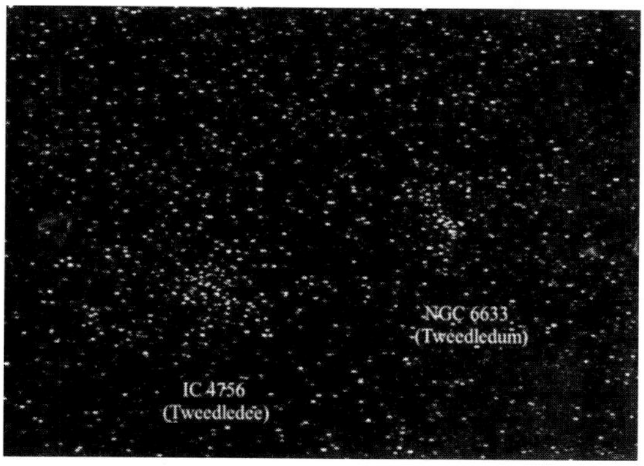

Tweedledee Cluster, Secret Garden Cluster
IC 4756
Type: Open Cluster
Con: Serpens Cauda

RA: $18^h\ 38.9^m$
Dec: $+05°\ 26'$
Mag: 4.3 (O'Meara); 4.6
Diam: 40.0′
Dist: 1,300 light-years
Disc: Solon Irving Bailey, listed in his *A Catalogue of Bright Clusters and Nebulae* (1908); the cluster was probably seen as part of the naked-eye Milky Way by Jean Philippe Loys de Chéseaux in 1745–46, and possibly by others prior

HERSCHEL: None.

IC: Cluster, compressed.

---

92 & 93

> They were standing under a tree, each with an arm round the other's neck, and Alice knew which was which in a moment

So BEGINS THE FOURTH CHAPTER of Lewis Carroll's classic children's book *Through the Looking Glass and What Alice Found There*. The chapter describes Alice's encounter with the "fat little," pugilistic twins called Tweedledum and Tweedledee. Since the publication of Carroll's book, the twins' names have become synonymous with any two people, places, or things worthy of comparison. And so it was in the fall of 2003, when I spied two magnificent naked-eye open clusters – NGC 6633 and IC 4756 – standing side by side on the banks of the river Milky Way. For the first time in four decades I was absolutely dumbfounded by what I was seeing. I was so taken by their appearance, and so stunned that they had escaped my gaze for so long, that I immediately changed my *Hidden Treasures* list to incorporate them. I did not want much time to pass without notifying others of these secret gems.

The two clusters do share some common traits. Both shine at magnitude 4.3 and can be seen without optical aid. They also lie at comparable distances from the Sun. Both are "fat" with stars: NGC 6633 spans 20′ in apparent diameter and IC 4756 is twice as large. When seen upside down from the Northern Hemisphere, the two clusters stand only 3° apart "under" a treelike asterism of stars, whose broad, 10°-wide canopy is formed by the stars Beta (β), Gamma (γ), 67, 68, 70, and 74 Ophiuchi, and whose trunk is marked by the stars 71 and 72 Ophiuchi. These same stars comprise the now defunct constellation Taurus Poniatovii, the Polish Bull (see Hidden Treasure 87).

In honor of Carroll's two chubby twins, I dubbed NGC 6633 and IC 4756, Tweedledum and Tweedledee, respectively. After I had logged these names, I found the two clusters in the *Millennium Star Atlas* and was surprised to see a 6th-magnitude star (Struve 2375) in Serpens Cauda, $1\frac{1}{2}°$ due west of IC 4756, with the name "Tweedledee and Tweedledum." Struve 2375 is a double star with a magnitude 6.3 primary and a 6.7 secondary 2.2″ apart (2003). Both stars are themselves spectroscopic binaries, so for every Tweedledee there is a Tweedledum and vice versa. This region of space, then, is almost as bizarre as Lewis Carroll's tale; it's as if Tweedledum and Tweedledee are peering into a looking glass that's being reflected by a looking glass, so the twins are seeing multiple and successively smaller images of themselves.

The Swiss astronomer and mathematician Jean Philippe Loys de Chéseaux (1718–1751) was the first person to discover NGC 6633. It was one of 20 such objects he found between the years 1745–1746, at least eight of which were genuine nebulae or clusters that had not been previously seen; NGC 6633 was one of them. De Chéseaux had prepared a list of these discoveries, which was read before the Academie Royale des Sciences on August 6, 1746. Unfortunately, the list was not otherwise published, though Guillaume le Gentil knew of it in 1759. But, as Kenneth Glyn Jones explains in his book *The Search for the Nebulae*, Messier was not aware of de Chéseaux's contributions to this field when he later rediscovered six of de Chéseaux's objects and included them in his 1771 catalog. And while Messier did not independently discover NGC 6633, Caroline Herschel did in 1783. While sweeping the heavens with a

4.2-inch reflector, which gave her a power of 24× and a field of view a little larger than 2°, she found, "About halfway from S Serpentarii [Ophiuchi] towards theta Serpentis, a Cluster of large stars. I counted about 80. Mess. has it not."

Alas, de Chéseaux's discoveries went unnoticed until the late nineteenth century, when Bigourdan included most of them in a "List of Nebulae and Clusters Discovered from 1700 to 1750," published in the 1884 Paris *Annales*.

Without question, de Chéseaux was a gifted observer. In a letter to his grandfather, de Chéseaux made this comment about his discoveries, "I have observed most of these nebulae and have found none, other than the first, – that of Andromeda, – which truly deserved the name. The last two [M17 and M13] are in fact nebulae to the naked eye but in the telescope they are no more than clusters of stars." This observation is testament to the great skill of this careful observer, who would later gain his fame for the great comet of 1744, whose tail was divided into six equal "rays." So brilliant and impressive was this comet that it impressed a 13-year-old Frenchman named Charles Messier, who would later make comets his lifelong passion.

Of NGC 6633, specifically, de Chéseaux wrote, "Near the tail of Serpens in which there is a small cluster of stars, a little separated from the rest towards the West." Although his description is simple, it is nonetheless intriguing and, perhaps, telling. After doing some initial research on NGC 6633, I had to wonder why de Chéseaux and other great observers after him – such as Charles Messier, William Herschel, James Dunlop, and John Herschel – had failed to see the equally bright, albeit larger, open cluster IC 4756 only 3° to the east. In fact, IC 4756 was not discovered visually at all. Harvard astronomer Solon Bailey found the cluster on plates taken with the 1-inch f/13 Cook lens at Harvard's Arequipa station in Peru, as part of a systematic search for bright clusters and nebulae. IC 4765 continued to be a problem child after Bailey's discovery of it. In *Star Clusters*, Brent Archinal and Steven Hynes explain in elaborate detail how the cluster was independently discovered by K. Graff in 1922 (IC 4756 = Graff 1), but various sources misidentified, confused, or erroneously listed Graff 1 and IC 4756 as separate objects.

That a 4th-magnitude open cluster would remain elusive to the greatest visual observers of the eighteenth and nineteenth centuries seemed odd – until I realized the importance of de Chéseaux's words, that NGC 6633 is "a little separated from the rest towards the West." What did de Chéseaux mean by "the rest?" A glance at the sky – or even a photograph – will reveal the answer. IC 4756 lies at the southwestern tip of a thin Milky Way star cloud, which is separated from NGC 6633 by a dark lagoon of obscuring cloud. While IC 4756 can be seen with the naked eye as a patch of diffuse light, it is seen against the splendor of the Milky Way. Couple this fact with another fact – that IC 4756 is twice as large as NGC 6633 – and the reason for its secrecy becomes clear. Had de Chéseaux turned either his 4.2-meter focal length, or 60-centimeter focal length, telescope toward IC 4765, he would have seen just a spread of unconcentrated stars in his tiny field of view. The magic of IC 4765 seen with the naked eye would have vanished just as magically when a telescope was trained on it, for the cluster would have diffused out and appeared no different telescopically to him than would any other rich star field in the Milky Way.

NGC 6633, on the other hand, stands apart from the milky band that includes IC 4756. Furthermore, its bright and concentrated core would have stood out from "the rest" of the Milky Way as a "cluster." Observers after de Chéseaux would have had the same problem with their limited fields of view. Walter Scott Houston concurred, writing, "The Herschels probably missed it because the small fields of their reflectors would have passed right over the group without revealing any concentration of stars. IC 4756 appears as a patch of Milky Way to the unaided eye." Here we have yet another irony of the night sky: until rich-field telescopes came into vogue, some of the bright yet diffuse deep-sky objects remained beyond the reach of telescopic observers and therefore defied imaging – even though they could be seen with the unaided eye. The effect is reminiscent of these words about "beauty" by the Elizabethan poet Anthony Munday (1553–1633): "When fond imagination / Seemed to see, but could not tell / Her feature or her fashion." Seeing IC 4756 in a powerful telescope with a narrow field is like trying to drive under a rainbow; the closer one sees it the more diffuse and obscure it becomes.

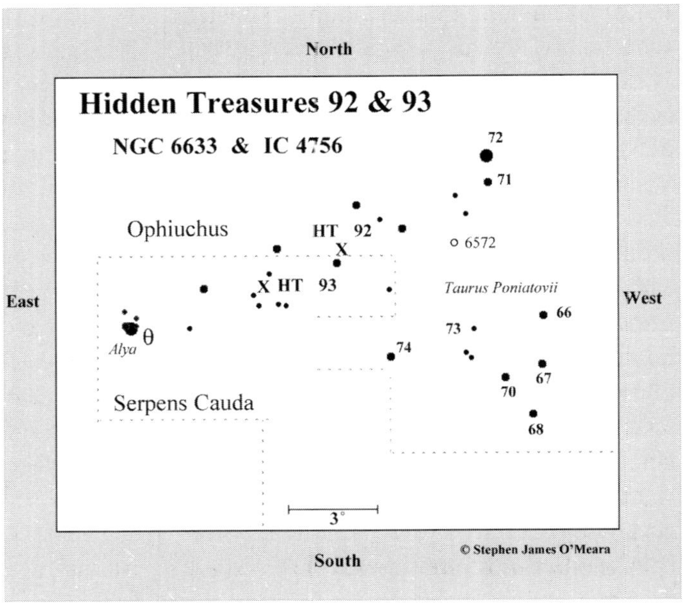

Because NGC 6633 and IC 4756 lie at similar distances and share a similar age (about 600 million years), and because both are relatively large in apparent diameter yet well defined, they are perfect candidates for studying the evolution of stellar activity between them and between other large and well-defined open star clusters of similar age – such as the Praesepe (M44), Coma Berenices (Hidden Treasure 62), and the Hyades (Caldwell 41). Open star clusters are key to studies of stellar evolution because they can usually be considered to contain stars that were all born at the same time. Observations of young open clusters like NGC 6633 and IC 4756 have been used to investigate theories about stellar evolution.

For instance, in a classical view, lithium is supposed to decay as a star ages. And observations of solar-type stars in open clusters show this to be the case. Indeed, in a 2002 *Monthly Notices* of the Royal Astronomical Society, Rob Jeffries (Keele University) and his colleagues found that the spectrum of solar-type stars in NGC 6633 shows a strong lithium depletion at approximately the same temperature range as solar-type stars in the Hyades. And at cooler temperatures, the lithium abundance patterns are remarkably similar to those in Hyades, Praesepe, and Coma Berenices. But cross comparisons of lithium depletion in star clusters are not entirely uniform. The stars of late-*G* and *K* spectral types in NGC 6633, have used up less lithium than their Hyades counterparts. To find out what causes these differences, astronomers have begun using the new generation of large telescopes (like the Kecks), coupled with efficient high-resolution spectrographs, to study solar- and later-type stars in open clusters covering a wide range of ages and metallicities. Since lithium is observed to be correlated both with stellar activity and rotation, studies of lithium depletion in solar-type stars of different ages may some day unlock the secrets to understanding the physical processes that drive stellar evolution both before and during a star's life on the main sequence.

Finding NGC 6633 and IC 4756 is simple. If you are under a dark sky, just look for two hazy 4th-magnitude globes halfway between magnitude 4.6 Theta (θ) Serpens Cauda – a glorious lemon-yellow double star resolvable in binoculars – and the similarly bright pair of stars, 71 and 72 Ophiuchi, which mark the tip of Taurus Poniatovii's western horn. From suburban locations, just raise your binoculars to this spot; the clusters will be unmistakable.

If you would like to see a perfect demonstration of how an object's apparent size affects its appearance, just look at Theta Ser and 72 Oph, both of which shine at magnitude 4.6. Next, compare these stars with NGC 6633, which has a catalog magnitude of 4.6 but is 20′ in apparent diameter, and then with IC 4756, which also has a catalog magnitude of 4.6 but is 40′ in apparent diameter. M44, the famous Beehive Cluster in Cancer, the Crab, is a little more than 1-magnitude brighter than IC 4756 and $\frac{1}{2}^{\circ}$ larger in apparent size. Beyond its greater brilliance, what makes M44 so much more obvious to see than IC 4756? The answer is the absence of Milky Way in Cancer. Still, I have to wonder how it is that I did not notice NGC 6633 and IC 4756 in almost a decade of deep-sky observing under dark skies. The clusters are just dim enough to avoid a casual sweep of the heavens. But once you know where to look, they are easy and quite obvious, especially with averted vision. The situation reminds me of seeing Venus in the daytime sky. If you know exactly where to look, Venus can be easy to see. But just try and find Venus blindly and, chances are, you'll fail.

On the evening of July 31, 2003, Larry Mitchell (Houston, Texas) and I drove to the nearly 14,000-foot-high summit of Mauna Kea in Hawaii and observed NGC 6633 and IC 4756 with our naked eyes and binoculars. Both have stars that the naked eye can resolve, or at least make the clusters appear patchy. The brightest stars in both clusters shine at 8th magnitude, and while it is possible to see these faint magnitudes from the summit of Mauna Kea, my guess would be that, for the most part, the naked-eye resolution is actually of star clumps not individual stars. Seeing these wonders together from such a magnificent location was like seeing the Double Cluster (Caldwell 14) in Perseus from a closer vantage point in space. Both clusters vie for attention and each has its own facets of visual splendor.

In true physical extent, NGC 6633 spans 5.8 light-years of space, while IC 4756 spans 15.2 light-years. Interestingly, the brightest component of the Double Cluster (NGC 869) shines at magnitude 4.5, about as bright as NGC 6633 and IC 4756. But this is an illusion. In reality, both components of the Double Cluster span 38 light-years of space, it's just that we see them seven times farther away. If we could place NGC 6633 and IC 4756 at the distance of the Double Cluster, they would vanish from naked-eye visibility. NGC 6633 would be an 8th-magnitude glow measuring a mere 3′ in apparent diameter, while IC 4756 would be an 8th-magnitude glow twice as large. Thankfully they are close, so we can fully appreciate their beauty.

Like some precious gem, NGC 6633 has a way of possessing the hearts of those who see it for the first time. For instance, Houston writes that when Canadian amateur Dunstan Pasterfield swept up NGC 6633 in binoculars and saw how large and impressive it was, he immediately turned his 8-inch reflector to it. "It's a lovely, great, straggling thing . . . of an absurd shape," Pasterfield writes. "I keep my eye on it annually. I'm convinced that some day something will happen in *my* cluster." [Emphasis mine.]

In the 4-inch at 23×, NGC 6633 displays a fine, 30′-long neck of stars oriented northeast–southwest. Look for a perpendicular row of some two dozen suns (between 8th and 11th magnitude) at the northeastern end of that neck, and for a bright and tight ellipse of starlight at the southwestern end. A dash of roughly 10th-magnitude suns gently curves away from the south end of that ellipse to the east, where it ends 20′ at a brilliant 6th-magnitude sun. Seen together, all these features describe a stunning hook, like the iron hook worn by James Barrie's infamous and fictitious pirate, Captain Hook – the principal antagonist in his delightful children's tale *Peter Pan*: "He has

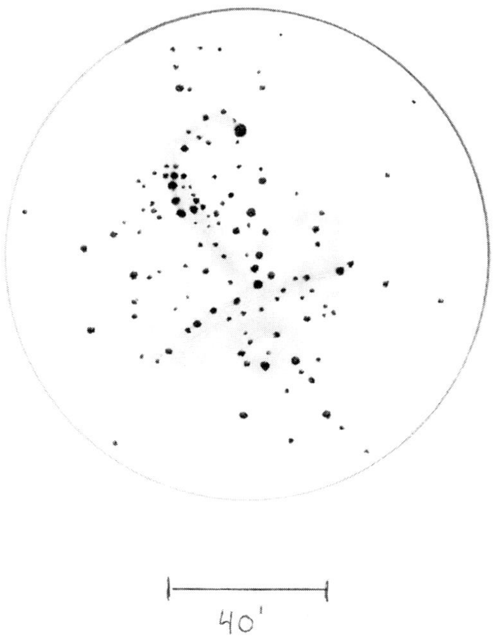

an iron hook instead of a right hand, and he claws with it . . . His eyes were of the blue of the forget-me-not, and of a profound melancholy, save when he was plunging his hook into you, at which time two red spots appeared in them and lit them up horribly."

NGC 6633's true dimensions are hard to determine visually because vast wings of starlight extend outward from the "hook" to the north, west, and east, making the cluster appear twice as large as it truly is. Seen with these wings, the cluster also looks like a flying squirrel. Seen in another way, I have also called NGC 6633 the Wasp-Waist Cluster, for the thin shaft of the hook can also be imagined as the narrow waist of a wasp, whose wings are formed by the perpendicular row of stars at the cluster's northeast end; the wasp's bulbous abdomen is marked by the southwestern ellipse; the wasp's stinger is the row of 10th-magnitude suns that lead

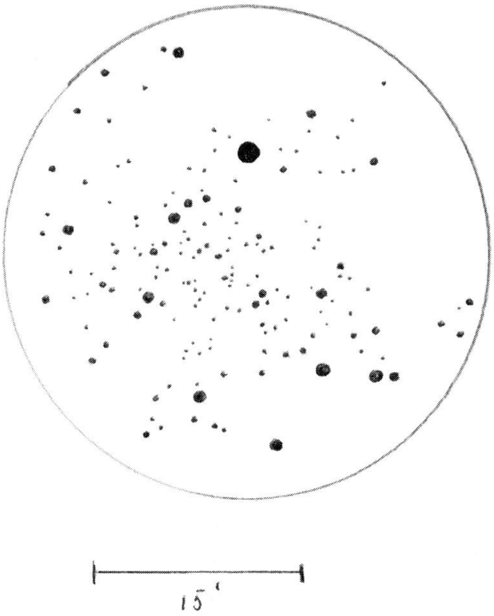

to the bright 6th-magnitude star mentioned above. While NGC 6633 incrementally loses its splendor with increasing magnification, be sure to examine the "wasp's" abdomen and stinger stars at high power; there are several fine pairs of stars worthy of attention. Most noteworthy, is a 5′-long, elongated gathering of eight stars at the southwest tip of the wasp's abdomen, and an even smaller mini-cluster of eight suns 7′ due west of the wasp's abdomen. If, like Dunstan Pasterfield, you want to keep your eye on NGC 6633 annually, I'll give you a hint as to what to look for. Nearly 15′ northeast of the 6th-magnitude stinger star is the red long-period variable T Ser. This star varies in brightness from magnitude 9.1 to 15.5 every 341 days. So you would indeed have to keep returning to NGC 6633 to catch this star in action. Imagine it as Captain's Hook's red eye winking – as he thrusts his iron hook toward you.

Like NGC 6633, IC 4756 is a superb binocular object, one so giddy with starlight that it has been called the Summer Beehive. In the 4-inch at 23×, IC 4756 matches NGC 6633 in splendor. What it lacks in concentration it makes up in number. While NGC 6633 has only about 160 known members, IC 4756 has more than three times as many. At first glance, IC 4756 appears as a large, sprawling swarm of homogeneous starlight, scattered across the field like the slivers of aluminum that fall to the floor from the action of a lathe. But take the time to sit back and relax. With time, you'll see order among the chaos. For instance, I see the brightest suns forming an elongated mass oriented northeast–southwest. Weak spirals of starlight seem to extend from this central bar. Two arms on the north side of the cluster appear to be rotating in a counterclockwise direction, while two arms in the southern side appear to be rotating in the opposite direction. At 72×, the cluster splinters into all manner of geometrical shapes, including a hook of multiple suns at the core. The cluster contains numerous mini-asterisms, which are best left to you to label. It is a veritable alphabet soup – with Os, Ws, Ts, Ls, Cs, Us, Vs, and Ys of starlight. It looks almost as if the cluster had failed to congeal and is slowly drifting apart like ashes in the wind. Use low power to trace out the faint extensions of stars that form the spirals, and be sure to take the time to let your imagination fly.

By the way, observers who own a late edition of Rev. T. W. Webb's *Celestial Objects for Common Telescopes* should first note that a description of NGC 6633 appears under Serpens Cauda, because prior to 1930, when the International Astronomical Union established our official constellation borders, the object did belong to that constellation. Second, in the description you'll find this sentence: "Between it and [Theta], nearer the former, is a beautiful large cloud of stars, chiefly of 8 and 9 [magnitudes.]" These words, which describe IC 4756, do not belong to Webb but to the successive editors who updated Webb's book over the years.

Finally, in *Through the Looking Glass*, the sight of Tweedledum and Tweedledee reminds Alice of the words to an old song that kept ringing through her head:

Tweedledum and Tweedledee
Agreed to have a battle;
For Tweedledum said Tweedledee
Had spoiled his nice new rattle.

Just then flew down a monstrous crow,
As black as a tar-barrel;
Which frightened both the heroes so,
They quite forgot their quarrel.

If you want to see the black crow, look 3° east of IC 4756 where you should see a 6°-long cloud of obscuring matter oriented north-northeast–south-southwest – the crow's extended wings.

Hidden Treasure 94

NGC 6709

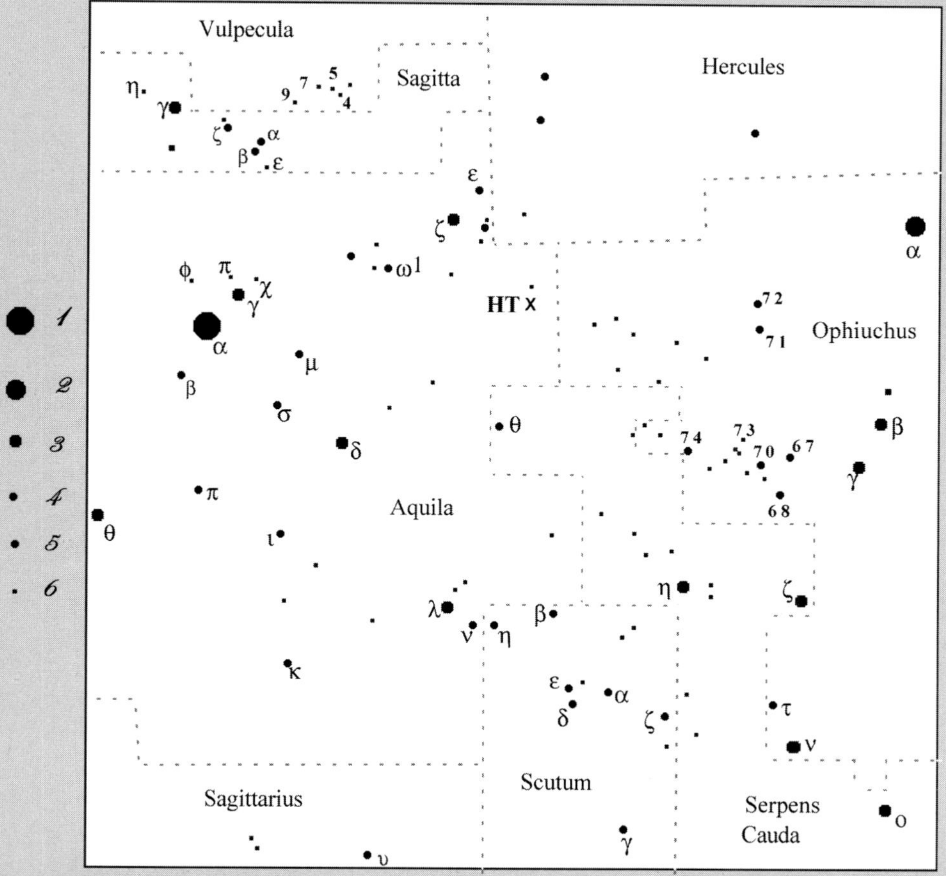

© Stephen James O'Meara

Tirion: Chart 16

Uranometria: Chart 205

94

Flying Unicorn Cluster
NGC 6709
Type: Open Cluster
Con: Aquila

RA: $18^h\ 51.5^m$
Dec: $+10^\circ\ 20'$
Mag: 6.8 (O'Meara); 6.7
Diam: 15.0′
Dist: 2,900 light-years
Disc: John Herschel, 1827

J. HERSCHEL: A pretty considerable cluster; 15′ diameter; irregular figure, 50 or 60 stars large and scattered, the place is that of the double star No. 870 of my third catalog. (h 2020) (h 2022)

NGC: Cluster, pretty rich, little compressed, irregularly faint.

OPEN CLUSTER NGC 6709 IS TO Aquila what open cluster NGC 6819 (Hidden Treasure 98) is to Cygnus – a lonely outcast on the sidelines of the Milky Way. In his 2002 book *Celestial Harvest: 300-Plus Showpieces of the Heavens for Telescope Viewing and Contemplation,* James Mullaney refers to it as "Aquila's only good – but often ignored – stellar clan." The cluster sits in a little nook in the northwestern boundary of the constellation, which is bordered by three inconspicuous "fence lines" – that of southeastern Hercules, northeastern Ophiuchus, and northeastern Serpens Cauda, which, by the way, contains the brightest deep-sky spectacle near NGC 6709: the equally neglected 5th-magnitude open cluster IC 4756 (HT 93). It's as if someone has cordoned off this area with yellow tape, forbidding anyone to enter the region without a search warrant.

If you sweep this area with your naked eye, you will see that this "insignificant" nook in Aquila is part of a bright and sudden enhancement of the Milky Way along the northern spur forming the Great Rift – a dark divide in the Milky Way between Cygnus and Sagittarius caused by a succession of large, overlapping clouds of interstellar dust in the equatorial plane of the galaxy. The section that contains NGC 6709 is an epaulet of stars on the stiff shoulder of the galactic bulge.

To find NGC 6709, first locate 3rd-magnitude Zeta (ζ) Aquilae – the Eagle's Tail in early times, and its northern wing tip in some modern depictions, such as H. A. Rey's *The Stars.* Now look $2\frac{3}{4}^\circ$ south-southeast of Zeta Aql for 5th-magnitude 18 Aquilae. It is

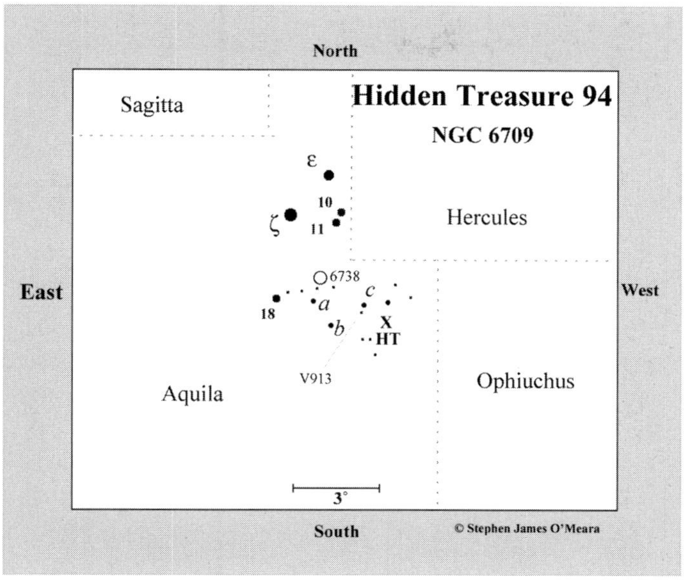

star cloud, comprised of all manner of dim stellar clumps, which bubble into view with averted vision. The entire scene is one of stellar effervescence.

When I relax my gaze but concentrate on the cluster at low power, I immediately see two noticeable forms: first, there's a stunning triangle of stars east-southeast of the cluster's hollow core (a dim fourth companion lies to the southeast), with two parallel rows of stars extending due south like a pair of legs. The second conspicuous grouping is an attractive string of stars on the west side of the cluster. It curves gently from the north-northwest to the south-southwest, becoming gradually fainter in the same direction, until it branches into

the only star of that brightness in the immediate vicinity. Center this star in your telescope. A little more than 1° west and a bit south is a magnitude 7.5 star (*a*); it marks the west end of an arc of three suns that curves northward from 18 Aql. Now move your scope about 1° to the southwest, where you will find a roughly 6.5-magnitude sun (*b*). A near $1\frac{1}{2}$° hop to the northwest brings you to a similarly bright star (*c*) with a 7.5-magnitude companion (V913 Aquilae) 10′ to the south-southeast. NGC 6709 is 50′ to the southwest of Star *c*.

The 7th-magnitude cluster is easily visible in 7 × 50 binoculars; with a steady hand, and a lot of time looking, you can partially resolve it. Telescopically, as the Rev. Thomas W. Webb says in his *Celestial Objects for Common Telescopes*, NGC 6709 is a "[b]eautiful wide group." In the 4-inch, it is a marvel at essentially every usable power. But it is especially dynamic when seen at 23× in a near-3° field of view. That's because the cluster sits right at the heart of a dense

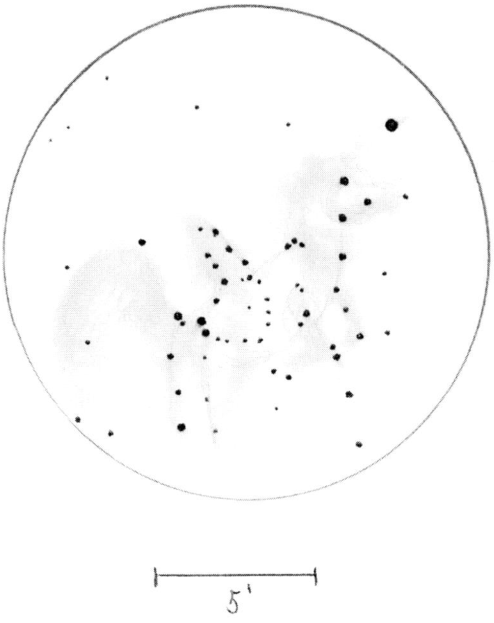

another pair of legs. The cluster's core is not completely hollow, there is a dim equilateral triangle of suns near the cluster's heart. Seen together, all these forms look like the head, neck, legs, and torso of a horse. Add the inverted "V" of stars immediately northwest of the bright triangle and you have a small wing emerging from a horse's back. And a bright star northwest of the gentle curve in the west part of the cluster looks like a sharp horn jutting from a horse's head, making the entire cluster look, with imagination, like a flying unicorn, thus the name I have given it. The cluster's perimeter is also flanked by delicate chains or wispy strings of stars, which, seen another way, makes the whole mish-mashed array look like a tangle of Christmas tree lights just taken out of storage.

Virtually everywhere I look at higher powers I see doubles and tiny clusterings of dim suns – which could be background suns in the star cloud. Indeed, NGC 6709 is a treasure trove of multiple star systems, and one could easily spend an evening with this object, surveying it for new groups at different magnifications. The star systems I see at low power are easily lost to the doubles I see at moderate magnifications, which are lost to those dim and elusive ones visible at higher magnifications. I cannot think of another open cluster with this alluring power; like the gingerbread house in the fairy tale "Hansel and Gretel," NGC 6709's tight multiple systems have the ability to draw the curious viewer in . . . closer and closer, until they cannot help but visually "nibble, nibble, gnaw."

In a 1999 *Astronomical Journal* article, Annapurni Subramaniam and Ram Sagar (Indian Institute of Astrophysics) made spectroscopic observations of 43 bright stars in NGC 6709 and determined a distance of about 3,900 light-years, which places it 1,000 light-years more distant than the value listed in Archinal and Hynes's *Star Clusters*. If we accept the distance in *Star Clusters*, NGC 6709 spans a diminutive 12.7 light-years across with 111 stars, the brightest of which shines at 9th magnitude. If we accept a distance of 3,900 light-years, the cluster's size swells to 17 light-years. Subramaniam and Sagar also found NGC 6709 interestingly young (315 million years), which is about as old as M39 in Auriga and much younger than the five-billion-year age of our Sun. NGC 6709 emerged from the void when reptiles first began to roam the Earth – 65 million years before the dinosaurs.

Hidden Treasure 95

NGC 6712

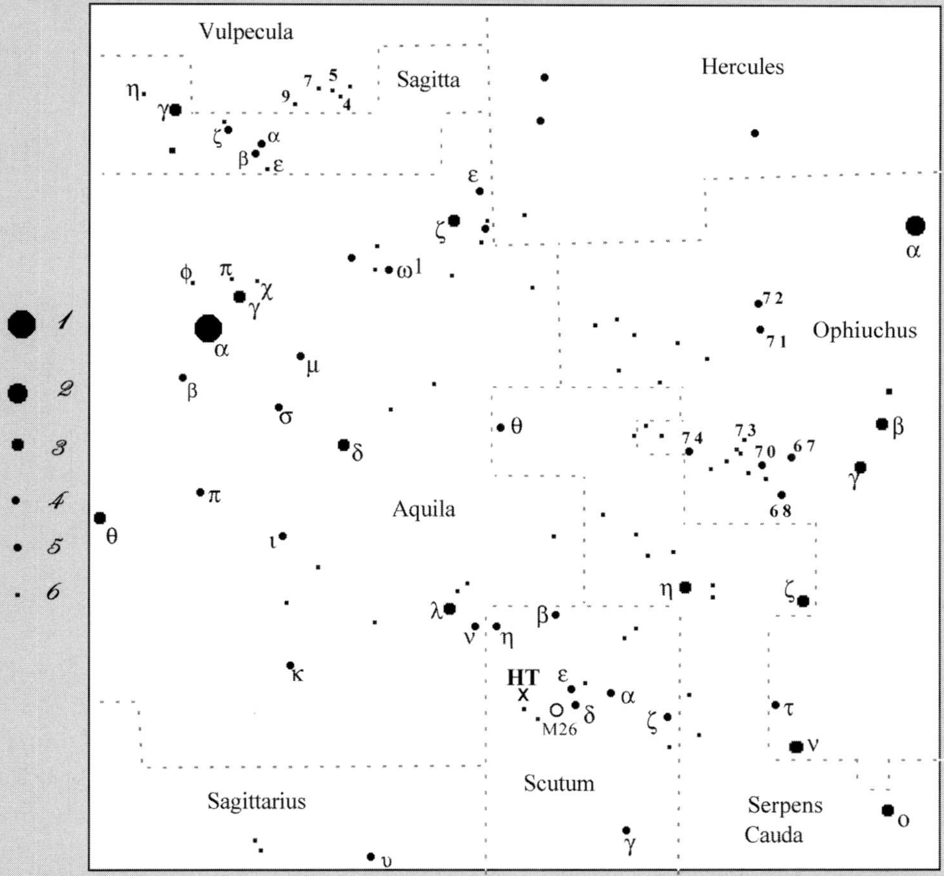

Tirion: Chart 16

Uranometria: Chart 295

95

NGC 6712
Type: Globular Cluster
Con: Scutum

RA: $18^h\ 53.1^m$
Dec: $-08^\circ\ 42'$
Mag: 8.3 (O'Meara); 8.1
Diam: 9.8′
Dist: 22,500 light-years
Disc: William Herschel, 1784

W. HERSCHEL: [Observed June 16, 1784] Bright, very large, irregularly faint, easily resolvable stars visible. (H I-47)

NGC: Globular cluster, pretty bright, very large, irregular, very gradually, then a little brighter in the middle, well resolved, clearly consisting of stars, 15 stars.

SCUTUM, ORIGINALLY SCUTUM Sobiescianum (Sobieski's Shield), is one of those "fill-in-the-blank" constellations. Johannes Hevelius created it in 1690 using seven (the magic number) unfigured 4th-magnitude stars in the Milky Way. It lies immediately southwest of Aquila, east of Serpens Cauda, and north of Sagittarius. The Milky Way rushes right through it. The Shield represents the Coat of Arms of the third John Sobieski, King of Poland. This brave defender of his homeland so successfully resisted an army of invading Turks in 1683 that a grateful priest praised him during church service as a "man sent from God." Interestingly, the constellation, in part, has long been known to Chinese skywatchers as *Tien Pien,* the Heavenly Casque – a divine piece of armor.

Without question, Scutum is one of my favorite constellations, for no other reason than that its dense star clouds hold my attention more than any other in the sky. Beyond the highly observed open clusters M11 and M26, I mainly enjoy seeing how the region's bright and dark nebulae play with the pale green shimmer of the Great Scutum Star Cloud, Barnard's Gem of the Milky Way. But there is one object in the region that has had a hold on me for years – NGC 6712, one of the most understated globular clusters in the heavens.

NGC 6712 is a relatively small (64 light-years) and sparse (~1 million suns) globular cluster located in our galaxy's halo about 11,400 light-years from the galactic center. Still, images of the cluster's core with the European Southern Observatory's Very

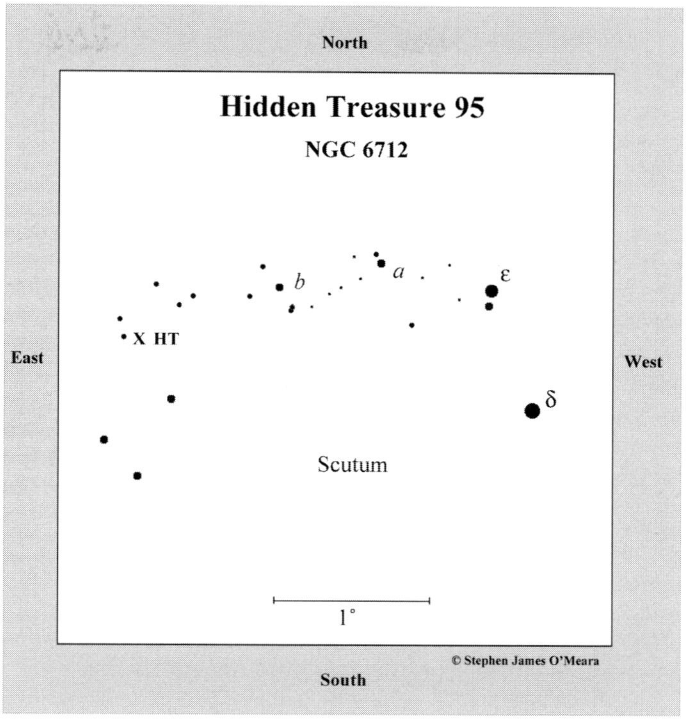

Large Telescope on Cerro Paranal in Chile reveal a blizzard of starlight. But the situation appears to be different in the outer envelope, where stars seem to be slowly dissolving – losing fainter, lower mass stars into our galaxy's halo. Francesco Paresce (European Southern Observatory in Garching, Germany) notes that none of NGC 6712's stars are less massive than the Sun, making it totally unlike any globular cluster surveyed by the Hubble Space Telescope. Most likely, NGC 6712 is unique only because no other globular cluster comes as close to the Milky Way's center as does NGC 6712; it penetrates very deeply into the galactic bulge, venturing to within 1,000 light-years of the galactic center. NGC 6712 is thought to have crossed through the crowded galactic plane only a few million years ago. Each passage not only robs low-mass stars from the cluster (because they tend to lie at the cluster's periphery) but also stretches out the cluster like a comet's tail.

To find NGC 6712, first locate 4th-magnitude Alpha (α) Scuti, which lies on the southwest edge of the Scutum Star Cloud, about $2\frac{1}{2}°$ northwest of M26. If you know how to find M26, NGC 6712 is just 2° northeast of that famous cluster. If not, look 2° east of Alpha Scuti, where you'll find 5th-magnitude Epsilon (ε) Scuti, which has a 7th-magnitude companion about 5′ to the south. Now move almost 50′ to the east-northeast to another 7th-magnitude star (*a*). Next, hop a little more than 40′ to the east-southeast to another 7th-magnitude star (*b*), which has an 8th-magnitude companion 10′ to the south-southwest. NGC 6712 is just 1° southwest of Star *b*.

Under a dark sky, the cluster is visible as a pale glow in 7 × 50 binoculars without difficulty. It is quite evident in my antique telescope, appearing like the head of a distinct cometlike glow, but that's only if I concentrate with averted vision. I would have easily swept over it during a comet search with such a small aperture. In the 4-inch at 23×, which gives a near-3° field of view, NGC 6712 and M26 can be seen in the same field of view, which offers the probing mind an opportunity visually to plumb the depths of three-dimensional space. Although M26 shines at about the same magnitude as NGC 6712, it is $4\frac{1}{2}$ times closer but five

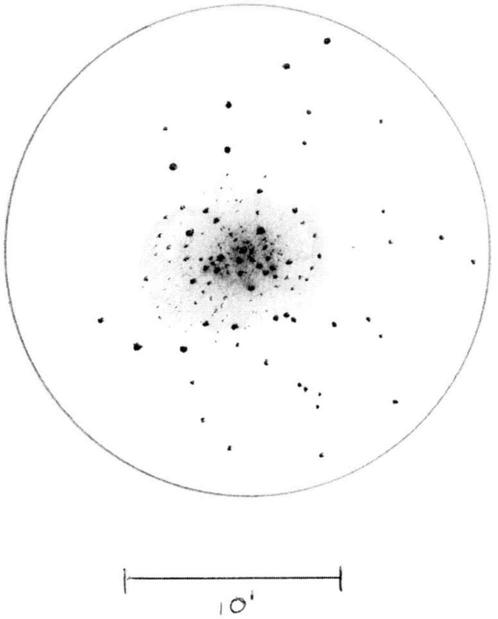

times smaller in true physical extent. Now, if NGC 6712 is placed at the southeast edge of the field, M11 appears at the extreme northern edge. M11 is three magnitudes brighter than NGC 6712, four times closer, and three times smaller in true physical extent.

NGC 6712 is very compact at 23× and the field is very rich, so the edges of the cluster appear to be resolved, but it's hard to judge which stars belong to the cluster and which to the Milky Way background. Resolution is certainly possible in a 4-inch, for the brightest stars in the cluster shine at magnitude 13.3. The core screws in with a direct gaze and looks irregularly round with averted vision. At 72×, the cluster's fine details remain elusive, but it seems to be on the verge of resolving – a distinct ball, bristling with fuzzy starlight – as if we are seeing it through a scrim. The cluster comes alive at 101×. The outer halo is speckled like flecked metal. I can't see any central concentration – no bright stellar pip, just a mantle and an outer halo. At the core, however, there appears to be a hollow surrounded by arcs of dense clumps of starlight. With averted vision the southeast flank of the mantle has two bright arms forming a "V." Other feathery arms spiral off to the north, west, south, and east. The mantle's northwest edge is capped by stars. And the outer halo, the one being stripped of stars that jitter about in the wavering sea like flecks on coffee. Arizona amateur Andrew Cooper observed NGC 6712 through an 18-inch f/4.5 reflector at 174× and saw the cluster fully resolved. He called it "dense, well defined, not very condensed at the core." Jere Kahanpää in Helsinki, Finland, using an 8-inch f/10 telescope at 160× said it looked "somewhat broadly concentrated . . . like M71, about 20 stars resolved against a grainy glow."

When you are done admiring this wonderful object, move your scope 25′ to the southeast. There you should see a large (1.7′) and uniformly round glow – planetary nebula IC 129. Under a dark sky, it is easy to see in a 4-inch and could easily be mistaken for a comet. And if you have a rich-field telescope, I'd suggest spending some time just sweeping the area at low power. If you do, see if you can't see the incredible web of dark nebulosity encircling M26 (center M26 and slightly, ever so slightly, defocus the telescope and tap the tube). If you see it, take a breath, walk away from the telescope, and return to the eyepiece, when you feel refreshed. Now, see if you can see not only dark (black) nebulosity but also gray. After enjoying these ribbons of darkness, return your gaze to our original target, NGC 6712, which should now look like a castaway on a beach of starlight, onto which the dark nebulosity washes like dark waves at night.

Hidden Treasure 96

NGC 6723

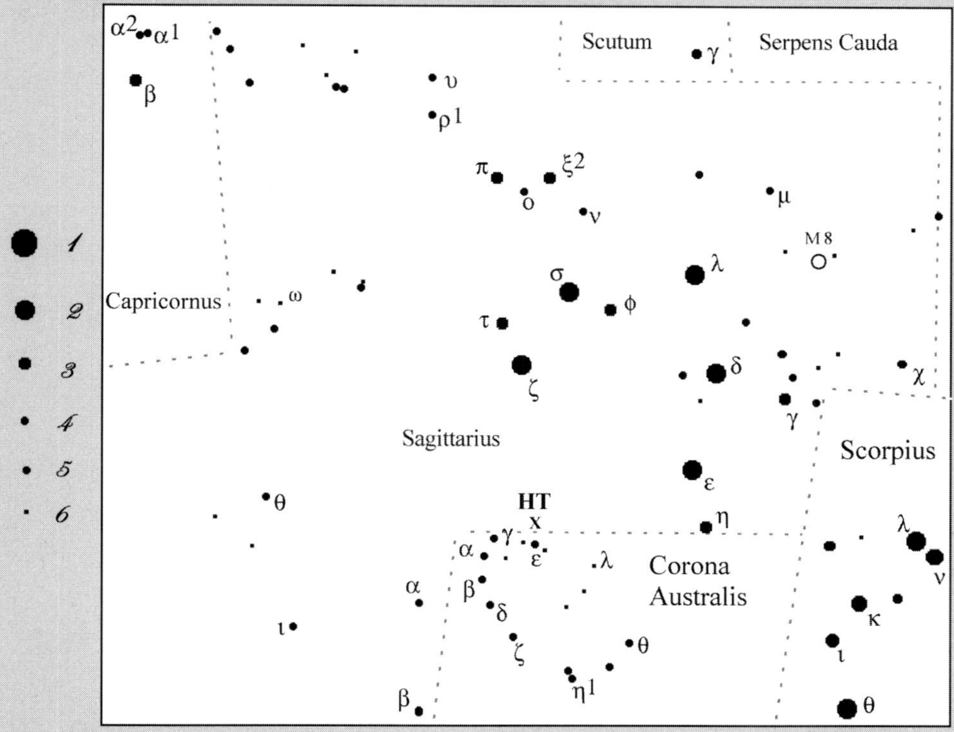

© Stephen James O'Meara

Tirion: Chart 22

Uranometria: Charts 378 & 379

96

NGC 6723
Type: Globular Cluster
Con: Sagittarius

RA: $18^h\ 59.5^m$
Dec: $-36°\ 38'$
Mag: 6.9 (O'Meara); 6.8
Diam: 13.0′
Dist: 28,400 light-years
Disc: James Dunlop, included in his 1827 catalog

J. HERSCHEL: Globular, bright, large, round or very little extended, very gradually brighter in the middle, diameter 5′, perfectly resolved into stars from magnitude 14 to 16, with stragglers extending to 8′ diameter. (h 3770)

NGC: Globular cluster, very large, very little extended, very gradually brighter in the middle, well resolved, clearly consisting of stars, stars from magnitude 14 to 16.

SOMETIMES OUR INTENT TO FOCUS on certain celestial treasures causes us to miss others nearby. Consider NGC 6729 (Caldwell 68), the R Coronae Australis Nebula. By selecting this tiny and rather obscure object for his Caldwell list, Patrick Caldwell-Moore missed the forest for a tree, because a mere 30′ northwest of NGC 6729 lies the glorious Sagittarius globular cluster NGC 6723 – one of the most spectacular objects of its kind in the night sky. (Of course, the Caldwell catalog is not a list of the sky's most *spectacular* non-Messier objects, but of Moore's *favorite* non-Messier objects, chosen either on the basis of visual appeal or astrophysical interest.)

Owing to its low southerly declination, NGC 6723 avoided the scrutiny of William Herschel's peering eye. Not until June 3, 1826, did James Dunlop, the "gentleman astronomer" of New South Wales, Australia, sweep it up with a 9-inch reflector. He listed it as the 573rd object in his *Catalogue of Nebulae and Clusters of Stars in the Southern Hemisphere*. Of it he wrote:

A beautiful bright round nebula, about $3\frac{1}{2}'$ diameter, moderately and gradually condensed to the centre. This is resolvable. The moderate condensation, the blueish colour of the stars which compose it, give it a very soft and pleasant appearance. This is rather difficult to resolve, although the condensation is not very great.

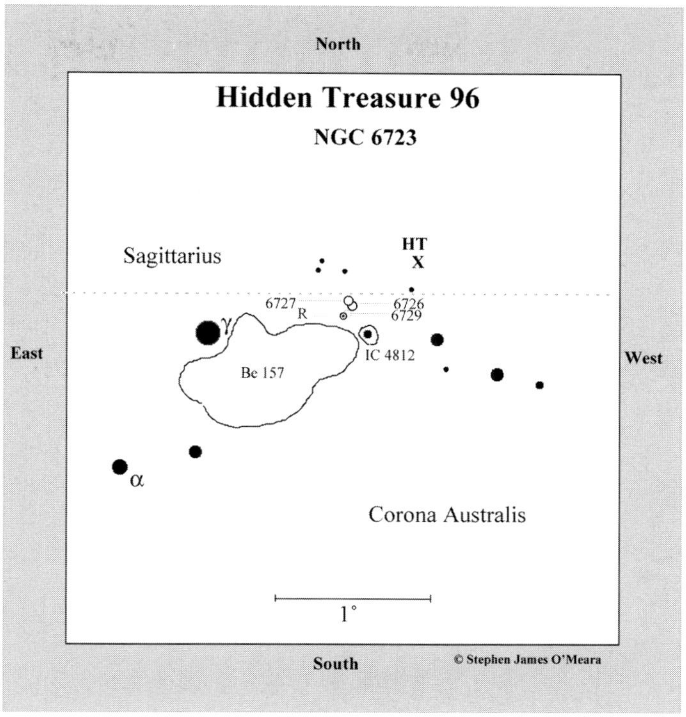

Although NGC 6723 appears very close to the vast swath of obscuring matter associated with the Corona Australis molecular cloud, it is seen through a relatively transparent window of sky. It is a moderate-sized globular spanning some 108 light-years of space. But it is an old system; its age is comparable to that of the universe's, which is now estimated to be about 13.7 billion years old. The age is also reflected in the cluster's low metallicity. Each of the cluster's stars contain, on average, about $\frac{1}{10}$ to $\frac{1}{16}$ as much iron (per unit hydrogen) as does our Sun. The amount of iron present in globular clusters ranges between 0.1 and 0.03 of the solar value. NGC 6723's is at the lower limit, suggesting, again, that this cluster formed at a time when the chemical evolution of the universe had not progressed very far. The cluster's integrated spectral type is $F9$ and its radial velocity has been measured at 87 kilometers per second in recession.

L. K. Fullton and B. W. Carney (University of North Carolina), used NGC 6723 as part of their study of globular clusters near the galactic center – to probe the formation history and chemical evolution of the galaxy's central regions out to 15,000 light-years. While NGC 6723 lies 28,000 light-years from the Sun, it is only 8,500 light-years from the galactic center, making it a perfect candidate for this study. The researchers found that the age and metallicity of NGC 6723 is similar to that of RR Lyrae stars found in the galactic bulge, suggesting that the bulge's RR Lyrae stars are at least as old as the oldest globular star clusters. If the bulge's RR Lyrae stars and NGC 6723 represent the first stages of bulge formation, then this result suggests that our galaxy's bulge required some 8 billion years to form.

To find NGC 6723 look for 5th-magnitude Epsilon (ε) Cornonae Australis, an exquisite binary star discovered by John Herschel in 1834. The binary comprises two dazzling yellow suns (magnitudes 4.5 and 6.4) separated by 1.3″. The stars, which are now opening toward a maximum separation of 2.5″ in 2060, will split on only the steadiest of nights in a 4-inch, whose Dawes Limit (4.5/aperture) is 1.1″. Epsilon CrA marks the apex of a near-perfect isosceles triangle with Zeta (ζ) and Epsilon (ε) Sagittarii, which lie about $7\frac{1}{4}^\circ$ to the north-northwest and $8\frac{3}{4}^\circ$ to the northwest of the cluster, respectively. Just $1\frac{1}{2}^\circ$ due west of Gamma CrA is 5th-magnitude Epsilon CrA in the northern section of the Southern Crown. Use binoculars to confirm the location of Epsilon, which is 6° due south of the midpoint in the base of the Sagittarius Teapot. NGC 6723 is only 30′ north-northeast of Epsilon CrA and forms the northern apex of an equilateral triangle with Epsilon CrA and a 7th-magnitude double star associated with IC 4812, which is located 30′ east-northeast of Epsilon CrA.

Shining at 7th magnitude, the cluster is a tad faint to be detected with the unaided eye, but it is so concentrated that someone with young eyes should consider taking up the challenge, especially if attending one of the more southerly star parties, such as the Texas Star Party. Needless to say the object is a cinch in 7 × 50 binoculars and a fine sight in the smallest of telescopes. In my antique telescope, NGC 6723 begins to show structure, namely a bright core surrounded by an irregular and mottled halo of diffuse light. The cluster's brightest star shines at magnitude 12.8 and its horizontal branch magnitude is only 15.5. At 23× in the 4-inch, it shines as a stunning globe of suns, like a jar of fireflies glowing on a hot summer's night. Christian Luginbuhl and Brian Skiff say that in a 60-mm refractor the cluster looks similar to M4 in Scorpius. I came to the same conclusion independently after I resolved the needlelike row of stars at the cluster's core.

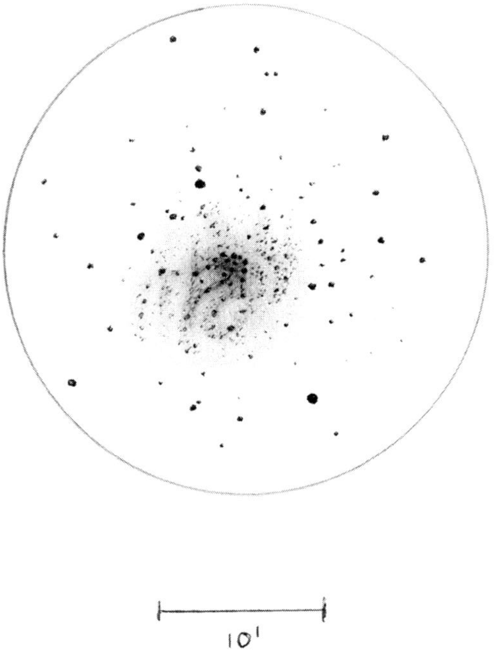

At 72×, the cluster turns to grains of salt and pepper. Indeed, with averted vision, the cluster is a sizzling haze of stars in three distinct layers: a globular core, an inner halo, and an outer aureole. Many of NGC 6723's outlying stars resolve into a web of blue-tinted starlight, while the core looks particularly frothy. It is as if some divine artist tried to depict these words of the nineteenth-century American poet William Hamilton Hayne: "Pure leagues of stars from garish light withdrawn / Behind celestial lace-work pale as foam." With any concentration the southern fringes of the cluster's halo fracture into long fingers of light, especially to the south, where four strong strings of stars radiate from a bright wedge of light at the core. Another prominent finger extends to the northeast and a few stubby nubs can be seen to the northwest. Overall, the cluster's southern side is brighter than the north. Ernst Hartung saw similar features, calling

NGC 6723 "irregularly round and looking like an almost hemispherical heap of star dust, the outliers extending to 4′ across, and being in irregular rays."

At high power, NGC 6723 looks like a blizzard of starlight. Look for dark lanes running through its core and fringes. One prominent lane lies just northwest of the inner core and is oriented northeast–southwest. It looks like the wound from a cat scratch. Now return to low power and just gaze at the cluster without concentrating. Just let the object's light settle on your eye. Isn't there something extremely elegant about this cluster? It has a delicate sheen to it, one that softens its silent majesty. NGC 6723 stands before the eye like a weathered mountain in the mist, aged yet beautiful, catching the last rays of Sun. It's a reminder that, like an old, cherished photograph, beauty does not wither with age.

Hidden Treasure 97

Collinder 399

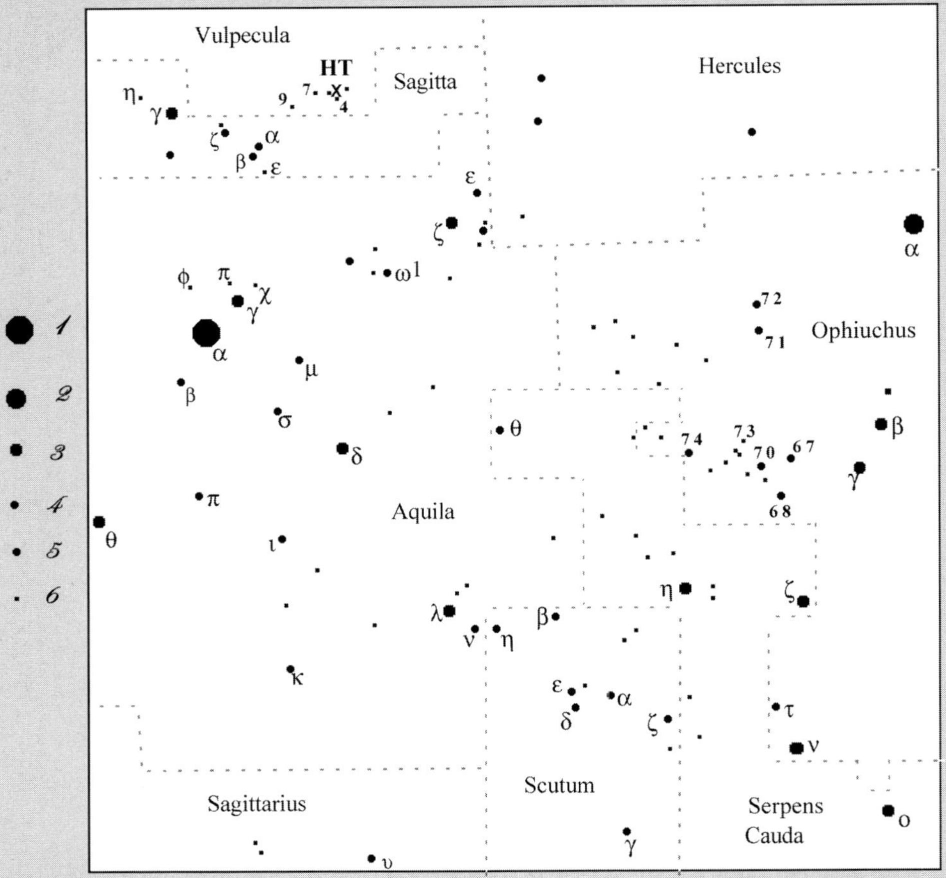

Tirion: Charts 8 & 16

Uranometria: Charts 161 & 162

97

Al-Sufi's "Nebula," Brocchi's "Cluster," Coathanger, The Snail
Collinder 399
Type: Asterism
Con: Vulpecula

RA: 19^h 26.2^m
Dec: +20° 06′
Mag: 3.5 (O'Meara); 3.6
Diam: 90.0′
Dist: 424 light-years
Disc: Al-Sufi, before AD 964

HERSCHEL: None.

NGC: None.

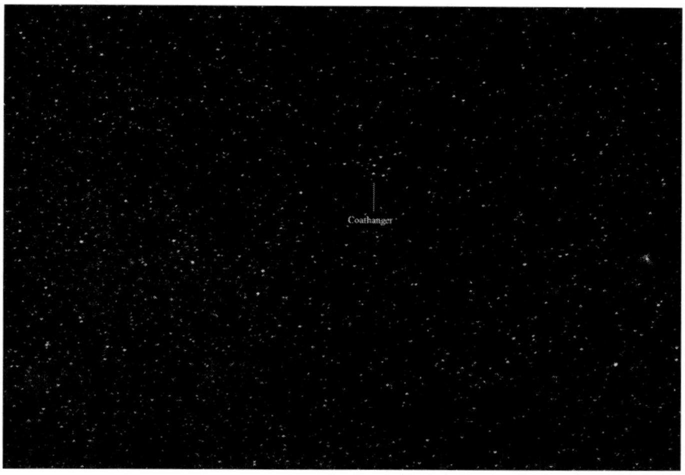

ON WARM AUGUST NIGHTS, RIDING high in the southern sky 4° northwest of the feathered shaft of Sagitta, the Arrow, is a remarkable gathering of suns in the vicinity of 4 and 5 Vulpeculae. The gathering is visible as a 3rd-magnitude mottled haze to the naked eye, while binoculars reveal it to be a conspicuous straight line of six stars (oriented east–west) with a southward-dipping hook. For this reason, the group is commonly known as the Coathanger and is adored by stargazers of all ages. It is one of the first objects Donna, my wife, looks for whenever she sweeps the summer Milky Way with her binoculars. The Coathanger has a magical charm that can bring novices into a lasting relationship with the night sky with just one look.

The group is quite large, spanning three Moon diameters across, with some 40-odd stars readily visible. These stars were long believed to be physically related, being first mentioned as a cluster in a 1903 proper motion study. But in a 1970 publication of the Astronomical Society of the Pacific, Douglas S. Hall (Vanderbilt University) and Franklin G. van Lanndingham (Kitt Peak National Observatory), revealed that only six of these stars form a physical system – five early Type-*A* stars and one red giant. They estimated its distance to be similar to that of the Pleiades but with an age of about 200 million years (between that of the Hyades [Caldwell 41] and M11). And in a 1989 *Astronomicheskii Zhurnal* article, Russian astronomers E. D. Pavlovskaya and A. A. Filippova, confirmed that the grouping is one of 10 clusters sharing a common motion in space – including the Pleiades (M45), NGC 6633 (Hidden Treasure 92), NGC 6709 (Hidden Treasure 94), NGC 6882 and 6885 (Caldwell 37), and IC 4665 (Hidden Treasure 83). The tables turned in 1998, however, when Brian Skiff (Lowell Observatory) described in *Sky & Telescope* magazine how he used Hipparchos satellite data to

determine that the colors, magnitudes, and proper motions of the stars reveal that the group is an apparent asterism rather than a true open cluster.

Although it is part of the visible naked-eye sky, the group was first singled out by the Persian astronomer Al-Sufi (AD 903–986)[1] who included it in his *The Book of the Fixed Stars* (circa 964). A French translation by H. C. F. C. Schjellerup published at St. Petersburg in 1874, says that he called the group "a little cloud [or cloudy patch] situated to the north of the two stars of the notch of Sagitta." Al-Sufi's discovery was not widely known until the late Kenneth Glyn Jones published his 1975 *The Search for the Nebulae.*

Several others have staked independent discovery claims. Chief among them was Giovanni Batista Hodierna (1597–1660), whose observation of it appeared in his *De Admirandi Coeli Caracteribus* – a catalog, published in 1654, of some 40 deep-sky objects, of which at least 19 are real or verifiable. Hodierna classified his objects into three groups: *luminosae* (stars visible to the naked eye), *nebulosae* (nebulous to the naked eye, but resolved in a simple Galilean telescope of 20×), and *occultae* (unresolved nebulosity). He classified the Coathanger asterism as the 8th object in the *nebulosae* category.

Dalmiro F. Brocchi, who for many years drew the AAVSO charts for variable stars, later independently discovered this grouping at some point during the 1920s or 1930s. Archinal and Hynes, however, could not find any publication that explains the circumstances or date of Brocchi's discovery, though he did create a map of the object for use in calibrating photometers. For this reason, his name has long been associated with the group, being referred to as the now erroneous Brocchi's Cluster. The most commonly used and recognized name for the asterism, however, is Collinder 399, based on Per Arne Collinder's yet one more possible independent discovery, which he lists in his 1931 *On Structured Properties of Open Galactic Clusters and Their Spatial Distribution.*

To find this popular object, simply locate the tail end of Sagitta's shaft – the 4th-magnitude stars Alpha (α) and Beta (β) Sagittae. You'll find the Coathanger just 4° northwest of Alpha Sge. If you're under a dark sky, look for a milky patch of light that seems to flare every now and again

[1] Al-Sufi's full name was Abd-al-rahman Bin Omar Muhammad Bin Sahl Abu'l-husain al-Sufi al-Raazi. He was court astronomer and teacher of astronomy to the powerful Prince Adhad al-Daulat, who eventually ruled most of what is now Iran and Iraq.

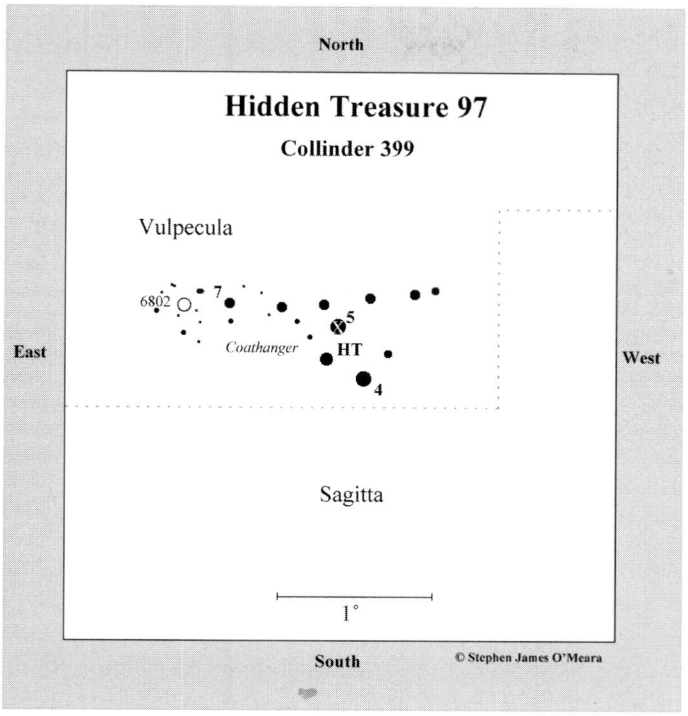

the Coathanger. Robert Burnham Jr. called it a snail-shaped asterism and it is also most appropriate, since the asterism's long axis and southern shell look like a snail "on the move."

At 23× in the 4-inch, the asterism is most fine. Not only is there an evenly spaced row of stars oriented east–west, but also a dimmer string of stars running parallel to them about 15′ to the north. If you relax your gaze and slightly defocus the image, you should see something equally fascinating – three definite parallel rows of dimmer suns oriented southwest–northeast, with

with flecks of sharp starlight. Be sure to spend some time scrutinizing this sparkling cloud, because five of the stars are between magnitude 5.2 and 6.3, and seven of them are brighter than 7th magnitude (see the chart above). I have seen the five brightest with effort and have suspected a sixth. The more time I take, the more distinct the stars appear. Make sure you are relaxed and comfortable, try lying supine on a blanket or recliner. Take breaks and be sure to breathe; inhale several times after each break before you start anew. It could take up to a half hour to achieve success. Of course, there is nothing like seeing the Coathanger in binoculars. As E. E. Barnard described in 1927, it is "a fine group of five bright stars equally distant in a straight line east and west . . . With two others to the south they make a remarkable figure." Of course, the two stars to the south are part of the distinct hook that describes

the center row being centered on 4 and 5 Vulpeculae. In fact, if you just relax, the entire asterism seems to be comprised of parallel rows of stars in typical escheresque visual confusion – just look and you will see. The magnitude 5.6 star 10′ southeast of the midpoint between 4 and 5 Vulpeculae in the Coathanger's hook is a nice telescopic double with a 10th-magnitude companion to the northeast. Once again, relax your gaze, slightly defocus the telescope, and look at the hollow of the hook; it is a pool of empty starlight.

Then, there's NGC 6802, the forgotten treasure within a treasure – a small (5′) 9th-magnitude open cluster less than 20′ due east of 6th-magnitude 7 Vulpeculae, the eastern end of the 2°-long straight row of stars in the Coathanger asterism. It is one of the ironies of the sky that the bright and obvious wonder that everyone adores

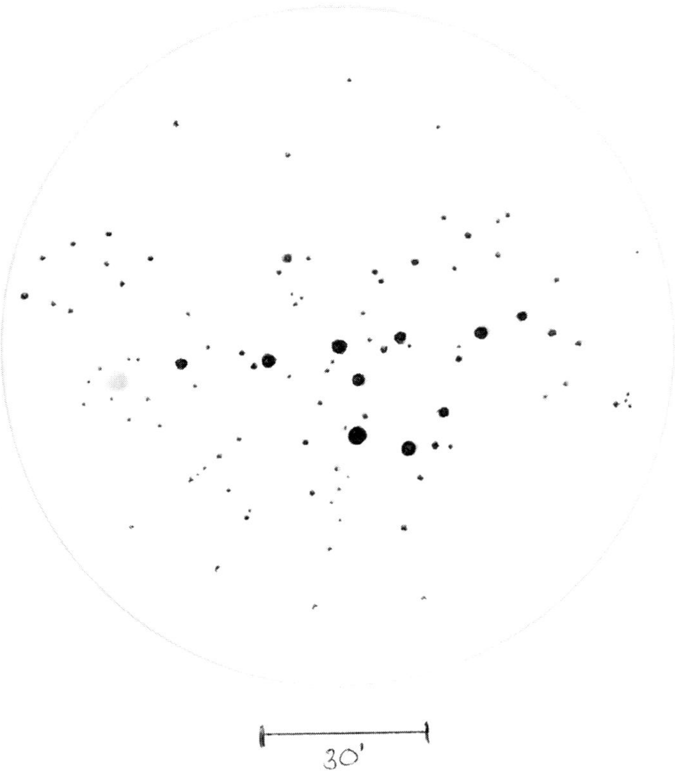

is not a true physical system but a random assortment of stars in the same line of sight. Meanwhile a dim and absolutely oblivious open cluster essentially hugs the asterism to the east, but hardly anyone notices. Walter Scott Houston called it a "delicate splash of celestial faint fire." Under higher magnifications, the cluster, which appears circular at low power, is actually extremely elongated north–south. I can partially resolve it in the 4-inch, though the brightest stars shine at magnitude 14. The cluster also breaks down into three segments: a little triangular core bordered to the north and south by opposing triangles, making it look like a little bow tie.

The beautiful 6th-magnitude Algol-type eclipsing binary star U Sagittae – which dips to a minimum of magnitude 9.2 for approximately 1 hour 40 minutes, every 3.4 days – lies about $1\frac{1}{2}^\circ$ due west and slightly south of 4 Vulpeculae; while the dim and distant (26,000 light-years) globular cluster Palomar 10, lies 1° due south and a tad west of U Sge. Palomar 10 is one of two globular clusters with the very lowest known galactic latitudes and highest reddenings (the other being NGC 6749).

So in this little region of space measuring 3° in angular extent, we have a brilliant asterism known since antiquity, a bright Algol-type variable star easily observed in binoculars, one of the most highly ignored open clusters in the heavens, and arguably the most mysterious globular cluster in our galaxy. It is as if the Coathanger is a decoy, forcing our attention away from these other wonders. But, to paraphrase T. S. Eliot, who am I to disturb the universe?

Hidden Treasure 98

NGC 6819

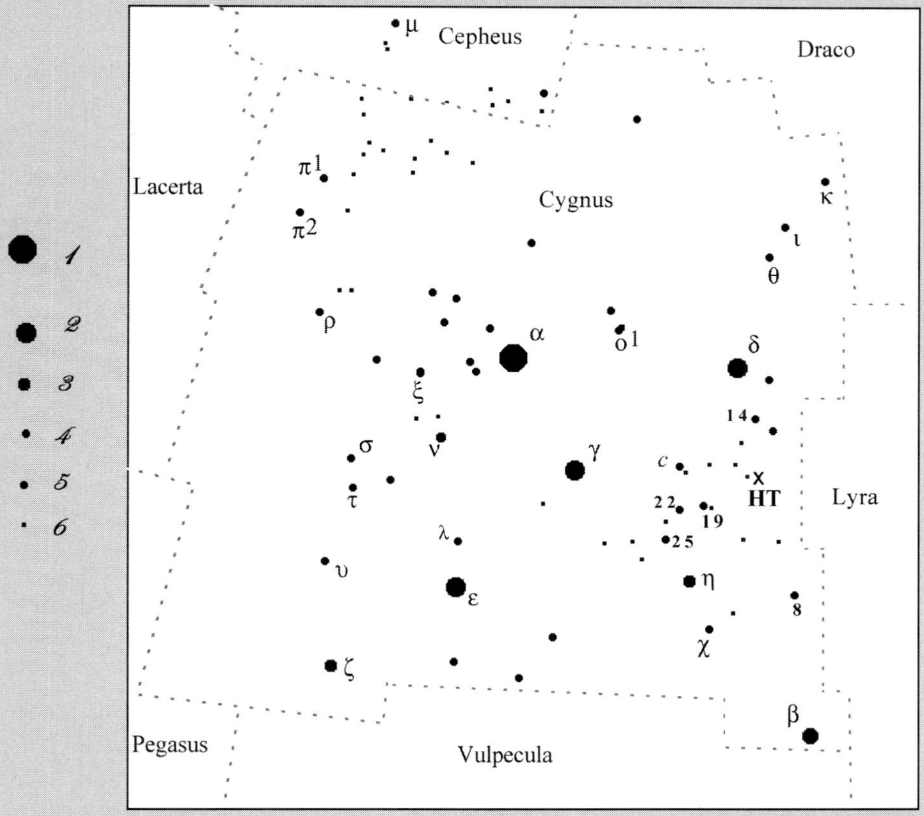

© Stephen James O'Meara

Tirion: Charts 8 & 9

Uranometria: Chart 84

98

98

Fox Head Cluster, Octopus Cluster
NGC 6819
Type: Open Cluster
Con: Cygnus

RA: $19^h\ 41.3^m$
Dec: $+40°\ 11'$
Mag: 7.3 (O'Meara); 7.3
Diam: 5.0′
Dist: 6,700 light-years
Disc: Caroline Herschel, 1784; Karl Ludwig Harding, independently discovered it between 1820–1824

W. HERSCHEL: [Observed December 31, 1801] There is a scattered cluster of small stars about $2°\ 10^m$ from 14 Cygni towards 21. My sister saw it May 12, 1784. [Added by Caroline:] my number 16.

NGC: Cluster, very large, very rich, stars from magnitude 11 to 15.

NGC 6819 IS A LITTLE AND LONELY open cluster located near the galactic plane in the vast Cygnus Milky Way. If that sounds like a paradox, it isn't. Not every object in the Milky Way stands out like a bursting firecracker. The fact is, NGC 6819 is just far enough away from the main stream of the celestial Mardi Gras parading through the long axis of the Northern Cross that it is as inconspicuous as a butterfly fluttering in a field near a busy highway.

It has long been wondered how so bright and tight an object escaped the gaze of William Herschel in his grand survey of the heavens. But it didn't: thanks to research done by Michael Hoskin and Owen Gingerich, we find that William's sister, Caroline, discovered the cluster on May 12, 1784, "halfway between delta Cyg & eta & theta Lyrae making an isosceles triangle downwards when in that situation." William reviewed it in detail on December 31, 1801. But, as Hoskin explains, "[t]his cluster was reobserved by William after 1790 when he established the existence of 'true' nebulosity and so had lost interest in clusters."

German astronomer Karl Ludwig Harding (1765–1834) independently found it sometime between 1820 and 1824 – probably with a 4-inch refractor at Göttingen Observatory, though that fact is uncertain. He listed it in the 1827 *Berliner Jahrbuch* along with another of his discoveries – planetary nebula NGC 7293 (Caldwell 63), the famous Helix Nebula in Aquarius – sometime between 1820 and 1824. Harding, however, is best known for his discovery of the asteroid 3 Juno on September 1, 1804; at the time he was employed. He also discovered three comets and published the *Atlas novus coelestis* which cataloged 120,000 stars.

NGC 6819 is a highly concentrated and rich cluster with stars of nearly uniform magnitude. While the most widely used diameter for the cluster is 5′ (as listed here), Korean astronomer Yung-Woo Kang (Pusan National University) and his colleagues in a 2002 *Journal of the Korean Astronomical Society* revealed that the cluster may be as large as 18′ round when stars to magnitude 20.5 are included. They also found the cluster to be dynamically relaxed, meaning that its high-mass stars (giants and upper main-sequence stars) tend to sink to the inner region while low-mass stars (lower main-sequence stars) tend to occupy the outer regions, though it appears that a large number of these low-mass stars has escaped from the cluster.

There has been another advance in knowledge: early photometric studies of NGC 6819 indicate that it is an old open cluster, similar to M67 (~4 or 5 billion years old), but recent estimates show its age at about 2.5 billion years, making it an intermediate-age, not an old, cluster. If we accept NGC 6819's distance to be 6,700 light-years, and its angular extent 5′, the cluster spans 10 light-years of space and has some 930 members, the brightest of which shines at 11th magnitude. If the cluster spans 18′, the true linear extent is about 35 light-years.

K. T. Hole (University of Wisconsin, Madison) and his colleagues at the 2003 American Astronomical Society Meeting presented results of a radial-velocity study of NGC 6819, in which they determined 5,140 measurements of 1,140 stars, obtained over the course of some 5 years with the WIYN Multiple-Object Spectrograph. They found the cluster's mean velocity to be about 22 kilometers per second. In a 2003 *Monthly Notices* of the Royal Astronomical Society, R. A. Street (University of St. Andrews, Belfast) presented results from a survey for planetary transits in NGC 6819. Street obtained high-precision time-series photometry for over 38,000 stars in the field and identified eight candidate stars showing

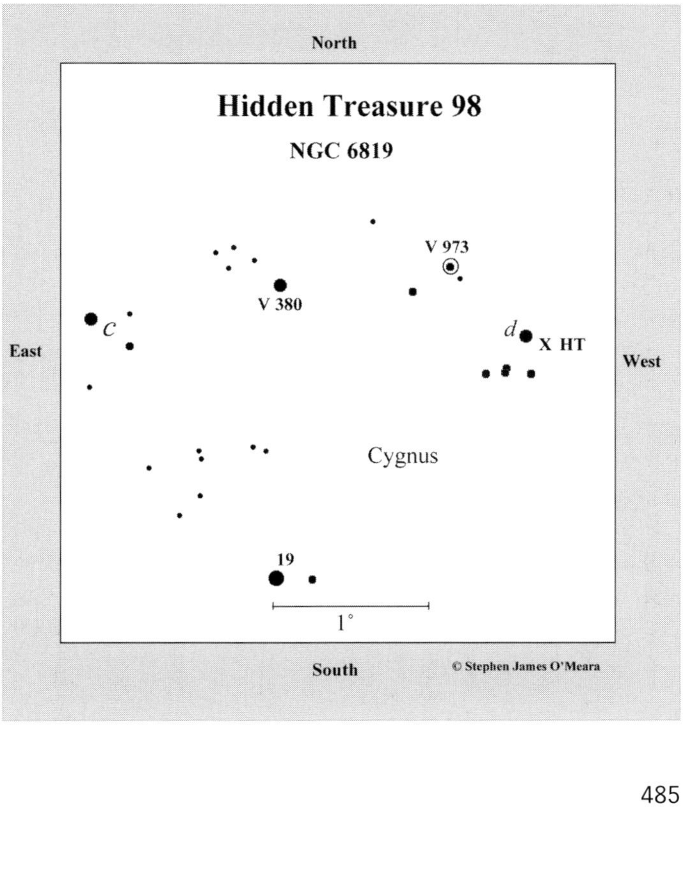

multiple transitlike events, plus 3 stars with single eclipses. On closer inspection, while most candidates are shown to be low-mass stellar binaries, some of them could be the result of brown-dwarf companions. By the way, Archinal and Hynes note that the Lund catalog incorrectly lists NGC 6819 as a globular cluster.

As with some of the other more obscure hidden treasures, NGC 6819 will require a fair amount of star hopping, perhaps with the aid of binoculars. Start with 4th-magnitude Eta (η) Cygni in the long axis of the Cross. About $3\frac{1}{4}^\circ$ to the northwest is 5th-magnitude 15 Cygni, which marks the southwesternmost star in a 4°-long line of three similarly bright stars – the other two are 19 Cygni (almost 2° to the northeast) and Star *b* (about 2° further in the same line). Almost $1\frac{1}{2}^\circ$ to the west-northwest is the $5\frac{1}{2}$-magnitude eclipsing variable star V380 Cygni which varies in brightness by only a few tenths of a magnitude about every $12\frac{1}{2}$ days. A little more than 1° in the same line is the 6th-magnitude semi-regular variable V973 Cygni. A little more than 40′ to the southwest is another 6th-magnitude Star *d*. NGC 6819 is less than 10′ further to the southwest.

NGC 6819 is indeed a true hidden treasure in a very lonely part of Cygnus. It is visible in both 7 × 50 binoculars – if you know exactly where to look; my antique telescope shows it much the same. At 23× in the 4-inch, the cluster looks like a condensed globe of starlight, like a globular cluster, so it is no surprise that Lund would have mistaken it for one of these more distant and older objects. With little effort, prominent streams and clumps of stars can be seen or traced within it. A lazy V-shaped pattern of stars is emblazoned on the diffuse background of dimmer suns, as if burned there with a hot branding iron. One arm of the V extends to the north, the other to the northwest. The longer I look, the more I see the apex of the "V" branch out to the southeast and the southwest, forming a weak Greek letter χ.

At low powers Skiff and Luginbuhl call it granular, and this is an excellent description, since it really is a globe of dim stellar "noise," like TV static. It also lies in a rich Milky Way field – a glorious sight, really, at low power. Increasing magnification brings out the "V" and diminishes the impression of granularity. It also brings out a knot of dim suns around a prominent double near the southern apex of the "V" in the eastern arm. Another double appears midway in the western arm, which is comprised of about six stars.

Another mini-cluster lies midway in the eastern arm directly opposite the double in the western arm. It's a "Z" of similarly bright suns. There's a very challenging double for a 4-inch under moderate viewing conditions. The southern knot is also part of a longer complex oriented southeast–northwest. The χ asterism, under low magnifications, has angular extensions giving the cluster a spiral shape. So, it's sort of like an octopus. The southern double is its eyes and the oval to the southwest is its head.

Interestingly, in the Webb Society's *Deep-Sky Observer's Handbook* (Volume 3), Guy Hurst describes the brightest portions in a 10-inch telescope at 120× as being "roughly U-shaped," which is a lazy "V," with resolved stars of 11th to 13th magnitude "set against unresolved nebulosity." No doubt, this nebulosity is unresolved Milky Way.

Hidden Treasure 99

NGC 6818

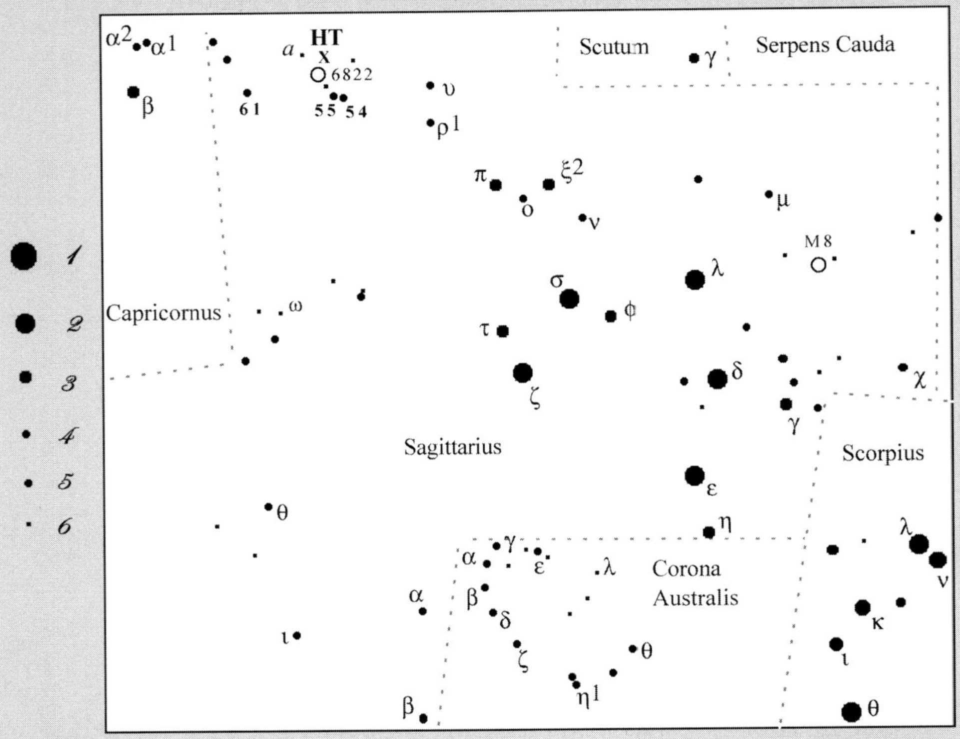

© Stephen James O'Meara

Tirion: Charts 16 & 22

Uranometria: Chart 297

99

Little Gem Nebula, Green Mars Nebula
NGC 6818
Type: Planetary Nebula
Con: Sagittarius

RA: $19^h\ 43^m\ 57.8^s$
Dec: $-14°\ 09'\ 11''$
Mag: 9.3 (O'Meara); 9.3
Dim: $22'' \times 15''$
Dist: 5,500 light-years
Disc: William Herschel, 1787

W. HERSCHEL: [Observed August 8, 1787] A considerably bright small beautiful planetary nebula; but considerably hazy on the edges, of a uniform light; 10″ or 15″ in diameter, perfectly round. "I shewed it to M. De La Lande." (H IV-51)

NGC: Planetary nebula, bright, very small, round.

NGC 6818 IS A BRIGHT TURQUOISE jewel neatly tucked away in the northeastern recesses of Sagittarius, the Archer. It is popularly known as the Little Gem, a name coined by the late John H. Mallas in 1963. (Note that NGC 6445 [Hidden Treasure 84] is sometimes mistakenly referred to as the Little Gem, but this is an error propagated by a mistake in *Sky Catalogue 2000.0*, Volume 2.) NGC 6818 is an attractive planetary nebula that lies about a Moon diameter north-northwest of one of the most celebrated dwarf galaxies visible from mid-northern latitudes – NGC 6822, Barnard's Galaxy (Caldwell 57). Normally, bright and popular galaxies can distract observers, making them blind to other interesting objects nearby. But if it weren't for Barnard's Galaxy, NGC 6818 would probably be even more obscure than it is. And while the nebula is considered a southern planetary, it can be seen from Alaska and northern Canada.

The nebula's telescopic size equals that of Mars at a perihelic opposition, which is about the apparent size of the annulus in NGC 6369 (Hidden Treasure 81) – the Ghost of Mars Nebula. For this reason, NGC 6818 is also known as the Green Mars Nebula. Robert H. Rubin (NASA/Ames, Orion Enterprises) and colleagues used the Hubble Space Telescope to image the Little

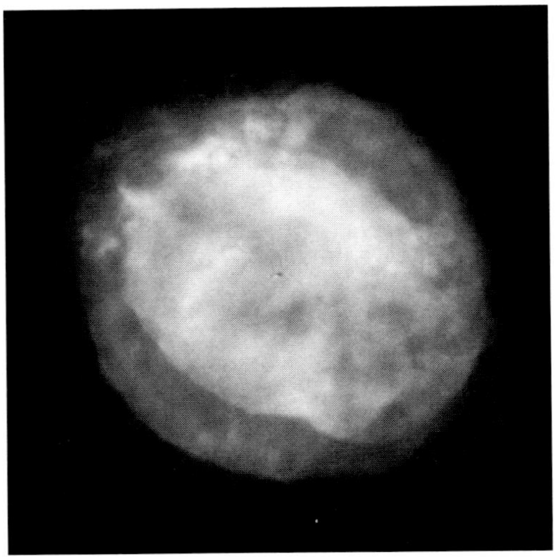

Gem in 1997. They discussed their findings at a 1998 American Astronomical Society meeting. The Hubble Space Telescope (HST) images showed the nebula as having a roughly spherical outer envelope as well as a brighter vase-shaped interior "bubble." The scientists believe that a fast wind from the hot central star is creating the elongated vase shape and, in fact, has caused a "blowout" at the two ends of the major axis.

Italian astronomer S. Benetti (Astronomical Observatory of Padova), and colleagues elaborated on NGC 6818's three-dimensional structure in a 2003 *Astronomy and Astrophysics* paper. The Little Gem appears to be a young (3,500 years), optically thin double-shell nebula seen almost equatorial on, with a tenuous and patchy spherical envelope (0.6 light-years across) encircling a dense and inhomogeneous triaxial ellipsoid with a hole along the major axis and a pair of equatorial, thick "moustaches."

The central star, which has a mass of about 0.6 Sun, is faint, shining around 17th magnitude. In a 1999 *Astrophyiscal Journal* paper, Seik Hyung and Lawrence H. Aller (University of California, Los Angeles) and Walter A. Fieibelman (NASA-Goddard Space Flight Center) note that the star's spectrum reveals that it may be a Wolf-Rayet type. It has a luminosity of about 3,500 Suns and a temperature of at least 140,000 K, suggesting an age of 9,000 years. And while NGC 6818's spectrum showed carbon and nitrogen appearing more abundant than in the Sun, other elements seem to have roughly a solar abundance. NGC 6818 may, then, have originated from a star resembling the Sun, at least in chemical composition.

NGC 6818 is at the start of what's known as the recombination phase, whereby electrons are being captured by protons following a luminosity decline of the central star, which is rapidly moving toward becoming a white dwarf. The nebula is destined to become thicker and thicker, with an increasing fraction of neutral, dusty gas in the outermost layers.

The exciting star of NGC 6818 is also a visual binary: a faint, red companion of about magnitude 17.7 appears at 0.09″ in position angle 190°, corresponding to a separation of about 150 astronomical units and to an orbital period of 1,500 years.

If you know how to find Barnard's Galaxy, you can find NGC 6818, which, again, is only 35′ to the north-northwest. If not, try star hopping to the galaxy, beginning with the bright double star Alpha^{1,2} ($\alpha^{1,2}$) Capricorni (Algedi). Just 4° to the west-southwest is the 6th-magnitude star 63 Sagittarii. A 3° hop due west brings you to another 6th-magnitude star (*a*). Barnard's Galaxy is only about $1\frac{1}{4}°$ to the southwest. It is quite large (15′ in diameter), but can be easily swept up under a dark

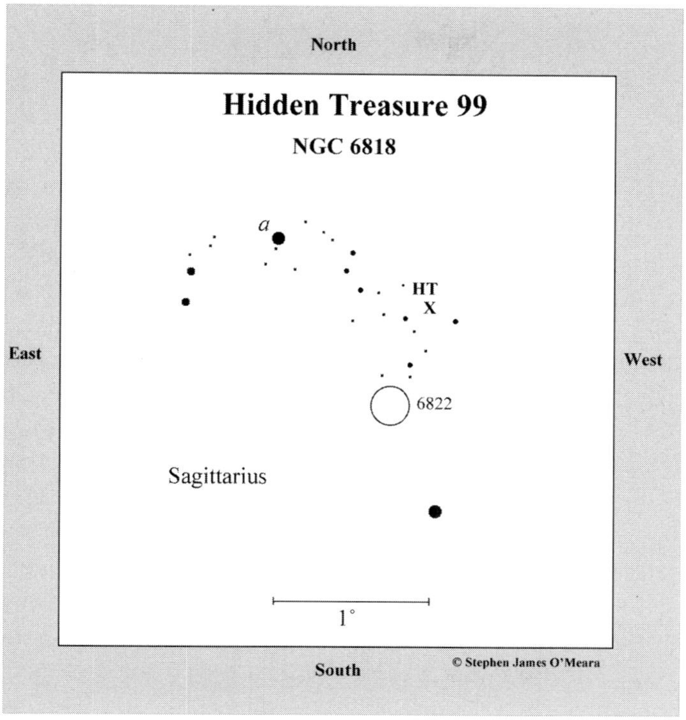

sky in a rich-field telescope: Edward Emerson Barnard discovered it visually with a 5-inch refractor in 1884. NGC 6818, as mentioned, is 35′ to the north-northwest, about 12′ east-northeast of an 8th-magnitude star.

The nebula is visible in 7 × 50 binoculars as a "star" among stars, which shows up more prominently in my antique telescope. At 23× in the 4-inch, one could easily pass over this 9th-magnitude glow because its light is so condensed. But if you stare long enough with averted vision, you can see that the planetary has some definition – a disk just ever so slightly non-stellar. The nebula is definite at 72×, appearing as a pale ashen globe, like Neptune. Again, with some concentration and averted vision, the Little Gem's color changes to a pale ashen blue or steel gray. Admiral William Henry Smyth saw the object's color as "pale blue." And Hartung sees it like me, "pale gray blue."

The disk does not appear uniform in brightness. There is a dense core surrounded by a more tenuous and slightly irregular crown. At 101×, it is a sparkling annulus – but the tiniest annulus you can imagine. The hole is not so definite as it is implied. I sense two bright beads or enhancements along the ring. With continued effort, I can see two dim stars, one to the northwest and one to the southwest. The one to the northwest is brighter and closer. Smyth saw these stars too, noting that they formed a square around the nebula.

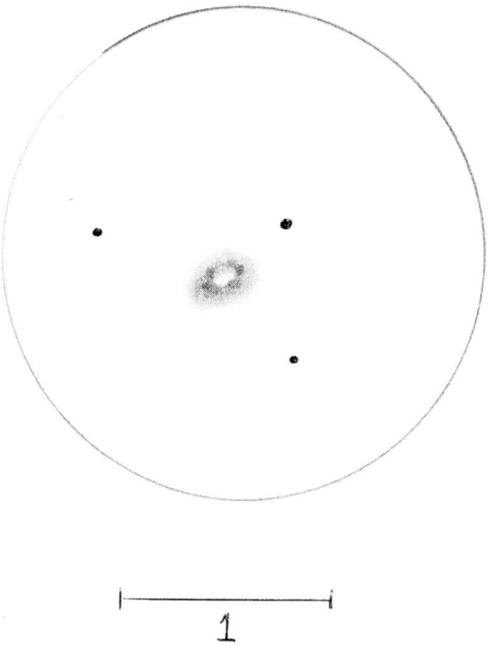

At 182×, the annulus is more apparent as is another much dimmer star to the northeast, and the annulus seems to be beaded in three places, to the south, northeast, and northwest. At 303×, the inner annulus has a sharp thread of light connecting the beads, some four of which stand out. The inner ring dominates the view. There is very little gas extending beyond it. I see no sign of a central star. When I read the brief description of this nebula in my *Deep-Sky Companions: The Caldwell Objects* (in my discussion on Barnard's Galaxy), it's obvious that I did not use sufficient power to see the annulus well.

Herschel notes that he showed the object to the French astronomer Joseph Jerome le Francais de La Lande (1732–1807), who must have been visiting the English astronomer at Slough. (La Lande is quoted as saying: "I am an oilskin for insults and a sponge for praise.")

By the way, in *Deep-Sky Companions: The Caldwell Objects*, I also mentioned that I did not see color in NGC 6818. Since this object was not in the Caldwell list, I did not spend much time observing it. It just goes to show you how making repeat observations and spending more time with an object can make you a better observer. The more we look, the more we see; the more we see, the more we learn – about the sky, the objects in it, and ourselves.

Hidden Treasure 100

NGC 6866

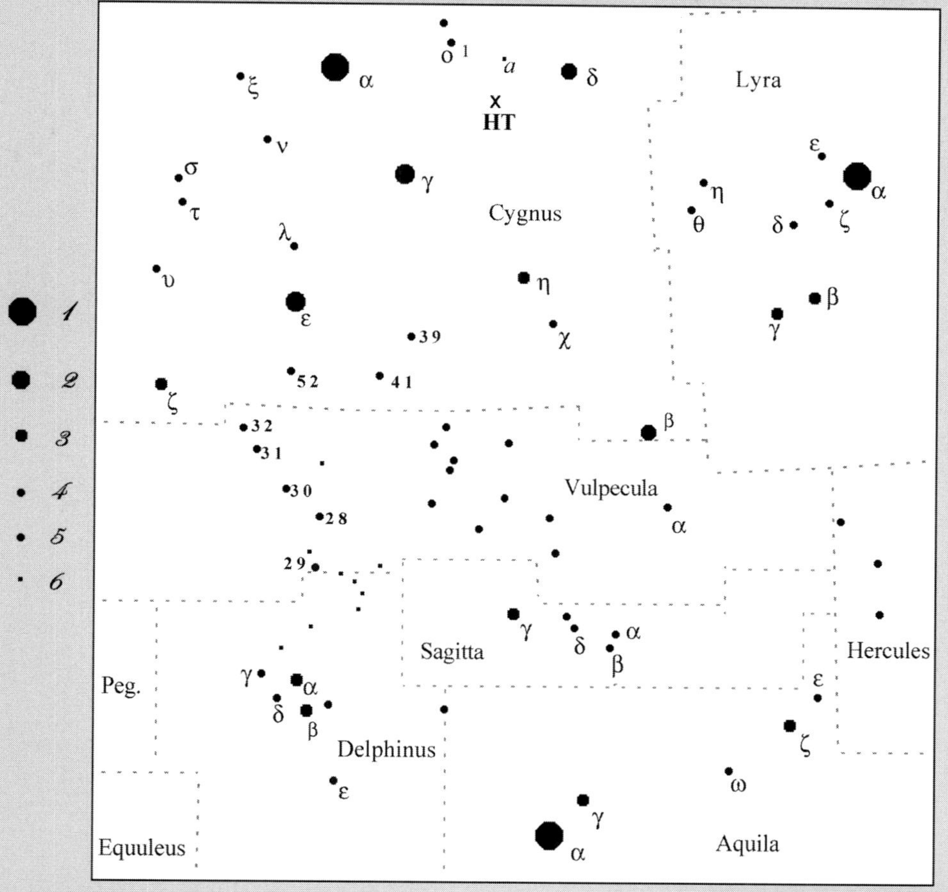

© Stephen James O'Meara

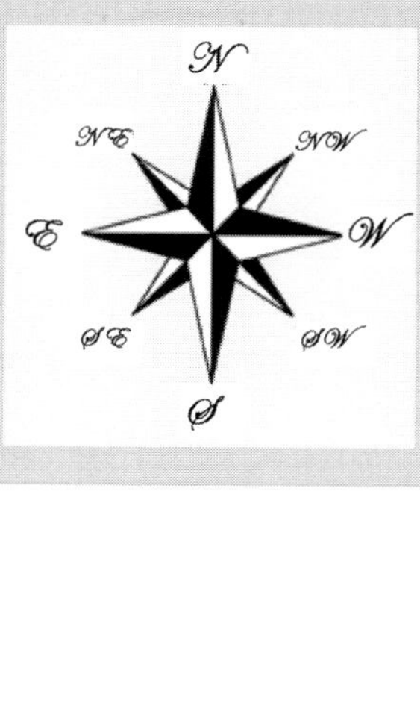

Tirion: Charts 9 & 16

Uranometria: Chart 163

100

Frigate (Pirate) Bird Cluster
NGC 6866
Type: Open Cluster
Con: Cygnus

RA: $20^h\ 03^m\ 55^s$
Dec: $+44^\circ\ 09.5'$
Mag: 7.6
Diam: 15′
Dist: 4,200 light-years
Disc: Caroline Herschel, 1783

W. HERSCHEL: [Observed September 11, 1790] A very rich cluster of large stars considerably compressed, above 15′ diameter by the size of the star. It is situated in the Milky Way, towards us. (H VII-59)

NGC: Cluster, large, very rich, considerably compressed.

NGC 6866 IS A MARVELOUSLY obscure, but visually enticing, cluster that Caroline Herschel discovered on the evening of July 23, 1783. This discovery went unnoticed until Michael Hoskin found the following note in Caroline's original journals for that date:

"Some small stars; or perhaps a Nebula. My Brother put, I believe, a power of 70 to the Sweeper, then what is call'd some small stars are about a hundred or more." Caroline saw the object again on October 16, 1783. "My Brother shewed me the spot in Cygnus, which I saw first July 23rd," she writes "the stars very numerous, and intermixed with strong nebulosity. Mess[ier] has it not."

The identity of this cluster is undeniably certain, for Hoskin also found the following description of the cluster in William Herschel's journal for the evening of October 16, 1783: "Cluster of stars making an isosceles triangle with delta & omicron Cygni, the vertex laying S following; about 3 or 4 degrees from delta or omicron. It contains very great variety of magnitudes; the large ones are scattered the smaller stars being mixed between them. 7ft [reflector] compound [eyepiece] – Lina [i.e. Caroline]." William's closing remark, "Lina" was his affectionate way of stating that Caroline had discovered this object, not him. William Herschel observed the object yet again on September 11,

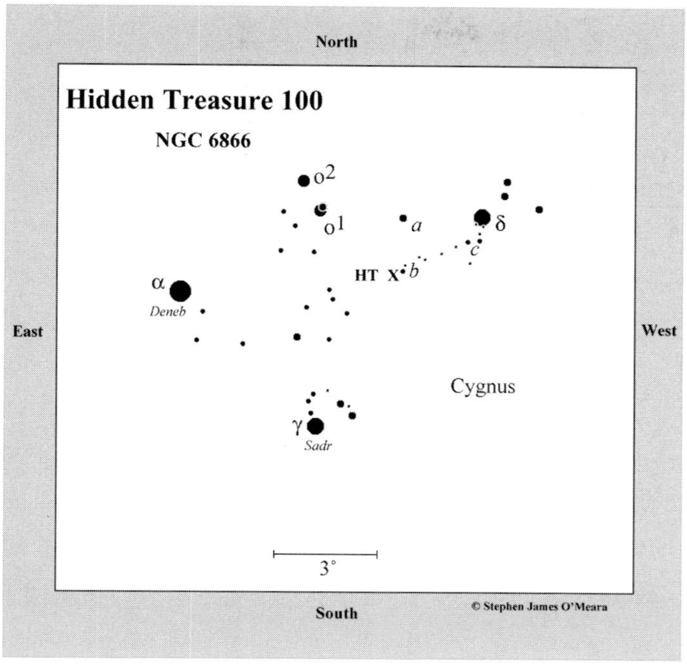

1790 and cataloged it as "VII.59," but as Hoskin says, "without either partner realizing that they had seen it before."

Barbara Wilson notes that the discovery was not made with her small refractor. In July 1783 Caroline exchanged that object glass for a Newtonian sweeper 4.2 inches in aperture and 2-feet long. It was a gift from her brother, William, who was delighted at the success she was having in her sweeps of the sky. "It was convenient," Wilson says, "as she could use it while seated, making vertical sweeps from the horizon to the zenith. It gave 24 power and a 2° 12′ field of view. This rich-field telescope was especially suited for the discovery of this magnitude 7.6 cluster."

To find it, first locate 3rd-magnitude Delta (δ) Cygni – the star that marks the western arm of the Northern Cross – and the 4th-magnitude star Omicron[1] (o^1) Cygni, about 5° east-northeast of Delta Cyg. NGC 6866 is about 3.5° east-southeast of Delta Cygni and makes an isosceles triangle with Delta and Omicron[1] Cyg. To star hop to it, center Delta in your telescope, then move about 1° to the south-southeast, where you will find two 7th-magnitude stars (*c*), which are oriented east–west and separated by about 25′. These stars are at the western end of a roughly $2\frac{1}{2}$° chain of near, similarly bright suns, the eastern end of which is marked by a star of magnitude 7.6 (*b*). NGC 6866 is only about 20′ southeast of Star *b*. (Note: the cluster is misplotted in the *Millennium Star Atlas*; the symbol is nearly 10′ south of the cluster, whose core of stars, ironically, is plotted in the correct position.) From a dark sky, the cluster is visible in 7 × 50 binoculars as a dim but concentrated glow near Star *b*, which shines at the same magnitude. The cluster becomes more apparent with averted vision.

You should take advantage of this opportunity to see how an extended object can *appear* fainter than a star of equal magnitude. If you memorize the appearance of NGC 6866, then defocus your binoculars until the star swells to the same size as NGC 6866, see if the defocused image of the star does not match the brightness of the cluster. If you spend some time viewing the cluster in binoculars with averted vision, you should see it irregular in form, amorphous, and somewhat mottled. In my antique telescope, the cluster is a diffuse but mottled glow with a partially resolved

affects of peripheral vs. direct vision, there must be suns, or groups of suns, here at or near the limit of resolution. With averted vision, the outlying regions of the cluster are very scattered with discrete patches of starlight strewn hither and yon to great extents. The cluster is, in fact, quite rich, having some 130 stars of 10th magnitude or fainter spread across an area of sky equal to that of half the diameter of the full Moon. In true physical extent, we are looking at an object spanning 18 light-years across. In size, distance, and stellar population, NGC 6866 is a dead ringer for M38 in Auriga. The difference is that M38 is about a magnitude brighter than NGC 6866 and its core is much looser. Try comparing the two on the same night.

NGC 6866 becomes even more dynamic at 101×, for the brightest stars stand out from the rich background to form an unmistakable outline of a frigate bird, Nature's "aerial pirate of supreme daring." The bird's body is the dense elliptical stretch of starlight at the cluster's core. A brilliant yellow star marks the tip of the bird's powerful beak to the north, while a zigzagging line of stars to the south marks its tail. Frigate birds have impressive wingspans that can stretch to eight feet, and these can be seen dramatically through the eyepiece as the sharp east–west extensions of starlight. With imagination, it looks as if this celestial frigate bird is soaring effortlessly overhead as we look upon its majestic undercarriage.

spine of stars oriented north-northwest–south-southeast.

At 23× in the 4-inch, a view that mimics what Caroline Herschel would have seen, the cluster is very nice, indeed, showing a small core of stars that appears a bit bulbous, fuzzy, and bright on the northern end. (Note that when Caroline Herschel discovered this object, she thought it might be a nebula.) This pouch of stars is surrounded by a dimmer crown of scattered suns that, with concentration, has sharp and long extensions to the east, west, and south.

The scene becomes more majestic at 72×, when the pouch transforms into a "pulsating" mass of starlight; if you sweep across this area with averted, then direct, vision, the pouch puffs out, then contracts, in regular fits and starts. Aside from the obvious

100

Having a frigate bird cluster in this book is most befitting. Frigate birds, whose diet consists of fish and other marine creatures, are also piratical feeders. They torment other, smaller, diving birds and force them to give up their booty. When you see a frigate bird circling in the air overhead, you can bet they are waiting to steal the meals from unsuspecting birds. And it is for this reason that the frigate bird is known throughout its domain as the pirate bird.

Hidden Treasure 101

NGC 6940

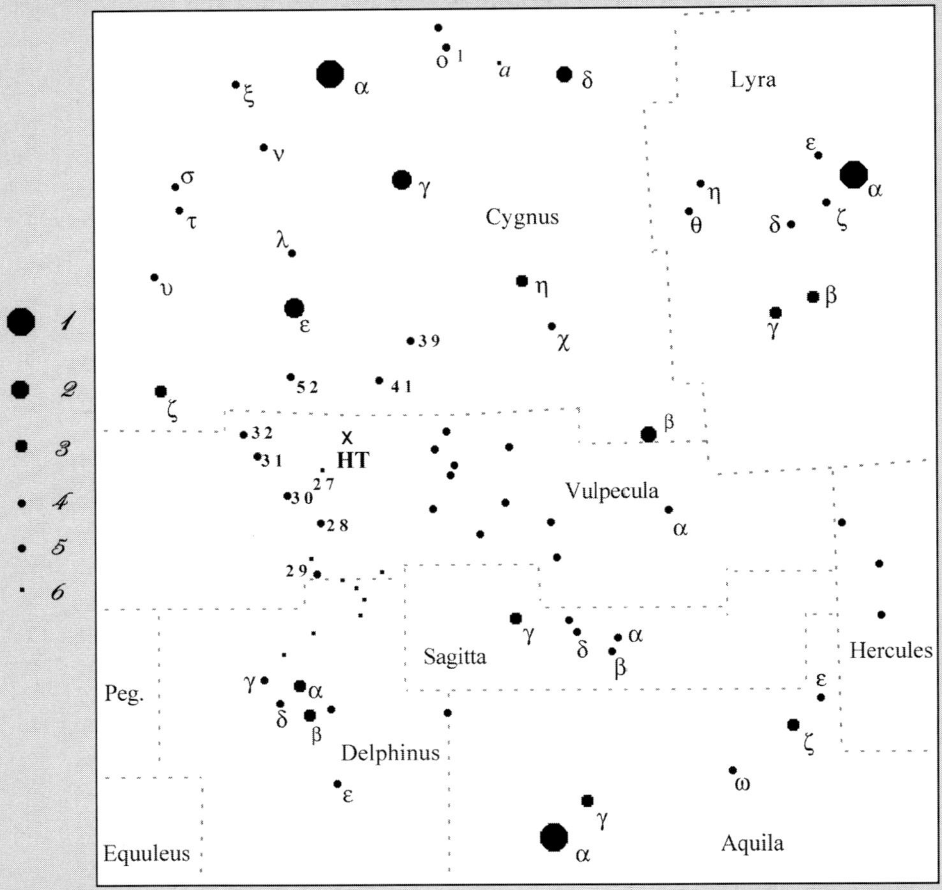

Tirion: Chart 9

Uranometria: Chart 120

101

Mothra Cluster
NGC 6940
Type: Open Cluster
Con: Vulpecula

RA: 20^h 34.5^m
Dec: $+28°$ 17′
Mag: 6.3
Diam: 25.0′
Dist: 2,600 light-years
Disc: William Herschel, 1784

W. HERSCHEL: [Observed July 17, 1784] A very rich cluster of pretty small scattered stars most of the same size, 20′ diameter. (H VII-8)

NGC: Cluster, very bright, very large, very rich, considerably compressed, stars pretty [bright].

WHEN YOU HEAR DEEP-SKY observers mention the constellation Vulpecula, the Little Fox, it is usually in connection with either the spectacular planetary nebula M27 (the Dumbbell Nebula), the Coathanger asterism (Collinder 399 [Hidden Treasure 97]), or the mysterious open cluster NGC 6885 (Caldwell 37). Vulpecula is also home to another visual gem, NGC 6940, a most abundant star cluster in the southern branch of the Milky Way bordering the Great Rift. But it is largely ignored. You'll find it lost in the void between the southern wingtip of Cygnus and the Little Fox's hind quarters. It is a binocular treat and a telescopic wonder of stellar riches – one whose variety and multitude almost defies imagining.

But first a word about Vulpecula. The original name of the constellation is *Vulpecula cum Ansere,* meaning the Little Fox (Vulpecula) with the Goose (Ansere). Johannes Hevelius created it to occupy a space between Sagitta, the Arrow, and Cygnus, the Swan. As Hevelius said, "I wished to place a fox with a goose in the space of the sky well fitted to it; because such an animal is very cunning, voracious, and fierce. Aquila and Vultur are of the same nature, rapacious and greedy."

Alas, Ansere, the Goose has disappeared from the constellation's title – probably a response to our ever increasing need for an economy of words (our fast-paced world spawns short attention spans), or as Patrick Caldwell-Moore jibes in the 1968 edition of his book *Amateur Astronomy,* "possibly the fox has eaten it." Ironically, there is no known mythology related to the fox, but one German myth describes how Antenteh and

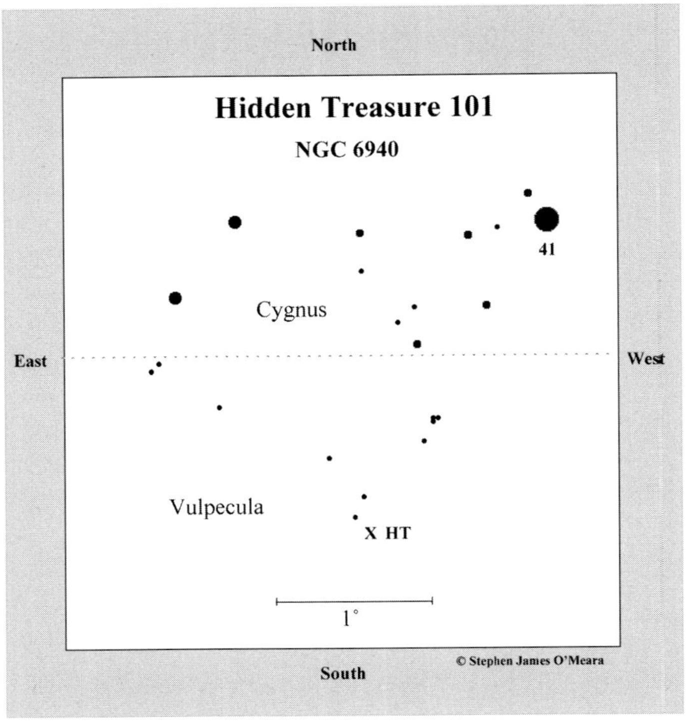

find excitement in trying to test their eyes with very faint and elusive stars." I must be among those excitable ones, because I take great liking to its form, though NGC 6940 lies several degrees east of the constellation's most prominent stars. In fact, the brightest stars near NGC 6940 are in the southern wing tip of Cygnus. To find it, look about $3\frac{1}{4}°$ southwest of 4th-magnitude 52 Cygni (and the western segment of the Veil Nebula [Caldwell 34]), which is about 3° south of magnitude 2.5 Epsilon (ε) Cygni. It also marks the southern apex of a near-equilateral triangle with 52 Cyg and 4th-magnitude 41 Cygni, which is about $3\frac{1}{2}°$ west and slightly south of 52 Cygni.

The cluster is easy to see in 7 × 50 binoculars, appearing as an elliptical swarm of dim suns shining with a total light of a 6th-magnitude star; the brightest members of NGC 6940 shine at 11th magnitude. Some brighter field stars are superimposed on the cluster. In my antique telescope, the cluster looks like a dwarf elliptical galaxy, elongated northeast–southwest – as if it were being pulled apart by tidal forces from a close encounter with the Milky Way.

Of course, open clusters are only weakly held together by gravity and fated to disperse over time. The oldest known open clusters tend to have ages that hover around 5 billion years. Dutch astronomer T. Belloni (Astronomical Institute "Anton Pannekoek") and his Italian colleague

his wife (a poor couple who lived by the sea in a cabin with only a tub) to keep warm, had to fill the tub with feathers and down from the Swan and the Goose. Since Hevelius intended Vulpecula to have a Goose, why not return it to the sky, so that its feathers and down can warm the hearts of those just learning the sky. Like some mythological god, the power to grant this request lies within you. There is another reason to bring back the Goose. While there is no known record of Vulpecula having any named stars, Richard Hinckley Allen notes that the *Standard Dictionary* says that the constellation's *lucida,* a magnitude 4.4 star just west of the Fox's head, is named "Anser."

As for the stars of Vulpecula, Julius D. W. Staal writes in his book, *The New Patterns in the Sky: Myths and Legends of the Stars,* "Vulpecula is of interest only to those who

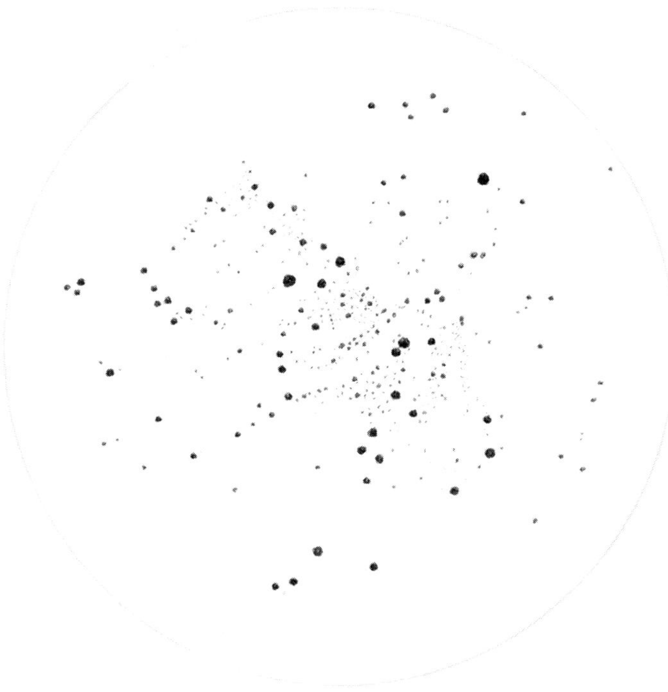

the cluster is at its best – a highly elliptical body of shimmering lights with many star clumps that give off a rich, mottled glow. A bright pair of stars lies just west of the cluster's heart; the brightest member is the semi-regular variable star FG Vulpeculae, which varies in brightness from magnitude 9.0 to 9.5 every 80 days. FG Vul has a warm orange hue, like that of the setting Sun, which is in stark contrast to the cool aquamarine hue of its companion. The rest of the cluster's stars are scatted across the field like diamond dust in a strong and erratic wind. Indeed, the cluster's core is splintered into three highly elongated groups, as if carried aloft by a jet stream. The cluster's core is marked by a long and narrow vein of stars running northwest–southeast immediately northeast of FG Vul. The brightest section of the core lies to the northeast and is a dense elliptical clutch of suns, oriented east–west. The third section of the core is an elongated crowd of stars extending southeast of FG Vulpeculae. Two elongated bands of darkness separate these groups. So the core of NGC 6940 is an incredibly busy and complex gathering of mini-clusters, whose overall shape (seen with north up) forms a backward epsilon (∍).

G. Tagliaferri (Astronomical Observatory of Brera) used the now defunct ROSAT satellite to study coronal activity in the stars of NGC 6940. Coronal activity in stars depends crucially on stellar rotation, which decreases with age due to magnetic breaking. Clusters with ages greater than 1 billion years are not expected to contain rapidly rotating late-type stars. ROSAT data show that stellar coronal activity still exists in NGC 6940, but it is confined to known late-type stars in binary systems. Therefore, Belloni and Tagliaferri determined that NGC 6940 has an estimated age of about 1 billion years, placing it at the threshold of coronal inactivity.

NGC 6940 is quite stunning in small telescopes under a dark sky. At 23× in the 4-inch,

The cluster is rich, having some 170 members, which are, in addition, projected against a rich Milky Way background, so it is most exhausting to observe. There is so much to see, so much to imagine. Increase magnification and more background stars can be seen. More and more pairings and groupings – lots of doubles and triples and quintuplets. With averted vision it's like seeing layer upon layer of clusters whose combined light measures nearly a full Moon diameter across.

The cluster's outlying regions contain two looping arcs like insect wings; one wing lies to the northeast, the other to the southwest of the elliptical core. A forked string of stars flaring to the northwest forms the insect's tail. With imagination the cluster looks like Mothra, that classic 1960s Japanese science-fiction creature – a moth-like goddess who spends eternity seeking a path to peace. The double star near the cluster's center, on the giant Moth's back, is, of course, the tiny telepathic twin princesses who ride the back of Mothra and sing pleas to their goddess whenever they need help.

Hidden Treasure 102

Northern Coalsack

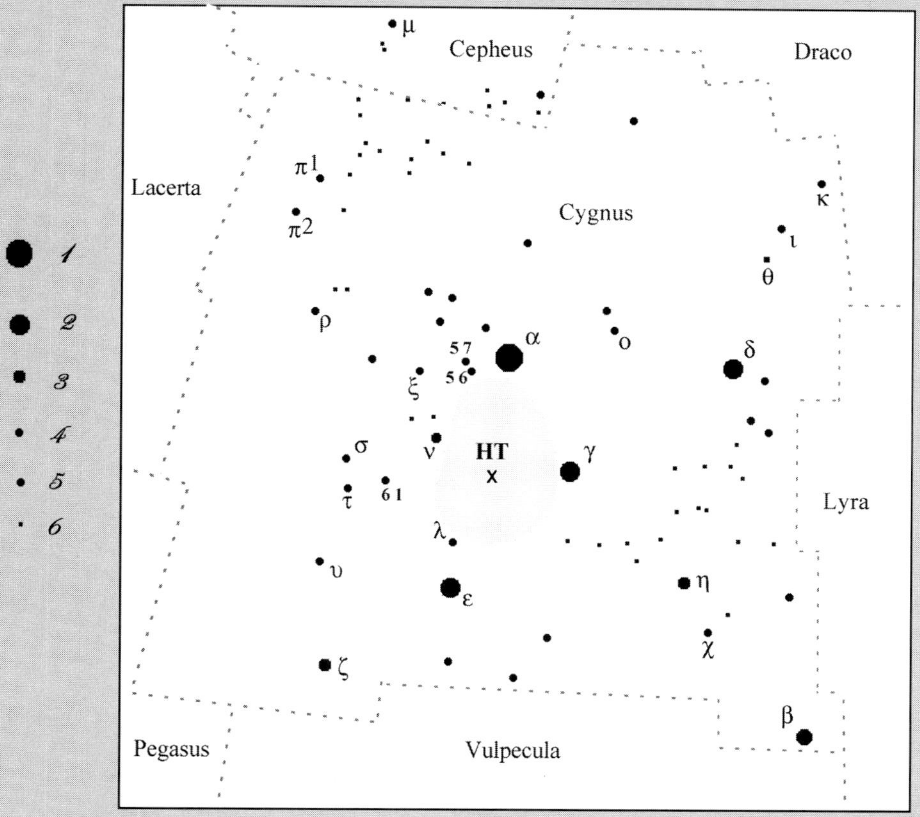

© Stephen James O'Meara

Tirion: Charts 3 & 9

Uranometria: Chart 56

102

Northern Coalsack
Type: Dark Nebula
Con: Cygnus

RA: $\sim 20^h\ 40^m$
Dec: $\sim +41°\ 00'$
Mag: –
Diam: $\sim 480' \times 300'$
Dist: ~520 light-years
Disc: Known since antiquity

HERSCHEL: None.

NGC: None.

I WISH I STILL HAD IT: ONE OF THE star charts I used in the early 1960s to learn the sky. On it was an object that both intrigued and mystified me. It is called the Northern Coalsack, and it resides in the Cygnus Milky Way. But I wonder how many amateurs are aware of it today. For some reason, the Northern Coalsack (at least in name) has been banished from almost all of today's popular star charts and atlases; I know of only one (*The Millennium Star Atlas*) that still refers to it, though it does not show the nebula's outline. The reason may be that confusion abounds over the Northern Coalsack's precise identity and location.

Part of the problem may be that any dark nebula visible to the naked eye was, at one time, called a "Coalsack." The name was created by the bold mariners who sailed the southern seas centuries ago. The first void to be named was the Coalsack next to the Southern Cross, and it was looked upon with superstitious wonder.[1] This dark cloud, sailors believed, possessed some occult connection with the Cross, so whenever they chanced upon it, they would "cross themselves" and "shudder." Little wonder then that the first Coalsack to draw astronomers' attention in the Northern Hemisphere was that deep, dark "shadow" of the Northern Cross.

"You see that dark opening in [the galaxy] near the Swan?" Swithin St. Cleeve says to his lover Lady Constantine. "There is a still more remarkable one south of the equator, called the Coal Sack, as a sort of nickname that has a farcical force from its very inadequacy." These words, from Thomas Hardy's 1882 novel *Two on a Tower*, are part of a fantastic monolog in which St. Cleeve, while pointing to the Northern Coalsack, tries to explain the unfathomable size of the universe. "Those are deep wells, for the human mind to let itself down into."

If you're wondering where this "dark opening" might be, do not fear, its position has been dutifully recorded by several astronomy popularizers over the last century and more. In the 1923 edition of *Astronomy with an Opera Glass* (first published in 1888), Garret P. Serviss includes this passage in his discussion of Cygnus the Swan: "Between the stars (α), (γ), (ε), is the strange dark gap in the galaxy called the Coal-Sack, a sort of hole in the starry heavens. Although it is not entirely empty of stars, its blackness is striking in contrast with the brilliancy of the Milky-Way in this neighborhood." And William Tyler Olcott gives this description in his 1911 *Star Lore of All Ages: A Collection of Myths, Legends, and Facts Concerning the Constellations of the Northern Hemisphere*: "Between (α), (γ), and (ε) Cygni is one of the vacant spaces in the Milky Way, a black and seemingly bottomless abyss, the brink over which man peers into the profound and mysterious depths of interstellar space. This wonderful region in the sky is known as 'the Northern Coal Sack'."

The reason the Northern Coalsack's identity has been a bit fuzzy of late (beside the fact that for more than a half century it has not been plotted on star charts) is that Cygnus harbors several naked-eye coalsacks; but only one of them – the one between Alpha, Gamma, and Epsilon – carries that proper name. The dark nebula most frequently mistaken for the Northern Coalsack is an intense, 5°-wide ink spot some 8° north-northeast of Alpha Cygni (Deneb), close to the Cepheus border (centered at right ascension: 21^h 06^m; declination $+52°$ $30'$). Another is B168 near the Cocoon Nebula (Caldwell 19). Some observers wrongly believe that the Northern Coalsack and the Great Rift are synonymous; they're not – at least not entirely. And some have confused it for Lynd's 896 – the 896th object in the 1962 *Catalogue of Dark Nebulae*; that object, however, is a tiny ($2'$-wide) opaque cloud centered at right ascension 20^h 41.9^m and declination $+39°$ $41'$, near the Northern Coalsack's visual center.

[1] The great Coalsack of the south was known to the ancient Aborigines and was an important figure in their myths and legends. For a more detailed history of dark nebulae, see my write-up of the Coalsack (Caldwell 99) in *Deep-Sky Companions: The Caldwell Objects*.

The Northern Coalsack has also been misidentified as Barnard 348 and Barnard 349 – two small dark nebulae also *within* the Northern Coalsack. In his 1927 *A Photographic Atlas of Selected Regions of the Milky Way*, Barnard identifies B348 as a narrow, 1°-long, dusky lane that "curves west, and then north around the star B.D.+41°3799 ($6^{m}.7$)." Barnard 349 is an even smaller commalike dust cloud, immediately south and west of the Pelican Nebula. He describes it as being a small (6′-long), curved, dusky lane, "elongated north and south, with dusky extension running 15′ northwest. A short line of [faint] stars close west." These features are but threads of darkness in the overall fabric of the larger hidden treasure you seek. Modern observers have literally missed the forest for the trees.

The Northern Coalsack has been known as a naked-eye object since antiquity; it appears, albeit indirectly, in the myth of Cygnus the Swan. In one story (and there are many), Cycnus [*sic*] – a brother of Phaethon (the mortal son of Helios, the Sun god) – let Phaethon test-drive the Sun chariot across the sky. It was a mistake. Phaethon, the proverbial impetuous youth, lost control of the steeds and became the world's first reckless driver. As the steeds roamed wild, the Sun threatened to scorch the Earth and its inhabitants. Zeus put an end to Phaethon's rakishness by striking him down with a lightning bolt. Although the Sun chariot righted itself, poor Phaethon was burned to a crisp; his charred remains fell from the sky like a meteor and plunged with a hiss into the river Eridanus. Heartbroken over the loss of his friend, Cycnus dove into the river, retrieved Phaethon's remains, and gave him a proper burial. Moved by such devotion, Zeus turned Cycnus into a Swan and changed his name to Cygnus. He positioned Cygnus high among the stars of the Milky Way, so he could look down upon the river where Phaethon fell. Today, when we look up at the Northern Coalsack, we see the very point at which Phaethon lost control of the Sun chariot, for it marks the northern end of the scorched path of Phaethon's disastrous ride – a feature we know today as the Great Rift, a long lane of obscuring dust that extends from Cygnus to Centaurus.

I have always been fond of the Northern Coalsack. It epitomizes the great unknown. When I looked into its depths as a child, it caused my mind to stir, to awaken to a sad reality: that there are things in this universe not meant for human eyes to see. It created a sense of helplessness, knowing that I could never peek behind that veil. What wonders lie in wait? Here was an object as impenetrable as the rock (molten or otherwise) between Earth's core and its crust. It's a secret that lay hidden in a darkened room.

This "window of absolute night," as Serviss calls it in his 1919 book *Curiosities of the Sky*, is a black mirror of infinite depth – something we gaze into with blank eyes and open minds. It is a gathering place for all thought and wonder. "Infinity seems to acquire a new meaning in the presence of these black openings in the sky," Serviss says, "for as one continues to gaze it loses its purely metaphysical quality and becomes a kind of entity, like the ocean." Serviss felt the name "coal-sacks" was hardly descriptive. "Rather," he said, "they produce upon the mind the effect of blank windows in a lonely house on a pitch-dark night, which, when looked at from the brilliant interior, become appalling in their rayless murk."

Serviss notes that the Northern Coalsack is "less perfect" than the one in the south.

"Instead of looking like an absolute void in the sky, it rather appears as if a canopy of dark gauze had been drawn over the stars."

We know today that there is nothing outrageously mysterious about these "voids." They are simply cold, dusty clouds of molecular hydrogen gas that cloak or dim the light of whatever lurks behind them. And while the Northern Coalsack in Cygnus does not appear as dark as its southern counterpart, it is. In a 2003 *Astronomical Journal* article, A. Fresneau (Astronomical Observatory, Strasbourg, France) and colleagues argue that not only are the Northern and Southern Coalsacks located at roughly the same distance, "about 160 ± 50 [parsecs] from the Sun," but they are equally as dark, exceeding "the mean interstellar extinction by more than 0.7 ± 0.1 [magnitude]."

The reason the Northern Coalsack appears less evident than the Southern Coalsack is a matter of contrast. The Milky Way surrounding the Southern Coalsack is one of the brightest and richest swaths of stars, clusters, and nebulosities in the heavens. The Southern Coalsack looks very dark because the Milky Way next to it is very bright. In fact, Sir John Herschel, who believed that the Sun was near (not at) the center of the Milky Way, suggested that the Sun was slightly closer to the south side of the Milky Way, considering the greater brightness of the stars in that direction.

The Northern Coalsack is a true hidden treasure. It has been largely "overshadowed" by the North America Nebula (Caldwell 20), which is ironic because the dark regions just west of the North America and Pelican Nebulae form the Northern Coalsack's eastern border. If you can get away from city lights in late summer and find the time to lay back on a blanket on a warm, starlit evening, the Swan will be flying overhead. But do not look at the stars that populate the Cygnus Milky Way. Instead, relax your eyes and focus on that magnificent gap of dark dust lying between Alpha, Gamma, and Epsilon Cygni. This gap forms the body of the Milky Way's Black Swan. The Swan's wings are comprised of two slightly less intense ellipses of darkness: one lies between Alpha, Gamma, and Omicron^1 (o^1) Cygni (including Barnard 46), while the other lies to the south of the North America Nebula, between the stars Xi (ξ), Nu (ν), and Tau (τ) Cygni. The long neck of the Black Swan is the Great Cygnus Rift that begins just south of Gamma and follows the galactic equator to a point between 35 and 39 Cygni, where you'll find the Swan's head.

Many ancient cultures saw creatures in the dark nebulae, and I can understand why. Creatures formed by dark nebulae have a more three-dimensional appearance to them than, say, the stick figure constellations of the two-dimensional sky. Interestingly, in *Star Names: Their Lore and Meaning*, Richard Hinckley Allen notes that while ancient Greek astronomers saw the stars of Cygnus as some kind of bird, most likely a hen, "Aratos described it as 'dark,' especially as to its wings." It seems then that Aratos is the first person known to call attention to the Northern Coalsack. Curiously, Hipparchos, believing that Aratos had made some sort of verbal error, "corrected" the situation.

The Northern Coalsack, like a petroglyph, is the dark imprint of human imagination. When I look at it with modern eyes, I see a "swallow" or "purple martin." The question is what fantastic creature does your mind chisel out in this region?

Finally, sticking to our treasure theme, I'd like to return your attention to the second most popular Coalsack in Cygnus: the 5°-wide ink spot some 8° north-northeast of Alpha Cygni (see the photograph). Here we have the negative imprint of the pirates' Black Flag, the one with the grinning skull and cross bones, the infamous Jolly Roger. Can you see these features with the naked eye?

Hidden Treasure 103
NGC 7008

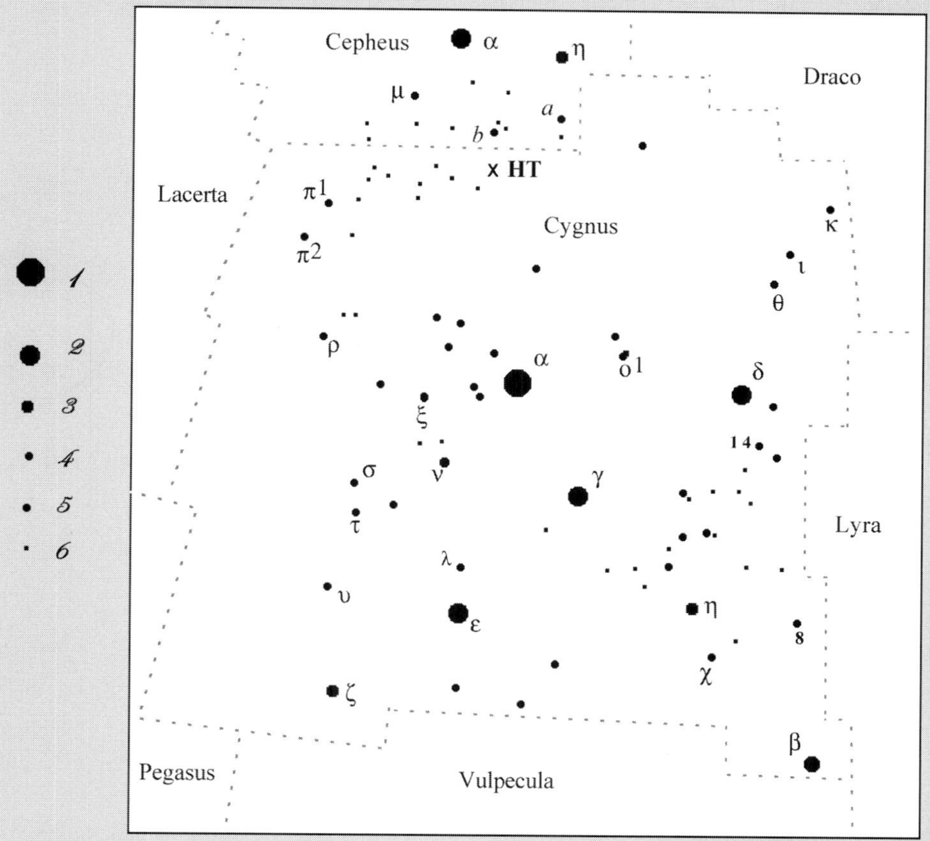

© Stephen James O'Meara

Tirion: Charts 3 & 9

Uranometria: Chart 56

103

Coat Button Nebula, Fetus Nebula
NGC 7008
Type: Planetary Nebula
Con: Cygnus

RA: 21^h 00^m 32.8^s
Dec: +54° 32′ 35″
Mag: 9.9 (O'Meara); 10.7
Dim: 98″ × 75″
Dist: 2,700 light-years
Disc: William Herschel, 1787

W. HERSCHEL: [Observed October 14, 1787] Considerably bright, irregularly faint, 3′ length, 2′ breadth, nebulosity. (H I-192)

NGC: Considerably bright, large, extended roughly toward position angle 45°, [mottled], double star attached.

NGC 7008 IS A BRIGHT AND PECULIAR planetary nebula all but lost in the "other" coalsack of Cygnus (see Hidden Treasure 102) – a 5°-wide dark nebula some 8° north-northeast of Alpha (α) Cygni (Deneb). The planetary lies in the non-descript northern shores of this black lagoon, less than 1° from the Cepheus border.

When William Herschel found the object during a survey sweep in 1787, he classified it not as one of his planetary nebulae but as a bright nebula. He must have been particularly fond of this curious object because he returned to it two more times.

Based on spectroscopic data obtained with the 72-inch reflector at the Asiago Astronomical Observatory at Cina Ekar, Italian astronomer F. Sabbadin (Asiago Astronomical Observatory) and colleagues describe the main part of the nebula as being "very inhomogeneous," consisting of two coaxial prolate spheroids (ellipses whose polar radii are greater than their equatorial radii, so the spheres are "pointy" rather than "squashed"). The internal ellipse is brighter than the surrounding elliptical crown, which is expanding at a velocity of 11 kilometers per second; the inner shell is expanding twice as fast. Assuming that NGC 7008 lies at a distance of 2,700 light-years, it has a true physical extent of 1.3 × 1.0 light-years.

Herschel was right, the "stars" superimposed on the nebula are indeed very interesting features. One of them is, in fact, the 13th-magnitude central star, which appears

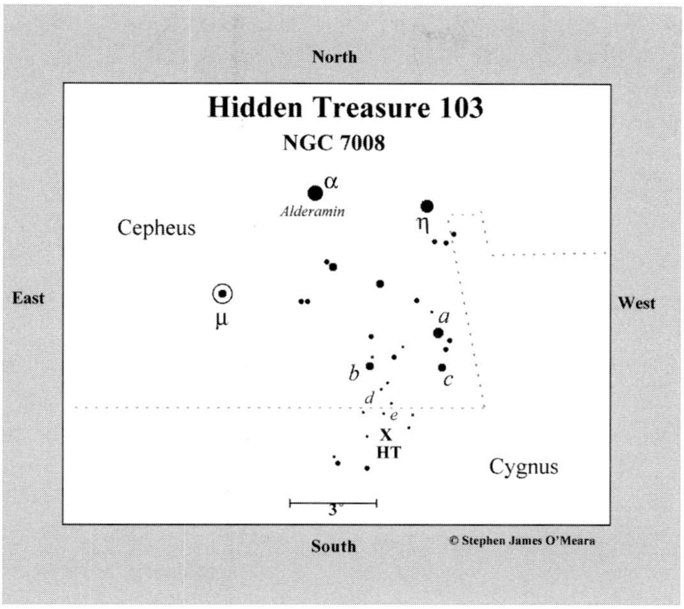

offset slightly from the center to the west. Spectral observations of three of the other "stars" reveal them to be condensations expanding with the surrounding shells. One of them, called Condensation A, is a star-like knot (5″-wide) about 22″ northwest of the central star. Lubos Perek and Lubos Kohoutek classify it as a distinct planetary nebula (K 4–44, 93 +5°1) in their 1967 *Catalogue of Galactic Planetary Nebulae.* But, as Sabbadin and colleagues' data have shown, it appears to be only a bright knot inside NGC 7008 – a possibility first suggested by M. C. Lortet-Zuckermann in 1976.

To find this peculiar object, it's best to start from the House of Cepheus. First locate 3rd-magnitude Eta (η) Cepheii, the west cornerstone of the House. Use your eye or binoculars to sweep about $4\frac{1}{4}°$ due south to a 4th-magnitude star (*a*). Star *a* is the northernmost star in a $2\frac{1}{2}°$-wide triangle with 6th-magnitude Stars *b* and *c*. You want to focus on Star *b*, which marks the southeastern corner of a roughly 50′-wide triangle with two slightly dimmer stars. Now look about $1\frac{1}{2}°$ to the south-southwest for a 1°-wide trapezoid of 7th-magnitude suns (*d*). NGC 7008 is 40′ south and a tad east of the southwesternmost star (*e*) in the trapezoid.

At 23× in the 4-inch, the planetary is kissing a 9th-magnitude star to the south–southeast. This star also has a 10th-magnitude companion 18″ to the south. Seen at a glance at low power, the planetary and these stars form a very small and fuzzy double, like M40 in Ursa Major. So it is easy to sweep over. At 72×, the planetary comes to life, being a very beautiful and charismatic object – a chiffon skirt of light clinging to the 9th-magnitude sun.

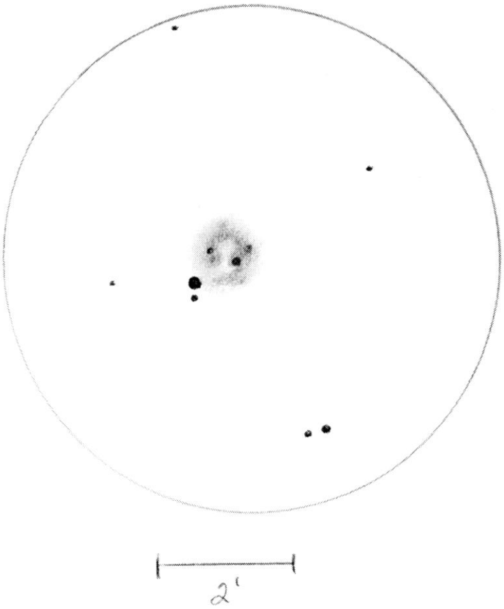

The longer I stare at it the more fascinating it becomes. The nebula appears mottled. It scintillates like sunlight on water and beckons for more power. At 101×, the planetary appears as an elliptical mass, pinched in the middle. It's oriented northeast–southwest and has two bright bowtie like knots at the ansae. The shell is dappled with starlight or glowing beads of light that form a misshapen, boxy annulus. Given the visual impression it's understandable as to why William Herschel classified this irregular object as a bright nebula rather than a planetary. But modern images tell a different story. The object is clearly a bright ellipse of light with an irregular ring measuring 65″ × 52″ and oriented north-northeast–south-southwest. It is surrounded by a fainter, irregular crown 98″ × 75″. Using more magnification on the object in a 4-inch is almost useless, since the nebula starts to fade into the background and I lose any sense of beauty. Of course, such high powers should be used, especially if you want to see the central star as a distinct object.

Using an 8-inch f/10 Schmidt-Cassegrain telescope, Central Coast (California) Astronomical Society member Kent Wallace observed NGC 7008 at 100× and had a similar experience, seeing the nebula being fan-shaped spreading out from the double star. He also notes that, as with many planetaries, NGC 7008 is enhanced by the use of an O-III filter, and it shows well with an ultra-high-contrast (UHC) filter. He found no response to the H-β filter. In a 2001 issue of *Amateur Astronomy* magazine, Eric Honeycutt called NGC 7008 the "Fetus Nebula," due to its appearance in his 22-inch telescope. The name is beginning to stick. Of the object's appearance, he writes, "The Fetus Nebula is a [planetary nebula] that has absolutely striking detail. The fetus shape is very obvious, even at low powers. [The c]entral star is an easy direct vision object (13.2v) and is near [the] 'abdomen region.' The [planetary] is very large in the 9[-mm] Nagler at 288×. [Two] medium-brightness stars are also within nebula; one as an 'eye' of the fetus and the other at the rear. [Two] stars of roughly 11th [magnitude] reside on [the north] side of [the planetary]. [G]reat object."

Since I already have a Fetus Nebula of sorts in this catalog – NGC 1333 (Hidden Treasure 15), the Embryo Nebula, I have decided to call NGC 7008 the Coat Button Nebula, based on its photographic appearance, which is reminiscent of the large buttons on a winter's coat, the irregular shape of the annulus's center also looks like two button holes, through which one needles the thread.

Hidden Treasure 104

NGC 7027

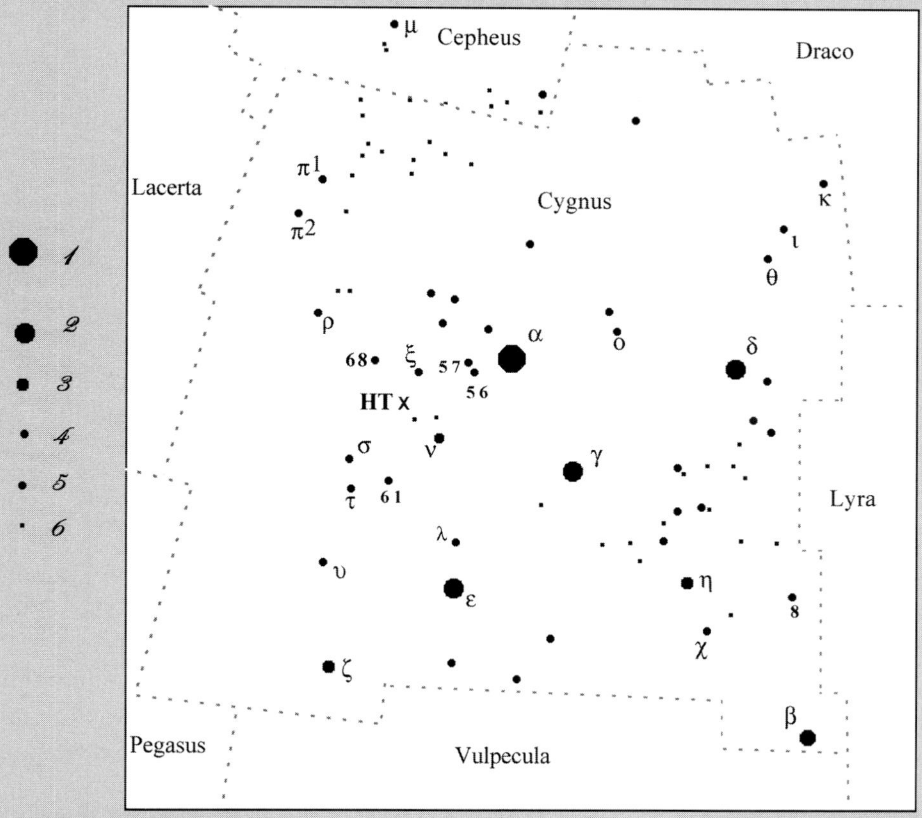

Tirion: Chart 9

Uranometria: Chart 85

104

Pink Pillow Nebula, Sugar Pops Nebula, Green Rectangle
NGC 7027
Type: Planetary Nebula
Con: Cygnus

RA: $21^h 07^m 01.7^s$
Dec: $+42^\circ 14' 10''$
Mag: 8.5 (O'Meara); 8.5
Dim: $18'' \times 10''$
Dist: 2,100 light-years
Disc: Edouard Jean-Marie Stephan, 1878; independently discovered by Rev. Thomas W. Webb in 1879

HERSCHEL: None.

NGC: Planetary nebula, stellar magnitude 8.5.

NGC 7027 IS ONE OF THE BRIGHTEST, smallest, most unusual, and arguably *the* most fascinating planetary nebula in the night sky. It is, hands down, *the* most extensively studied – both observationally and theoretically. As Steven J. Hynes notes in his 1991 book *Planetary Nebulae*, NGC 7027 has inspired more papers in professional journals than any other object of its kind. It has been at the forefront of planetary-nebula research in almost every region of the spectrum, and these studies have helped to advance our understanding of the fundamental processes involved in the creation of these intriguing objects.

Although NGC 7027 is bright (magnitude 8.5), it avoided detection until 1878 when the renowned French astronomer Edouard Jean-Marie Stephan (1837–1923) spied it with the 31-inch reflector at Marseille Observatory – one of the first large reflectors to use a silvered-glass mirror. NGC 7027 was one of many new nebulae Stephan discovered with that instrument (see Hidden Treasure 35). Alas, despite all of the attention NGC 7027 receives in the professional community, it is one of the least observed planetaries in amateur circles. And while NGC 7027 is so bright that it can be spied in an opera glass (if one knows where to look), amateurs are more eager to seek out Stephan's most popular find: Stephan's Quintet (Arp 319) – a gaggle of five dim galaxies about $\frac{1}{2}^\circ$ south-southwest of the stunning spiral system NGC 7331 (Caldwell 30) in Pegasus.

Actually, Stephan's discovery of NGC 7027 is a bit ironic. In 1873–1874, Stephan had used the 31-inch reflector at Marseille in the first attempt to measure the apparent diameter of stars. Although he was unsuccessful (he could only prove that they were smaller than 0.16″), his later discovery of NGC 7027, proved to be, in fact, that of a star's outer atmosphere, which is indeed measurable. Of course, at the time Stephan did not know the true nature of the nebula he had discovered.

The Rev. Thomas W. Webb – whose classic handbook, *Celestial Objects for Common Telescopes*, is to this day a handy reference guide for amateur astronomers – independently discovered NGC 7027 on November 14, 1879: "Planetary, like a [magnitude] 8.5 star, about 4″, found by me independently." This discovery is quite remarkable, since Webb notes that "I had at one time projected a survey of the wonders of [Cygnus] with a sweeping power; but want of leisure, an unsuitable mounting, and the astonishing profusion of magnificence, combined to render a task hopeless for me which, I trust, may be carried through by some future observer." Even he was not aware of what a wonder he discovered. His calling it a "planetary nebula" had no astrophysical understanding behind it. He was simply classifying it in the same manner of William Herschel before him. What he classified was simply a round nebula shaped like a planet.

Peter Collins, one of the world's greatest visual nova discoverers, introduced me to NGC 7027, in the early 1970s. He showed it to me through the 9-inch f/12 Clark refractor at Harvard College Observatory in Cambridge, Massachusetts. At the time, astronomers were just beginning to understand the nature of this enigmatic beast. A decade earlier, astronomers had begun using radio telescopes to study planetary nebulae, which turned out to be powerful radio sources, and NGC 7027 was one of the brightest. NGC 7027's radio emissions were so intense that astronomers determined it was one of the densest, and therefore youngest, planetary nebulae known – about a thousand years young. And while NGC 7027 is too dim to be noticeable with the unaided eye, the nebula formed coincidentally around the same time as the ancient skywatchers chronicled in 1006 the appearance of a supernova in Lupus. This coincidence, as you will soon see, is also ironic, because the more professional astronomers studied NGC 7027, they more they learned about the ultimate fate of stars more massive than our Sun.

One early finding, in 1967, was that NGC 7027 is not only a strong radio source but also a strong emitter of infrared radiation, an indication that NGC 7027 is, and probably all planetary nebulae are, very dusty places – though universal acceptance of that truth had to wait until the launch of the Infrared Astronomical Satellite in January 1983. The spectrum of NGC 7027 is rich atomic lines and complex molecules, including some complex organic compounds. But while the nebula was steadily giving up its secrets, NGC 7027's central star remained elusive to study, until the Hubble Space Telescope (HST) imaged NGC 7027 in 1997 with its near-infrared camera and multi-object spectrometer (NICMOS).

As principal investigator William B. Latter (SIRTF Science Center/IPAC/Caltech) and colleagues describe, NICMOS clearly revealed the central star, which has a temperature of some 200,000 °C, making it one of the hottest stars known. It sits at the center of a white ring of gas and dust cast off by the star. This region is highly ionized – atoms are ripped apart into nuclei and electrons by radiation from the hot central star. This ring

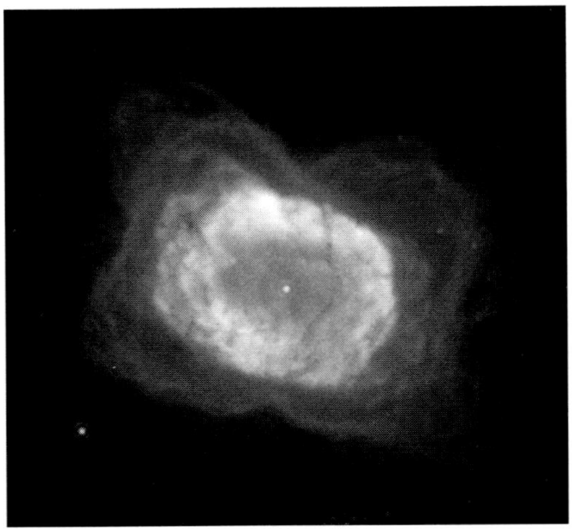

has a temperature of several tens of thousands of degrees. The ring, which is reminiscent of M57, the Ring Nebula in Lyra, is surrounded by a fluorescent pink pillow of molecular hydrogen. Here is the boundary layer between the ionized shell and the surrounding molecular envelope; the region glows pink because the molecules here are being violently split into separate hydrogen atoms by the stellar ultraviolet radiation. The pillow is actually two cones of light seen at an angle of 45°. Another interesting feature of these data is the appearance of a disturbance that might be caused by an as yet unseen jet of material that is inclined from the main axis of the object.

Prior to the HST observations, NGC 7027 was believed to be a proto-planetary nebula (one with a central star still too cool for it to ionize the core region), but we now see it as a planetary nebula in the early stages of evolution. Indeed, the nebula is exceptionally small. While most planetary nebula we see in the sky extend about 1 light-year in true physical extent, NGC 7027 measures only 0.2×0.1 light-years. The central star has a mass of about 0.7 Sun and appears to have left the asymptotic giant branch of the Hertzsprung–Russell diagram about 700 years ago, which is a mere breath in the lifetime of a star.

The gas in the circumstellar envelope is mostly made up of simple molecules such as molecular hydrogen and carbon monoxide, combined with several other gases such as cyanide, sodium chloride, and possibly water vapor. Complex hydrocarbon molecules are also known to be present in circumstellar envelopes. Most importantly, the material cast off during this phase of stellar evolution includes a large abundance of the key elements to the origin of terrestrial life – carbon, nitrogen, and oxygen. These elements are created through nuclear fusion of hydrogen and helium in the stellar core.

Since planetary nebulae were believed to be formed from the remnants of circumstellar envelopes of asymptotic giant branch stars, the discovery of molecules in NGC 7027's envelope (and in those of other planetary nebulae) is important; these molecules have the same spectral characteristics as those detected in well-studied asymptotic giant branch stars, such as CW Leonis.

The estimated molecular mass in the expanding halo of NGC 7027 is about 3 Suns, or about 100 times the mass contained in the ionized shell. That discovery is of vast importance. In 1930, Subrahmanyan Chandrasekhar applied Einstein's theory of special relativity to the structure of white dwarfs and demonstrated that the maximum mass for white dwarfs is 1.4 solar masses (known as Chandrasekhar's limit). If we accept this limit, a star born with a mass greater than 1.4 solar masses could not become a white dwarf; it would have to end its life in a supernova explosion. If true, supernovae should be a frequent occurrence in our galaxy. But they are not. Clearly, something is amiss.

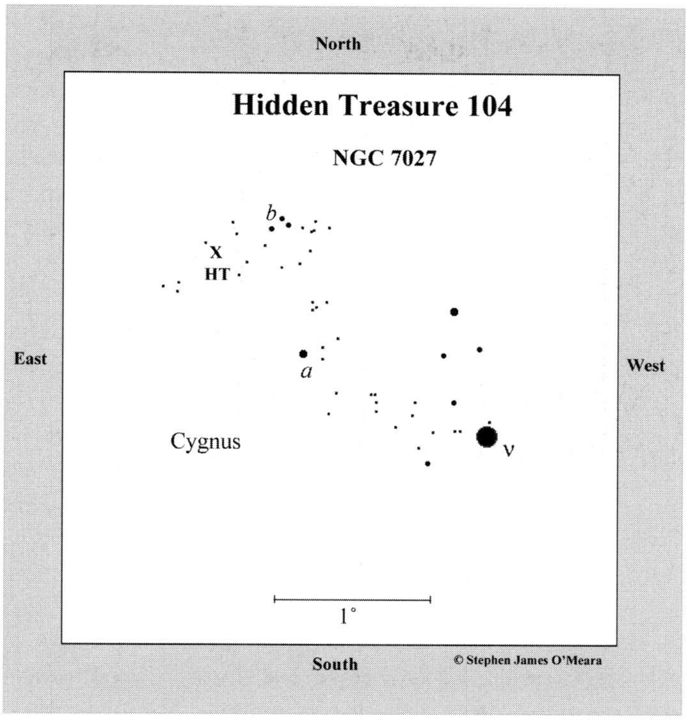

The combined mass of NGC 7027's central star and that of its molecular envelope implies that this planetary nebula must have descended from a star with an initial mass of at least 4 solar masses. These data and others became essential evidence in a theory that stated stars born with masses several times higher than the Sun will not become supernovae, but end their lives as planetary nebulae instead. "While stars maintain a constant mass through most of their billion-year lifetime," Sun Kwok explains, "they undergo a rapid weight loss in the last one million years of their lives. The last minute weight loss allows a star to avoid a violent death as a supernova. In spite of the glamorous status of supernovae and black holes in the minds of many people, they are relevant to the lives of only a small fraction of all stars. Planetary nebulae, not supernovae, are the fate of most stars. Unlike old soldiers, old stars do die. But, like them, most just quietly fade away."

Even though little of this was known to me at the time I first saw NGC 7027, I had much to think about as I peered into the eyepiece and listened to Peter describe the significance of the exciting early radio and infrared discoveries. I was soon swimming in a visual mire. The swollen disk of NGC 7027 was not so much a visual wonder, at least not at the time, but one with a secret and seductive invisible nature. Now when I look at NGC 7027, I have an even greater appreciation for the significance of the scene. As Latter says, "We have caught NGC 7027 at a very important time for study . . . This is an object that will survive in its present state for only the blink of an eye (in cosmological time)."

To find this tiny marvel, start at Alpha (α) Cygni (Deneb). Just 5° to the southeast is 4th-magnitude Nu (ν) Cygni. A short 1° hop to the northeast brings you to a 1°-wide acute triangle of three 6th-magnitude stars pointing to the southeast. Center the southeast apex of that triangle (Star *a*), and move 50′ to the north-northeast, to a roughly 7′-wide triangle of 8th- and 9th-magnitude suns (*b*). NGC 7027 is 30′ to the southeast of that triangle. The planetary is absolutely brilliant and easily detectable in 7 × 50 binoculars. It is flanked to the northwest and south-southeast by two pairs of stars. The nebula shines alone and is the brightest object between these pairs.

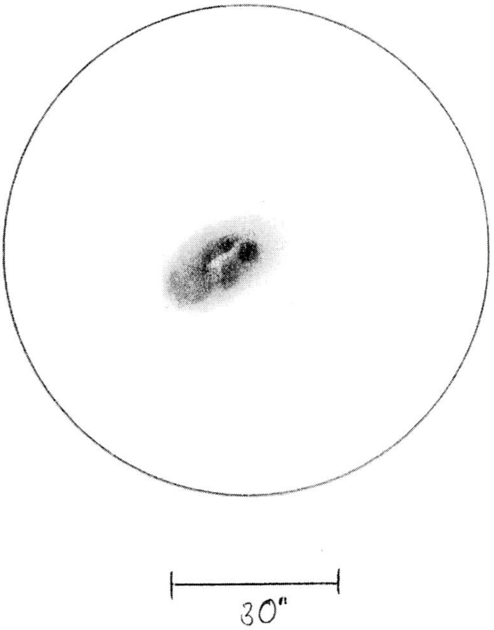

At 23× in the 4-inch, the nebula is a slightly swollen "star" with a blue-gray tint, a delicate ashen shade, almost stone-cold slate, a fitting color for this near-death star. Although all manner of moderate magnifications will make the planetary disk more obvious from the surroundings, high powers are needed to make sense of it. Moderate powers give the illusion that the planetary has a bright central star surrounded by an asymmetrical crown of light. But the longer I look at it at moderate power, I get the distinct feeling that something is wrong: that there are two objects here.

High power (the highest possible), in my case 504×, reveals a most puzzling sight. The planetary looks like a dumbbell oriented north-northwest–south-southeast, with the north-northwest side being the brightest; that bright section also contains a bright knot not centered on the nebula but on the northwest edge. The south-southeast part of the dumbbell is mystifying, looking like an emerging puff of smoke or an unfinished afterthought. I wondered, however, if this was nothing more than a slightly swollen disk of a dim star nearly touching the nebula. The entire dumbbell also appears to be surrounded by a dim glow, but this is so tenuous that I cannot, or was not able to confirm its existence to my satisfaction.

But the most fascinating aspect of the bright north-northwest section, especially after many long minutes of observing at high power is that it breaks up into a cruciform of knots of various brightnesses. Together, these knots form an annulus or appear to do so. But the knots do not stand out so clearly from the nebulosity. And, during times of imperfect observation, all the knots merge and separate in a turbulent dance, like a *do-si-do*. When connected, the annulus looks boxy – a Southern Cross asterism in a donut. Finally, now, I have to add the odd extension to the south-southeast.

The knots, I found, can be followed into bright twilight. So, if you are having difficulty seeing them, observe the group, as I did, in the early morning and let the brightening sky wash away the nebulous glow and help you see these fine details.

NGC 7027 has a variety of nicknames. The first two mentioned in the table are mine, and apply to the NICMOS and wide-field planetary camera (WFPC) color images, respectively, taken by the HST. The nebula has two other popular names, the Green Rectangle and the Magic Carpet Nebula. Both were given by Kent Wallace (Navaho Flats). The Green Rectangle is the only one based on the object's visual appearance through a 20-inch telescope at high powers. I also find that name appropriate for high-power views through small telescopes. See what you think.

Hidden Treasure 105

Trumpler 37

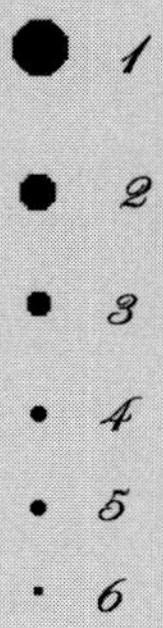

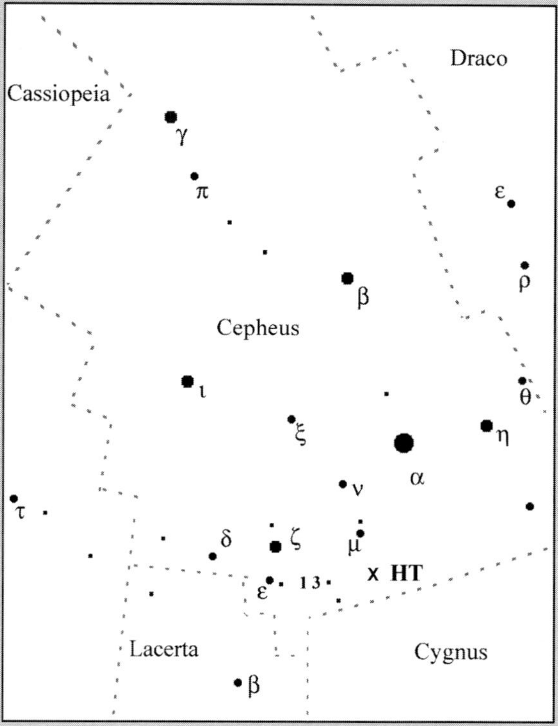

© Stephen James O'Meara

Tirion: Chart 3

Uranometria: Chart 57

105

The Misty Clover Cluster
Trumpler 37 (Cluster)
Type: Open Cluster
Con: Cepheus

RA: 21^h 39.0^m
Dec: $+57°$ $30'$
Mag: 3.8 (O'Meara); 3.5 cluster
Diam: 90.0′, cluster
Dim: 170′ × 140′, nebula
Dist: 2,600 light-years
Disc: Trumpler, 1930; E. E. Barnard discovered the nebula presumably on his 1890 images of the Milky Way

HERSCHEL: None.

IC: Nebulous part of Milky Way.

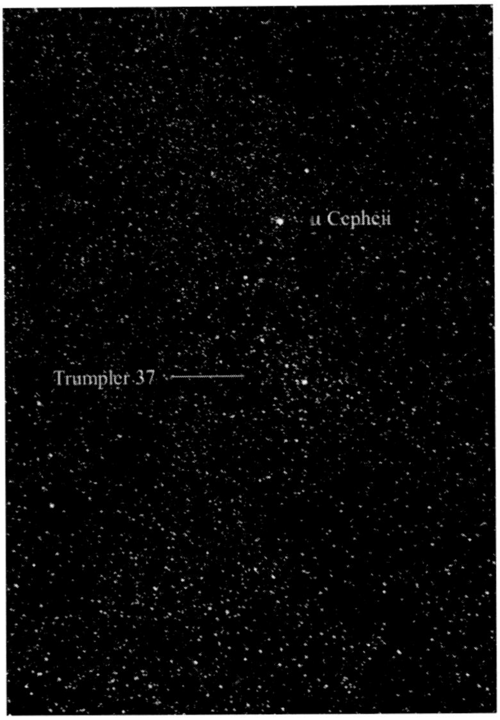

CEPHEUS, THE KING, IS RICH IN celestial treasures. Riding high on the crests of the northern Milky Way, the constellation is home to a variety of galactic and extragalactic wonders, including NGC 188 (Caldwell 1), one of the oldest open star clusters in the heavens; the beautiful reflection nebula NGC 7023 (Caldwell 4); the hauntingly dark and elusive Cave Nebula (Caldwell 9); the bright mixed spiral galaxy NGC 6946; and the visually appealing open cluster NGC 6939. It also contains one of the most famous and important variable stars Delta (δ) Cepheii – the classical example of a high-luminosity star that undergoes very regular physical pulsations with periods of 1 to 135 days – and Mu (μ) Cepheii, Herschel's Garnet Star, one of the most brilliant of all the red-giant stars known and that's perhaps the reddest star visible to the naked eye in the northern sky, where it shines like a spot of blood on dusty black velvet.

This small but exhaustive list is but a single layer of deep-sky riches in the constellation, which contains all manner of nebulous forms and tiny clusters. The irony is that the constellation's largest and brightest deep-sky wonder is also one of the least admired. It is the 3rd-magnitude open cluster Trumpler 37 (Tr 37) and its associated nebulosity IC 1396. What's more, this dynamic duo lies less than $1\frac{1}{2}°$ south and slightly west of Mu Cepheii, which is seen against the northern skirt of IC 1396. The cluster is visible to the naked eye under a dark sky as an enormous round glow three Moon diameters in extent. It looks like a blossoming flower in a moonlight garden covered in frost.

Trumpler 37 and IC 1396 are part of an extended H-II region situated near the galactic plane at the core of the Cepheus *OB*2 Association – an extensive region rich in bright-rimmed clouds (globules) distributed over an area of 3°. The Cepheus *OB*2 Association is divided into two subgroups: the 3-million-year-young Trumpler 37 (480 stars populating 46 light-years of space), and a more dispersed, 6- or 7-million-year-old group (about 75 Type *O*- and *B*-stars spanning 554 light-years of space). The region also contains more than 200 faint Type *F–G* stars with masses below 3 Suns, all of which are in their pre-main-sequence stage. In fact, an analysis of the color-magnitude diagram of the brightest members of Trumpler 37 indicated that low-mass pre-main-sequence stars should exist in the cluster itself.

One thing is for certain, the nucleus of Trumpler 37 (one of the youngest star clusters known in our galaxy) is a wonder unto itself. Here is the star complex HD 206267, whose hottest component, a magnitude 5.6 Type *O*6.5V star, is the main source of excitation of IC 1396; in many respects this complex is similar to the Orion Trapezium, though twice as distant, having three fainter stars: companion B at 1″.6 (V = 13.6), companion C at 11″.7 (V = 8.1), and companion D at 19″.9 (V = 8.0). The main component itself has long been classified as a spectroscopic binary with an orbital period of 3.7 days. Its ultraviolet spectrum also reveals a third eclipsing Type *O* component. The two brightest components have a mass of about 40 Suns and an additional speckle component may also exist.

The surrounding nebula, IC 1396, is quite rich in bright-rimmed globules. Such globules are usually found in presumably old H-II regions and resemble compact and massive areas of H-I interacting with the surrounding neutral and cooler gas. In particular, two bright-rimmed clouds IC 1396A and B, which are centered about 30′ to the west of HD 206267, have been the focus of previous investigations. Near the center of IC 1396A, two very young T Tauri stars (about 100,000 years old) have been identified. Both stars may have been born simultaneously and are still in the pre-main-sequence stage. The central part of this globule may resemble a cavity that is formed by interaction of the stellar wind from the pre-main-sequence stars and the globule.

In December of 2003, the Spitzer Space Telescope (formerly known as the Space Infrared Telescope Facility) penetrated the interior of one of the most famous globules in IC 1396 (the Elephant's Trunk Nebula) and found bright reddish proto-stars within. And in a 2005 *Astronomy and Astrophysics* paper, Irish astronomer D. Froebrich (Dublin Institute for Advanced Studies) and colleagues present data on a large-scale study of the IC 1396 region using new deep near-infrared and optical images, complemented by 2MASS satellite data. For ten globules in IC 1396, the researchers identified the young stellar population within. Five of these globules contain a rich population of reddened objects, most of them probably young stellar objects. Two new Herbig–Haro were also identified. Using star counts based on 2MASS data, the researchers identified 25 globules whose masses show a significant increase with the distance from the exciting star (HD 206267). They explain this correlation

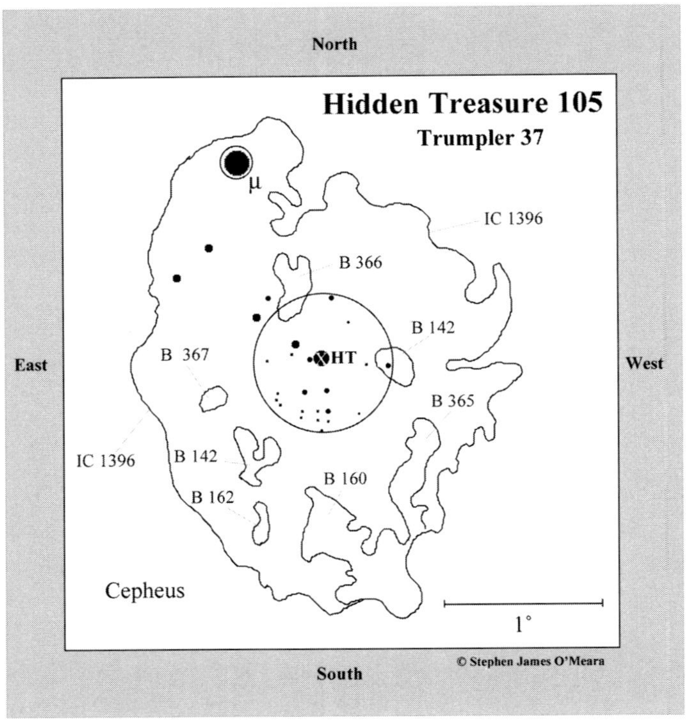

by the enhanced radiation pressure close to this star, leading to evaporation of the nearby clouds and hence smaller globule masses. The radiation from HD 206267 then has a major impact on the star-formation activity in these globules.

To find this stunning region, you simply need to locate 4th-magnitude Mu Cephei, which is about $4\frac{3}{4}°$ southeast of 2nd-magnitude Alpha (α) Cepheii (Alderamin) or $3\frac{1}{2}°$ west-northwest of Zeta (ζ) Cepheii. The cluster/nebula complex is $1\frac{1}{2}°$ south and slightly west of Mu Cep. On one extremely clear December night in 2003, I got the distinct impression that I could see the nebula with the naked eye, but I was not so certain; after all, how does one truly distinguish the nebulous appearance of an unresolved cluster from the nebulosity enshrouding it? The nebula looked a tad more distinct in 7 × 50 binoculars, surrounding an X-shaped asterism of stars, but I had the same nagging doubt.

As for the cluster, the view in the 4-inch is outstanding. The first and most obvious object to strike the eye is the magnitude 5.6 star at the core of the cluster, HD 206267, the striking triple star system otherwise known as Struve 2816. At 23×, it appears only as a double, but it is immediately a triple at 72×. What's most intriguing is that, while the star is classified as a Type *O* star, it appears to have a most definite yellow hue. Randy Moench (Northern Colorado Astronomical Society) had the same impression with a 7-inch Maksutov. As for the two companions, one has a yellowish tinge with a dusty rose sheen, and the other has a slight greenish hue. Moench cuts to the quick and says that "each star in the triplet offers a different color making it a real gem."

Beyond the mesmerizing core, long looping arms extend radially about the cluster, forming petals that give the cluster the appearance of a four-leaf-clover. Add the mistiness of the surrounding nebulosity and you arrive at my coined name for the cluster. It's quite a spectacle, given the richness of the cluster and the Milky Way background light. It's easy to get lost among the masses, but overall, the arms extending from the core mimic the arms and legs of the constellation Hercules, or that of an Alaskan King Crab.

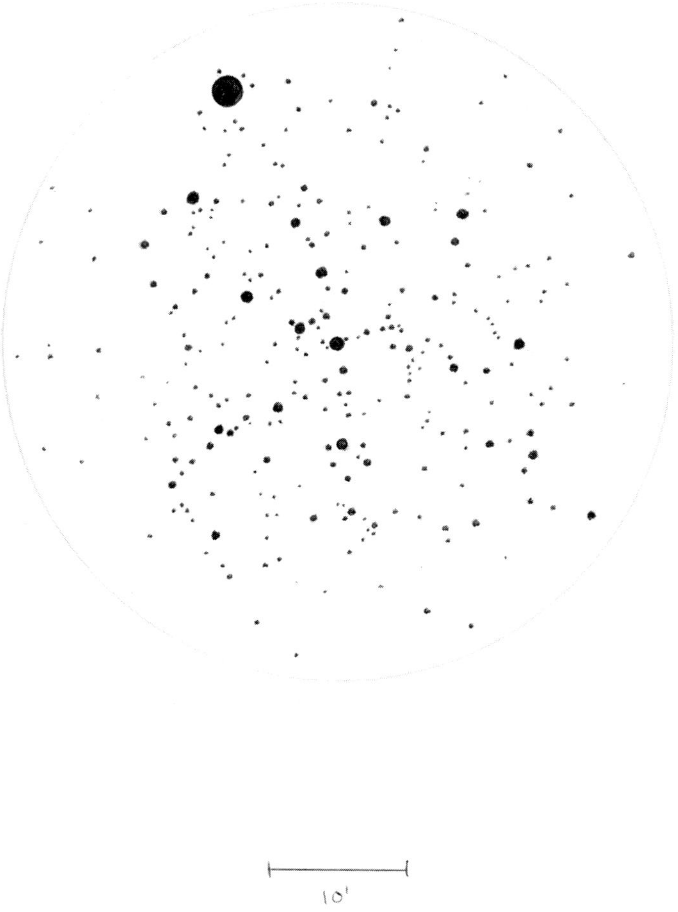

cluster and nebula, one sightseer went so far as to call it a 'magnificent object.' The view of this entire amazing region of sky through the eyepiece of a rich-field telescope leaves no doubt that this is indeed an apt assessment."

I can see a multitude of dark nebulae, without a filter, within the nebulosity. (But remember, this is under a very dark sky and with a nearly 3° field of view.) The dark nebulae become enhanced if I slightly defocus the image then tap the telescope tube gently to set the object in motion, which helps the eye–brain system to detect faint light. The most obvious lagoon of darkness is Barnard 161, located 30′ north and slightly east of HD 206267. Barnard 160, about 1° to the south of HD 206267, is also seen very well. But the entire nebula is covered in an intricate web of dark, dusty cobwebs. Barnard 367 (50′ east-southeast of HD 206267) and Barnard 163 (50′ southeast of HD 206267) appear as two wide arcs of darkness whose boundaries melt into one another in a dim wash of gray. There is, in fact, a lot of gray, smoky nebulosity (not black, not white) here as well.

If you live under suburban skies, try using a filter to enhance the nebula. David Knisley (Prairie Astronomy Club), who used a 10-inch f/5.6 Newtonian reflector, at 59×, 70×, and 141× recommends using an ultra-high-contrast or deep-sky filter which will not

The nebula is a definite telescopic glow – but only in the sense that, after much effort, on the very clearest of nights under a dark sky, an unsuspecting observer might marvel at its subtle grandeur. In a 1999 *Sky and Telescope* magazine, James Mullaney eloquently notes that the "huge nebulous expanse demands a very dark, transparent night and fully dark-adapted eyes. Careful sweeping back and forth across the nebula's position should reveal a faint gossamer glow – it can be seen even in a 3-inch scope at 30×. Various observers have described the nebula as a 'large haze' and 'like a very faint Rosette Nebula.' Referring to the combined only make the nebulosity more visible but also the dark inclusion of Barnard 161 more definite, "but the glow is still faint." Then again, the matter is still up for debate. As Moench says, "[n]o nebulosity was detected in either the wide field 3-inch refractor or the 7-inch Maksutov using Skyview, UHC or O-III filters," though there is no reason for this emission nebula to appear bright in an O-III filter.

Return to Mu Cephei. At 23× in the 4-inch, I was immediately stunned by its gory glory – such a rich red color, like freshly spilled blood; I have seen a leopard in Africa lap up the blood of prey and it is this memory that comes to mind whenever I look at this bleeding star. This may be the largest star visible to the naked eye – it is roughly 2.4 billion miles across. If it replaced our Sun, it would extend beyond the orbit of Saturn. Mu Cep is a slow semi-regular variable that is easily visible to the unaided eye in suburban skies when at its maximum brightness of magnitude 3.4, but may be a little more challenging at its minimum of 5.1. The life expectancy of a star like Mu Cephei is only around a few million years – short for astronomical time scales. Mu Cepheii will eventually explode leaving a supernova remnant with a neutron star or a black hole.

Hidden Treasure 106

NGC 7380

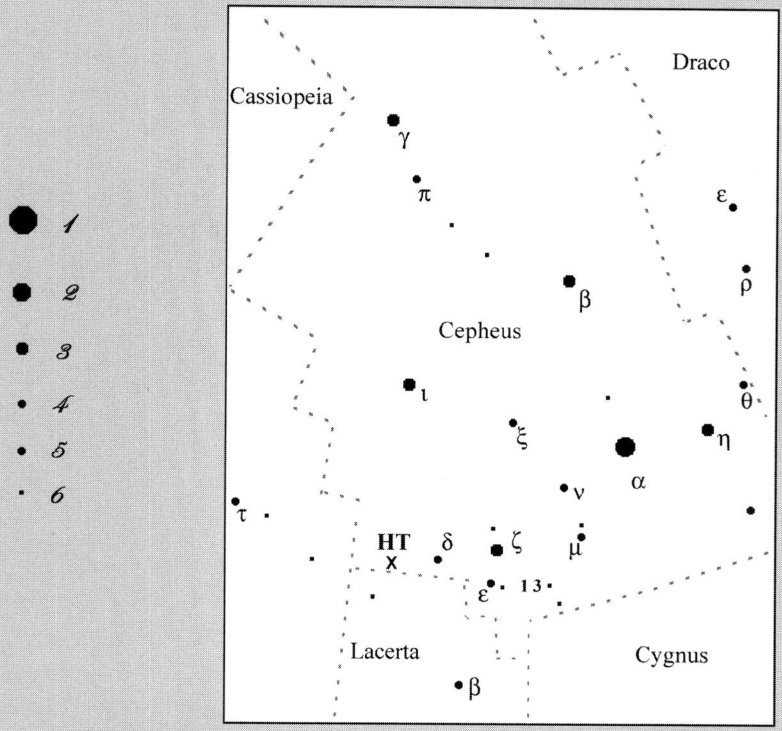

© Stephen James O'Meara

Tirion: Chart 3

Uranometria: Chart 57

106

Harry Potter and the Golden Snitch
NGC 7380 (Cluster)
Type: Open Cluster
Con: Cepheus

RA: $22^h\ 47.3^m$
Dec: $+58°\ 08'$
Mag: 7.4 (O'Meara); 7.2 cluster
Diam: 20.0′, cluster
Dist: 9,700 light-years
Disc: Caroline Herschel, 1787

W. HERSCHEL: [Observed November 1, 1788] A cluster of coarse scattered stars, 8′ diameter, C. H. 1787. (H VIII-77)

NGC: Cluster, pretty large, pretty rich, little compressed, stars from magnitude 9 to 13.

IT WAS A BEAUTIFUL SIGHT. ON THE evening of September 18, C/2002 C1 (Ikeya-Zhang), in a long and flowing evening dress, slipped by Caroline Herschel's open cluster in Cepheus and gave it a kiss. It was a scene straight from Walter Hartright's narrative in Wilkie Collin's novel *The Woman in White*: "In the time when my dread of losing her was most present to my mind, I had always kissed her when she left me at night and when she met me in the morning." It was also a fitting romantic interlude between the two objects Caroline Herschel loved the most – nebulae and comets. Her kinship for these new finds are reflected in a December 22, 1788, letter to the "Rev. Dr. Maskelyne," in which she reports her discovery of a new comet and pens this plea, "I beg the favour of you to take it under your protection." She signed the letter "Your most obliged, humble servant."

When Caroline Herschel discovered NGC 7380 in 1787, things were rather hectic: "The discovery of the Georgian satellites caused many breaks in the sweeps which were made at the end of 1786 and beginning of 1787," she writes in her memoirs, "by leaving off abruptly against the meridian passage of the planet, which occasioned much work." She also noted that the discovery of these satellites "brought many nocturnal visitors to Slough . . . and many times I have listened with pain to the conversation my brother held with his astronomical friends when quite exhausted by answering their numerous questions." Caroline had also begun work on a temporary "index to Flamsteed's observations," and "calculating the beginning and ending of sweeps and their breadth." What's more, her brother Jacob was visiting "from April till October 1787," the 40-foot reflector was being completed, William was preparing for marriage, and Missus Herschel was ill. (She died in January of 1788.) Still, as Caroline writes, "The fine nights were not neglected, though observations were often interrupted by visitors." But the year of 1787 was also a special one for Caroline. It's the year King George III gave her a £50 per year salary to continue as William's assistant, making her the first woman officially recognized for a scientific position. Her discovery of NGC 7380 that year, then, came when she was a salaried astronomer.

To Caroline and her brother William, NGC 7380 was merely a coarsely scattered cluster of stars. But what a marvel of nebulosity surrounds it. NGC 7380 is situated about 4° from the nucleus of the extensive Cepheus *OB*1 Association – an "old" star-forming region 2,000 light-years-wide and 10,000 light-years distant. Although NGC 7380 is young (between 2 and 4 million years) star formation in the cluster either has stopped or is extremely low. Indeed, Infrared Astronomical Satellite (IRAS) data reveal 18 sources indicating that stars may still be forming in the complex. NGC 7380 has 125 members of 9th magnitude and fainter. Most are Type *O* and early *B* stars still embedded in their natal cocoon of dust and gas. The nebula (Sharpless 2–142) is being excited mainly by the 8.6-magnitude eclipsing binary DH Cepheii (HD 215835).

Color images of the region show a turbulent sea of pink gas clouds washed by waves of dark dust whose sharply chiseled crests reveal the eroding action of the region's strong stellar winds. Hydrogen-alpha light imagery of Sharpless 2–142 shows it to be composed of several bright arcuated clouds, rims, and knots interlaced with local dust lanes. Mexican astronomer Carlos Chavarria (Institute of Astronomy, UNAM, Mexico) and his colleagues note in a 1994 *Astronomy and Astrophysics* paper, "Sharpless 2–142 is but a 'blister' on the surface of the more extensive molecular which is seen roughly edge on." Radio observations reveal two components to this cloud: one to the northeast, one to the southwest. The northeastern component is a 'sphere'

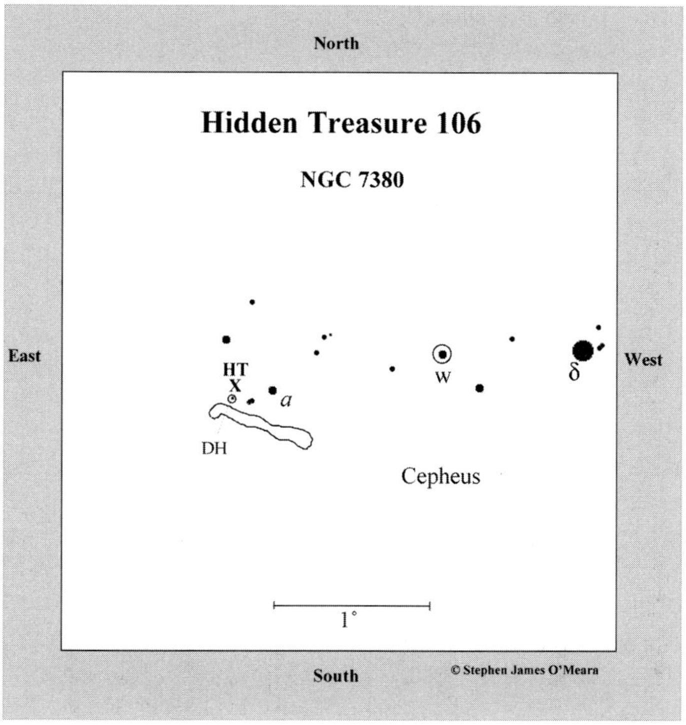

with a roughly 30 light-year-wide diameter containing about 70 percent of the cloud's mass.

To find this almost mystical complex start at the famous variable star Delta (δ) Cepheii, which is also a beautiful double star for small telescopes. As James Mullaney chimed, "Sadly, it's often overlooked as an attractive double due to its notoriety as a variable." The magnitude 6.3 green companion is 41″ distant from the yellow primary. The pair's colors are just as striking, if not more so, than those in Beta (β) Cygnii (Albireo). Of course, the colors you see may be different. Various observers have reported the colors as "pale orange and blue," "blazing yellow and bluish white," and "very yellow and blue." NGC 7380 is simply 2° east of Delta. You can first sweep 1° to the east, where you'll find the 8th-magnitude irregular variable star W Cepheii. Hop a little more than 1° to the east-southeast and you'll see a magnitude 6.5 star (*a*). NGC 7380 is about 15′ east and a little south of Star *a*, nearly centered on DH Cepheii. It's easy to see in 7 × 50 binoculars as a small condensed glow of uniform brightness. And through my antique scope it appears as a delicate glow with a central condensation and faint asymmetries.

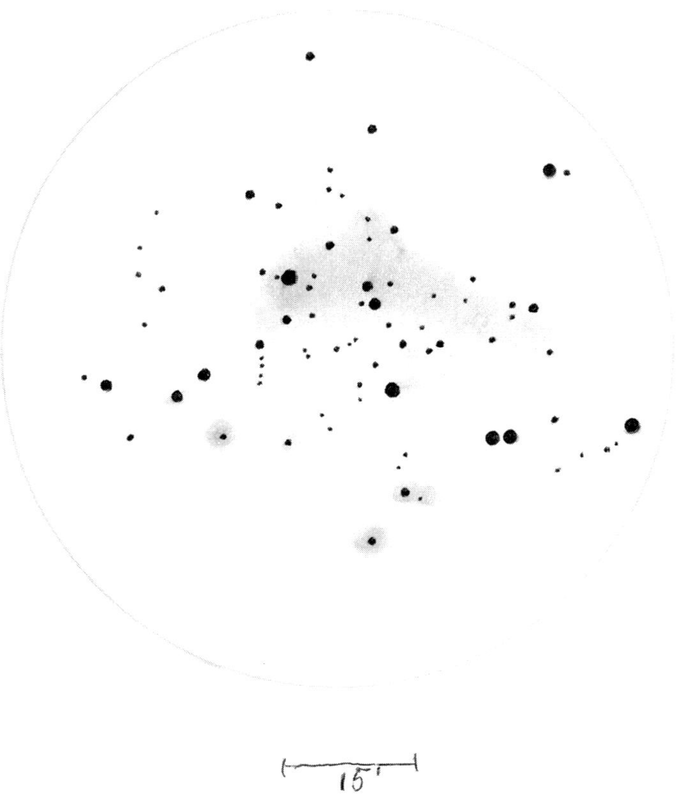

I immediately suspected bright nebulosity within the triangular-shaped cluster at 23×. Mario Motta also noticed the dark nebulosity and saw a broad band slicing east–west just north of the cluster's main body. I was fighting to convince myself that the bright nebulosity was real, because this Milky Way region is so replete with veins of dark nebulosity which cuts up the Milky Way like separate tectonic plates – islands of starlight floating on an ocean of ink. It's almost maddening to fathom. However, the cluster had a ghostly quality to it – a glow that differs ever so slightly from that of neighboring stellar patches. This is something that has to be appreciated from a dark sky.

At a glance, about 50 stars pop out, and my drawing shows about 70 that span nearly 60 light-years of space.

At 23×, the nebula is a hyper fine glow. The brightest section surrounds a triangular-shaped wedge of starlight that lies at the northeast end of an arc of three bright stars – the central one of which is a fine double whose components shine at magnitude 7.7 and 8.6. Overall, it's a pretty area, with lots of dark nebulosity. The most prominent dark feature is a 30′-long lane, oriented east-northeast–south-southwest, that is immediately south of the main arc of three bright stars. As with IC 1396, the entire region is patchy with starlight. If I relax my gaze, I can see a little dipper, or a ladle, comprised of the triangular wedge and a handle of stars to the east.

At 101×, I could detect definite condensed patches of nebulosity around at least four stars. I know these nebulous glows are real because stars of similar magnitude near them do not share the glow. Furthermore, these patches coincide with dense nebulous regions as seen in long-exposure photographs. Most prominent is a knot of nebulosity about 5′ southeast of the double in the bright arc mentioned above. The other patches have been recorded on the drawing shown here. Also notice in the photo the complex arrangement of dark nebulosity, as if it is some form of hieroglyphs.

Finally, try using your imagination to see the brightest stars in the region as Harry Potter on his Firebolt broomstick, playing Quidditch and trying to grab the Golden Snitch.

Hidden Treasure 107

O'Meara 1

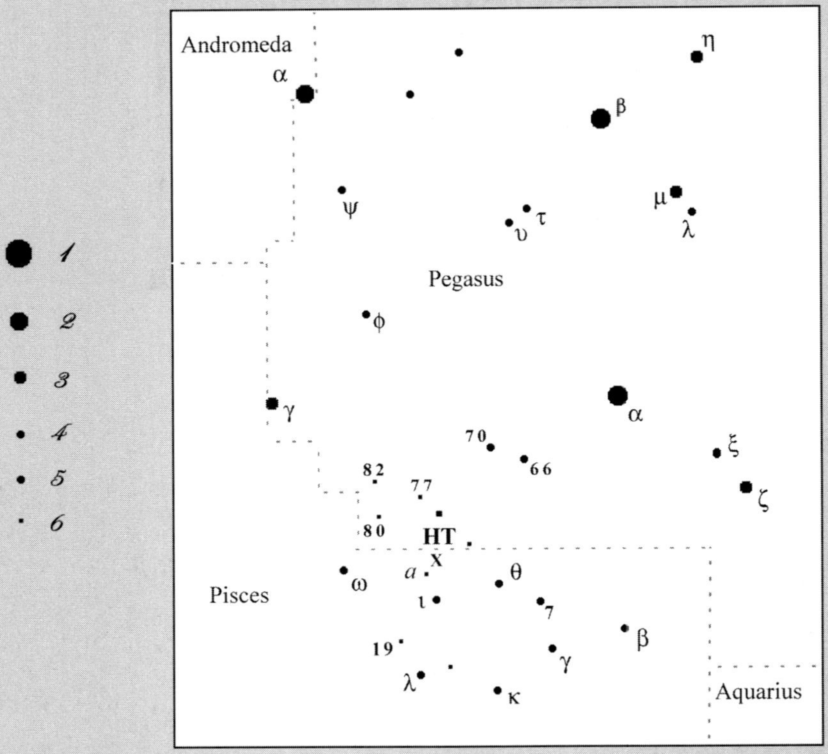

© Stephen James O'Meara

Tirion: Chart 17

Uranometria: Chart 214

107

Little Ladle
O'Meara 1 = Alessi J23407 +0757
Type: Asterism
Con: Pisces

RA: $23^h\ 40.7^m$
Dec: $+07°\ 57'$
Mag: –
Diam: 15.0′
Dist: –
Disc: Bruno Alessi, 1997–1998; Stephen James O'Meara independently discovered it in January 2003

W. HERSCHEL: None.

NGC: None.

THE NIGHT SKY IS FILLED WITH hidden treasures, more than we can imagine. To see them, all we have to do is keep our eyes and minds open. A celestial treasure does not have to be a galaxy, nebula, or cluster. It can be any wonder that sparks emotion or stirs the imagination. It can be as simple as a solitary star with an "adorable" hue or a chance grouping of stars that form an attractive or bewitching pattern. The treasure's worth is a matter of personal appeal. Does it bring a smile to our face when we see it? Does it kindle even the smallest of flames in our hearts? The 107th treasure in this catalog may be one of those objects. You will have to judge. You might not have heard of it, because its discovery is relatively new. But that's the purpose of this catalog: to introduce you to new wonders that will keep you company in the night. Here's the story behind my first encounter with the asterism I call the Little Ladle.

On the evening of January 20, 2003, I was driving helter-skelter across Volcano, Hawaii, looking for a clear spot between clouds to get a glimpse of Comet 2002 V1 (NEAT). But the weather failed to cooperate. Giving up the chase, I turned off the Volcano Highway and started to return home. Suddenly, I saw a hole in the clouds to the west. Quickly, I pulled the car over and turned off the engine. Grabbing my 7 × 50 binoculars from the passenger seat, I slipped outside and promptly found the 6th-magnitude comet in southeastern Pegasus near the Pisces border. Quickly, but carefully, I scanned the region for comparison stars, so I could estimate the comet's brightness. That's when I spied a large diffuse object some $1\frac{1}{2}°$ south of the comet and $2\frac{1}{4}°$ north of Iota (ι) Pisces, in the famous Circlet of Pisces. The new object looked like a comet with multiple nuclei strung out in a line. Suddenly I recalled the famous "squashed comet" Carolyn Shoemaker discovered in 1993 while scanning photographic plates taken by her husband, Eugene, and David Levy; this comet, Shoemaker-Levy 9, had fragmented into more than 20 pieces before it ultimately smashed into Jupiter the following year, giving observers around the world one of the most memorable astronomical experiences of their lives.

For a moment I wondered if Comet 2002 V1 had broken apart in similar fashion. Better yet, had I stumbled upon a new comet, perhaps a faint one that had flared in brightness as it split apart? I had witnessed several comets perform in this manner over the years. Was this an unknown event? Before I could finish my thoughts, fate turned against me; the hole in the clouds closed and I lost sight of the stars. Fortunately, I had memorized the new object's position. Without hesitation, I returned to the car, found a pen, and recorded the new object's position as best I could on a paper napkin.

The skies remained overcast for two days. So you can imagine my state of mind and heart – especially after I had confirmed that no known comets (other than Comet 2002 V1) – were in the area. I also checked all my star charts for deep-sky objects in or near the suspect's position. Oddly enough, two galaxies (NGC 7742 and NGC 7743) were within 2° of comet NEAT and my suspect's mean position. What's more, these galaxies shared a similar north–south orientation and separation. I grilled myself with questions. Did I record the object's correct position? Was I confused by the hole in the clouds? Did I really see the comet, or was I looking at these two galaxies? A look in the *Deep-Sky Field Guide* provided an immediate answer – *impossible* – these galaxies shine at magnitudes 11.6 and 11.5, respectively, much too faint to be seen in 7 × 50 binoculars; besides, the comet was shining at 6th magnitude. What deep-sky objects in Pisces are *that* bright?

I had little to do but wait until the evening of January 22, 2003. That night I brought my binoculars and my 4-inch telescope into the front yard under a crystal clear sky. I found Comet 2002 V1 in binoculars and estimated its brightness – magnitude 5.8; it was the object I saw on January 20. Next I scanned the sky with binoculars for the "strange object." And there it was, *exactly* where I had plotted it. Alas, it was not a comet, which would have moved by even a modest amount in two days (unless it was on a collision course with Earth). So, I wondered, what is it?

The telescope revealed the truth. My squashed comet is a lovely gathering of nine 8th- to 11th-magnitude suns in an area measuring a mere 15′ across. (The Coathanger asterism (Hidden Treasure 97) in Vulpecula, by comparison, measures 90′ across.) The stars are arranged in such a way that they resemble a ladle, the kind you use to scoop hot fudge for an ice-cream sundae. The asterism looks that delicious.

To find the Little Ladle, look for two larger ladles. Start with Alpha (α) Pegasi, or Algenib; it marks the tip of the handle to a large ladle, which is comprised of Alpha Pegasi, magnitude 4.0 Omega (ω) Pisces, and the Circlet of Pisces. Next use your telescope to star hop from Iota Pisces, the 4th-magnitude star marking the northeastern flank of the Circlet. Iota also marks the southern apex of a $1\frac{1}{2}$°-wide triangle of stars (the scoop of another ladle), oriented north–south; the star at the northeast corner of this scoop is a nice pair of stars (magnitudes 7.7 and 8.5). Just 1° due north of this pair of stars

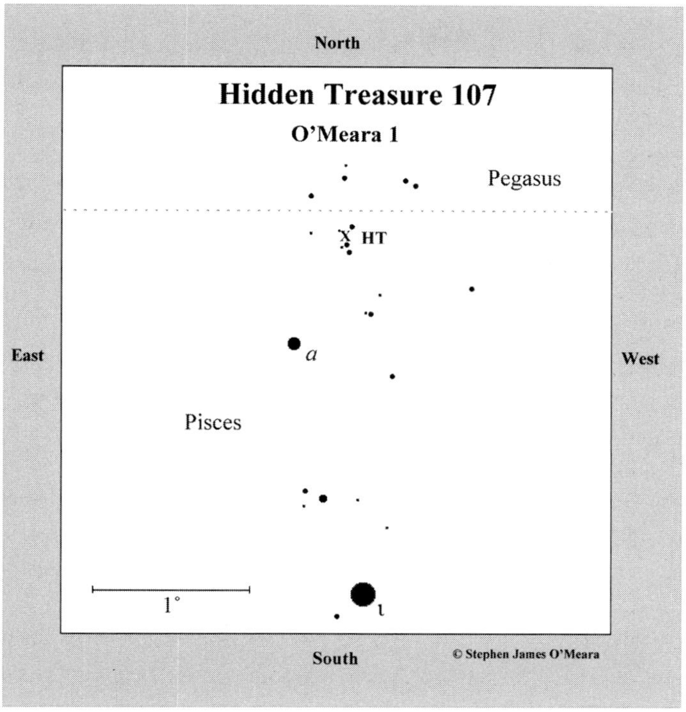

Sao Paulo, Brazil, who had alerted Archinal to 41 possible star clusters. He had discovered these objects between 1997 and 1998 while looking at plots for clumps of galactic star clusters or by examining Digitized Sky Survey images. Stellar parallax data suggest no real link between the stars, so they are not a physical group but a chance alignment. Archinal includes Alessi's discovery of this object in his book, so its official designation is Alessi J23407 +0757. Mine is just an independent discovery.

That close pairings of stars look fuzzy in small apertures is not new. Consider the asterism M40, a close

is a solitary magnitude 5.7 sun (*a*) – the handle of this second ladle. The ladle you seek is about $\frac{3}{4}°$ northwest of Star *a*.

Four of the Little Ladle's stars – three in the handle and one at the western tip of the scoop – shine between magnitude 8.2 and 8.8 and comprise a 12′-long arc oriented north–south; the middle star in the handle is a fine visual double (magnitudes 8.2 and 10.2). It is this tight-knit group of stars that appears fuzzy in binoculars. The three other main stars in the Ladle shine around 10th magnitude. The faintest star in the group is magnitude 11.2. All should be easy to see in small telescopes.

Shortly after finding this object, I contacted Brent Archinal to see if he knew about this quaint grouping of stars; he did. The asterism was first discovered by Bruno Alessi, an advanced amateur astronomer in

double star in Ursa Major. Charles Messier included it in his now famous catalog of

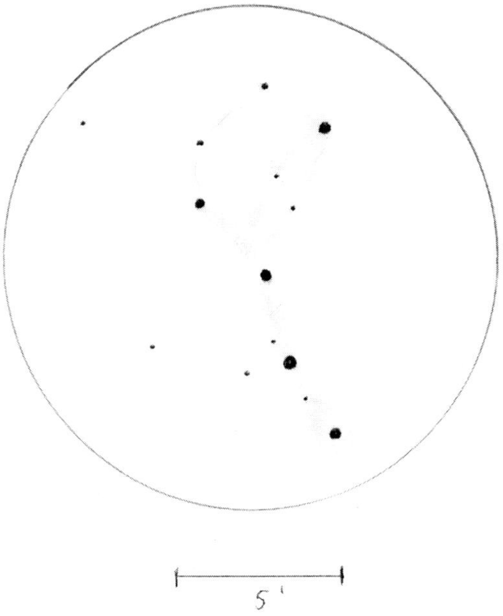

deep-sky objects not because he mistook it for a comet but because John Hevelius mistook it for a nebula in 1660. By including this object in his catalog, Messier was sending a warning to comet hunters that close pairings of dim stars can masquerade as comets in small apertures. It's the same with the naked-eye appearance of the famous Coathanger; what looks like a star-studded nebula to the naked eye is really a wonderful asterism of binocular stars in the shape of a coathanger.

So my discovery tale is nothing new. It is part of a never-ending story, one that Messier and his contemporaries knew all too well. It was quite common in Messier's day for comet hunters to make discoveries of deep-sky objects while they were searching for, or observing, known comets. Indeed, what inspired Messier to create his catalog of non-comets was his independent discovery of the Crab Nebula (M1) in Taurus while observing the comet of 1758. Jean-Dominique Miraldi II had a similar experience; his discovery of M2, a globular cluster in Aquarius, occurred while he was looking for de Chéseaux's comet in 1746: "I took, at first, this nebula for the comet."

And so the story goes, on and on, hopefully until eternity. I encourage you to go out and find your own hidden treasure. Consider how popular Kemble's Cascade (HT 21) has become. Perhaps you have also heard of Pazmino's Cluster (Stock 23) in Cassiopeia; New York amateur John Pazmino independently discovered this magnitude 6.5 open cluster on the Camelopardalis–Cassiopeia border in the late 1970s with a 4.3-inch refractor. Although Pazmino is not the official discoverer of this cluster, his name has become intimately attached to it ever since he announced his finding in the March 1978 issue of *Sky & Telescope*. So do not fear to put a formal stamp on your personal discovery, even if it only appears as such in your logbook or diary. There is nothing more valuable in life than your own words, thoughts, and discoveries.

Hidden Treasure 108

NGC 7789

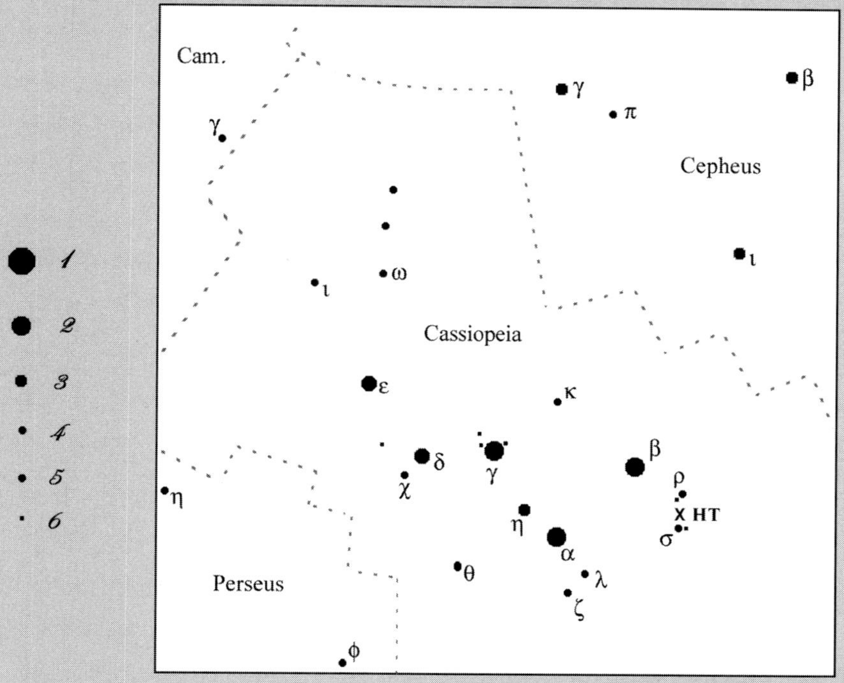

© Stephen James O'Meara

Tirion: Charts 1 & 3

Uranometria: Chart 35

108

Herschel's Spiral Cluster, Crab Cluster, Screaming Skull Cluster
NGC 7789
Type: Open Cluster
Con: Cassiopeia

RA: $23^h\ 57.5^m$
Dec: $+56°\ 43'$
Mag: 6.6 (O'Meara); 6.7
Diam: 25.0′
Dist: 6,000 light-years
Disc: Caroline Herschel, 1783

W. HERSCHEL: [Observed October 18, 1787] A beautiful cluster of very compressed small stars, very rich, C. H. discovered it in 1783. (H VI-30)

NGC: Cluster, very large, very rich, very much compressed, stars from magnitude 11 to 18.

JUST 3° SOUTHWEST OF BETA (β) Cassiopeia, or Caph (the westernmost star in Cassiopeia's famous W asterism), and midway between Rho (ρ) and Sigma (σ) Cassiopeia, lies the large uniform glow of NGC 7789, one of the finest and richest open star clusters in the sky. The Rev. T. W. Webb, in his *Celestial Objects for Common Telescopes,* calls it a "[b]eautiful large faint cloud of minute stars," while John Herschel called it a "most superb" cluster.

Caroline Herschel discovered this beautiful object in the fall of 1783. And it was upon this discovery that, as Adm. William Henry Smyth notes in his *Cycle of Celestial Objects,* she "realized the poet's description: *Some sequestered star / That rolls in its Creator's beams afar, / Unseen by man; till telescopic eye, / Sounding the blue abysses of the sky, / Draws forth its hidden beauty into light, / And adds a jewel to the crown of night.*"

Shining at magnitude 6.6, NGC 7789 lies at the limit of unaided vision, making it a great naked-eye challenge under dark skies. While taking the time to investigate its naked-eye visibility in October 2003, I found a way to help you confirm whether or not you truly believe that you see it. The trick is first to study the brightness of the sky between Rho and Sigma Cassiopeia using keen averted vision. Next, compare that impression with the sky brightness between Rho and Tau (τ) Cassiopeia. If you see the sky between Rho and Sigma looking decidedly brighter, you've detected the tight core of NGC 7789.

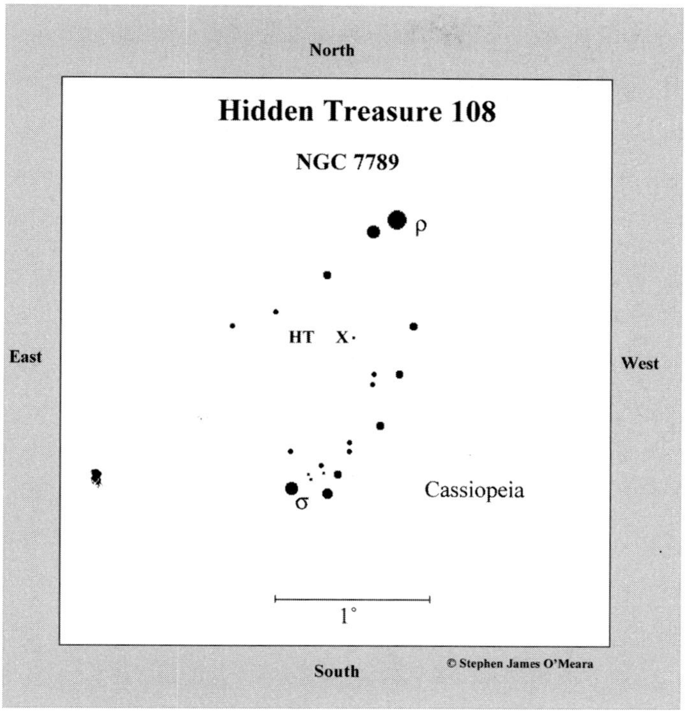

especially with averted vision. In fact, no matter what instrument you use, this cluster, with its great range between bright and dim members, forces the observer to use peripheral vision.

There are some 580 stars in the cluster – the brightest shines at magnitude 10.0 while the faintest hover around magnitude 18. A pleasing number of them, perhaps 60 or so, can be spied in the 4-inch at 23×, with which the cluster may just be at its best. That number increases to about 150 with 72×. With imagination, the central core of suns appears three-dimensional at 23×; it looks as if some pirate, eager to eye his bounty, had poured a fistful of diamond dust into his blackened hands and squeezed them hard with greed. In an equally sinister way, the core resembles a skull, with two oval rings of stars marking its eye sockets, and a curving dark lane of obscuring matter marking the skull's open jaw. For this reason I call NGC 7789 the Screaming Skull Cluster; the skull is most apparent in photographs, like the one above, when viewed with north up.

To see the eye sockets, look about 10′ west-northwest of the 7th-magnitude star just east of the cluster's core. I've shaded them in the drawing. Midway between the easternmost eye socket and the 7th-magnitude star is a solitary 9th-magnitude star, just west of which is a neatly defined horseshoe asterism of stars. The western side of the horseshoe forms the western outline of

It's a magnificent spectacle in 7 × 50 binoculars, being nearly as large as the full Moon and just as round. In fact, through binoculars, the cluster looks like a tailless comet crossing the rich star fields of the northern Milky Way. With just a few minutes of scrutiny, though, the cluster should begin to sparkle – a phenomenon caused by the light of dim suns popping in and out of view in your peripheral vision. Concentrate now on the cluster's core with averted vision. Do you see a central core of light surrounded by a larger and fainter halo of largely unresolved suns?

As the cluster was a bit too large to make a decent magnitude estimate with binoculars (the eyepieces did not rack out far enough), I had to rely on the sliding tubes of my antique telescope. Through that telescope, the cluster appears compressed at the core, partially resolved, and particularly clumpy,

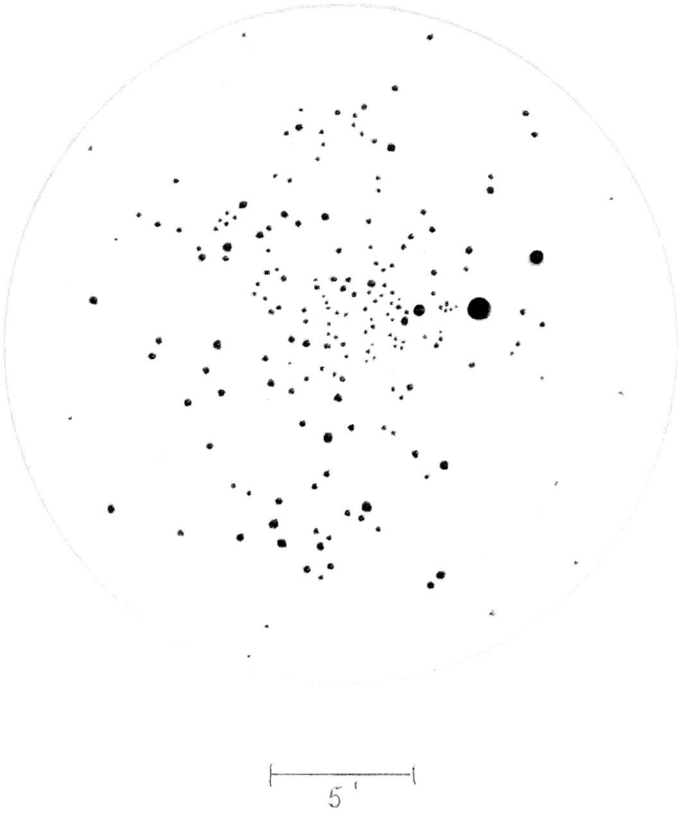

brighter than the main body. With averted vision, the east side has a brighter assortment of suns. Now look for curved extensions off these arms, making it appear to have a counterclockwise spiral structure. These arms give rise to another common nickname: Herschel's Spiral Cluster.

Smyth was most impressed with the cluster, which he said was comprised of "minute stars, on a ground of star-dust." He also saw the cluster as "having spangled rays of stars which give it a remote resemblance to a crab . . . The crab itself is but a mere condensed patch in a vast region of inexpressible splendour, spreading over many fields."

At 72×, the cluster indeed takes on a most crablike form. Look for parallel rows of wavy suns intermixed with large loops of stars. Now, turn the focus knob at this power and place the cluster ever so slightly out of focus. The easternmost eye socket transforms into a beautiful "smoke ring." It really requires a delicate touch. Using powers greater than 100× transforms the tight core into all manner of patches of starlight, arcs, ellipses, trapezoids, clusterings, and pairings.

With every change of magnification, the cluster takes on a new appearance. It is, without question, the Lon Chaney[1] of star

the Skull's eye socket and cheekbone. NGC 7789's horseshoe asterism bears an uncanny resemblance to the main form of NGC 663 (Caldwell 10) in Cassiopeia, which is known as the Horseshoe Cluster. Be sure to compare the two clusters on the same night.

Now look midway between the 7th- and 9th-magnitude stars. Do you see a small puff of "nebulosity?" This is a minute clustering of six dim suns in the shape of a fishhook, which is visible as such only at high power.

It's very difficult to judge the cluster's boundaries; the stars fly off in long arms, all

[1] Lon Chaney was a silent movie actor frequently cast as twisted and deformed characters. He is most famous for his roles as the Hunchback in *The Hunchback of Notre Dame* and the Phantom in *The Phantom of the Opera*. Chaney was so masterful at creating characters through a combination of elaborate makeup, contorted postures, and sensitive performances, that he earned the nickname of "The Man of a Thousand Faces."

clusters; I never tire of looking at this celestial marvel. At the accepted distance of 6,000 light-years, the cluster's angular diameter corresponds to a respectable physical size of 44 light-years. The cluster's age is about 1.6 billion years, making it intermediate in age – a contemporary of NGC 752 (Caldwell 28) in Andromeda, which is 2 billion years old. The brighter and more massive stars in NGC 7789 have already exhausted the hydrogen fuel in their cores, so they have evolved from hydrogen burning stars like our Sun into red giants and subgiants, which, if you look carefully, reveal themselves with their tepid orange hue. The cluster is also rich in variable stars, 45 of which were identified by Polish astronomers in 1999, including 35 eclipsing binaries, two RR Lyrae stars, and one Delta Scuti star (which is a blue straggler – a star that appears to linger on the main sequence – belonging to the cluster). Some of the variables, like the two RR Lyrae stars, are line-of-sight objects in the rich Milky Way background.

As pointed out by Smyth, the Milky Way surrounding NGC 7789 is equally impressive, especially in wide-field telescopes. Most noteworthy is the complex of dark nebulae in and around the cluster. These impressive veils are not outlined on any star atlas that I'm aware of.

Two stars in the immediate vicinity of NGC 7789 are of particular interest. Alex J. R. Lobel and Andrea Dupree (Harvard-Smithsonian Center for Astrophysics) suspect that Rho Cassiopeia, a hypergiant star as hot as the Sun but roughly a million times more luminous, is in the very last stages of its evolution and could go supernova in as little as 50,000 years. In 2000, the star underwent the largest stellar mass ejection ever recorded, ejecting 50 Earths per day for 200 days. About 5 percent of a solar mass was expended during that event, which is roughly a thousandth of Rho Cas's mass. Since the star cannot withstand many more such mass ejections and remain stable, it's believed the end is near for Rho.

The second star, Sigma Cassiopeia, pales by comparison – at least in potential drama. In color, though, it's one of the best – a close pair of blue-green gems in a rich field teeming with jewels. The primary shines at magnitude 5.0; the secondary shines at magnitude 7.1; the stars were separated by 2.4″ in 2002, making them a decent challenge for small telescopes. Of course, no two people see the colors of these stars the same. Admiral Smyth, for instance, saw the primary as "flushed white" and the secondary as "smalt blue." He also says, "The colours are clear and distinct, though less fine than those of ε Bootis, of which this is a miniature." You be the judge.

Hidden Treasure 109

NGC 7793

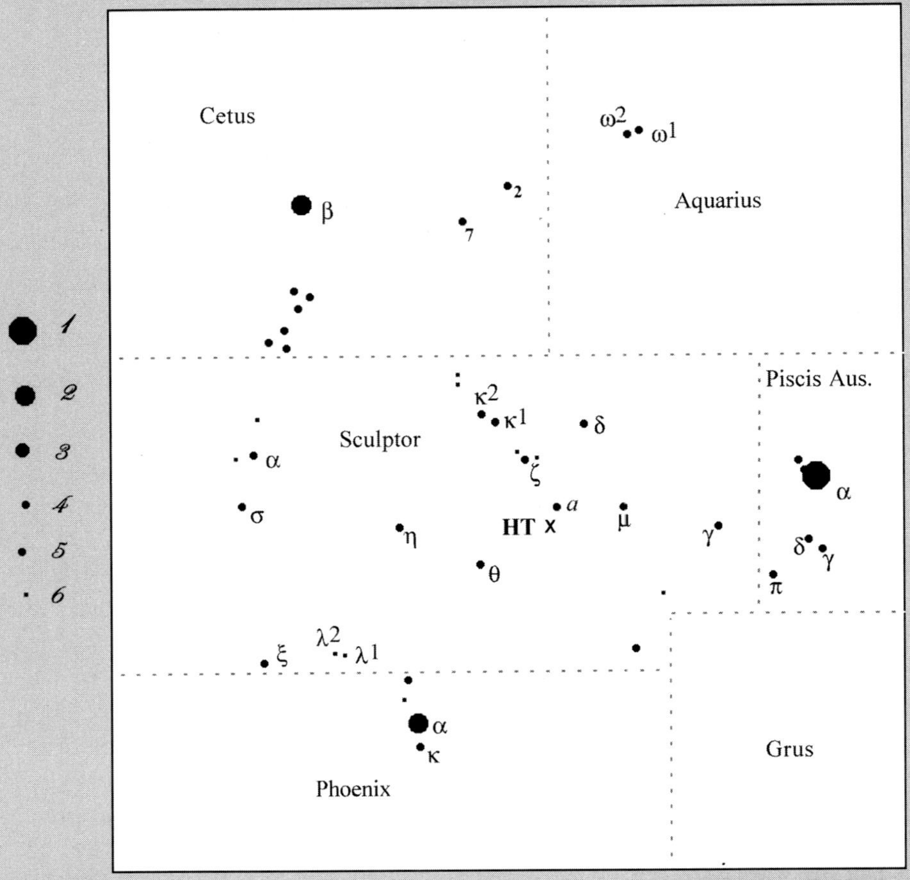

Tirion: Charts 18 & 23

Uranometria: Chart 350

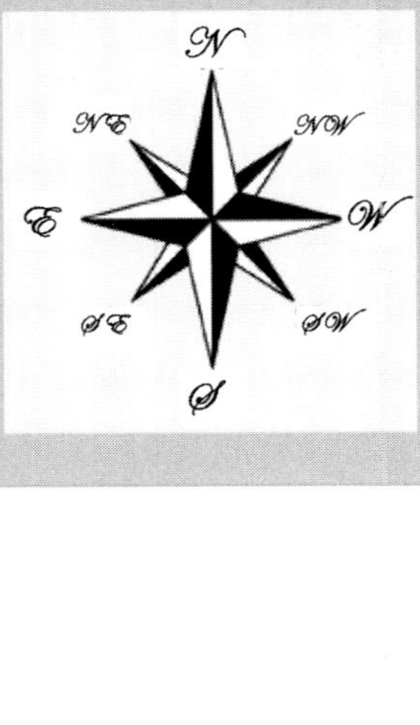

109

Bond's Galaxy
NGC 7793
Type: Dwarf Spiral Galaxy (Sdm)
Con: Sculptor

RA: $23^h\ 57.8^m$
Dec: $-32°\ 35'$
Mag: 9.2
Dim: $9.2' \times 6.4'$
SB: 13.6
Dist: 9 million light-years
Disc: James Dunlop included it in his 1827 catalog; independently discovered by George Phillips Bond, 1850

HERSCHEL: None.

NGC: Like a comet (1850).

THE LAST HIDDEN TREASURE IS tucked away in the star-poor regions of southwestern Sculptor. It lies some 13° east-southeast of brilliant Fomalhaut, Alpha (α) Piscis Austrini, and is near the apex of a nearly perfect isosceles triangle whose other corners are the magnitude 4.4 Beta (β) Sculptoris ($7\frac{1}{4}°$ to the southwest of the galaxy) and Gamma (γ) Sculptoris (8° to the west). The 9th-magnitude galaxy will be about the same distance above your horizon as the spout in the Teapot of Sagittarius.

The visibility of NGC 7793 in a small telescope from mid-northern latitudes is a good yardstick of your area's light pollution. That's because the object was independently discovered in 1850 by Harvard astronomer George Phillips Bond in Cambridge, Massachusetts. He found it while hunting for comets from observatory grounds while using a 4-inch f/8 Merz comet seeker. Bond was quite successful in his comet searches, finding at least 11 in the course of a few years. He also discovered five other "nebulae" with the great 15-inch Merz & Mahler refractor. All are small and dim 13th- and 14th-magnitude galaxies: NGC 219, NGC 223, and NGC 391 in Cetus; NGC 5366 in Virgo; and NGC 7692 in Aquarius. These were the days when the Milky Way still stretched from horizon to horizon over the city, and the zodiacal light blazed high into the night like the broad sweep of a comet's dusty tail.

The 1888 *New General Catalogue*'s description of NGC 7793 is based on Bond's discovery observation, in which he described the object as appearing like a comet. And though the *NGC* also credits Bond with the discovery, he was not the

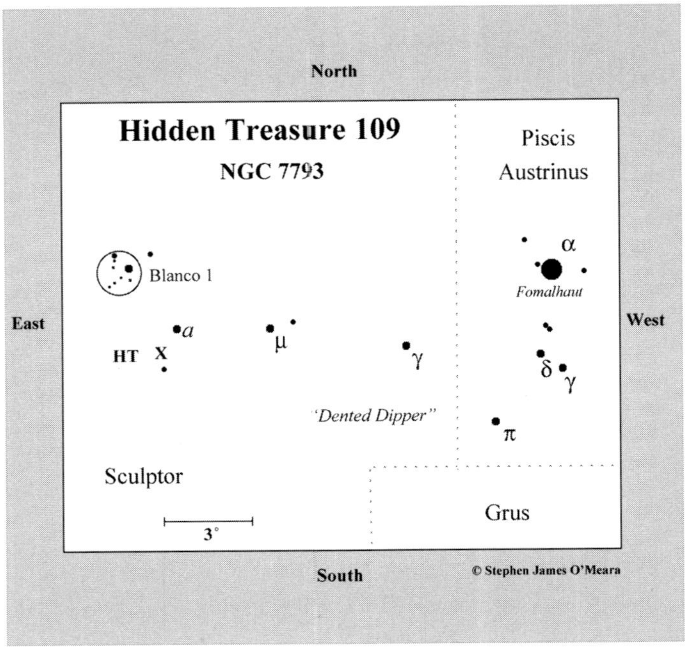

first. That credit goes to James Dunlop, who swept it up with a 9-inch f/12 refractor at Paramatta in New South Wales. Dunlop listed it as the 608th object in his 1827 catalog of southern nebulae and clusters and described it as, "[a] faint round nebula, about 2′ diameter, with very slight condensation towards the centre; a double star is north preceding."

Today, we know NGC 7793 is a prototypical late-type spiral galaxy with a total luminosity of 1.9 billion Suns. We see its disk inclined 40° from edge-on, so its tiny nuclear bulge and dominating spiral arms are prominently revealed. If we accept a distance of 9 million light-years for NGC 7793, its true linear diameter is a modest 23,000 light-years. This dwarf belongs to the Coma–Sculptor Cloud of Galaxies – also known as the South Polar Group of Galaxies – which includes the fantastic spiral system NGC 253 (Caldwell 65), also in Sculptor. Both NGC 7793 and NGC 253 are more distant than the group's other members, NGC 55 (Caldwell 72) and NGC 300 (Caldwell 70) in Sculptor, and NGC 247 (Caldwell 62) in Cetus. NGC 7793 is the faintest of the five members of the group and is receding from us at a speed of 230 kilometers per second.

NGC 7793 is highly resolved into individual stars beginning at a blue magnitude of 18. A spiral pattern is seen, but it is more disorganized than that of M33 or NGC 300. The object's ultraviolet spectrum is dominated by H-II regions; at optical wavelengths, the spectrum is dominated by early Sun-type stars. The brightest stars are massive red supergiants that are located in the central regions. Also present is a rich population of asymptotic giant branch stars with ages near 10 billion years. This indicates the disk component contains an underlying old population of stars which is what one would expect.

An optical study in 1997 using charge coupled device (CCD) imagery and interference filters indicated 28 objects that have morphologies consistent with supernova remnants. These are the first large samples of supernova remnants identified in galaxies outside the Local Group. A later radio study in 2002 using the Very Large Array added five new candidate supernova remnants to the 28 already detected by optical methods. One of these remnants, identified as N7793-S26, has been detected at three wavelengths: X-ray, optical, and radio. It has a long filamentary morphology and its radio luminosity exceeds that of the Milky Way's Cassiopeia A: a supernova remnant. In fact, N7793-S26 is so energetic that astronomers believe it may be the product of multiple supernova explosions rather than a single supernova event.

The galaxy is a cinch to find under a dark sky, especially from the more southerly latitudes of the USA. It is a little less than 1° southeast of a 6th-magnitude star (*a*), which marks the tip of the handle of a roughly fist-wide "dented dipper" asterism oriented east–west; Fomalhaut marks the end of the dented bowl. City and suburban observers should try star hopping along the dipper, starting with Fomalhaut. Star *a* is a beautiful double star: a pair of yellow suns of near equal magnitude. While I could not see the galaxy in 7 × 50 binoculars, it did appear as a dim glow in my antique telescope.

Through the 4-inch at 23×, the 9th-magnitude galaxy appears quite large, and it looks very much like the smoothly textured head of a comet with no tail. A tight arc of three similarly bright stars lies nearly 10′ to the south. At this power, the galaxy does not have a sharp or distinct core, just an amorphous glow that seems to get denser gradually toward the center. It also appears mottled.

At 72×, the galaxy and the nearby arc of stars look like a fly used for fishing. Interestingly, the galaxy's core appears sharply stellar now; the bright haze immediately surrounding the core is dappled with fuzzy clumps that would cause any small-telescope user to believe he or she has chanced upon a globular cluster near the limit of resolution. Seeing these details

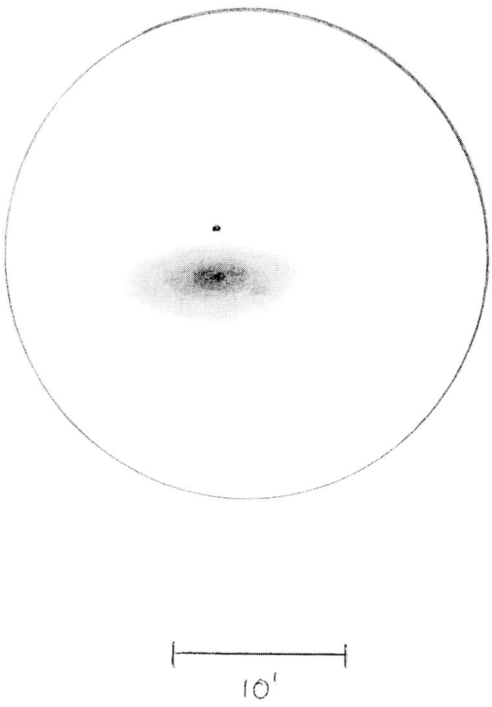

is maddeningly enticing. It's like standing on the lip of a lit stage with the audience in near-complete darkness. All manner of shapes and forms can be seen, but they blend together like the fine brush strokes in an impressionist painting. The galaxy's fainter outer envelope is slightly elliptical and shows less mottling. But it is nonetheless mystifying. At 101×, the core is a sharp pip of light surrounded by a condensed coma of dappled light both patchy and fragmentary.

Although NGC 7793 is the faintest of the five major members of the Sculptor Group, the detail it displays in small instruments makes it a superb yet visually challenging, target for beginners. It is a fitting object to end with. Many of the deep-sky objects available to small-telescope users beyond the scope of the *Hidden Treasures* list (not including those in the Messier and Caldwell lists) will only further test your hunting and observing skills. They will draw you in to the inner sanctums of our universe, where new island universes filled with unseen treasures await the watchful eye of the celestial pirate.

> The bar of silver and the arms still lie, for all that I know, where Flint buried them; and certainly they shall lie there for me . . . "Pieces of eight, pieces of eight."

So ends Robert Louis Stevenson's *Treasure Island*; so ends this chapter, which closes the lid on our main chest of astronomical riches.

APPENDIX A
Caroline Herschel: no ordinary eighteenth-century woman

BARBARA WILSON

February 26, 1783. Night falls cold and clear. A slight young woman of 32 slips out of her house in Datchet, England, and sets up a small refractor on a grass plot covered with frost. Seated in the frigid air, wrapped in a wool shawl and cap, the woman points her telescope just above the southern horizon and begins sweeping the heavens. Her desire is to discover a comet, but she cannot help but stop to record every remarkable object she sees. Time passes uneventfully at first, until she spies a "very faint nebula" near Gamma (γ) Canis Majoris not in Messier's list of known objects. A spark of warmth ignites in her body. Caroline Herschel has just made the first of several discoveries that, arguably, will alter the course of astronomical history.

CAROLINE LUCRETIA HERSCHEL WAS born in Hanover, Germany, on March 16, 1750. She was the eighth of 10 children born to Isaac and Anna and nearly 12 years younger than her revered older brother, Friedrich Wilhelm (later William). She had, like her father and William, a penchant for music and was a talented soprano. In 1778 she was offered an engagement for the Birmingham Festival, but she declined, having resolved to sing in public only where her brother, William, was conductor.

The strong attachment and affection between Caroline and William began as soon as Caroline could show or express her feelings and continued throughout their lives. Caroline (affectionately known as Lina) was tiny, standing only about 4 feet 9 inches tall. She suffered smallpox at the age of 3 (which stunted her growth) and typhus at the age of 11; she nearly succumbed to that disease. As Caroline recalled, the typhus had reduced her strength to such a degree that "for several months after I was obliged to mount the stairs on my hands and feet like an infant; but here I will remark that from that time to this present day (June 5, 1821) I do not remember ever to have spent a whole day in bed."

From infancy to old age, William was the "best and dearest of brothers." Indeed, in August 1772, William brought Caroline to his home in Bath, England, saving her from a dreary life as a "household drudge," and she was clearly grateful to him: "I saw that all my exertions would not save me from becoming a burden to my brothers, I was not qualified to become a governess for want of knowledge in languages. My father cautioned me against all thoughts of marrying, saying as I was neither hansom nor rich."

Blue-eyed Caroline was not only physically small in height, but she had disfiguring scarring on her face resulting from her bout with smallpox. Her appearance was not unusual, as smallpox scarring was a common occurrence at the time. In fact, much of the population suffered disfigurement from the more severe and most common form of the disease (variola) until 1798, when vaccination campaigns started. Commenting on her appearance Caroline states: "Although I recovered [from smallpox], I did not escape being totally disfigured and suffering some injury to my left eye."

When Caroline arrived in Bath, William had already established himself as a teacher of music. He was also organist of the Octagon Chapel and frequently composed anthems, chants, and entire services for the choir. But Caroline also arrived at the turning point in her brother's career – from music master to master observer of the night

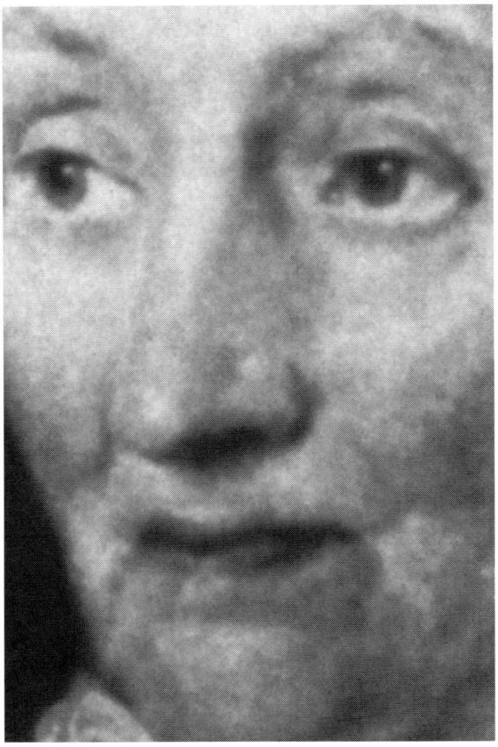

sky. And Caroline's service to her brother was never more needed.

As Mrs. John Herschel recorded in her 1879 *Memoir and Correspondence of Caroline Herschel,*[1] "No contrast could be sharper than that presented by the narrow domestic routine she had left to the life of ceaseless and inexhaustible activity into which she was plunged . . . For ten years she persevered at Bath, singing when she was told to sing, copying when she was told to copy, 'lending a hand' in the workshop, and taking her full share in all the stirring and exciting changes by which the musician became the King's astronomer and a celebrity."

Her brother's rise to fame was, of course, owing to the discovery of the planet we now call Uranus. King George III, William's patron (and for whom William wanted to name the new world), arranged for William to give up his musical career and settle in the vicinity of Windsor Castle, where he would be obliged, on occasion, to show the Royal Family celestial objects of interest through his telescope. In return, William would receive a salary of £200 a year. The King subsequently made an allowance of £50 a year to Caroline as her brother's assistant, making her the first woman professional astronomer. "I received 12 pounds being the first quarterly payment," Caroline writes, "and the first money in all my lifetime I ever thought myself at liberty to spend to my own liking."

Caroline's career as an astronomer started on August 2, 1782, once she and her brother moved to Datchet – a tiny village about 3 kilometers east of Windsor Castle, on the north bank of the Thames. Caroline's days were filled with writing down her brother's discoveries and calculating their positions. She prepared all his papers for publication by the Royal Society of London. She ground mirrors, hired the gardeners, kept the household running, did needlework, cleaned the brass-work for the telescopes her brother used, and entertained their distinguished guests. She was the consummate multitasker.

William was also training Caroline to be an assistant observer. And on August 22, 1782, she began sweeping the skies with a small telescope built by her brother and adapted for "sweeping." The telescope consisted of a "tube with two glasses, such as are commonly used in a 'finder'." Her task was "'to sweep for comets,' and I see by

[1] Available as a reprint from the William Herschel Museum, 19 New King St., Bath, BA1 2BL, UK.

my journal that I began . . . to write down and describe all remarkable appearances I saw in my 'sweeps,' which were horizontal." William also instructed her to search for double stars, clusters of stars, and nebulae. He gave her instructions on how to describe them by lines from certain stars and figures drawn on paper. During these sweeps, she kept Messier's list of nebulous objects by her side. The list, published in the 1784 *Connaissance des Temps,* was a gift to her brother from the London businessman and prominent amateur astronomer, Alexander Aubert, soon after its appearance.

As Caroline swept the heavens for comets and other remarkable objects, William focused his attention on his "third review of the heavens," which he had begun at Bath with 6.2-inch and 12-inch reflectors. Herschel was reviewing all Flamsteed stars, to find and measure doubles. Little was known about double stars, their nature, and their motions, and William's examination of them was the first systematic survey. Observations continued from dusk to dawn, many hours at a time, and he often examined 400 stars a night.

It's important to note here that William's single-minded focus at the time was on double stars and not nebulae. In fact, until their move to Datchet, and until Caroline had begun her comet sweeps, William had seen only four nebulae.

CAROLINE'S TREASURES

When Caroline first started sweeping the night sky, she "knew too little of the real heavens to be able to point out every object so as to find it again without losing too much time by consulting the Atlas." This was most disturbing when her brother was away: "I was, of course, left solely to amuse myself with my own thoughts, which were anything but cheerful . . . But all these troubles were removed when I knew my brother to be at no great distance making observations with his various instruments on double stars, planets, etc., and I could have his assistance immediately when I found a nebula or cluster of stars, of which I intended to give a catalogue."

Caroline's "finds" were relatively few at first. On her first night of observing (August 22, 1782) she found one of William's double stars. Then, on September 30, she encountered a few more double stars and her first comet masquerader – the 27th object in Messier's list. By October she had added M36 and M13 to her repertoire. By January 23, 1783, she was clearly beginning to enjoy her time with the stars, for on that night she had spent 30 minutes looking at M44, noting that she did not find any "nebulas" in or around it, just distinct stars.

It was on the evening of February 26, 1783, that Caroline's luck changed. On that night she recorded a nebula not recorded in Messier's list, and showed it to her brother. The object, known today as open cluster NGC 2360 (Caldwell 58), was the first of 14 deep-sky objects she would discover or independently discover, 11 of which had never before been seen by humans. The identity of some of these objects have mystified modern astronomers, prompting much controversy over which objects truly exist and which truly belong to Caroline and not her brother. The matter has been satisfactorily cleared up by astronomical historian Michael Hoskin in the November 2005 *Journal for the History of Astronomy* (Churchill College, Cambridge, UK). A list of each of Caroline's discoveries, adapted with permission from Hoskin's article, appears in chronological order below. Each entry includes a description from Caroline's original journal. The modern designation of each discovery is listed in parentheses at the end of each description. In some cases, additional comments follow from Hoskin, Stephen James O'Meara, and me.

CAROLINE'S NEBULAE

(1) 26 February 1783.
Following gamma Canis majoris a very faint Nebula, my Brother's observation upon this Nebula: about $3\frac{1}{2}$ degrees following gamma Canis maj Amas d'étoiles [the term used by Messier for a cluster]. It is about $\frac{1}{2}$ deg following a star of the 7th or 8th magnitude. With 460 [power] there are about 15 or 16 stars. which are all excessively obscure, and seem a little nebulous; but I think it owing to the low situation and high power. with 227 [power] about 40 or 50 small stars. with the compound piece, a cluster of small stars closer than those in the foregoing Nebula [M 93]. Messier has it not. (NGC 2360 [Caldwell 58])

HOSKIN: William's notebook contains the observation as cited by Caroline. In Caroline's fair copy there is a marginal note by John Herschel for some reason declaring this to be William's VIII.45 [NGC 2358], but William is clearly correct in saying it is VII.12 [NGC 2360 = GC 1512, a galactic cluster later seen by William in sweep 366 on February 4, 1785]. William credits the discovery to Caroline in his first catalog, but his explanatory note at the end of the catalog reads VII.13 [NGC 2204], and this slip by Caroline in preparing the text for publication has misled modern writers into thinking this cluster too should be credited to Caroline. In John [Herschel's] *General Catalogue* Caroline is rightly credited with VII.12 (and with neither VII.13 nor VIII.45 [NGC 2358], the subject of a similar slip).

(2) 8 March 1783.
At an equal distance from 29 [Zeta (ζ)] & 30 Monocerotis, making an equilateral triangle with those two stars is a nebulous spot. By the telescope it appears to be a cluster of scattered stars, it is not in Mess. catalogue.

William wrote:

Lina. equilateral triangle with 29 & 30 Monocerotis, to the south, a Cluster of scatter'd stars. (NGC 2548 [M48])

HOSKIN: The cluster NGC 2548 = GC 1637, the "missing" M48 of Messier's list [VI.22, seen by William in sweep 519 on February 1, 1786]. In 1959, T. F. Morris of the Royal Astronomical Society of Canada showed that Messier had made an arithmetical error in the reduction, mistaking the sign of the right ascension difference.

O'MEARA: The importance of Caroline's discovery of this object is understated. It was only through William Herschel's inclusion of it in his catalog that astronomers became aware of its existence.

(3) 23 July 1783.
Some small stars; or perhaps a Nebula. My Brother put, I believe, a power of 70 to the Sweeper, then what is call'd some small stars are about a hundred or more.

16 October 1783.
My Brother shewed me the spot in Cygnus, which I saw first July 23rd, the stars very numerous, and intermixed with strong nebulosity. Mess. has it not. (NGC 6866 [Hidden Treasure 100])

HOSKIN: William's entry for 16 October in his Journal No. 7 reads: "Cluster of stars making an isosceles triangle with delta & omicron Cygni, the vertex laying S following; about 3 or 4 degrees from delta or omicron. It contains very great variety of magnitudes; the large ones are scattered the smaller stars being mixed between them. 7ft [reflector] compound [eyepiece] – Lina [i.e. Caroline]." William's first observation of this object antedates the commencement of his first catalog. In all probability this is the cluster NGC 6866, observed again by William on September 11, 1790 and cataloged as VII.59 but without either partner realizing that they had seen it before.

WILSON: In July 1783 Caroline exchanged her small refractor for a Newtonian sweeper 4.2 inches in aperture and 2-feet long. William was delighted with her success and built this special telescope just for her. It was convenient, as she could use it while seated, making vertical sweeps from the horizon to the zenith. It gave 24 power and a 2° 12′ field of view. This rich-field telescope was especially suited for the discovery of this magnitude 7.6 cluster.

O'MEARA: Given the clear description of the object's nature and its position by William, and given the appearance of the object in my 4-inch refractor at 23×, it's certain that NGC 6866 is the object discovered by Caroline Herschel.

> **(4) 31 July 1783.**
> About halfway from S Serpentarii [Ophiuchi] towards theta Serpentis, a Cluster of large stars. I counted about 80. Mess. has it not.

William wrote:

> About halfway from S Serpentarii towards theta Serpentis a Cluster of large stars visible to the naked eye, about 70 large with many extremely small between. (NGC 6633 [Hidden Treasure 92])

> **(5) 31 July 1783.**
> From beta Serpentarii towards S, $1\frac{1}{3}$, a cluster of stars. (IC 4665 [HT 83])

WILSON AND O'MEARA: There is little doubt that Caroline discovered this object but reversed the sign, as IC 4665 lies $1\frac{1}{3}$° north of Beta Ophiuchi. This is an independent discovery, though, since Philippe Loys de Chéseaux included it in his list of 21 nebulous objects, which was compiled in 1745–1746. Interestingly, while de Chéseaux's written description of the object and its position are fine ("Above the shoulder, β Ophuichi, a cluster of stars"), he provides the precise position of the two brightest stars in the cluster but places them in a southern declination rather than a northern one. We believe that, at least in Caroline's case, she most likely mistakenly recorded the object as being south of Beta, because in her rich-field telescope, which simply inverted the image, Beta would have appeared at the top of the field and the cluster "below" it.

WILSON: If you change the sign to north instead of south this would be IC 4665, a quite large and coarse cluster of stars at 40′ in size, that is visible to the bare eye and would be visible in her telescope's eyepiece along with Beta Serpentarii. I experimented with an eyepiece overlay of the size field Caroline was using of 2° 12′. This would put Beta Ophiuchi in the field and completely enclose IC 4665.

> **(6) 27 August 1783.**
> About $\frac{1}{2}$ deg preceding & a little north of Mess 31st a nebula. There are many stars besides in the field, but these two are the largest. (NGC 205 [M110])

HOSKIN: This, the second companion to M31, was discovered by Messier on August 10, 1773, and is listed as M110, but his discovery was not published until 1807. Caroline was therefore an independent discoverer, and William's was the first publication.

WILSON: Messier never added this object to his 1781 catalog but he published a description in *Observations Astronomiques* (1770–1774) and in *Connaissance des Temps* for *1801* (Paris 1798, p. 461). In 1807 a drawing was published from his August 10, 1773, observations. The addition of M110 to Messier's catalog was made by Kenneth Glyn Jones in 1966, in his book *Messier's Nebulae and Star Clusters* (Faber and Faber, 1968). Caroline Herschel's independent discovery was 10 years after Messier's unpublished observation, so she is an independent discoverer. William numbered it H V.18 on October 5, 1784.

(7) 23 September 1783.
A faint nebula below the 2d Triangle under beta Ceti in a direction parallel to theta & tau, under the two preceding stars of the 2d triangle nearly at an equal distance. Mess. has it not. (NGC 253 [Caldwell 65])

(8) 27 September 1783.
Delta and epsilon Cassiopeae & chi Persei making a trefoil. A cluster of stars in the middle . . . Mess. has them not.

30 October 1783.
I saw the cluster which is placed between delta & epsilon Cassiopeae and chi Persei (a crouded place). (NGC 659 [Hidden Treasure 7])

(9) 27 September 1783.
About 2 degrees from gamma Cassiopeae making an Isosceles triangle with gamma & kappa, a small cluster of stars, seeming to be intermixed with nebulosity . . . Messier has [it] not.

30 October 1783.
$1\frac{1}{2}$ deg. from gamma towards kappa Cass, (by the finder) the first cluster of Septr 27th.

23 February 1784.
I look'd for the two clusters of stars near gamma Cassiopeae (see Sept 27th 83) but could not find them; some clouds interfering . . . I saw a cluster of obscure stars about $1\frac{1}{2}$ south preceding [sic] kappa Cassiopeae. My Brother observed it with the 7 feet [reflector], and counted abt 20 stars.

8 March 1784.
When kappa Cassio: is just going out of the field, the cluster is a little advanced as in the Fig[ure]. By the delineation it seems to be 47′ S following kappa Cassiopeae at the distance of abt $1\frac{1}{4}$ degree.

11 March 1784.
I saw the Cluster near kappa Cassiopeae and found it instantly by the description of the 8th of March. (NGC 225 [Hidden Treasure 2])

HOSKIN: It is [erroneously] said in William's catalog to have been discovered by Caroline in 1784 (rather than 1783).

(10) 27 September 1783.
About 1° south of the above cluster [NGC 225] a faint nebula surrounded with a great number of both large and small stars, there are more large stars in the field than are marked here [in a diagram] but I took particular notice of the two between which the nebula is situated . . . Mess. has them not. (NGC 189 [Hidden Treasure 1])

HOSKIN: The object must precede Gamma Cas, and the description is indeed that of NGC 189. But it is William's VIII.64 (NGC 381) that is credited to Caroline, even though this follows Gamma Cas. It would seem therefore that this attribution to Caroline is a mistake (perhaps because VIII.64 is positioned between two 6th-magnitude stars 1° apart), and that in fact she should be credited with NGC 189, an object that William never saw.

O'MEARA: It has long been my contention that NGC 381 is too faint for Caroline to have noticed in a sweep with a 4.2-inch telescope and a speculum mirror. The cluster shines at a dim magnitude 9.3 and has a dull surface brightness of 13.5. While faint, NGC 189 is 0.5-magnitude brighter and more condensed than NGC 381, giving it a greater surface brightness of 12.3. Also, NGC 189 is near Dolidze 12, a slight concentration of Milky Way adding to the region's dynamicism, especially in a rich-field telescope, which Caroline used. Hoskin's discovery of Caroline's original journal entry clearly dismisses NGC 381 as being one of Caroline's discoveries.

(11) 29 September 1783.
About 3° south of gamma Andromedae, a fine Cluster of Stars. Mess. has it not. (NGC 752 [Caldwell 28])

HOSKIN: Although the bright cluster NGC 752 is nearly 5° from gamma [Andromedae],

Owen Gingerich proposes that this is what Caroline saw. This is highly likely, in that Caroline's sweeper had a field of view of some 2° 15′, and so she could have made only a rough estimate of the angle between the two objects. NGC 752 was observed by William as VII.32 in sweep 599 on September 21, 1786 but it is not credited to Caroline [and was] intentionally omitted from William's catalogs.

(12) 30 October 1783.
Between sigma & rho Cassiopeae a fine nebula, very strong. (NGC 7789 [Hidden Treasure 108])

(13) 12 May 1784.
Halfway between delta Cyg & eta & theta Lyrae making an isosceles triangle downwards when in that situation.

From my Brother's Review or &c. Nr. 2. Decr 31, 1801: There is a scattered cluster of small stars about 2° 10′ from 14 Cygni towards 21. My Sister saw it May 12, 1784. [Added by Caroline:] my number 16. (NGC 6819 [Hidden Treasure 98])

HOSKIN: This cluster was reobserved by William after 1790 when he established the existence of "true" nebulosity and so had lost interest in clusters. It is the bright "Foxhead Cluster," NGC 6819 = GC 4511. I thank Owen Gingerich for this identification.

(14) 7 August 1787.
I saw a nebulous patch in a line from epsilon Cephei continued thro' delta towards 1st and 2nd Fl. Cassiopeae. (NGC 7380 [Hidden Treasure 106])

Hoskin notes that several deep-sky objects have also been erroneously attributed to Caroline. A clerical error by Caroline's hand, erroneously attributes V.19 (NGC 891 [Caldwell 23], a 10th-magnitude edge-on galaxy in Andromeda) to her, and on March 4, 1783, Caroline recorded that she found a "very obscure" nebula near 19 Monocerotis, William wrote: "Lina near 19 Monocerotis very obscure." Either this observation is very confused or Caroline had found an unclaimed comet. Hoskin explains the confusion:

> William credits Caroline with first sighting of the cluster VII.27, seen by him in sweep 529 on 24 Feb 1786, immediately after he had viewed M50. The position he gives for M50 in his sweep record is $6^h53^m11^s$, –8° 6′, and the position for VII.27, $7^h1^m17^s$, –8° 16′. VII.27 was therefore some 8^m following M50 and slightly south of it, whereas 19 Mon was nearly 4° north of M50. Therefore, whatever it was that Caroline saw in 1783, it was not VII.27.

Although there is no deep-sky object in the position of Herschel's VII.27, there is a "very obscure" deep-sky object a little more than 1° west and slightly south of 19 Monocerotis – NGC 2311, a 9th-magnitude open cluster that William discovered on November 26, 1786, and cataloged as VIII.60. But this object would have been much too faint for Caroline to see in her small refractor (her reflector had not yet been completed). O'Meara notes that even with a modern 4-inch telescope at 23× under a dark sky, NGC 2311 is visible only if one knows where to look with averted vision. It's possible then that Caroline had either seen (and missed) her first comet, or picked up a small splash of stars in this Milky Way region – perhaps even the roughly 12′-wide arc of four 8th-magnitude stars immediately east of NGC 2311.

The latter hypothesis is not far-fetched. Several mysterious observations appear in Caroline's journals, mostly of nebulous stars. For instance, on the evening of

October 13, 1783, she recorded that "chi Aur or near it is nebulous." Most likely, the budding observer was recording the fuzzy appearance of Chi Aurigae and its close attendant stars. She made a similar observation while sweeping through Cetus on December 1, 1783, seeing some faint stars "which I cannot help thinking make a nebulous appearance." It's important to note that at times Caroline suffered the fate of many modern observers when her telescope dewed over so fast she could not record the locations of her finds. It is possible then, that some of the "nebulosity" the blossoming observer suspected was caused by moisture collecting on the objective or eyepiece.

One of these "mysterious" observations may not be all that mysterious after all. On July 30, 1783, Caroline recorded the following observation: "In the neck of Equuleus or head of Aquarius, a rich spot near 3, 4, 7 Equ." The star 7 Equ is not near 3 and 4 Equulei, but if one ignores the reference to 7 Equ, there is a remarkable asterism approximately 54′ northeast of 4 Equ documented in Archinal and Hynes's book *Star Clusters.* David Levy and Wendee Wallach-Levy found it in December 2000 and call it the Equuleus S owing to its location and S-shaped appearance – a sinuous, 22′-wide, S-shaped stream of some 25 stars of 9th magnitude and fainter suns that extend in an east–west direction (right ascension 21^{h} 08.8^{m}; declination +06° 20′; epoch 2000). With an eyepiece overlay of 2°, I found that both 3 and 4 Equulei and the Equuleus S would be in the same eyepiece that Caroline used for sweeping. Seen at low power, the Equuleus S could be her "rich spot," as she was using a new telescope with a deeper limiting magnitude.

O'Meara also believes it's possible that Caroline confused the two horses (Pegasus and Equuleus), since 3, 4, and 7 Pegasi (which are grouped together) lie only about 4 fields east and slightly north of 3 and 4 Equulei, near the head of Aquarius. He notes that when Caroline made her discovery, she was only just learning how to use her new sweeper, which gave her wider vistas and more stars to see, so any grouping of stars in this sparsely populated region would have given her reason to pause and reflect.

Regardless of these understandable observations, Caroline's penchant for discovery did not go unnoticed. In remarking about her discovery of NGC 7789, Adm. William H. Smyth, an admirer of Caroline's career, states in his *Bedford Catalogue* that this discovery "would alone justify Miss Herschel's nomination [for tribute.]"

Equally important is the inescapable contention that Caroline's rash of deep-sky discoveries are what prompted William to turn his attention away from double star observing and to start his greatest endeavor – a systematic search for nebulae with his large 20-foot telescope, which began shortly after Caroline made her 12th deep-sky discovery on October 30, 1783. As Caroline recalled, "In the beginning of December [1783] I became entirely attached to the writing-desk, and had seldom an opportunity after that time of using my newly acquired instrument." William was certainly eager to start his own hunt for deep-sky objects, as Caroline recalls:

> My brother began his series of sweeps when the instrument was yet in a very unfinished state, and my feelings were not very comfortable when every moment I was alarmed by a crack or fall, knowing him to be elevated

fifteen feet or more on a temporary cross-beam instead of a safe gallery. The ladders had not even their braces at the bottom; and one night, in a very high wind, he had hardly touched the ground before the whole apparatus came down.

That my fears of danger and accidents were not wholly imaginary, I had an unlucky proof on the night of the 31st December. The evening had been cloudy, but about ten o'clock a few stars became visible, and in the greatest hurry all was got ready for observing. My brother, at the front of the telescope, directed me to make some alteration in the lateral motion, which was done by machinery, on which the point of support of the tube and mirror rested. At each end of the machine or trough was an iron hook, such as butchers use for hanging their joints upon, and having to run in the dark on ground covered a foot deep with melting snow, I fell on one of those hooks, which entered my right leg above the knee. My brother's call, "Make haste" I could only answer by a pitiful cry, "I am hooked."

A new fire was raging in William, and Caroline, with her noble discoveries, undoubtedly sparked it. Yet Caroline did not give up on her desire to discover. And in August 1786 she took advantage of some "free time" in her brother's absence to sweep the sky with her 4-inch. It was a moment that would open a new chapter in her life.

CAROLINE'S COMETS

August 2, 1786. Nearly four years have passed since Caroline began sweeping the heavens for comets without success. It has been raining all day, and she is worried that the skies will not clear. She spends the day calculating 150 nebulae discovered by her brother, who was away on a trip, but her mind is on the object she saw last evening at 10 p.m. – an object "very much resembling in color and brightness the 27 nebula of the *Connaissance des Temps* [M27], with the difference, however of being round. I suspected it to be a comet; but a haziness coming on, it was not possible to satisfy myself as to its motion. I made several drawings of the stars in the field of view with it."

Caroline's hopes rise as the sky clears in the afternoon. After sundown she sets up her 4.2-inch reflector and refinds the field. Suddenly she spies the object, which looks "like an out of focus star," and notices that it has moved. The night before the object had made a perfect isosceles triangle with two stars, but no longer. Taking up her pen, she redraws the field and plots the motion in relation to the two stars. By 1:00 a.m. she has followed her object long enough to declare that it is a comet. Rather than sleep, she writes a letter to Alex Aubert and describes the discovery. In that letter, the modest woman apologizes, saying, "I hope, sir, you will excuse the trouble I give you with my wag [meaning, vague] description, which is owing to my being a bad (or what is better) no observer at all . . . Lastly, I beg of you, sir, if this comet should not have been seen before, to take it under your protection.

What an exciting time it must have been for this first lady of comet hunters. She was also surrounded by friends who wished her well. In a letter dated August 7, 1786, Aubert writes to her, "I wish you joy, most sincerely, on the discovery. I am more pleased than you can well conceive that you have made it, and I think I see your *wonderfully clever* and *wonderfully amiable* brother, upon the news of it, shed a tear of joy."

Comet hunting was quite unusual for a woman of the 1800s. In fact, it was unprecedented. Yet this is what Caroline loved the most; to this day her comet finds are

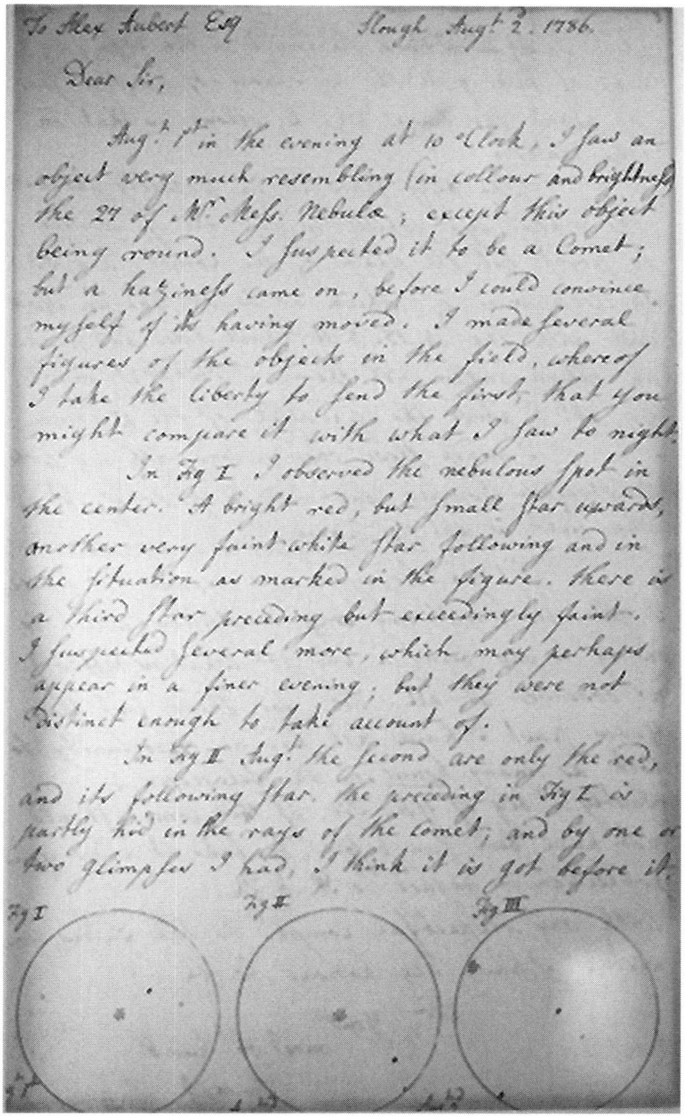

To Alex Aubert Esq. Slough Aug.t 2. 1786.

Dear Sir,

Aug.t 1st in the evening at 10 o'Clock, I saw an object very much resembling (in colour and brightness) the 27 of Mr Mess. Nebulæ; except this object being round. I suspected it to be a Comet; but a haziness came on, before I could convince myself of its having moved. I made several figures of the objects in the field, whereof I take the liberty to send the first, that you might compare it with what I saw to night.

In Fig I. I observed the nebulous spot in the center. A bright red, but small star upwards, another very faint white star following and in the situation as marked in the figure. there is a third star preceding but exceedingly faint. I suspected several more, which may perhaps appear in a finer evening, but they were not distinct enough to take account of.

In Fig II Aug.t the second are only the red, and its following star. the preceding in Fig I. is partly hid in the rays of the comet; and by one or two glimpses I had, I think it is got before it.

Fig I Fig II Fig III

considered her most important astronomical contributions, though this is very subjective. And it was the 10 years from 1788 to 1798, that she enjoyed the most fruitful period of astronomical discovery, for during that time she had discovered seven more comets.

Interestingly, in 1788, William married Mary Pitt, daughter of a wealthy merchant and a widow of John Pitt, Esq., whose first child had died. Mary later gave birth to William's only child, John Herschel, who later continued his father's investigations of the heavens and was adored by Caroline. Although Caroline continued to work as William's assistant, she also began to strike out on her own. Her independence blossomed further on December 21, seven months after her brother's marriage at age 49, when she discovered her second comet, 35P/1788 Y1 a little more than 1° southwest of Beta Lyrae. Her brother called it a "pretty visible object," noting that it was slow moving and much larger than M57. At the time of discovery, Caroline was searching for the return of the Great Comet of 1661, which was speculated upon by Halley as being possibly identical with the Comet of 1532.[2]

[2] The Comet of 1661 returned to the inner solar system 341 years later, when it was discovered by Kaoru Ikeya of Japan and Daqing Zhang of China and independently discovered by Paulo M. Raymundo of Brazil. The comet's modern designation is 153P/2002 C1 (Ikeya-Zhang). Interestingly, on July 28, 1939, Roger Rigollet of France discovered an 8th-magnitude comet that L. E. Cunningham found to be identical with Caroline's comet of 1788. This periodic comet is now known as 35P/Herschel-Rigollet.

William described the circumstances of the discovery to the Earl of Salisbury, Hertfordshire:

> My Lord, the Great Comet (Comet of 1661) which is expected has not yet made its appearance but we are in hopes every day to meet with it. Your Lordship will find the small comet which we are now observing, which was discovered by my sister the 21st Dec when she was on the lookout for the expected one in the Constellation of the Lyre, and the enclosed figure will point it out. Etc. . . .
>
> Wm Herschel Slough, Jan 8, 1789.

William also wrote to Sir Harry Englefield, as Englefield had once pointed out a comet to William that Messier had found. Englefield replied on Christmas Day 1788:

> I am much obliged to you for your account of the Comet and beg you to make my complements to Miss Herschel on her discovery. She will soon be the great Comet finder and bear away the Prize from Messieurs Messier and Mechain.

Caroline's reputation was soaring. She was being congratulated, recognized, and compared to other great astronomers of her day – ones who had gained great reputations as comet hunters and discoverers.

Caroline discovered two of the three comets found in 1790: C/1790 A1 on January 7 and C/1790 H1 on April 17. Charles Messier observed C/1790 A1 on January 19, describing its appearance as nebulous with a bright condensation; he likened its brightness to the globular cluster M15 in Pegasus. The comet was last seen only two nights later. At discovery, C/1790 H1 was a small, faint object in Andromeda, without a tail.

It's a compliment to Caroline that she found this comet, given the circumstances, which she describes in a letter to Alexander Aubert dated April 18, 1790:

> I am almost ashamed to write to you, because I never think of doing so but when I am in distress. I found last night at 16h 24′ sidereal time a comet, and do not know what to do with it, for my sweeper is not half finished; and besides I broke the handle of the perpendicular motion in my brother's absence (who is on a little tour into Yorkshire) . . . All my hopes were that I should not find anything which would make me feel the want of these things in his absence; but, as it happens, here is an object in a place where there is no nebula, or anything which could look like a comet.

Aubert spied the suspect, which he called "*something faint,*" on the morning of January 21. His happiness for Caroline is certain: "You cannot, my dear Miss Herschel, judge the pleasure I feel when your reputation and fame increase; everyone must admire your and your brother's knowledge, industry, and behaviour. God grant you many years of health and happiness." By early May, Caroline's comet had grown in brightness and was slightly fainter than M31; it also had the beginnings of a tail. Indeed, by May 20, the tail had grown to 4° in length. On May 28, Caroline wrote, "I saw the comet thro' the clouds, when I could not see a single star in the heavens, with the naked eye." The comet was last seen on June 29.

Caroline's fifth comet, C/1791 X1, was the only comet discovered in 1791. She found it in Lacerta on December 15 using a telescope William built for her – a 9-inch Newtonian reflector, 5-feet long, that magnified 25–30 times with a 1° 49′ field of view. By now, Caroline also knew all the Messier objects on sight, for William had advised her years ago, that, as a comet hunter, she needed to know them. Caroline simply described the object as a "pretty large, telescopic comet." But her brother, William noted that it was 6 arcminutes in diameter with a central condensation 5 or 6 arcseconds across. He also recorded a faint, ill-defined ray extending nearly 15 arcminutes to the north. The comet slowly faded from view and was last seen by Messier on January 28.

Caroline found her sixth comet on October 7, 1793, near Delta (δ) Ophiuchi. Charles Messier, however, had found it 10 days earlier. And though the comet is officially known as C/1793 S2 (Messier), Caroline is credited as an independent discoverer in cometographies.

A little more than two years passed before Caroline made another comet find. She announced her seventh discovery in a modest letter to Sir Joseph Banks, dated November 8, 1795:

> Last night in sweeping over a part of the heavens with my five-foot reflector, I met with a telescopic comet... It will probably pass between the head of the Swan and the constellation of the Lyre, in its descent towards the sun. The direction of its motion is retrograde . . . As the appearance of one of these objects is almost become a novelty, I flatter myself that this intelligence will not be uninteresting to astronomers, and therefore hope, sir, you will, with your usual kindness, recommend it to their notice.

Indeed, this was a great discovery. The comet was seen for 23 days, with carefully measured positions obtained. But mathematicians struggled with determining its orbit, trying to fit a parabolic, rather than an elliptical orbit, and subsequently it was lost. This comet is now popularly known as Encke's comet, for the German mathematician Johann F. Encke who firmly established its periodicity and predicted its 3.3-year return to Earth's vicinity. Indeed, Caroline's discovery of Encke's comet (2P/Encke) was the second known observed return to perihelion of this comet. (Pierre Mechain first discovered it on January 17, 1786, while sweeping for comets in the Aquarius region.) Comet 2P/Encke is still the comet with the shortest known orbital period.

Caroline's last discovery, C/1797 P1/ Bouvard-Herschel, was the only comet discovered in 1797. She discovered it with her naked eye on August 14, almost at the same time as Eugene Brouvard (Paris). She must have known it would be found soon by others, for Caroline climbed onto a horse to deliver her letter of discovery to Dr. Maskelyne herself, as she explains in a letter dated August 17, 1797, to Joseph Banks:

> I have so little faith in the expedition of messengers of all descriptions that I undertook to be my own . . . but unfortunately I undertook the task with only the preparation of one hour's sleep, and having in the course of five years never rode above two miles at a time, the twenty to London, and the idea of six or seven more to Greenwich in reserve, totally unfitted me for any action.

Caroline's account of the discovery of her eighth and last comet closes the discovery period of her life:

> August 14, 1797, at 9 hr 30′ common time, being dark enough for Sweeping I began in the usual manner with looking over the heavens with the naked eye, and immediately saw a comet nearly as bright as that which was discovered by Mr Gregory, Jan 8, 1793. I went down from the observatory to call my Brother Alexander, that he might assist me at the clock. In my way in the garden I was met and detained by Lord

Storker and another gentleman who came to see my brother and his telescopes. By way of preventing too long an interruption I told the gentleman that I had just found a comet and wanted to settle its place. I pointed it out to them and after having seen it they took their leave. 19 h 30′ correct Sid. time = 9 h 59′ 13 Mean time.

August 16, 1797. Memorandum. Tuesday morning the 15th, I went to Greenwich and carried my memorandums to Dr. Maskelyne. When I arrived there, he had no intelligence yet of the comet. In the evening Dr. M received a letter from Stephen Lee, Esq., who had also seen it on the 14th. I had the pleasure to find that my observations agreed perfectly with Mr. Lee's.

After Caroline left the cottage she was living in, and moved to temporary lodgings in October 1797, she gave up her practice of regular sweepings for comets. But her *Book of Observations* shows she continued to observe and watched for reports of any comets discovered by other observers. In 1807 when she was living at Upton she wrote the following news:

Jan 27, 1807, I brought this morning a memorandum with me from Windsor, which I copied from the Hamburgher papers at the Castle. The place of a comet for the following dates:

Jan 15. R.A.	25 14′ Decl.	39 18	Discovered by Mr Ponse,
Jan 25	19 40	29 34	assistant to Mr. Schroeter zu Lilienthal.
Feb 2	17 33	23 58	

This being a part of the heavens which I could get sight of at the house at Upton I took my sweeper down in the garden and began to sweep as soon as daylight was gone and at 6h 52′ I saw it in the field of view; It is one of the smallest comets I have seen, but not the smallest and would undoubtedly have appeared to be very considerable if it had been observed at a higher altitude; for it was actually setting at the time I was looking. It appeared like an oblong nebula, rounded at the corners, shaped like an egg.

Jan 28, I had my sweeper carried to Slough and went there with the intention of showing the comet to my Brother, but the heavens were totally overcast.

Jan 31, I was at Slough time enough to prepare my sweeper. The weather coming on very fine I kept on sweeping near the place and found the comet again. The wind was very high and the weather very cold but my brother (though very ill) came out to make proper observations on the comet and its situation for which see his Review.

She observed another comet in October 1807, with her brother's 7-foot telescope in his absence. This comet must have been the Great Comet of 1807, which was discovered on September 19 of that year and was still 1st or 2nd magnitude in October. And in September 1811, she made some lengthy observations of a comet of which her friend Mr. Pigot had given her notice. She described it as having:

a nucleus of white light, which appeared to me equal in size to 47 Ursae (which is of reddish light); then to the nucleus adhered a circular white nebulosity, which I estimate about 3 times the breadth of the nucleus.

The September 1811 comet she observed was clearly the Great Comet Flaugergues. That month it was 1st magnitude with a 12°-long tail. The Comet was at perihelion on September 12, and was visible all night long below the handle of the Big Dipper in the beginning of October.

Caroline's last observation was of The Great Comet C/1823 Y1 (De Breaute-Pons) which was visible from December 23, 1823, to April 1, 1824, and was the only comet found in 1823. This final entry was written in her *Book of Observations* at age 73:

> Hanover Jan 31, 1824, Eyedraft of the situation of the Comet which has been from some times past (since Jan 6) observed here in Hanover. It is at present surrounded with a diffused light of not very great extent and with difficulty seen by the naked eye.

CAROLINE THE INDEFATIGABLE

Besides her regular daily chores, Caroline, at the request of her brother, undertook the enormous task of indexing and cross referencing Flamsteed's *Observations of the Fixed Stars,* along with a list of 560 stars that were not included in the original catalog. It took her 20 months to complete her work, which was read to the Royal Society in March 1798.

William's health was failing by 1819. In a feeble hand, on a yellow slip of paper William wrote: "LINA, There is a great comet." The Great Comet of 1819 (Comet Tralles) ascended above the northwest horizon as a conspicuous naked-eye object in Europe in the first days of July with an 8°-long tail. The earliest observation of its position was made by Professor Tralles of Berlin on July 1. The comet transited the Sun a few days earlier. Possible observations of the transit by Pastoroff were presented by none other than William's son John Herschel to the Royal Society many years later.

"I want you to assist me. Come to dine and spend the day here. If you can come soon after one o'clock we shall have time to prepare maps and telescopes. I saw its situation last night – it has a long tail. July 4, 1819." Caroline annotated this carefully preserved note: "I keep this as a relic," she wrote, "Every line now traced by the hand of my dear brother becomes a treasure to me."

After William died in 1822, Caroline returned to Hanover and, working independently at age 75, completed William's and her catalog of 2,500 nebulae reordered in zones from the North Pole. Caroline explained to her nephew John Herschel that "If you see an "L" and find a cluster of stars thereabout, I shall claim it as one of those I mentioned in my last letter." The unpublished manuscript (the *Zone Catalogue),* was mailed to John who was continuing his father's work on the nebulae. The *Zone Catalogue* was of immeasurable assistance to John in completing the *General Catalogue of Nebulae* published in 1863. In the introduction to his *General Catalogue* John honored her astounding contribution: "I learned fully to appreciate the skill, diligence, and accuracy which that indefatigable lady brought to bear on a task which only the most boundless devotion could have induced her to undertake or enabled her to accomplish." For this crucial contribution to astronomy she was awarded the Gold Medal of the Royal Astronomical Society of London on February 8, 1828.

In the presentation of this medal John South, Esq., addressed the Royal Society outlining her many endeavors, a small portion is quoted here:

> Soon after the death of him to whom she had given up so much of her life, her best energies, and her ripest faculties, she returned to Hanover, – unwilling, however, to relinquish the astronomical researches which had been so pure and permanent a source of pleasure. She undertook and completed the laborious reduction or registration of the places of 2500 nebulae, down to the 1st of January 1800; thus presenting in one view the results of

all the observations Sir William Herschel had made upon those wonderful bodies, and triumphantly bringing to a close half a century of scientific toil. For this more immediately and to mark their estimation of services rendered during a whole life to astronomy your Council resolves to confer on her the distinction of a medal of this Society.

She went on to receive other honors including honorary membership, along with Mary Somerville, into Britain's Royal Astronomical Society of London in 1835, election into the Royal Irish Academy in 1838, and awarded the Gold Medal for Science by the King of Prussia in 1846 on her 96th birthday.

Living to age 97, her mental faculties intact, Caroline was indeed a remarkably accomplished astronomer. The sheer scale of her contributions to astronomy are exhausting. Dedicated to her brother and her work, Caroline would walk to and from her brother's observatory through his own grounds – a distance quite half a mile, which must have been trying in bad weather. There is a tradition in Slough that the son of the owner of one of the houses where she lodged for a short time, in the High Street, would tell how, when a boy, he used sometimes to be roused in the night by her rap on the wall, when the weather, which had at first been cloudy, cleared up. Knowing the signal, he would get up, light a lantern, and going downstairs find her ready dressed and awaiting him. He perfectly recalled her gentle manner of saying to him: "Please will you take me to my Broder."

Caroline wrote her own epitaph, which was engraved on her tombstone:

Here rests the earthly shell of Caroline Herschel, Born at Hanover, March 16, 1750, Died January 9, 1848.

The gaze of Her who has passed to glory was, while below, turned to the starry Heaven: her comet discoveries, and her share in the undying work of her Brother, William Herschel, shall tell of this to all time.

The Royal Irish Academy of Dublin, and the Royal Astronomical Society of London counted her among their members.

At the age of 97 years 10 months she fell asleep in perfect peace, and in full vigor of mind, following into a better life; her father, Isaac Herschel, who lived to the age of 60 years 2 months 17 days, preceded her in 1767, and lies buried hard by.

In 1889, Caroline received a final honor for her achievements when a minor planet was named "Lucretia."

Barbara Wilson Houston, Texas, July 7, 2005.

Author's Note: This history is based on the following works:

1. William Herschel, *Philosophical Transactions* **76**, 457 (1786); **79**, 212 (1789); **92**, 477 (1802).
2. Michael Hoskin, *The Herschel Partnership* (Cambridge: Science History Publications, 2003).
3. Michael Hoskin, editor, *Caroline Herschel's Autobiographies* (Cambridge: Science History Publications, 2003).
4. Mrs. John Herschel (Mary Cornwallis Herschel), *Memoir and Correspondence of Caroline Herschel* (London: John Murray, 1879; reprinted in 2000 by the William Herschel Society, 19 New King Street, Bath).
5. J. L. E. Dreyer, *Scientific Papers of Sir William Herschel*, Vol. 1 and 2 (London: 1912).

6. Constance A. Lubbock (William Herschel's Granddaughter), *The Herschel Chronicle* (Cambridge: Cambridge University Press, 1933; reprinted in 1997 by the William Herschel Society, 19 New King Street, Bath).
7. Larry Mitchell, "William Herschel: the greatest visual observer of all time," in Stephen James O'Meara, *Deep-Sky Companions: The Caldwell Objects* (Cambridge, MA: Sky Publishing / Cambridge: Cambridge University Press, 2002).
8. Sir John Frederick William Herschel, *Catalogue of Nebulae and Clusters of Stars*, published in the *Philosophical Transactions*, 1863.
9. Gary W. Kronk, *Comets: A Descriptive Catalog* (Hillside, NJ: Enslow Publishers, 1984).
10. John Bortle, *Bright Comet Chronicles*, http://encke.jpl.nasa.gov/bright_comet.html#pre1850

APPENDIX B
Hidden treasures: basic data

HT[a]	NGC/IC[b]	Type[c]	Con.[d]	RA[e]	Dec.[f]	Mag.[g]	Size (′)[h]	Notes
1	189	OC	Cas	00^h 39.6^m	+61° 06′	8.8	5	Caroline Herschel
2	225	OC	Cas	00^h 43.6^m	+61° 46′	7.0	15	Caroline Herschel
3	281	BN	Cas	00^h 52.8^m	+56° 37′	7.8	35 × 30	Pacman Nebula
4	288	GC	Scl	00^h 52.8^m	−26° 35′	7.9	13	
5	404	GX	And	01^h 09.4^m	+35° 43′	9.8	6.6 × 6.6	Mirach's Ghost
6	584	GX	Cet	01^h 31.3^m	−06° 52′	10.4	3.8 × 2.4	Little Spindle
7	659	OC	Cas	01^h 44.4^m	+60° 40′	8.2	6	Caroline Herschel
8	772	GX	Ari	01^h 59.3^m	+19° 00′	9.9	7.1 × 4.7	Fiddlehead
9	908	GX	Cet	02^h 23.1^m	−21° 14′	10.4	6.1 × 2.7	
10	1023	GX	Per	02^h 40.4^m	+39° 04′	9.3	7.5 × 3.0	Perseus Lenticular
11	1232	GX	Eri	03^h 09.8^m	−20° 35′	9.5	7.2 × 5.7	Eye of God
12	1291	GX	Eri	03^h 17.3^m	−41° 06′	8.6	11.1 × 10.1	Snow Collar Galaxy
13	1316	GX	For	03^h 22.7^m	−37° 12′	8.2	13.5 × 9.3	Fornax A
14	Mel 20	OC	Per	03^h 24.3^m	+49° 52′	2.3	300	Alpha Per. Moving Cluster
15	1333	BN	Per	03^h 29.3^m	+31° 25′	5.7	6 × 3	Embryo Nebula
16	1360	PN	For	03^h 33.2^m	−25° 52′	9.1	9 × 5	Comet Planetary
17	1365	GX	For	03^h 33.6^m	−36° 08′	8.3	11.2 × 5.9	
18	1399	GX	For	03^h 38.5^m	−35° 27′	8.8	8.1 × 7.6	
19	1398	GX	For	03^h 38.9^m	−26° 20′	9.5	7.6 × 5.3	
20	1404	GX	For	03^h 38.9^m	−35° 36′	9.7	4.8 × 3.9	
21	Kem 1	AST	Cam	03^h 57.4^m	+63° 04′	–	180	Kemble's Cascade
22	1501	PN	Cam	04^h 07.0^m	+60° 55′	10.6	56″ × 48″	Oyster Nebula
23	1502	OC	Cam	04^h 07.8^m	+62° 20′	6.0	20	Jolly Roger Cluster
24	1535	PN	Eri	04^h 14.3^m	−12° 44′	9.1	48″ × 42″	Cleopatra's Eye
25	1528	OC	Per	04^h 15.3^m	+51° 13′	6.2	18	m & m Double Cluster
26	1545	OC	Per	04^h 20.9^m	+50° 15′	6.6	12	m & m Double Cluster
27	1647	OC	Tau	04^h 45.7^m	+19° 07′	6.2	40	Pirate Moon Cluster
28	IC 418	PN	Lep	05^h 27.5^m	−12° 42′	9.1	14″ × 11″	Spirograph Nebula
29	Cr 69	OC	Ori	05^h 35.0^m	+09° 56′	3.0	70	Lambda Orionis Cluster
30	1981	OC	Ori	05^h 35.2^m	−04° 26′	3.7	28	Coal Car Cluster
31	Cr 72	OC	Ori	05^h 35.4^m	−05° 55′	3.0	15	Lost Jewel of Orion
32	1977	BN	Ori	05^h 35.5^m	−04° 52′	6.3	20 × 10	Mermaid's Purse Nebula
33	1999	BN	Ori	05^h 36.5^m	−06° 42′	9.5	2 × 2	13th Pearl Nebula
34	2024	BN	Ori	05^h 41.9^m	−01° 51′	7.2	30 × 30	Lips Nebula
35	2163	BN	Ori	06^h 07.8^m	+18° 39′	~11	3 × 2	Cederblad 62
36	2169	OC	Ori	06^h 08.4^m	+13° 58′	5.9	6	Shopping Cart Cluster
37	2175	BN	Ori	06^h 09.7^m	+20° 30′	6.9	40 × 30	
38	2264	OC	Mon	06^h 41.0^m	+09° 54′	4.4	40	Christmas Tree Cluster
39	2301	OC	Mon	06^h 51.8^m	+00° 28′	6.0	15	Hagrid's Dragon
40	2353	OC	Mon	07^h 14.5^m	−10° 16′	6.0	18	Avery's Island
41	2440	PN	Pup	07^h 41.9^m	−18° 12′	9.1	74″ × 42″	Albino Butterfly Nebula
42	2451	AST	Pup	07^h 45.4^m	−37° 57′	2.7	50	Stinging Scorpion

HT[a]	NGC/IC[b]	Type[c]	Con.[d]	RA[e]	Dec.[f]	Mag.[g]	Size′[h]	Notes
43	2467	BN	Pup	07^h 52.5^m	−26° 24′	8.0	16 × 12	
44	2547	OC	Vel	08^h 10.2^m	−49° 13′	5.0	25	Golden Earring Cluster
45	2539	OC	Pup	08^h 10.6^m	−12° 49′	6.5	15	"The Dish" Cluster
46	2546	OC	Pup	08^h 12.4^m	−37° 37′	5.3	70	Heart & Dagger Cluster
47	2683	GX	Lyn	08^h 52.7^m	+33° 25′	9.7	9.1 × 2.7	UFO Galaxy
48	2655	GX	Cam	08^h 55.6^m	+78° 13′	10.2	5.9 × 5.3	
49	2841	GX	UMa	09^h 22.0^m	+50° 59′	9.0	6.6 × 3.4	Tiger's Eye Galaxy
50	IC 2488	OC	Vel	09^h 27.4^m	−56° 57′	~6.0	18	Lacaille
51	2903,05	GX	Leo	09^h 32.2^m	+21° 30′	9.0	11.6 × 5.7	
52	3184	GX	UMa	10^h 18.3^m	+41° 25′	9.4	7.5 × 7.0	Little Pinwheel Galaxy
53	3228	OC	Vel	10^h 21.4^m	−51° 43′	6.0	5	Lacaille
54	3293	OC	Car	10^h 35.8^m	−58° 13′	4.6	5	Lacaille
55	3344	GX	LMi	10^h 43.5^m	+24° 55′	9.3	6.7 × 6.3	Sliced Onion Galaxy
56	3521	GX	Leo	11^h 05.8^m	−00° 02′	9.1	11.7 × 6.5	
57	3621	GX	Hya	11^h 18.3^m	−32° 49′	8.5	9.8 × 4.6	Frame Galaxy
58	3628	GX	Leo	11^h 20.3^m	+13° 35′	9.5	14.8 × 3.3	King Hamlet's Ghost
59	4214	GX	CVn	12^h 15.6^m	+36° 20′	9.1	9.6 × 8.1	
60	4216	GX	Vir	12^h 15.9^m	+13° 09′	10.0	7.9 × 1.7	Silver Streak Galaxy
61	4361	PN	Crv	12^h 24.5^m	−18° 47′	10.2	1.9 × 1.9	
62	Mel 111	OC	Com	12^h 25.1^m	+26° 07′	1.6	300	Coma Berenices Cluster
63	4490	GX	CVn	12^h 30.6^m	+41° 39′	9.5	5.6 × 2.8	Cocoon Galaxy
64	IC 3568	PN	Cam	12^h 33.1^m	+82° 34′	10.3	18″ × 18″	Theoretician's Nebula
65	4526	GX	Vir	12^h 34.0^m	+07° 42′	9.9	7.4 × 2.7	Hairy Eyebrow Galaxy
66	4605	GX	UMa	12^h 40.0^m	+61° 36′	10.3	5.7 × 2.5	Fabergé Egg Galaxy
67	4656	GX	CVn	12^h 44.0^m	+32° 10′	10.5	18.8 × 3.2	"Messier's" Hockey Stick
68	4699	GX	Vir	12^h 49.0^m	−08° 40′	9.5	3.1 × 2.5	Vinyl LP Galaxy
69	4725	GX	Com	12^h 50.4^m	+25° 30′	9.2	10.5 × 8.1	
70	5102	GX	Cen	13^h 22.0^m	−36° 38′	9.3	8.3 × 3.5	Iota's Ghost
71	5281	OC	Cen	13^h 46.6^m	−62° 55′	6.1	8	Lacaille
72	5363	GX	Vir	13^h 56.1^m	+05° 15′	10.1	4.7 × 3.2	
73	5662	OC	Cen	14^h 35.5^m	−56° 40′	5.5	30	Lacaille
74	5746	GX	Vir	14^h 44.9^m	+01° 57′	9.3	8.1 × 1.4	Blade & Pearl Galaxy
75	5866	GX	Dra	15^h 06.5^m	+55° 46′	9.8	7.3 × 3.5	Fool's Gold Galaxy
76	5897	GC	Lib	15^h 17.4^m	−21° 01′	8.2	11	Ghost Globular
77	5986	GC	Lup	15^h 46.1^m	−37° 47′	7.8	9.6	
78	6210	PN	Her	16^h 44.5^m	+23° 48′	8.4	48″ × 8″	Turtle Nebula
79	6242	OC	Sco	16^h 55.5^m	−39° 28′	6.2	9	Lacaille
80	6281	OC	Sco	17^h 04.8^m	−37° 53′	5.8	8	Moth Wing Cluster
81	6369	PN	Oph	17^h 29.3^m	−23° 46′	11.4	58″ × 34″	Little Ghost Nebula
82	6400	OC	Sco	17^h 40.2^m	−36° 58′	8.3	12	Phantom Cluster
83	IC 4665	OC	Oph	17^h 46.2^m	+05° 43′	4.7	70	Caroline Herschel
84	6445	PN	Sgr	17^h 49.2^m	−20° 01′	11.2	3.1 × 0.9	Box Nebula

(cont.)

HT[a]	NGC/IC[b]	Type[c]	Con.[d]	RA[e]	Dec.[f]	Mag.[g]	Size′[h]	Notes
85	6503	GX	Dra	$17^h 49.4^m$	+70° 09′	10.4	5.9 × 2.2	Lost-In-Space Galaxy
86	6441	GC	Sco	$17^h 50.2^m$	−37° 03′	7.2	9.6	Silver Nugget Cluster
87	–	HPMS	Oph	$17^h 57.8^m$	+04° 40′	9.5	–	Barnard's Star
88	6520	OC	Sgr	$18^h 03.4^m$	−27° 54′	7.6	5	Dead Man's Chest
89	6544	GC	Sgr	$18^h 07.3^m$	−25° 00′	7.5	4.6	Starfish Cluster
90	6572	PN	Oph	$18^h 12.1^m$	+06° 51′	7.3	16″ × 13″	Emerald Eye Planetary
91	6624	GC	Sgr	$18^h 23.7^m$	−30° 22′	7.8	8.8	
92	6633	OC	Oph	$18^h 27.2^m$	+06° 30′	4.3	20	Caroline Herschel
93	IC 4756	OC	Ser	$18^h 38.9^m$	+05° 26′	4.3	40	Tweedledee Cluster
94	6709	OC	Aql	$18^h 51.5^m$	+10° 20′	6.8	15	Flying Unicorn Cluster
95	6712	GC	Sct	$18^h 53.1^m$	−08° 42′	8.3	9.8	
96	6723	GC	Sgr	$18^h 59.5^m$	−36° 38′	6.9	13.0	
97	Cr 399	AST	Vul	$19^h 26.2^m$	+20° 06′	3.5	90	Coathanger
98	6819	OC	Cyg	$19^h 41.3^m$	+40° 11′	7.3	5	Caroline Herschel
99	6818	PN	Sgr	$19^h 44.0^m$	−14° 09′	9.3	22″ × 15″	Little Gem Neb
100	6866	OC	Cyg	$20^h 03.9^m$	+44° 09′	7.6	15	Caroline Herschel
101	6940	OC	Vul	$20^h 34.5^m$	+28° 17′	6.3	25	Mothra Cluster
102	—	DN	Cyg	$20^h 40.0^m$	+41° 00′	—	480 × 300	Northern Coalsack
103	7008	PN	Cyg	$21^h 00.5^m$	+54° 33′	9.9	98″ × 75″	Coat Button Nebula
104	7027	PN	Cyg	$21^h 07.0^m$	+42° 14′	8.5	18″ × 10″	Pink Pillow Nebula
105	Tr 37	OC	Cep	$21^h 39.0^m$	+57° 30′	3.8	90	IC 1396 (Nebula)
106	7380	OC	Cep	$22^h 47.3^m$	+58° 08′	7.4	20	Caroline Herschel
107	OMe 1	AST	Psc	$23^h 40.7^m$	+07° 57′	–	15	Alessi J23407 +0757
108	7789	OC	Cas	$23^h 57.5^m$	+56° 43′	6.6	25	Caroline Herschel
109	7793	GX	Scl	$23^h 57.8^m$	−32° 35′	9.2	9.2 × 6.4	Bond's Galaxy

[a] HT, hidden treasure.

[b] *NGC/IC, New General Catalogue/Index Catalogue*.

[c] Type: AST, asterism; BN, bright nebula; DN, dark nebula; GC, globular cluster; GX, galaxy; HPMS, high proper motion star; OC, open cluster; PN, planetary nebula.

[d] Con., constellation.

[e] RA, right ascension (equinox 2000.0).

[f] Dec., declination (equinox 2000.0).

[g] Mag., magnitude.

[h] Size (′), in arcminutes.

APPENDIX C

Twenty additional hidden treasures

HT[a]	*NGC/IC*[b]	Type[c]	Con.[d]	RA[e]	Dec.[f]	Mag.[g]	Size (′)[h]	Notes
1	134	GX	Scl	$00^h\ 30.4^m$	−33° 15′	10.0	7.6 × 1.5	Giant Squid Galaxy
2	1245	OC	Per	$03^h\ 14.7^m$	+47° 14′	8.4	10	
3	1300	GX	Eri	$03^h\ 19.7^m$	−19° 25′	10.4	5.5 × 2.9	
4	1491	BN	Per	$04^h\ 03.4^m$	+51° 19′	–	25 × 25	
5	1514	PN	Tau	$04^h\ 09^m$	+30° 47′	10.9	2.3 × 2.0	
6	2022	PN	Ori	$05^h\ 42.1^m$	+09° 05′	11.0	28″ × 27″	
7	3114	OC	Car	$10^h\ 02.7^m$	−60° 06′	4.2	35	
8	3918	PN	Cen	$11^h\ 50.3^m$	−57° 11′	8.1	12″	
9	IC 4406	PN	Lup	$14^h\ 22.4^m$	−44° 09′	10.3	1.7 × 0.6	Retina Nebula
10	5617	OC	Cen	$14^h\ 29.7^m$	−60° 42′	6.3	10	
11	5846	GX	Vir	$15^h\ 06.5^m$	+01° 36′	10.0	4.9 × 4.4	
12	5907	GX	Dra	$15^h\ 15.9^m$	+56° 20′	10.9	11.2 × 1.9	Splinter Galaxy
13	5927	GC	Lup	$15^h\ 28.0^m$	−50° 40′	8.0	6	
14	IC 4603	BN	Oph	$16^h\ 25.6^m$	−24° 28′	–	35 × 20	
15	6356	GC	Oph	$17^h\ 23.6^m$	−17° 49′	8.2	10	
16	6388	GC	Sco	$17^h\ 36.3^m$	−44° 44′	6.8	10.4	
17	6664	OC	Sct	$18^h\ 36.5^m$	−08° 11′	7.7	12	
18	6781	PN	Aql	$19^h\ 18.5^m$	+06° 32′	11.4	1.9 × 1.8	Ghost of the Moon
19	6905	PN	Del	$20^h\ 22.4^m$	+20° 06′	11.1	42″ × 35″	Blue Flash Nebula
20	6939	OC	Cep	$20^h\ 31.5^m$	+60° 40′	7.8	10	Ghost Bush Cluster

[a] HT, hidden treasure.

[b] *NGC/IC*, *New General Catalogue/Index Catalogue*.

[c] Type: AST, asterism; BN, bright nebula; DN, dark nebula; GC, globular cluster; GX, galaxy; OC, open cluster; PN, planetary nebula.

[d] Con., constellation.

[e] RA, right ascension (equinox 2000.0).

[f] Dec., declination (equinox 2000.0).

[g] Mag., magnitude.

[h] Size (′), in arcminutes.

APPENDIX D

Deep-sky lists: comparison table[a]

HT	NGC/IC	H400	OME	SAC	TAAS	CH	SC	RASC	ICC	JBC	SAAO	CHer	H
1	189											*	
2	225	*			*				*			*	
3	281					*	*	*	*	*			
4	288				*	*		*	*	*	*	*	
5	404	*	*		*	*	*						
6	584	*	*										
7	659	*										*	
8	772	*	*		*			*					
9	908	*	*		*				*				
10	1023	*		*	*		*	*	*				
11	1232			*	*		*		*	*	*		
12	1291		*		*		*			*			
13	1316		*		*	*	*			*		*	
14	Mel 20		*				*	*		*			
15	1333		*										
16	1360		*		*	*	*			*			
17	1365		*		*	*	*						
18	1399		*							*			
19	1398		*							*			
20	1404		*							*			
21	Kem 1		*		*	*							
22	1501	*	*	*	*	*	*	*	*				
23	1502	*	*		*	*	*						
24	1535	*		*	*	*	*	*	*	*	*		
25	1528	*	*		*	*	*						
26	1545	*	*										
27	1647	*	*		*		*						
28	IC 418		*										
29	Cr 69		*			*					*		*
30	1981		*			*							
31	Cr 72		*										
32	1977		*	*	*	*		*					
33	1999	*	*										
34	2024	*	*	*	*	*	*	*					
35	2163		*										
36	2169	*	*	*									
37	2175		*		*								
38	2264	*	*		*	*			*		*		
39	2301	*	*		*		*						
40	2353	*	*			*							
41	2440			*	*	*	*						
42	2451	*				*			*	*			*

HT	NGC/IC	H400	OME	SAC	TAAS	CH	SC	RASC	ICC	JBC	SAAO	CHer	H
43	2467		*		*				*	*			
44	2547		*				*				*	*	
45	2539	*	*	*		*	*						
46	2546												*
47	2683	*		*	*	*	*	*	*				
48	2655	*	*					*					
49	2841	*	*	*	*	*	*	*	*				
50	IC 2488												*
51	2903,05	*	*	* *	*	*	*	*		*			
52	3184	*	*	*	*	*	*	*					
53	3228												*
54	3293		*										*
55	3344	*	*	*	*	*	*						
56	3521	*	*	*	*		*	*					
57	3621	*	*		*		*		*	*			
58	3628	*	*	*	*	*	*	*					
59	4214	*		*	*	*	*						
60	4216	*		*	*	*	*	*					
61	4361	*	*	*	*	*		*					
62	Mel 111		*	*	*	*		*					*
63	4490	*	*	* *		*	*	*					
64	IC 3568		*		*								
65	4526	*		*	*		*	*	*				
66	4605	*	*	* *			*	*					
67	4656	*	*	*	*								
68	4699	*		*	*		*	*		*			
69	4725	*	*	*		*	*	*			*		
70	5102		*			*							
71	5281												*
72	5363	*	*										
73	5662												*
74	5746	*	*	*	*								
75	5866	*	*										
76	5897	*	*		*				*				
77	5986		*		*		*		*				
78	6210	* *		*	*	*	*						
79	6242												*
80	6281		*				*						
81	6369	*			*	*	*						
82	6400		*										
83	IC 4665		*			*	*				*		*
84	6445	*	*	*	*			*		*			

(cont.)

HT	NGC/IC	H400	OME	SAC	TAAS	CH	SC	RASC	ICC	JBC	SAAO	CHer	H
85	6503		*	**			*		*				
86	6441		*			*	*			*			
87	Barnard's Star		*			*							
88	6520	*	*	*	*		*						
89	6544	*	*							*			
90	6572		*	*	*	*	*	*	*		*		
91	6624	*	*		*					*			
92	6633	*		*	*	*	*	*				*	
93	IC 4756		*		*	*	*						
94	6709		*			*	*						
95	6712	*	*	*		*	*			*			
96	6723		*		*			*			*		
97	Cr 399		*		*		*	*					*
98	6819			*	*	*	*						
99	6818	*	*	*	*	*	*	*			*		
100	6866	*					*						*
101	6940	*		*	*	*		*		*			
102	N. Coalsack		*										*
103	7008	*	*		*								
104	7027		*	*	*		*		*				
105	Tr 37		*			*	*						
106	7380	*			*							*	
107	O'Meara 1		*										
108	7789	*	*	**			*	*	*			*	
109	7793		*			*			*	*	*	*	

[a] Selected comparison lists: H400, Herschel 400; OME, O'Meara (Hawaii); SAC, Saguaro Astronomy Club (Arizona); TAAS, The Albuquerque Astronomical Society (New Mexico); CH, Celestial Harvest (Pennsylvania); SC, Supplementary Catalogue (UK); RASC, Royal Astronomical Society of Canada; ICC, Index Caldwell Catalog (New Zealand); JBC, Jack Bennet Catalogue (South Africa); SAAO, South African Astronomical Observatory (South Africa); CHer, Caroline Herschel (UK); H, Historical (prior to Caroline Herschel).

APPENDIX E
Photo credits

Photographs are credited in the order in which they appear within the book.

PREFACE	
	Stephen James O'Meara
CHAPTER 1	
	Stephen James O'Meara
	Stephen James O'Meara
	Charlene Meyers
	Stephen James O'Meara
	Stephen James O'Meara
	Stephen James O'Meara
CHAPTER 2	
Hidden Treasure 1	Digitized Sky Survey[1]
Hidden Treasure 2	Digitized Sky Survey[1]
Hidden Treasure 3	Digitized Sky Survey[1]
	Dr. Jean-Charles Cuillandre / Canada-France-Hawaii Telescope Corporation
Hidden Treasure 4	Digitized Sky Survey[2]
Hidden Treasure 5	Digitized Sky Survey[1]
	R. E. Griffiths / N. Homeier / J. Gallagher / NASA
Hidden Treasure 6	Digitized Sky Survey[2]
Hidden Treasure 7	Digitized Sky Survey[1]
Hidden Treasure 8	Digitized Sky Survey[1]
Hidden Treasure 9	Digitized Sky Survey[2]
	(Cetus) Michael Oates / Manchester Astronomical Society
Hidden Treasure 10	Digitized Sky Survey[1]
Hidden Treasure 11	Digitized Sky Survey[2]
Hidden Treasure 12	Digitized Sky Survey[2]
Hidden Treasure 13	Digitized Sky Survey[2]
	C. Grillmair / NASA
Hidden Treasure 14	Stephen James O'Meara
	Stephen James O'Meara
Hidden Treasure 15	Digitized Sky Survey[1]
Hidden Treasure 16	Digitized Sky Survey[2]
Hidden Treasure 17	Digitized Sky Survey[2]
Hidden Treasures 18/20	Digitized Sky Survey[2]
Hidden Treasure 19	Digitized Sky Survey[2]

Hidden Treasures 21/23	Stephen James O'Meara
	Digitized Sky Survey[1]
Hidden Treasure 22	Digitized Sky Survey[1]
Hidden Treasure 24	Digitized Sky Survey[2]
Hidden Treasures 25/26	Digitized Sky Survey[1]
Hidden Treasure 27	Stephen James O'Meara
	Digitized Sky Survey[1]
Hidden Treasure 28	Digitized Sky Survey[2]
	Hubble Heritage Team (STScI / AURA / NASA)
Hidden Treasure 29	Digitized Sky Survey[1]
Hidden Treasures 30–32	Digitized Sky Survey[2]
Hidden Treasure 33	Digitized Sky Survey[2]
	Hubble Heritage Team (STScI / AURA / NASA)
Hidden Treasure 34	Adriana Sherman
	(Grim Reaper) Dan Norder
Hidden Treasure 35	Digitized Sky Survey[1]
Hidden Treasure 36	Digitized Sky Survey[1]
Hidden Treasure 37	Digitized Sky Survey[1]
Hidden Treasure 38	Digitized Sky Survey[1]
Hidden Treasure 39	Digitized Sky Survey[1]
Hidden Treasure 40	Digitized Sky Survey[1]
Hidden Treasure 41	Digitized Sky Survey[2]
	Hubble Heritage Team (STScI / AURA / NASA)
Hidden Treasure 42	Digitized Sky Survey[2]
Hidden Treasure 43	Digitized Sky Survey[1]
Hidden Treasure 44	Digitized Sky Survey[2]
	(Argo Navis) Michael Oates / Manchester Astronomical Society
Hidden Treasure 45	Digitized Sky Survey[2]
Hidden Treasure 46	Digitized Sky Survey[2]
Hidden Treasure 47	Digitized Sky Survey[1]
Hidden Treasure 48	Digitized Sky Survey[1]
Hidden Treasure 49	Digitized Sky Survey[1]
	HST / NASA
Hidden Treasure 50	Digitized Sky Survey[2]

Hidden Treasure 51	Adriana Sherman
	Almudena Alonso-Herrero *et al.* / HST / ESA / NASA
Hidden Treasure 52	Digitized Sky Survey[1]
Hidden Treasure 53	Digitized Sky Survey[2]
	(Charles' Oak) Michael Oates / Manchester Astronomical Society
Hidden Treasure 54	Digitized Sky Survey[2]
Hidden Treasure 55	Digitized Sky Survey[1]
Hidden Treasure 56	Digitized Sky Survey[1]
Hidden Treasure 57	Digitized Sky Survey[2]
Hidden Treasure 58	Digitized Sky Survey[2]
Hidden Treasure 59	Digitized Sky Survey[1]
	Hubble Heritage Team (STScI / NASA)
Hidden Treasure 60	Digitized Sky Survey[2]
Hidden Treasure 61	Digitized Sky Survey[2]
Hidden Treasure 62	Stephen James O'Meara
	Stephen James O'Meara
	Digitized Sky Survey[1]
Hidden Treasure 63	Digitized Sky Survey[1]
Hidden Treasure 64	Digitized Sky Survey[1]
	Howard Bond / STScI / NASA
Hidden Treasure 65	Digitized Sky Survey[1]
	High-Z Supernova Search Team / HST / NASA
Hidden Treasure 66	Digitized Sky Survey[1]
	HST / NASA
Hidden Treasure 67	Digitized Sky Survey[1]
Hidden Treasure 68	Digitized Sky Survey[2]
Hidden Treasure 69	Digitized Sky Survey[1]
Hidden Treasure 70	Digitized Sky Survey[2]
Hidden Treasure 71	Digitized Sky Survey[2]
Hidden Treasure 72	Digitized Sky Survey[1]
Hidden Treasure 73	Digitized Sky Survey[2]
Hidden Treasure 74	Digitized Sky Survey[1]
Hidden Treasure 75	Digitized Sky Survey[1]
	(Draco) Michael Oates / Manchester Astronomical Society
Hidden Treasure 76	Digitized Sky Survey[2]
Hidden Treasure 77	Digitized Sky Survey[2]
Hidden Treasure 78	Digitized Sky Survey[1]
Hidden Treasure 79	Digitized Sky Survey[2]
Hidden Treasure 80	Digitized Sky Survey[2]
Hidden Treasure 81	Digitized Sky Survey[1]
	Howard Bond / STScI / NASA
Hidden Treasure 82	Digitized Sky Survey[2]
Hidden Treasure 83	Stephen James O'Meara
Hidden Treasure 84	Digitized Sky Survey[1]
Hidden Treasure 85	Digitized Sky Survey[1]
Hidden Treasure 86	Digitized Sky Survey[2]
Hidden Treasure 87	Digitized Sky Survey[1]
Hidden Treasure 88	Digitized Sky Survey[2]
Hidden Treasure 89	Digitized Sky Survey[2]
Hidden Treasure 90	Digitized Sky Survey[1]
Hidden Treasure 91	Digitized Sky Survey[1]
	Ivan King / NASA / ESA
Hidden Treasures 92/93	Stephen James O'Meara
	Stephen James O'Meara
Hidden Treasure 94	Digitized Sky Survey[1]
Hidden Treasure 95	Digitized Sky Survey[2]
Hidden Treasure 96	Digitized Sky Survey[2]
Hidden Treasure 97	Stephen James O'Meara
	Stephen James O'Meara
Hidden Treasure 98	Digitized Sky Survey[1]
Hidden Treasure 99	Digitized Sky Survey[2]
	Robert Rubin / Reginald Dufour / Matt Browning / Patrick Harrington / NASA
Hidden Treasure 100	Digitized Sky Survey[1]
Hidden Treasure 101	Digitized Sky Survey[2]
Hidden Treasure 102	Stephen James O'Meara
Hidden Treasure 103	Digitized Sky Survey[1]
Hidden Treasure 104	Digitized Sky Survey[1]
	William B. Latter / NASA
Hidden Treasure 105	Stephen James O'Meara
Hidden Treasure 106	Digitized Sky Survey[1]
Hidden Treasure 107	Digitized Sky Survey[1]
Hidden Treasure 108	Digitized Sky Survey[1]
Hidden Treasure 109	Digitized Sky Survey[2]

APPENDIX A
All by Stephen James O'Meara

1. From the Digitized Sky Survey, Northern Hemisphere, courtesy of the Palomar Observatory and NASA/AURA/STScI.
2. Use of these images is courtesy of the UK Schmidt Telescope (copyright in which is owned by the Particle Physics and Astronomy Research Council of the UK and the Anglo-Australian Telescope Board) and the Digitized Sky Survey created by the Space Telescope Science Institute, operated by AURA, Inc., for NASA, and is reproduced here with permission from the Royal Observatory Edinburgh.

Index

The treasure chest

HT	NGC/IC	Date observed	Location (Altitude)	Telescope & magnification	Seeing & transparency	Notes
1	189					
2	225					
3	281					
4	288					
5	404					
6	584					
7	659					
8	772					
9	908					
10	1023					
11	1232					
12	1291					
13	1316					
14	Melotte 20					
15	1333					
16	1360					
17	1365					
18	1399					
19	1398					
20	1404					
21	Kemble's Cascade					
22	1501					
23	1502					
24	1535					
25	1528					
26	1545					
27	1647					
28	IC 418					
29	Collinder 69					
30	1981					
31	Collinder 72					
32	1977					
33	1999					
34	2024					
35	2163					
36	2169					
37	2175					
38	2264					
39	2301					
40	2353					
41	2440					
42	2451					

HT	NGC/IC	Date observed	Location (Altitude)	Telescope & magnification	Seeing & transparency	Notes
43	2467					
44	2547					
45	2539					
46	2546					
47	2683					
48	2655					
49	2841					
50	IC 2488					
51	2903,05					
52	3184					
53	3228					
54	3293					
55	3344					
56	3521					
57	3621					
58	3628					
59	4214					
60	4216					
61	4361					
62	Melotte 111					
63	4490					
64	IC 3568					
65	4526					
66	4605					
67	4656					
68	4699					
69	4725					
70	5102					
71	5281					
72	5363					
73	5662					
74	5746					
75	5866					
76	5897					
77	5986					
78	6210					
79	6242					
80	6281					
81	6369					
82	6400					

(cont.)

HT	NGC/IC	Date observed	Location (Altitude)	Telescope & magnification	Seeing & transparency	Notes
83	IC 4665					
84	6445					
85	6503					
86	6441					
87	Barnard's Star					
88	6520					
89	6544					
90	6572					
91	6624					
92	6633					
93	IC 4756					
94	6709					
95	6712					
96	6723					
97	Collinder 399					
98	6819					
99	6818					
100	6866					
101	6940					
102	Northern Coalsack					
103	7008					
104	7027					
105	Trumpler 37					
106	7380					
107	O'Meara 1					
108	7789					
109	7793					